THE ANTIPODES OF THE MIND

THE ANTIPODES OF THE MIND

Charting the Phenomenology of the Ayahuasca Experience

BENNY SHANON

OXFORD
UNIVERSITY PRESS

OXFORD
UNIVERSITY PRESS

Great Clarendon Street, Oxford OX2 6DP

Oxford University Press is a department of the University of Oxford.
It furthers the University's objective of excellence in research, scholarship,
and education by publishing worldwide in

Oxford New York

Auckland Cape Town Dar es Salaam Hong Kong Karachi Kuala Lumpur
Madrid Melbourne Mexico City Nairobi New Delhi Shanghai Taipei Toronto

With offices in

Argentina Austria Brazil Chile Czech Republic France Greece
Guatemala Hungary Italy Japan Poland Portugal Singapore
South Korea Switzerland Thailand Turkey Ukraine Vietnam

Published in the United States
by Oxford University Press Inc., New York

First published 2002

British Library Cataloguing in Publication Data

Data available

Library of Congress Cataloging in Publication Data

Shanon, Benny
The antipodes of the mind : charting the phenomenology of the Ayahuasca experience /
Benny Shanon.
p. cm.
Includes bibliographical references.
1. Ayahuasca–Psychotropic effects. I. Title.
BF209 A93 S53 2002 615'.7883–dc21 2002027770

ISBN 0-19-925292-0
ISBN 0-19-925293-9 (pbk.) ; 978-0-19-925293-0 (pbk.)

Typeset in Ehrhardt MT
by Kolam Information Services, Pvt. Ltd, Pondicherry, India
Printed in Great Britain
on acid-free paper by
CPI Antony Rowe, Chippenham, Wiltshire

CONTENTS

FIGURES

The images presented on the cover and in the book are reproductions of the work *Planos* (Portuguese for 'planes', 'levels', and also 'plans') by the Brazilian artist Céu who travelled throughout the Amazonian rainforest and has partaken of Ayahuasca there; he is involved in ecological and educative projects both in the Amazon and elsewhere in Brazil. The art work is partly based on visions the artist has had with Ayahuasca. Figure 1 shows the entire work; Figures 2 (and the cover), 3, and 4 depict details from it. The author deeply thanks the artist for making his work available for reproduction here.

LIST OF TABLES

The light that the Holy One, blessed be He, created on the first day, one could see thereby from one end of the world to the other; but as soon as the Holy One, blessed be He, beheld the generation of the Flood and the generation of the Dispersion, and saw that their actions were corrupt, He arose and hid it from them. . . . And for whom did he reserve it? For the righteous in the time to come.

The Talmud (tractate *Hagiga*)

God wanted to hide his secrets in a secure place. 'Would I put them on the moon?' He reflected. 'But then, one day human beings could get there, and it could be that those who would arrive there would not be worthy of the secret knowledge. Or perhaps I should hide them in the depths of the ocean,' God entertained another possibility. But, again, for the same reasons, he dismissed it. Then the solution occurred to Him—'I shall put my secrets in the inner sanctum of man's own mind. Then only those who really deserve it will be able to get to it.'

A tale recounted to the author by an ice-cream vendor in the Peruvian Amazon

Prologue

In the summer of 1983 I took a walking tour in the jungle of the Oriente of Ecuador. I joined four other people, two couples, and together we hired an Indian woman as our guide. The two couples kept very much to themselves, and I, being alone and much more interested in the locals than in my fellow tourists, paired with the guide. We walked and talked. When the woman found out that I was a psychologist, the topics she discussed became more and more personal. She recounted her life, critically examining her relationships with her parents when she was young. The conversation went on into the night. In the hut at the riverside, the other four members of the group were sound asleep and we two went on talking. There was one dream, the woman told me, that had haunted her for more than thirty years. 'I thought it must mean something. Indeed, the dream made me realize that what I was thinking was the case could not have been so.' Thus, with increasing boldness and trust, the woman began to reveal to me thoughts and ideas she had never told anyone before. She spoke of a mental life unknown to us, of repression, of symbols and symbolic transformations, of dreams and their interpretation. What I was listening to was, in effect, an exposition of an entire theory, invented unawares by a person who hardly knew how to spell her own name. At one point, the woman inadvertently mentioned hallucinations. She referred to a concoction made out of a plant. I became interested but she did not want to delve too deeply into the subject. I pressed her, but still she insisted that it was not for *gringos* (foreign white people). I pressed further, and she said she had to ask the local sorcerer, the *brujo*.

When the jungle trip was over we went to the guide's village and looked for the *brujo*. He was not to be found. For a while I thought the woman did not really want me to find the man. But then she gave me an address where I could obtain the plant; 'Ask for Ayahuasca,' she told me. I spent a whole day on a bus going to the little town she had indicated and there I headed for the grocery store. I told the vendor the name of my guide and she turned to the back of the room. There, behind a curtain, was a big heap of logs. The vendor picked up one about a foot long and returning to the counter told me to make a soup out of it. 'Ayahuasca,' I reiterated the name looking at the piece of wood on the counter, 'Ayahuasca,' she nodded, 'pound it, make a soup, and drink it at night.'

Several months later, at my home in Jerusalem, I prepared the soup. My (erroneous) understanding was that I should use only the bark. I cut it into pieces which I immersed in water to boil. The water was evaporating and there was still solid material in the pot. Again and again, I added water and let the liquid simmer. The process took several hours. Eventually, I was left with a small schnapps-size cup of a concoction of dark purple-brown. I drank it.

For years I believed that nothing had really happened then. I danced and sang, and stars scintillated all around me. I was very happy. Yet dancing in this manner was not novel for me and it was difficult for me to fathom how real my experience of the stars had been. It was a pleasant, almost euphoric, experience, but there was nothing that felt or seemed supernatural about it.

Indeed, I had done it wrong. The entire piece of wood has to be used, not only the bark, and there are leaves of another plant, about which I knew nothing, that have to be added to the brew. In retrospect, however, it is not clear to me that absolutely nothing had happened.

I told the story—my encounter in the Ecuadorian Oriente and my experiment at home—to a friend who had recently returned from a trip to South America. My friend was startled. The story reminded him of one related to him by another traveller, a young Israeli girl he had met in Peru just before she began an excursion into the jungle. When he returned home to Jerusalem, he found a letter from her. The girl recounted an experience she characterized as the most extraordinary of her life. She had been in an Indian village and had participated in a nocturnal ceremony in which a certain potion was drunk. She felt she was losing her individual self and that her thoughts were being read and controlled by the leader of the ceremony. The world around her, she wrote, was transformed into something that she could not describe in words and the entire experience was horrible. Apparently, it was as if her body had been invaded by a foreign agent. All her insides rocketed and she knew she had to get out of it. She forced herself to throw up and then she crawled into a corner and went to sleep. In the morning she left. 'It was clear to me, however,' my friend read, 'that I would go back. These people know something we do not. It was just that I was not ready for it. One day I shall return and study with them.' 'Study,' she explained, was the word the Indians used—'study the wisdom of the Ayahuasca'.

A few weeks later my friend called me. The woman, his friend, was now back in Jerusalem. They had met and he had told her about my experiences. He said she had been furious. 'One is not supposed to prepare the soup and drink it by oneself. One can do so only with permission, in a native setting.' It was very clear, she said, that nothing had happened to me—my experiment was cursed. I was lucky that nothing worse had happened. My common sense rejected these hostile comments. After all, what is involved, I surmised, is but a plant specimen with a hallucinogenic compound in it. It could not be that the biochemical effect was contingent on blessings, or curses, or sacred permission. It was clear to me that my friend's friend had undergone a terrifying experience and that her reaction to me, a person she had never met, was overshadowed by it.

Eight years later, in 1991, I was invited to Brazil to participate in a scientific conference. Preparing for this trip, I consulted a friend about which places and things would be of interest in this vast land that I had never visited before. Having drawn up a long list of recommended places to visit and things to do, my friend recalled that he had once heard of a strange mystical group that uses some kind of infusion in their ceremonies. He knew nothing more about it. When he mentioned this, the thought suddenly crossed my mind that the infusion in question was the very one I had heard about in Ecuador. I had no specific reason for believing this, but it seemed to me pretty certain that this was the case. My friend referred me to a Brazilian friend of his in Rio de Janeiro. This person had not experimented with the concoction either but his brother, my friend remarked, was deeply involved with the mystical group he had mentioned. When I concluded my academic engagements in São Paulo I contacted this brother on the phone and he suggested that I go to the Colônia Cinco Mil, a farming compound at the outskirts of Rio Branco, the capital of the Amazonian state of Acre. It was a journey from one side of a continent to the other. I took a plane and went.

On my arrival at the Colônia, I was led to the leader of the community, the *Padrinho* (an affectionate diminutive for father, also godfather). With his smiling, yet deep and piercing eyes, the man was inspecting me, and even though no words were used—I knew no Portuguese then—I felt I was having a personal interview. He looked at me straight in the eyes and said 'My home is your home.' There was a session the next evening and I was invited.

My first encounter with the *Daime* (the local name for the Ayahuasca potion) was grounded in utter ignorance. As noted, I did not speak the language nor did I have any knowledge of Amerindian culture. When I first visited the Colônia I knew nothing about the Church of Santo Daime, its practices, or its doctrine. Indeed, I did not even know that the ceremonies in which I participated were religious rituals and I received no instructions as to what to expect or what to do. While in many respects this was very fortunate, as my initial experiences were not contaminated by prior knowledge or external influences, there is one aspect that I somewhat regret. For some reason it seemed to me that I should keep my eyes open as much as possible. This, I gather, reduced the time that I was subject to full-fledged hallucinations.

Keeping my eyes open, I inspected the room and the people in it. What struck me was that the rather dilapidated room became exceedingly beautiful. The colours shone, the simple paper decorations were brilliant, and giant butterflies—or were they birds?—seemed to be floating in the inner space of the hall. I experienced no distortions in perception as such, but it was all otherworldly. What I felt might be likened to what one feels when snorkelling and inspecting the world through glass goggles. It seemed to me that I was distanced from my normal self. At times I also felt fear.

When I did close my eyes a most wonderful arabesque presented itself to me. It was a visually rich tapestry whose dominating colours were blue and purple. At first it was a two-dimensional symmetrical pattern. Then the symmetry broke and there was no apparent division in what was seen: the entire visual space constituted one image. And, in oblique lines stretching from the lower left-hand corner of my visual field to the upper right-hand corner, there appeared reptiles—yellow-green lizards that were coming out of the arabesque in a manner very similar to that encountered in Escher's drawings. I know I had another vision, but I cannot recall it. What I do remember is the modified sensation of time. When my eyes were closed and I was hallucinating, time seemed to be an eternity and I was frightened. I opened my eyes and, watching the clock on the wall, I realized that only two minutes had elapsed.

The man with the shining eyes who had received me the previous day was offering another serving of the brew. My initial impression was that he was a kind person but now it also seemed that there was something malicious about him. In this Amazonian farmer, descendant of serfs on rubber plantations, I saw, in effect, a Mediterranean, Italian, or southern French man, a bartender in a cheap tavern. This perception, I should emphasize, did by no means degrade either the man or the situation. Yet it gave it all a special, strange flavour.

I also observed other participants in the ceremony. I noticed that I could detect the most minute interactions between people, and even though these were persons I had never seen before and knew nothing about, it seemed to me that I understood them well.

Throughout the session a small band was playing electric guitars. The music, I felt, was sublime. My hearing was very acute and there was something enchanting about the sounds. At one moment a woman (I later discovered she was the leader's wife) read out names of people. This part of the ceremony was very solemn. It seemed that the spirits of the dead were being contacted. At the end of the session, five hours later, I looked through the window and the world appeared fresh and immensely beautiful. I contemplated it and felt cleansed.

I should say that had it not been for the next session I was invited to, the impact of the experience on my life would have been rather limited. It was a remarkable experience, but it did not change me as a person. My world-view had not been altered and no shift in my plans or further interests would have occurred.

That second session to which I was invited was for a *feitio*, the ritual in which the *Daime* is prepared. I was told that this ritual takes place once every two or three months, so I should regard myself lucky to be able to participate in it. My original plan was to leave the day after the first session but, encouraged to prolong my stay in the village, I delayed my departure ticket for a couple of days. Again, I was ignorant. I knew I would be participating in the preparation of the brew, but I did not know that I would drink it myself that time.

The ritual began around 2:30 a.m. On an open platform, bounded by a low stone fence and surrounded by a small grove of trees I and seven other men sat on small wooden trunks that served as individual benches, and we all pounded twigs of wood. This we did to the rhythm of the chanting. I did not understand the words, but still the music governed the movements of my arm. The tempo was fast and the motor action was far from easy for me. And then, at one point the leader of the community came and offered each of us a cup of *Daime*. Along with the other men, I drank and I went on with the pounding. All of a sudden, the forest was aflame. I closed my eyes, I opened them—and it was just the same: fire all around. From the fire emerged dragons. They were terrifying, with glaring eyes, erupting tongues, shining teeth, and triangular protrusions on their backs. At first I did not know what to do, but then it occurred to me that the dragons were not evil. Admittedly, they looked ugly and menacing; this, however, was because of their nature, not because they meant any real harm. Rather than be afraid of them, the thought passed through my mind, I should respect their existence and accept them as they too are God's creatures. It further occurred to me that I have my place and they have theirs, and as long as I acknowledged and respected this, nothing bad would happen to me. And indeed, when I looked at the dragons it seemed to me that they were even smiling. I was on guard, but there was no longer fear in my heart.

Then something else happened—the twigs I was pounding were turning into snakes. The experience was horrendous. I knew my task was to pound, but now the pounding became an act of killing. And then the association dawned on me—Cain, the first murderer. I reflected: What was Cain's lot? God did not kill him. Rather, Cain was ordered to roam the earth and his descendants were the builders of human civilization. Cain's first son was the founder of the first city and amongst his later descendants were the first musician and the first 'artificer of brass and iron' (Genesis 4: 22). Thus, the thought crossed my mind, the reparation for crime was civilization. There was no way of undoing the evil deed, no way of going back. The only option was to go forward and to construct a new, and, it was to be hoped, a better, world.

With this, I started to see human history unfold before me. I remember only a few of the scenes I saw. In one, there were slaves rowing Roman boats. The slaves were chained to their seats, and the labour was hard. Yet, it suddenly occurred to me, in their debased position in the lower deck the slaves were, in some sense, freer than the masters on the upper deck who were trapped in their games of power and social status. Fixed to one place and confined to one action that was repeated again and again, the slaves had more liberty to think and reflect. Most significantly, they were closer to the Divine than their masters were.

Another scene depicted the looting of America by the Spaniards. It was cruel. Yet, all the evil notwithstanding, there was something else about the whole process, something that was above and beyond the awful catastrophe, something with respect to which the ostensible victors in the bloody human drama were only instruments. The proud and the brave greedily accumulated personal, material wealth, but in doing so they also contributed to the advancement of culture and the arts. Indeed, they built not only palaces, but also churches. Much of the South American gold ended up being used for the praise of the Lord. However evil and petty human beings are, I thought, they are also the creators of some of the most beautiful things that exist in the universe. With culture and art, as well as with religion and spirituality, humankind can be redeemed.

Still another scene was of the Jewish Holocaust in the Second World War. In this conjunction, I had many insights about evil and destiny, justice and redemption. In particular, it seemed to me that I was gaining a new understanding of both the killers and the victims, and also had conjectures regarding the more general, perhaps metaphysical, significance of the disaster. In particular, it seemed to me that the victims were given the opportunity to reach the highest levels of faith. Despite all the suffering, until the very last moment, human beings are free to maintain their dignity and not to lose the link with their Creator.

And then I realized it was day. The pounded material had accumulated and was brought to the big cauldron on the other side of the platform. At the bottom of the hill I saw a man preparing wood for the fire. The man was cutting trees and then moving the branches uphill to where the cauldron was. It was clear to me that I should help. Even though movement and motor co-ordination were difficult for me, I negotiated the low fence and started carrying the branches from the place where they were being cut to the fire. Doing this, it occurred to me that I was involved in holy work, precisely this: work that was holy. I should note that in my native Hebrew, the expression 'holy work' (*avodat kodesh*) is a rather common one, but never before had I sensed its meaning in such a literal fashion.

At one point I gazed at the forest, which was now washed by the fresh morning light. It seemed that this was the first day of creation. I also had a glimpse of a heavenly scene. It seemed to me that the righteous were there who, according to Jewish legend, eat the meat of the great Leviathan. I also had a glimpse of Jesus on the cross.

Some philosophical thoughts crossed my mind as well. In particular, it occurred to me that process theories, such as psychoanalysis or Darwinian evolutionary theory, miss the point. The essences of persons and of species are not determined by the processes that these theories specify. The essences at hand are determined on another level. The processes only define the realization of these essences in the physical world and in time.

It also occurred to me that what I was experiencing had to do with the world of the Kabbalah. It dawned upon me that the various heavenly scenes described in esoteric Jewish texts are not fictitious. I found myself entertaining the idea that reality as we normally perceive it is only a veil, a cover that hides the secret, hidden essences of things. With prayer, I further reflected, human beings can gain access to these essences and even influence what is happening in the Higher Realms. By way of association, the prince, from the tales of the Brothers Grimm, who was transformed into a toad, came into my mind. I felt that, in a certain sense, the story this fairy tale recounts was true, literally so. It goes without saying that all these ideas were totally new to me and very foreign to the views I normally held in the context of my life.

The third time I drank the *Daime* potion was the afternoon of the same day. I was offered a cup of the brew I had helped prepare. It was warm and honey-coloured. I was sitting on a small bench looking at the forest. A group of women were singing on the side. When the potion had its effect, I found myself presented with pure enchantment. The forest was full of animals—both natural and phantasmagoric: notably there were dragons, felines, and big birds. The dominant colours were green and blue. I was sitting viewing the forest as if it were a stage. It was as if a screen were raised and another world made its appearance. At moments, however, it seemed to me that even though I was sitting here on the bench, my own self was over there in the forest and I was dancing with the various creatures in it. It was all blissful, and very real. And I saw it all with open eyes.

With the second cup I was offered, the scene changed. What now presented itself to me was an enchanted city, a city of gold and precious stones. It was of indescribable beauty. And again, so very real.

The next morning I woke up early and I felt a desire to return to the place where the brew had been prepared. I thought there would be people there, but there were none. So I entered the forest, and soon I found a clearing. I stood and I started to chant. There was no planning or reason in this act, it was simply the thing to do. It seemed that a Bach cantata was coming out of my mouth. I closed my eyes. My visual field was all blue, then it changed into white. I held my hands clasped in a praying position, and I felt as if I were rising. My eyes were closed and I was singing. Then I saw a triangle and in it was circumscribed a hand. Moved, I lowered my posture and kept silent. I stayed in the place for a couple of minutes, and then felt that I should leave. Four years later I discovered that this place was considered, by the members of the Santo Daime Church, to be the holy spot of the Queen of the Forest (*a rainha da floresta*).

And lastly, before leaving the village, I drank one cup alone. I was sitting on a balcony overlooking a garden. As the brew was having its effect, I started to sing. What poured out of my mouth was a biblical text—the beginning of the book of Genesis. I chanted, in Hebrew, the creation of the world. And as I did so, the world was being created in front of my eyes: there was water, and then land surfaced, and trees came out of the earth and flowered, and then the world was populated with birds. I stopped there, before the creation of animals.

When I first took Ayahuasca I did so out of mere curiosity, as a traveller. As noted, the second, third, and fourth time I partook of the brew, I experienced very powerful visions that affected me profoundly and cast serious doubts on my world-view. Yet, I was even more impressed when I returned back home and started to read about the brew I had consumed.

Professionally, what drew me first and foremost to the mystery of Ayahuasca was reading that with this brew many people see snakes and jaguars. As a cognitive psychologist, I was baffled: How could it be that different people see the same things in their visions? I could understand, of course, visions having similar forms or structures. I could understand that visions manifest, in a pictorial manner, basic wishes, such as flying, or basic conflicts, such as those having to do with key stages or central figures in one's life. However, commonalities in specific contents defied my understanding. Snakes and jaguars seem to be just too specific to define cognitive universals.

What struck me even more was the discovery that many of my own experiences with the brew were similar to those of the indigenous Amerindians. In particular, in my first vision I saw large lizards, in the second snakes, and in the third jaguars. All this had occurred before I consulted the literature on Ayahuasca and without knowing anything about it. Puzzled, I decided to return to South America and study Ayahuasca further.

My first return was a short visit to southern Colombia, during which I was introduced to Ayahuasca in the native, Indian context. Then I decided to devote a sabbatical to the study of Ayahuasca, and I stayed in Brazil and Peru for over a year. During that period I consumed Ayahuasca many times and in various settings. Soon I realized that my involvement with the brew was neither a mere matter of curiosity, nor could it be confined to the professional interests that had guided it all along. Increasingly, the quest became personal. First and foremost, it dawned on me that what I was actually entering was a school. There were no teachers, no textbooks, no instructions; yet there was definitely structure and order to it. The teacher was the brew, the instruction was conducted during the period of intoxication without the assistance of any other person. And what was quite remarkable—there were grades. Each series of sessions centred on a topic or a problem. At times, I realized what the topic was only in retrospect. But there was always an order. I have heard the same impressions from other people. Indeed, one attribute by which *Daime* is referred to in the hymns of its church is the 'teacher of all teachers'.

Second, there was the element of struggle. Ayahuasca sessions are by no means easy, purely enjoyable experiences. One has to endure moments which may be very, very harsh, physically as well as mentally. Often I would take the brew and say to myself, 'Never again!' But I was driven to continue, hence I had to confront the difficulties and learn how to handle them. This was a training process in itself. And like any struggle that one overcomes, it does not leave one the same. One's inner constitution is affected, and in the process one undergoes significant personal transformations.

Third, I realized that one aspect of the Ayahuasca experience is a profound self-analysis. One is cruelly confronted with one's self and one finds oneself having no other option but to address issues that are often neither easy nor pleasant to handle. As one good friend told me, based on her own experiences, with Ayahuasca there is no way to cheat.

Last, but definitely not least, is the spiritual level. Sooner or later this is unavoidable. Ayahuasca introduces one to realms that pertain to religion, to faith, to the Divine. A significant number of the Ayahuasca sessions in which I participated were conducted in the context of groups with specific doctrines and religious beliefs. However, from the beginning I decided to keep to my own personal path. This was because I am Jewish and these groups are Christian (or semi-Christian), because I wanted to conduct objective research, and most of all because by my very nature I am fundamentally an individualistic freethinker. For years I have characterized myself as a 'devout atheist'. When I left South America I was no longer one. I did not, despite strong encouragement, become a member of any of the groups I associated myself with nor do I have any intention of doing so in the future. But my *Weltanschauung* has radically changed.

I decided to write a book about all this.

PART I

Preliminaries

1

General Background

The Soma is a god. He cures the sharpest ills that man endures. He heals the sick, the sad he cheers; he nurses the weak, dispels their fears. The faint with material ardour fires, with lofty thought the bad inspires. The soul from earth to heaven he lifts, so great and wondrous are his gifts. Men feel the god within their veins, and cry in loud exalting strains.

O soma-drinker, drink of the soma-wine!
The intoxication of thy rapture gives indeed the Light.

We have drunk the soma and become immortal!
We have attained the light, we have found the Gods.

The Rig Veda

Ayahuasca is a psychoactive brew consumed throughout the entire upper Amazon region. The term is a Quechua compound word meaning 'vine of the (dead) spirits'. Depending on the region and the context of use, the brew is also known by a variety of other names; the most well-known are *caapi*, *yagé* (*yajé, yahé*), *natem*, *cipo*, *mariri*, *Daime*, *hoasca*, and *vegetal*.[1] Typically, Ayahuasca induces powerful visions as well as hallucinations in all other perceptual modalities. Pronounced non-perceptual cognitive effects are also manifest. These include personal insights, intellectual ideations, affective reactions, and profound spiritual and mystical experiences. Moreover, Ayahuasca introduces those who partake of it to what are experienced as other realities. Those who consume the brew may feel that they are gaining access to new sources of knowledge and that the mysteries and ultimate truths of the universe are being revealed to them. All this is often coupled with what drinkers[2] describe as an encounter with the Divine.

In this book I present the case for the cognitive-psychological study of Ayahuasca. Before doing this, I would like to review some background information pertaining to Ayahuasca and the traditions of its use. This is by no means intended to be a comprehensive survey of the literature. Here I cite only information deemed directly pertinent to the analyses and discussions that follow. For further information the reader should consult the ethnobotanic, physiological, and anthropological literature.

[1] I have partaken of Ayahuasca in contexts using all these names except for the first.

[2] Some readers might feel that the term 'drinker' is somewhat unprofessional. In point of fact, however, it is one that is often used in the traditional contexts of Ayahuasca use. I prefer it to the New Age neologism 'psychonaut' which seems to me more pretentious and less neutral.

First and foremost are the numerous scientific publications of the dean of Amazonian ethnobotany, Richard Evans Schultes—for instance, Schultes (1972, 1982); other publications about the botany and pharmacology of Ayahuasca are listed below. Of the anthropologists working with indigenous tribes, the one who is most prominent, and who will be cited many times throughout this book, is Reichel-Dolmatoff—of his works let me mention *Amazonian Cosmos*, *The Shaman and the Jaguar*, *Beyond the Milky Way*, and *The Forest Within* (Reichel-Dolmatoff, 1971, 1975, 1978*a*, 1996, respectively), as well as the collection of essays compiled in Reichel-Dolmatoff (1997); other anthropological studies are referred to below. A unique work is the joint publication of the anthropologist Luna and the shaman-turned-painter Pablo Amaringo which is a collection of reproductions of paintings in which Ayahuasca visions are displayed (Luna and Amaringo, 1993). Excellent general, multidisciplinary reviews of the literature are presented in Ott (1993, 1994). A summary bibliography of the scientific research on Ayahuasca is found in Luna (1986*a*). Several interesting popular, non-scientific books on the topic appeared in recent years—of these let me mention T. McKenna (1991), Wolf (1992), and Narby (1998).[3]

In the vast region encompassing western Brazil and the eastern areas of Ecuador, Peru, and Colombia, as well as parts of the basin of the Orinoco, where Brazil and Venezuela meet, Ayahuasca has long been a key constituent of culture. Indeed, it appears that the indigenous peoples of this region have used the brew for millennia (for pertinent historical and archaeological data, see P. Naranjo, 1986). In the past, Ayahuasca was used in the making of all major decisions of a tribe, notably locating game for hunting and declaring war. It was also believed that the brew made it possible to see distant places and foretell the future. Even today, Ayahuasca is the basic instrument of shamans in the entire region. On the one hand, the brew is said to enable the shaman to see the inner constitution of his patients, and thus establish a diagnosis; on the other hand, it is said to bring the shaman in contact with wise beings and guiding entities that pass information to the shaman so that he knows how to perform the appropriate treatment. In addition, Ayahuasca is purported to allow the shaman to be in touch with the spirits—the beings of other worlds and the dead. For many, Ayahuasca is not merely a potion or a plant but also a being with special, unique qualities or even a deity. In the words of Schultes (1982: 206):

[3] Since these lines were written, the world witnessed the Internet revolution. This has dramatically affected Ayahuasca as well. Now the Web is full of sites and discussion groups dealing with the brew. Personal reports of Ayahuasca abound in them. Furthermore, just as the writing of this book was being completed, there appeared a book by Metzner (1999), which contains a collection of firsthand reports, most of them by North Americans who are mental health practitioners and/or associated with the spiritualism of the New Age genre. Thus, in the span of just a couple of years, Ayahuasca turned from an exotic, almost obscure, Amazonian brew to the prime psychedelic substance of the new millennium. But while the social scene has radically changed, not so the scientific psychological one. All statements made in this book regarding the scientific cognitive-psychological study of Ayahuasca remain valid to this day. Also to be noted is the recently published anthology by Luna and White (2000) which includes many firsthand reports of the Ayahuasca experience.

> Probably no other new world hallucinogen—even peyote—alters consciousness in ways that have been so deeply and completely evaluated and interpreted. Caapi (i.e. Ayahuasca) truly enters into every aspect of living. It reaches into prenatal life, influences life after death, operates during earthly existence, plays roles not only in health and sickness, but in relations between individuals, villages and tribes, in peace and war, at home and in travel, in hunting and in agriculture. In fact, one can hardly name any aspect of living or dying, wakefulness or sleep, where caapi hallucinogens do not play a vital, nay, overwhelming, role.

The following statement made by an indigenous shaman from the Sibundoy valley in Southern Colombia supports Schultes's assessment; this statement is cited in Ramírez de Jara and Pinzón Castaño (1992: 289): 'Yagé is a force that has power, will and knowledge; with it we can reach the stars, enter the spirit of other people, know their desire to do good or bad; we can foresee the future of ours and others' lives, see illnesses and cure them, and with it we can travel to heaven or hell.' Likewise, Taussig (1987: 140) observes that: 'The Indians ... say that yagé is a special gift from God for the Indians. ... "Yagé is our school, yagé is our study," they would say, and yagé is conceived of as something akin to the origin of knowledge and society. It was yagé that taught the Indians good and evil, the properties of animals, medicine and food plants.'

Normally, the Ayahuasca potion is made out of two plants. While there may be variations in the plants being used, usually one plant is *Banisteriopsis caapi*, a liana of the *Malpighiaceae* family, whereas the other is *Psychotria viridis*, a bush of the *Rubiacaea* family. In common parlance, the term 'Ayahuasca' is used to refer not only to the potion but also to the first of the two constituent plants. In the Amazonian context, *Banisteriopsis* is often referred to as *jagube* or *mariri*, whereas *Psychotria* is generally referred to as *chacrona* or *chacruna* (in Portuguese and Spanish, respectively). In the various contexts of Ayahuasca use, it is often said that the vine gives 'power' whereas the leaves give light; usually, the former is characterized as male and the latter as female. When preparing the brew, the *Banisteriopsis* is cut into twigs and pounded; thereafter the pounded fibers are boiled in water together with the leaves of *Psychotria*. At times, other ingredients are added to the brew. These are said to affect the quality and power of the visions that are induced; so do variations in the *Banisteriopsis* itself—the wood's distance from the root, its size, age, freshness, and the like.

Chemically, the main active constituents in the Ayahuasca potion are the alkaloids N,N-Dimethyltryptamine or DMT, and the so-called harmala alkaloids—notably, harmine, harmaline, and tetrahydroharmine; these latter ones are all beta-carbolines which are MAO (mono-amine oxidase) inhibitors. DMT is a potent hallucinogen enhancing the activation of the neurotransmitter serotonin. When taken orally, however, DMT is deactivated by MAO. This deactivation is, in turn, deactivated by the MAO inhibitors which protect the DMT from deamination by MAO and thus render it orally active. For details regarding the psychopharmacology of DMT the reader is referred to Strassman (1994, 1996); information regarding the effects of DMT is found in Strassman (2001) as well as Spinella (2001). The effects of harmaline were examined some thirty years ago by C. Naranjo (1973*a*, 1979, 1987; see also Naranjo, 1973*b*).

Significantly, the active ingredients of Ayahuasca are found separately in the two plants of which the brew is made. Normally, DMT, the substance inducing the hallucinations, is found in the *Psychotria* leaves, whereas the MAO inhibitors are found in the *Banisteriopsis* vine. Thus, the consumption of each constituent of the Ayahuasca potion alone results in hardly any psychoactive effect.[4] For such effect to obtain it is necessary to join the two indicated plants (or their functional equivalents) so that both DMT and MAO inhibitors are present. When alternative constituent plants are used, the basic principle is always maintained: one plant contains DMT, whereas the second contains the MAO inhibitors.[5] For further information regarding the botany and pharmacology of Ayahuasca, the reader is referred to Chen and Chen (1939), Der Marderosian, Pinkley, and Dobbins (1968), Holmstedt and Lindgren (1979), Schultes (1972, 1982, 1986), Schultes and Winkelman (1995), Schultes and Hofmann (1980), and Ott (1993, 1994). Especially to be noted are the investigations of Grob *et al.* (1996) and Callaway *et al.* (1999). A recent review of the literature is found in Spinella (2001).

When one thinks about it, the discovery of Ayahuasca is indeed amazing. The number of plants in the rain forest is enormous; the number of their possible pairings is astronomical. The common sense method of trial and error would not seem to apply. This appraisal, to which many drinkers of Ayahuasca arrive on their own, is aptly presented in the following lines that P. Naranjo (1983: 352–3) cites from a conference presentation by Davis (1991; see also Davis, 1996, as well as Furst, 1976, and Narby, 1998):

> When we attempt to account for these discoveries, the phrase that is inevitably employed is 'trial and error'. It is a reasonable term and may well account for certain processes and transformations. But at another level it is but an euphemism disguising the fact that we actually have very little idea of how Indian people come up with their insights.
>
> The experimental process that originally led to the manipulation and combination of these morphologically dissimilar plants, and the discovery of their unique chemical properties, is far more profound than the term 'trial and error' suggests. The patterns that any researcher—and the shaman most certainly has earned that title—observes in nature depend on cognitive constructs, an intellectual synthesis, and reflect in turn culturally patterned thoughts and values. Sensitivity to nature is not an innate attribute of Indian peoples.
>
> It is a consequence of adaptive choices that have resulted in the development of highly specialized perceptual skills. But those choices in turn spring from a comprehensive view of nature and universe, in which man is perceived as but an element inextricably linked to the whole. It is this unique, cosmological perspective that has enabled the shaman to comprehend implicitly the intricate balance that is the Amazon forest.

[4] Usually, consumption of *Banisteriopsis* alone results in a feeling of inebriation, but not in visions.

[5] Another plant in which MAO inhibitors appear in high concentration is the Syrian rue, *Peganum harmala*, a shrub growing in the Middle East (for further details, see Flattery and Schwartz, 1989; Festi and Samorini, 1996; as well as Shanon, 1999).

The mystery pertaining to the discovery of Ayahuasca is reflected in the various Amerindian mythologies that present accounts of how the knowledge of this brew originated. In general, these accounts attribute the discovery of Ayahuasca to some special, unusual event. Supernatural agents or forces are usually involved. Often the discovery is linked to the first moments in the existence of the human species, to the definition of the relationships between human beings and the natural world, to the creation of culture, and to the very identity of the ethnic group in question (see Reichel-Dolmatoff, 1975, 1978, 1990; Chaumeil, 1983, and Lagrou, 1998).

Three Examples

As attested to by all who have had any significant exposure to Ayahuasca, the effects of this brew can be stupefying to the utmost. Both in the literature and in reports of people I have interviewed, again and again I have found Ayahuasca visions to be characterized as exhibiting a beauty that is beyond imagination. Invariably, the visions impress their viewers as marvellous, and when powerful they introduce drinkers to what seem to be enchanted realities that fill them with wonder and awe. The affective and spiritual impact that the Ayahuasca experience may have on people can be very profound. Often, people say that their exposure to Ayahuasca has radically changed their lives; many say that after this exposure they were no longer the same person. But even when it does not have radical ramifications on people's lives at large, the Ayahuasca inebriation is a wondrous experience that those who have been subject to it almost invariably describe as like nothing they had experienced before. I am saying all this by way of apology, for in a deep sense the effects to be discussed here defy verbal description. In order to be fully appreciated they have to be experienced firsthand. Yet, in order to give the non-initiated reader some taste of what will be talked about here, I shall try to do what I have just said cannot be done, namely, I shall resort to description by means of words and cite three reports of Ayahuasca visions that have appeared in the literature. Many more descriptions of visions, based on my own personal experiences and those of people I interviewed, will be presented in the chapters that follow.

I annotate the reports with two different marks. The asterisk indicates content elements that I have experienced in the same manner myself; the † symbol indicates content elements that appear in the reports of other people I have interviewed. The reader will appreciate that almost all the content elements in the cited texts are marked. In other words, taken as a whole, the annotations highlight the great cross-personal commonalities that are exhibited in Ayahuasca visions. This topic will be the subject of focal discussion in later parts of this work.

The first report is the first firsthand account that ever appeared in print—that of the Ecuadorian civil servant Villavicencio (1858; also see below); the following quote is taken from Reichel-Dolmatoff (1975: 30):

When I have partaken of Ayahuasca, my head has immediately begun to strain, then I have seemed to enter an aerial voyage*†, where I thought I saw the most charming landscapes*†, great cities*†, lofty towers*†, beautiful parks*†, and other delightful things*†. Then all at once I found myself deserted in a forest*† and attacked by beasts of prey*† against which I tried to defend myself. I began to come around, but with a feeling of excessive drowsiness, headache, and sometimes general malaise.

The second report is that of Kusel (1965: 64–5), a trader who lived in the upper Amazon for seven years and, in his words, 'was very skeptical and not interested in these low-class local manners'. Twice he partook of Ayahuasca and nothing happened; then there was the third time:

The first visual experience was like fireworks*†. Then a continuously creating power produced a wealth of simple and elaborate flat patterns in colour*†. There were patterns that consisted of twining repeats, and others geometrically organized with rectangles or squares that were like Maya designs or those decorations which the Chamas [the Indians with whom Kusel partook of Ayahuasca, B.S.] paint on their thin, ringing pottery*†. The visions were in constant flux*†. First intermittently, then successively, the flat patterns gave way to deep-brown, purple or green depths, like dimly lighted caves in which the walls were too far away to be perceived*†. At times snake-like stems of plants were growing profusely in the depths*†, at others these were covered with arrangements of myriads of lights that like dewdrops or gems adorned them*†. Now and then brilliant light illuminated the scene as though by photographic flash*†, showing wide landscapes with trees*† placed at regular intervals or just empty plains*†. A big ship*† with many flags*† appeared in one of these flashes, a merry-go-round*† with people dressed in highly coloured garments in another*†.

At a certain point I felt, helplessly, that [the person administering the session] and his song could do anything with me, which made me slide . . . deeper and deeper into a place where I might lose consciousness*†. If, to reassure myself, I opened my eyes, I saw the dark walls of the jungle covered with jewels as if a net of lights had been thrown over it*†. Upon closing my eyes again, I could renew the procession of slick, well-lighted images.

The colour scheme became a harmony of dark brown and greens. Naked dancers*† appeared turning slowly in spiral movements*†. Spots of brassy lights played on their bodies which gave them the texture of polished stones. Their faces were inclined*† and hidden in deep shadows. Their coming into existence in the centre of the vision coincided with the rhythm of [the] song*†, and they advanced forward to the sides, turning slowly. I longed to see their faces. At last the whole field of vision was taken up by a single dancer with inclined face covered by a raised arm*†. As my desire to see the face became unendurable, it appeared suddenly in full close-up with closed eyes. I know that when the extraordinary face opened them, I experienced a satisfaction of a kind I had never known. It was the visual solution of a personal riddle.

The third report was written by a traditional Shipibo healer from Peru, Guillermo Arévalo Valera (1986: 156–7); it is taken from an article in which the process by which one is trained to be an *ayahuasquero* (an Ayahuasca shaman) is described; the translation from the Spanish is my own:

The [master and the initiate] enter a marvellous world*†, passing through a path of gold†, until reaching a very beautiful city*† with most good-looking men and women*†. They are introduced to a factory whose machines operate perfectly†. . . . They exit through a secret door† and they enter a bar*† where they drink a cup of black wine. . . . Then they reach the coast of a river*† or the sea*†, and from the distance they see arrive a white boat*†, with a white flag*†. When the boat has reached the place where they stand, they climb up and they salute the chief. Thereafter they leave the boat and they continue walking. There are many beautiful flowers*† and pretty hummingbirds†. . . . They continue walking. Here there are birds that know how to speak mysteriously*† and that know one's name. When they see that someone is coming they start to pass the word that so-and-so is coming, and soon they discover a small city whose forms are that of a circle†. In the middle there is a lagoon surrounded by flowers*†, nothing less than the horn of plenty*†. In order to return, one has to pass through a subterranean path*†. In the centre of the city one encounters a very big castle*† where there live the God of ayahuasca along with some other gods*†. He who enters through the main gate of this castle will never return back*†. In this city the master and the initiate are now led to a bar. There they drink a large cup of chicha that brings about the effects of the *toé* [datura, B.S.]. Soon they meet the gods of medicine†. The initiate is recognized as a future healer of high category and the gods reveal to him how much is left for the conclusion of his diet.[6] They bid farewell and follow the road back.

The Various Contexts of Ayahuasca Use

For several centuries the indigenous people of the Amazon managed to keep Ayahuasca from the eyes of the white man (see Taussig, 1987). The first European to encounter the brew was the English botanist Richard Spruce, in 1851 (see Schultes, 1972). Spruce participated in an indigenous Ayahuasca ritual and was offered to partake of the brew, but he only sipped a small amount and then did whatever he could to counteract the effects of the intoxication. The first written report on a firsthand experience with the brew was that of the Ecuadorian civil servant Villavicencio (1858); this is the first report cited in the previous section. Until the middle of this century, the number of foreigners who had direct access to the brew was very small. In addition to the works cited throughout this chapter, the old reports of Koch-Grünberg (1921), Reinburg (1921), Rouhier (1924), and Karsten (1935) will be noted, as well as the early investigations of Fischer Cardenas (1923) and Mallol de Recasens (1963). A summary review of the early encounters with Ayahuasca is presented in Reichel-Dolmatoff (1975); a collection of excerpts from the early literature is found in Schleiffer (1973).

In this century[7] there have been two developments that significantly changed the situation. First is the appearance of new contexts of Ayahuasca use. The rubber

[6] In the Amazonian shamanic context, a diet (*dieta* in Spanish) is an integral part of Ayahuasca consumption. The diet consists of rigid restrictions on what foods may and may not be eaten. Sexual abstention is also demanded. The diet may last from three days to a week before the consumption of Ayahuasca as well as a couple of days subsequent to it.

[7] The twentieth century.

boom provoked by the First World War brought white entrepreneurs and mercenaries as well as many black manual workers to the Amazon regions. Contacts between rubber plantation workers, *siringueiros*, and indigenous persons resulted in the creation of several new, non-indigenous contexts of Ayahuasca consumption. In particular, three syncretic religious sects established in western Brazil are to be mentioned. The oldest and best known is the Church of Santo Daime, founded in the 1930s. This church brings together indigenous and Christian elements, and the sessions it administers are essentially Christian-oriented religious services in which Ayahuasca, referred to as *Daime*, serves as a holy sacrament. The second sect, the União do Vegetal, was founded in the 1960s. It likewise incorporates some Christian elements, yet its sessions are more akin to teachings in an esoteric school. The third and least known religious sect, the Barquinha, brings together the use of Ayahuasca with the Afro-Brazilian tradition of the Umbanda. Until about fifteen years ago, these groups were by and large confined to the Amazonian regions of Brazil. In the 1980s they began to expand throughout the urban centres of Brazil. Nowadays, there are *Daime* and União do Vegetal communities in practically all major Brazilian towns, and their members are people from all walks of life, with a growing number of professionals and members of the upper middle class. Recently, these groups have also established communities in Europe and the United States. Since details pertaining to the rituals of these groups are relevant for the data to be discussed here, more information about these will be given below.

Another modern development reflects the drastic change in travelling patterns throughout the world. Regions which previously were extremely inaccessible are no longer so. The easier facilities of international travel along with the socio-cultural changes in the United States and Europe have brought travellers pursuing psychedelic experiences to the Amazon. The first of these were the novelist William Burroughs and his companion, the poet Allen Ginsberg, who described their experiences with the brew in *The Yagé Letters* (1963). Since then, various personal-impressionistic reports on Ayahuasca have followed suit. In the last couple of years an increasing number of such reports appeared in the New Age literature as well as on various sites on the Internet. A very recent phenomenon is organized tours focusing on shamanism, as well as seminars and workshops for personal growth, in which Ayahuasca sessions directed by traditional healers are administered for payment to Europeans and North Americans who travel to South America especially for this purpose.

These developments did not leave the traditional, indigenous use of Ayahuasca unaffected. First, it shall be noted that the traditional societies are constantly menaced and they rapidly lose their cultural traditions. This includes Ayahuasca and its rituals. Often, there are no young persons who wish to become shamans, and with the passing away of the older shamans, the tradition, which is exclusively oral, simply dies out. Also, the psychedelic tourism is not a neutral phenomenon. When a healer caters to foreigners, for sums of money which for him are very significant, he is bound to be less available to the persons from his own community that are in real need of his curing services. I expect this effect to be more pronounced in the near future. I shall also note that in recent years the Santo Daime Church has had its impact on the indigenous populations of the Brazilian Amazon, and in some Indian communities Ayahuasca is now being used in the framework of this Christian church.

My own exposure to Ayahuasca has been both in the traditional indigenous context and in the newer syncretic ones. As the various details of these settings are necessary for understanding the data and analyses that follow, I shall now present some background information regarding each. Again, let me note that in line with the perspective of this book, which is psychological, not anthropological, my intention is not to present a complete socio-cultural survey of the various settings in which Ayahuasca is used. The following presentation contains only information that is directly relevant to the discussions entertained later in this book.

The Church of Santo Daime

The Church of Santo Daime ('holy *Daime*', with the latter word being a colloquial form for the Portuguese phrase 'give-me') is a syncretic religion bringing together indigenous traditions and the Catholic religion, as well as some Afro-Brazilian elements. In its rituals *Daime* (the name given to Ayahuasca in this context) is consumed as a sacrament. The term '*Daime*', however, refers not only to the brew, but also to the animated force believed to reside within the brew. Moreover, the *Daime* is regarded as a divine being or even as the Divine itself, and is believed to be the source of life, vitality, and health on the one hand and all knowledge and supreme wisdom on the other hand. Thus, *Daime* is central in the rituals of the church that carries its name not only as a sacrament but also as the source of the doctrine that is believed to emanate from it.

The Church of Santo Daime was founded in the early 1930s by Raimundu Irineu Serra, a black *siringueiro* who came to the Amazon from the north-east of Brazil in the 1920s and who learnt about Ayahuasca from the native Indians. Alone, he consumed the brew repeatedly for eight days, during which he experienced a vision where a figure he identified as the Queen of the Forest revealed herself to him and ordered him to name Ayahuasca '*Daime*' and to create a church in which this brew would be administered. When Mestre Irineu passed away there was a split between his family and his disciple Sebastião Mota; nowadays most members of the Santo Daime Church are affiliated with the second line, that of Padrinho Sebastião.

Originally, the centre of the Santo Daime Church was at the Colônia Cinco Mil, a ranch on the outskirts of Rio Branco, the capital of the Brazilian Amazonian state of Acre. However, with the growing urbanization of the area, the centre was moved further into the forest, to the settlement of Mapia, along the Rio Purus. For the first few decades of its existence, the Church of Santo Daime was confined to the Amazonian region and its members were exclusively simple farmers, many of them being the descendants of rubber plantation workers. However, as noted above, in the past two decades the church has expanded dramatically. Travellers encountered the *Daime* and brought it to the urban centres of Brazil, in the so-called *sul*, the South. Nowadays, there are *Daime* congregations in all major cities in the country, and recently congregations have also been established overseas. This expansion has many ramifications. These are an issue for a sociological (or anthropological) investigation and I will not dwell on them here (for recent studies, the reader is referred to Castillo, 1996, as well as to a work currently being conducted by Groisman).

The *Daime* sessions are called *trabalhos* (works); in all, consumption of the brew is pivotal. Except for the rituals in which the *Daime* is prepared, all sessions are held in a closed, fully illuminated hall. At the centre of the hall there stands a table which is usually star-shaped, at the head of which sits the leader of the session, the *padrinho*—a person who is both the leader of the community and the one in charge of the session. To the right of the *padrinho* are rows of women, and to his left rows of men. Either behind the *padrinho* or at the other side of the table, facing him, are the musicians. In addition, around the table are seated key members of the community, among them the *madrinha* (the mother, the *padrinho*'s wife).

Five types of sessions are administered. The regular sessions are referred to as *concentração* (concentration or meditation sessions) and are held twice a month. The sessions open with the serving of the *Daime* to all present. Usually, there is another serving later on in the session. The central part of each session is a period of one and a half to two hours in which the participants sit in silence. Before this, and also afterwards, the participants pray and sing. The sessions usually begin around eight in the evening and last for about six hours. Second are the festive sessions, held on fixed dates. These include Christmas, New Year's Day, and various saints' days, as well as special dates in the life histories of the founders of the church; festive sessions may also be held to celebrate the anniversaries of leaders of particular communities as well as marriages. In the festive sessions, the participants stand aligned in ordered formations, and they dance and chant, accompanied by music performed by a group of players. In general, guitars are used but other instruments may be added as well—an electric organ or even classical instruments such as a flute or a clarinet. In addition, many participants use the *maraca*, a rattle. This is a derivative of an indigenous instrument made from a gourd; in the *Daime* setting, this instrument is made from tin or aluminum cans with seeds inside. Usually the festive sessions take place at night and they last for about eleven hours, so as to terminate after the break of day. At times, though, for practical reasons, the sessions are held during daytime. At the festive sessions the *Daime* is served four or five times. After the second serving there is usually an intermission during which the participants leave the hall where the session is being held to go outside and relax. A third type of session is that of the *trabalhos de estrela* (works of the star, after the star-shaped table at the centre). These are intimate sessions with a small number of participants that are held on special occasions decided on by the local *padrinho*. Usually they serve as healing sessions and, as such, they focus on one person who is ill. In these sessions there is chanting, but no dancing. The fourth type of session is held very rarely—it is a *missa* (mass) for the dead. Last but not least, there is the *feitio* (preparation), the ritual in which the *Daime* is prepared. The *feitio* is the only type of session that is held outside, in nature. Usually, this ritual starts in the late hours of the night and lasts until the late afternoon of the following day. Typically, the dosages of *Daime* served during the *feitio* are larger than those served in all other types of works; more on the preparation of the *Daime* is said below.

The core of *Daime* ritual sessions are the *hinos* (hymns) which together comprise *hinarios* (hymn books), of which there are several. The hymns are said to have been 'received' under the effect of the *Daime* by *padrinhos* and other prominent members of the community. The totality of the hymns are the only written teachings of the church and are said to present the doctrine of Santo Daime. Objectively, the hymns are rather simple texts consisting of prayer, supplication, and praise, as well as descriptions of visionary scenes and beings. Under the intoxication, however, additional readings are invested upon these texts and they are viewed as presenting hidden esoteric wisdom. Typically, hymns are several stanzas long and there are about 120 hymns in a *hinario*. Overall, the hymns manifest a progression that parallels the progression exhibited by the different stages of the Ayahuasca inebriation. Usually, the first hymns are benedictions and blessings for a good session, then come hymns expressing difficulty and struggle. Later in the sequences are hymns describing visions and religious devotion. The concluding hymns are usually songs of praise and jubilation. Musically, the hymns resemble popular and religious songs of the Brazilian north-east, and they combine European and African elements. Personally (and I am definitely not alone in this regard), I have found the *Daime* hymns exceptionally beautiful.

A few further details regarding the preparation of the *Daime* brew are in order here. The work is divided between men and women. The men collect the vines, cut them into pieces about a foot long, and pound the *jagube* (*Banisteriopsis*) lianas; the women collect and clean the leaves of the *chacrona*, but they do not participate in the *feitio* itself. During the *feitio*, the pounding is conducted by eight to fourteen men sitting on small, individual wooden logs. Hymns are chanted and the pounding (which is physically quite demanding) is done in unison to the rhythm of the chant, but otherwise complete silence is maintained. The *padrinho* is in charge of the preparation of the brew—the collection of the two plant ingredients, the addition of water, the determination of quantities and of cooking times. Other people help with the preparation of the fire, the wood, and the water.

For further details regarding the history, doctrine, and rituals of the Santo Daime Church the reader is referred to Monteiro da Silva (1983), Cunha (1986), Froes (1986), MacRae, (1992), Groisman and Sell (1995), Castillo (1996), and Groisman (1999); some of these books have texts of various hymns sung in the *Daime* rituals. It is pertinent to note that while the authors indicated are all anthropologists and social scientists, all of them are, or have been, members of the church as well. Of special interest are two books written by Alex Polari, a poet, journalist, and an ex-radical Marxist activist, who after nine years in jail encountered the *Daime* and converted; now Polari is one of the two heads of the Santo Daime Church.[8] In his first book, *O Livro das Mirações* ('The Book of the Visions'), Polari (1984) recounts his personal history and his experiences with the brew; this book presents what in my opinion are the best written descriptions of Ayahuasca visions by a modern Western person, and citations from it will be made on several occasions in places throughout this book. In a second book, Polari (1992) recounts the biography of Padrinho Sebastião and presents further information about the Santo Daime Church and his experiences in it.

União do Vegetal

Like the Church of Santo Daime, the *União do Vegetal* (the 'plant union'; henceforth, UdV) is a syncretic religious sect. *Hoasca* (or the *vegetal* [plant], or *cha* [tea]), as the brew is referred to in this context is, again, consumed as a sacrament and is regarded as the fountain of the ultimate knowledge constituted by the teachings of the UdV. These teachings are not written and are revealed only to the adepts.

The UdV was founded in 1961 following an encounter of a *siringueiro*, Gabriel da Costa, with the native inhabitants of the Amazon. Originally, the UdV was based in Porto Velho, in the western Brazilian state of Rondônia, but later it expanded. Like the Church of Santo Daime, it now has congregations in all major cities of Brazil. The demographic profile of its members has changed accordingly and it includes people from all walks of life. Recently, congregations have also been established in Europe and North America.

Formally, the UdV is a church too, but in many respects its sessions are like classes in an esoteric school. The sessions take place in closed, illuminated halls with the participants sitting in armchairs, each one by him/herself. Presiding over the session is a *mestre dirigente*, a leading master. In contrast to the practice in the Santo Daime Church, the person directing the session is not necessarily the highest-ranking person present in the session. That person, the *mestre representante* (representing master), sits behind the administering master, supervises him, and offers advice and assistance whenever needed.

In the UdV a rigid hierarchy is maintained. The lowest grade is that of affiliates or simple members, then there are members of what is called the *corpus instructivo* (the instructive body), next are the counsellors, and at the top are the masters. The organization is divided into communities referred to as 'nuclei'. Presiding over each nucleus is a representing master. At the top of the hierarchy in the capital city of Brasilia is the general representative master. The epithet 'representative' indicates a relationship to the one, supreme head of the organization, the founding—now deceased—Mestre Gabriel.

[8] More accurately, he is the head of the organization that oversees the activities of the various Santo Daime congregations.

Regular UdV sessions—called *trabalhos de escala*, works of the ladder—are held twice a month. In addition there may be extra sessions convoked by the leader of the group. Throughout the year there are also several festive occasions which take place on fixed dates. All regular sessions are held at night and last exactly four hours. The sessions open with the serving of the brew to all persons present. In addition to these sessions, which are open to all members of the community, there are also sessions confined to only some members, according to their rank in the organization. Especially to be noted are instructive sessions; these are held during daytime. As with the *Daime*, there is also a special ritual for preparing the brew, the *preparo*, during which *hoasca* is served.

Central in the UdV sessions are the teachings by the presiding master. Participants are also invited to ask questions or make comments. These may pertain to the presentation given by the leading master or any other topic. Among the topics I have heard discussed are the history of the UdV and its mission, the nature of the universe, love, health, and faith. Several times during a session the leader and possibly other participants engage in singing *chamadas* (calls)—special, unwritten hymns of the UdV. At times, regular music, usually popular Brazilian music, is played from records and tapes. No other chanting or playing of music takes place during the session. For further general information the reader is referred to Henman (1986), Centro de Memoria e Documentação (1989), and De Souza (1996/7).[9]

Barquinha

Essentially, this is a syncretism between the Afro-Brazilian religion of Umbanda and the *Daime*. The name Barquinha is the diminutive of *barco*, a boat. This is because the founder of the sect, Frei Daniel Pereira de Mattos, had a vision in which he saw a boat, which he interpreted as a metaphor for the spiritual voyage. The Barquinha was founded in 1945 in the Amazonian town of Rio Branco in Brazil. Often, the members dress in seamen's uniforms. By many accounts, the brew served at the Barquinha sessions is more concentrated than that used in any of the other aforementioned groups.

There are various types of sessions. All told, the number of sessions administered during the year is much greater than in any other of the sects mentioned in this survey. Some sessions are based on Umbanda rituals in which participants enter into trance and 'incorporate' spirits. In non-festive sessions the participants are seated. Hymns, referred to as *salmos* (psalms), are sung by a cantor and the audience joins in for the refrains. Typically, these hymns have many stanzas and are much longer than the hymns chanted in the rituals of the Church of Santo Daime. Catholic prayers are interspersed between the hymns. In some sessions there is a *gira* (merry-go-round), a dancing ritual also taken from the Umbanda

[9] Just as this book was being completed, I was presented with a copy of a Master's thesis by Brissac (1999), which is an ethnography of the UdV.

tradition. The dance begins with a slow movement in a circle, and gradually participants may fall into a trance and enter the centre of the circle. Three times a year there are *romarías* (pilgrimages) during which sessions are held nightly over a long period of time (in the case of the pilgrimage of St Francisco, thirty-four days in a row).

Nowadays, the Barquinha is split into three different groups, all centred in Rio Branco with one also having a branch in Rio de Janeiro. These groups are independent of one another but they maintain very similar rituals. Anthropological research on the Barquinha has just begun. Master's theses have been written by Balzer (1999) and Sena Araújuo (1999).

Lastly, let me mention that there are other small groups that bring together the traditions of the *Daime* and the Umbanda and that do not pertain to the Barquinha. One in which I participated is that of the Baixinha (small one, the nickname of the leading woman in that congregation) near Rio de Janeiro.

Traditional Healers

Nowadays, in many indigenous and *mestizo* (mixed race) communities of the Amazon the only usage of Ayahuasca is medicinal. Normally, sessions are held at night, on fixed days of the week, at the house or 'clinic' of the *curandero* (healer). Sessions usually begin around 8 or 9 o'clock in the evening and they normally last for about six or seven hours. The participants are members of the community who wish to receive treatment. Typically, four to eight patients would be present; often, an apprentice-assistant is present as well. The sessions are conducted in complete darkness, and there is only one serving of Ayahuasca in them. The *curandero* drinks in order to see what is wrong with his patients and thus make a diagnosis, to consult with supernatural beings and entities on how to treat the patients, to be in touch with the good energies of nature and serve as a vehicle to pass them on to the patients and cure them. The patients, depending on their stamina and wishes, may or may not partake of the brew. All throughout the session the *curandero* chants *icaros*, 'power songs'. These are quiet, usually repetitive chants especially associated with Ayahuasca. At a certain stage of the session, patients are invited to come and sit with the *curandero* and he will sing especially for each one of them. The song is accompanied by the blowing of tobacco, the sucking of the negative energies of disease, the beating with a bunch of leaves (the *chapada*), and the spraying of perfumes and scents. On the day following the session, there are usually individual treatments with various medicinal plants, but without Ayahuasca.

There is a large anthropological literature describing the use of Ayahuasca in indigenous and mestizo contexts. Early reports regarding Ayahuasca use by indigenous people are found in Waisbard (1958/9), Prance (1970), Rivier and Lindgren (1972), Ayala Flores and Lewis (1978), and Taussig (1987), as well as in the non-scientific books of Lamb (1971, 1985). General works on indigenous Amazonian cultures and their shamanistic practices are presented in Harner (1972),

Deltgen (1978/9, 1993), Goldman (1979), Chaumeil (1983), Chiappe, Lemlij, and Millones (1985), and Lagrou (1998). Theoretical works focusing on Ayahuasca and its place in the indigenous Amerindian cultures are presented in Langdon (1979*a*, 1979*b*, 1992), Luna (1984*a*, 1984*b*, 1986*b*), Lagrou (1991, 1996, 1998), and Mader (1999). Accounts of indigenous Ayahuasca rituals very similar to those in which I have participated are found in Siskind (1973*a*), Weiss (1973), Harner (1973*a*), Kensinger (1973), Arévalo Valera (1986), Gebhart-Sayer (1985, 1986), Taussig (1987), Illius (1992), Hampejs (1994), and Fericgla (1998). A systematic investigation of Ayahuasca use by *mestizo* healers in a context very similar to those in which I have participated myself are presented in Dobkin de Rios (1972, 1973, 1992). Specific studies of the musical aspects of indigenous Ayahuasca rituals are found in F. Katz and Dobkin de Rios (1971), Siskind (1973*a*), Dobkin de Rios and Katz (1975), Bellier (1986), Buchillet (1992), Hill (1992), and Luna (1992). The reader will note that some of these items appear in the anthologies edited by Harner (1973*b*) and Langdon and Baer (1992), as well as the special volume of the journal *America Indígena* (volume 46 of 1986). Various more specific works are cited throughout this book when relevant.

Other Contexts

In closing, let me mention some other, more recent contexts of Ayahuasca use. An especially interesting one is that developed by the French physician Jacques Mabit. This European-trained medical doctor has had extensive experience with Ayahuasca as traditionally employed by indigenous healers. Bringing together this experience with his professional medical training, he founded a rehabilitation centre in which Ayahuasca is employed for the treatment of drug addicts. The centre, called *Takiwasi*, is located at the outskirts of Tarapoto in the upper Peruvian Amazon region. Ayahuasca sessions directed in the traditional indigenous manner are central in the therapeutic work administered in this centre. Other traditional medicinal herbs are employed as well. The traditional healing practices are coupled with individual and group psychotherapy sessions as well as Western medical supervision. Most patients in Takiwasi are locals who have been using coca paste but the clientele has also includes Europeans. Overall, Takiwasi reports a significant rate of success in its rehabilitation work. For further information the reader is referred to Mabit (1988, 1992, 1996) and Nakazawa (1996).

Another context—or rather, cluster of contexts—of Ayahuasca use is what I refer to as independent. Private independent Ayahuasca sessions are held by individuals who have vested interest in the brew but wish not to be affiliated with any institutionalized group (for further details see Ch. 3). Such sessions vary in their form and in the manner in which they are conducted. Some are held in private homes, some in natural settings. Some are strictly organized and directed whereas others are totally free in form. The music employed in these sessions may consist of *Daime* hymns, indigenous *icaros*, classical music, popular music, or New Age spiritually oriented music. The number of participants in them is usually small.

Finally, as already mentioned above, in recent years various contexts of Ayahuasca use especially catering to Europeans and North Americans have opened up. These include traditional Ayahuasca sessions especially administered to tourists visiting the Amazon region as well as the so-called shamanistic excursions and psychological workshops directed by Westerners in which Ayahuasca is administered. Somewhat unofficially, some similar activities are now conducted outside of South America.

A comment regarding the legal status of Ayahuasca is called for. In the countries of the Amazon, the use of Ayahuasca, whether in the indigenous context or in the contexts of the various modern syncretic religious groups indicated, is legal. In Brazil, during the regime of military dictatorship (in the 1980s), there was an attempt by the authorities to change this legal status, but eventually this was not upheld. In the Western world, the situation is not clear. In itself, DMT is a Schedule I substance, and as such its administration and consumption in the USA (and by implication, in many other countries) are illegal. Yet, it is not clear that this makes the consumption of the brew Ayahuasca illegal. Recently, following a police raid on a *Daime* ritual in the Netherlands, the Church of Santo Daime won a lawsuit, and consequently, achieved the formal legalization of the use of Ayahuasca in the context of its religious rituals. Similar litigations are now underway in other European countries as well as in the USA.

A Comment on Terminology

I conclude this survey of background information with a terminological comment. In the literature agents such as Ayahuasca are referred to by different epithets—notably: *psychoactive*, *psychotropic*, *psychedelic*, *hallucinogenic*, *psychotomimetic*, and *psychotogenic* (see Masters and Houston, 1966; Stafford, 1993). Of these terms, I prefer (and therefore use here) the first, which is the most neutral. Etymologically, the term psychedelic (mind expanding) is just as fine; yet sociologically, I find, it is too linked with the New Age culture and therefore I prefer not to use it. The term 'hallucinogenic' is problematic in two respects. First, while the generation of hallucinations is a major effect of Ayahuasca (and other agents), surely it is not the only one. Indeed, one principal message conveyed by the phenomenological survey sketched in this book is that the Ayahuasca experience is by far too rich and complex for the brew to be characterized merely as a generator of non-ordinary perceptual effects. The second reason for my not opting for the term 'hallucinogenic' is that often it carries with it derogatory connotations. In particular, in the clinical literature hallucinations are regarded as perceptual perturbations that reflect a malfunctioning of the brain and/or mind (see, for instance, West, 1962; G. Reed, 1972; and Siegel and West, 1975). This value judgement does not reflect the attitudes held by responsible users of Ayahuasca, traditional and syncretic alike—for them, the brew is taken in order to increase one's sensibilities, not diminish them. Significantly, some hymns of the Church of Santo Daime speak of the brew as freeing its partakers from the world of illusion (that is, the ordinary state of human existence) and bringing them to the world of true knowledge. As for the last term mentioned above, psychotomimetic (mimicking psychosis), it is clearly both grounded in prejudice and phenomenologically wrong. While, admittedly, some aspects of the Ayahuasca inebriation resemble some aspects of psychosis, the two experiences are fundamentally different. Psychosis is neither voluntary nor wished for: It is a sad condition imposed upon those who experience it, one placing them in a disorderly mental state over which they have no control. No one ever enjoys being psychotic or sincerely wishes to be one. In contrast, Ayahuasca, when taken in a responsible manner, induces a state of being characterized by great internal order and brings about feelings of personal growth, intellectual insight, spiritual uplift, and happiness. Furthermore, when experienced, the Ayahuasca drinker has significant control over what is happening to him or her under the intoxication. As will be exemplified throughout this book, the experienced Ayahuasca drinker may be regarded as a competent navigator or a co-dancer with the brew. Nothing of the sort is true of the psychotic (see Walsh, 1990). Finally, a new term has been introduced recently and is currently gaining hold in the literature—'entheogen' (i.e. that generating the god within; see Ruck, Bigwood, Wasson, and Ott, 1979). This term does justice both to the role Ayahuasca plays in the traditional indigenous Amazonian cultures and to its religious function in all the modern, syncretic groups mentioned in this chapter. I am very sympathetic to the perspective regarding Ayahuasca as a sacrament and salute it; yet, in this scientific book I prefer the more neutral terms indicated above. For a most illuminating discussion of the terminological issue, see Ott (1996).

Let me end with a practical note. Above I have used the adjective 'responsible'. Indeed, it is crucial that Ayahuasca be used in a responsible manner. Traditionally, the brew has never been used for hedonistic reasons and its consumption was always embedded within a well-structured ritual led by responsible individuals with substantial experience and attended by persons sharing similar attitudes and beliefs. Further, except for shamans on some occasions, people never took Ayahuasca alone. I strongly believe that these wise ancient provisos should be applied to all modern usages of the brew as well. And then, if Ayahuasca is to be taken as an instrument for healing, a mind-expander, a vehicle for the gaining of knowledge, or, indeed, as an entheogen, the likelihood is significantly increased that the experience induced by this potent brew will be an enriching and beneficial one. For a related discussion in the context of psychoactive agents in general, the reader is referred to Grinspoon and Bakalar (1979) (also see Ch. 19 below).

2

Theoretical Foundations

> There is a story, which I have read somewhere, to the effect that Mohammed once compared a scholar or philosopher who writes about mysticism without having had any mystical experience to a donkey carrying a load of books.
>
> Stace, *Mysticism and Philosophy*

The Scientific Study of Ayahuasca

Most scientific research on Ayahuasca falls into two categories. The first is that of the natural sciences—botany and ethnobotany, pharmacology, biochemistry, and brain physiology. The second category is that of the social sciences—most notably anthropology. The disciplines of the first category try to determine the identity of the plants from which Ayahuasca is made, analyse the active chemical constituents in them, and discover the pharmaceutical action these generate and the physiological effects they produce in human beings. It seems that by now scientists have gained substantive knowledge with regard to all these questions (for an especially clear discussion, see Ott, 1994). Anthropologists, in their turn, are primarily concerned with societies and cultures. In the context of Ayahuasca, usually their main concern is with the rituals— religious and medicinal—in which the brew is consumed and with the cultural traditions, beliefs, artistic creation, and customs related to the brew and its use.

To my mind, important though they are, neither type of investigation addresses the core of the matter. Both view Ayahuasca from the outside, so to speak. Ayahuasca is intriguing, I think, because of the extraordinary experiences it generates, but experiences as such are outside the province of both the natural and the social sciences. The various natural sciences tell us what Ayahuasca is made of and what neurophysiological processes it may produce, but they say nothing—indeed, they can but say nothing—about the special experiences associated with the brew. The social sciences look at things from the outside too. Usually, anthropologists describe how people of a given social or ethnic group (as a rule, other than their own) use Ayahuasca. They focus more on the context of consumption and less on the subjective experiences that the brew generates in the minds of individual persons. Indeed, as far as the anthropologist is concerned, the consumption of Ayahuasca and the rituals and other activities associated with it are socio-cultural phenomena. These are interesting, but they are not categorically different from any other ritualized activities that a community may perform—initiation rites, religious procedures, divination, and the like.

Admittedly, Ayahuasca would not have been known to us in the West had it not been for the daring adventures of botanists and anthropologists. Yet it is not the plants of which it is made that make Ayahuasca so special, and the riddles the brew presents are not in the rituals associated with it. What is special about Ayahuasca is the extraordinary subjective experiences this brew generates in the mind. Thus, to my mind, the real puzzles associated with the brew pertain neither to the brain nor to culture but rather to the human psyche. As such, the study of Ayahuasca belongs first and foremost to the domain of psychology, and more specifically, cognitive psychology—that is, the discipline engaged in the study of the mental life of human beings. The paradigmatic topics of concern of this discipline are language, memory, thought processes, problem solving and reasoning, creativity, and last but not least, the phenomenon of consciousness.

Between cognitive psychology and anthropology there are some fundamental differences in perspective. The anthropologist's point of departure is the study of societies and culture; the psychologist's is the individual. The anthropologist studies the other and attempts to understand his or her foreign culture. The cognitive psychologist investigates the common structures and mechanisms that comprise the human mind. These general differences in perspective also entail significant differences in the study of Ayahuasca. For the anthropologist, Ayahuasca is interesting for the new (as far as the scientific community is concerned) socio-cultural practices it presents, and the variety of the human experience that these reveal. In contrast, I as a cognitive psychologist focus on the individual person. My scientific work is couched in the appraisal that different though our cultural envelopes may be, we human beings are all essentially the same. In essence, the cognitive endeavour is guided by a perspective analogous to that underlying biological research. Just as all of us, members of the species *Homo sapiens*, have hands, livers, bones, and the like, we also have what have been referred to as 'mental organs'—language, memory, reasoning, and the like. For me, Ayahuasca is a tool to discover new, heretofore unknown territories of the human mind—be it the mind of an Indian or of a Western person, myself included. As I see it, the subject matter at hand is not the other but rather the self in general (hence, also my own self).[1]

Cognitive psychology should also be contrasted with other subdisciplines of psychology, notably medical psychology, clinical psychology, and psychiatry. The very few psychological studies on Ayahuasca in existence belong to these fields (see

[1] I appreciate that this presentation of the contrast between anthropology and psychology is crude and simplistic. While it may be too radical as a generalization, it does, I think, capture too the state of affairs in the study of Ayahuasca when the present project was conceived. Such a contrast also represents the rationale that has led me to the research project presented in this book. In Ch. 19, in a section written more than five years after this early chapter, I present a more complex, more conciliatory, and more updated, characterization of the relationship between the psychological and the anthropological (or rather, socio-contextual) as they pertain to the study of Ayahuasca.

Grob *et al.*, 1996; McKenna, Callaway, and Grob, 1998; Callaway *et al.*, 1999; as well as the early studies of the effects of harmaline conducted by C. Naranjo, 1973*a*, 1973*b*, 1987). While I will have some comments on Ayahuasca and healing, my primary interest here is different.

I will close this introductory section with a methodological issue that I find paramount. Above, I have characterized the study of Ayahuasca by both natural and social scientists as external. This is due not only to the general perspectives of the disciplines at hand, but also to the specific research methodologies taken by the individual investigators studying the brew. When I started to study the scientific literature on Ayahuasca I was surprised to discover that in most cases the information scientists have on the experiential phenomenology of Ayahuasca is secondhand.[2] Even when they do partake of the brew, scientists usually do so rather cautiously and as something peripheral to what they consider their principal object of investigation. This, I believe, has important ramifications for what can be learnt from the existing literature on the topic.

My own firm belief is that there is no alternative to studying phenomenology from within. The experiences that Ayahuasca induces are extraordinary in the full sense of the term, and many are ineffable. There is no way to really appreciate what they are without experiencing them firsthand. After all, would anyone venture to study music without actually experiencing how music sounds? Moreover, for a serious study of the Ayahuasca experience a cursory, explorative exposure to the brew is not sufficient. The spectrum of phenomena pertaining to the Ayahuasca experience is extremely broad and there is simply no way these can be captured in a small number of probes. Again, the analogy with music is instructive: in order to appreciate what classical music is, it is not enough for one to go to a couple of concerts or to listen to a dozen discs. And as with music, learning to know a field and to appreciate what is interesting about it requires longitudinal, cumulative experience. What happens to one under the Ayahuasca intoxication is determined not only by the brew itself but also by one's attitude and stance, and these, in turn, change over the course of time. In sum, then, any serious study of Ayahuasca requires not only firsthand experience, but also substantive, long-term familiarity—indeed, training.

General Theoretical Background

The present work is guided by the assessment that bringing together Ayahuasca research and cognitive psychology should be most fruitful. The benefits of this are, I think, twofold. On the one hand, cognitive psychology presents a new, and to my mind, most pertinent perspective for the study of Ayahuasca. On the other hand,

[2] This was almost a decade ago. Since then the number of investigators who have had firsthand experience with Ayhuasca has been steadily growing. However, as shall be indicated throughout this book, Ayahuasca is a school, not a one-time affair. It is just impossible to gain a serious appreciation of the Ayahuasca experience and to understand it having partaken of the brew only a handful of times.

Ayahuasca, with the unusual psychological phenomena it generates, opens new vistas for the study of the key manifestation of the human cognitive system, namely, consciousness.

My general approach to the study of cognition is not the standard one in the field today, and therefore I would like to say something about it. The view that dominates the cognitive sciences today is based on current thinking in computer science. Essentially, it is maintained that the human mind is a kind of very sophisticated computer. More specifically, it is assumed that the mind operates by means of applying computational operations upon underlying mental semantic representations. Mainstream cognitive psychologists conceive of their endeavour as the definition of the structure of these underlying representations and the dynamics of the procedures that are associated with them. This programme of research is grounded in what is known as the *representational-computational* view of mind (or, for short, representationalism). In Shanon (1993*a*), I have extensively criticized this view and argued that it is fundamentally wrong. My critique is based on both a conceptual philosophical analysis and a comprehensive examination of human behaviour as manifested in many different domains—language and thinking, learning and memory, perception and action, child development, and cultural evolution. In my opinion, the basic capability of the human cognitive system is neither the manipulation of symbols in underlying representations nor the computational processing of information but rather being and acting in the world. The argumentation is extensive, and this is not the place for its elaboration. Here, I would like to point out only a main methodological ramification of my view. If cognition is not based on computational operations applied upon covert underlying representations, then cognitive theory does not consist of the specification of such underlying models. Rather than the uncovering of underlying structures and mechanisms, I maintain that the primary goal of cognitive investigation is the charting of mental geography and the marking of regularities in it (for a detailed exposition of this position, see Shanon, 1993*a*).

My position is a radical phenomenological one. Technically, what I propose is that cognitive research be centred around what I call 'natural cognitive domains'. A cognitive natural domain exhibits the following characteristics. First, it is part of the cognitive expression of the human mind. Second, for all human beings it occurs naturally and spontaneously, without prior concentrated deliberation or active intended involvement on the part of the cognitive agent. Third, it is phenomenologically distinct, manifesting specific characteristics not shared by other domains. Fourth, it is well defined in the sense that for any specific phenomenological token of the domain, it is straightforward to determine whether or not this token is a member of that domain. Fifth, it is well demarcated in the sense that the totality of the types pertaining to the domain is distinct. Sixth, as a whole the domain manifests intrinsic regularities and substantial richness and complexity. The number of cognitive natural domains thus defined is very small. The prime natural domain is natural language. Other natural cognitive domains include spontaneous trains of mentations, dreams and daydreams, spontaneous gesturing, certain modes of musical expression, but not many more.

By way of clarification, I will review the six criteria in conjunction with the specific domain of natural language. Language is a natural expression of the human cognitive system. It is an activity in which we all engage without being cognizant of the rules that govern it, without having been explicitly taught how to perform, and—except for those who intently scrutinize language or study it professionally—without really appreciating the great complexity that it involves. As a whole, the system of language exhibits structural characteristics that are specific to it—the rules of syntax, phonology, and semantics. Indeed, the appreciation that these characteristics are specific to the domain of language and cannot be reduced to general functional principles pertaining to human psychology is one of the fundamental insights of modern linguistic science (see Chomsky, 1980). Further (here we arrive at the fourth of the six criteria indicated in the previous paragraph), for each token of language—e.g. a word or a sentence—it is clear that it pertains to this particular domain and not to any other. Specifically, the determination that a sentence is a token of language and not of music is direct; so is, in fact, the determination that any given word or sentence is an expression in one specific language and not another. Further still, as a whole, the linguistic domain is well differentiated from other domains of human expression, for instance singing or motor performance. Finally, the structural regularities exhibited in language manifest a coherent system of significant richness and complexity; technically, this system is usually referred to as 'grammar'.

Guided by the belief that the way to appreciate the 'natural history' of mind is to map the various cognitive natural domains and to define the regularities they exhibit, I have devoted much of my previous work to the systematic study of what I refer to as 'thought sequences', namely, trains of verbal-like expressions that spontaneously pass through people's minds (see Shanon, 1989*a*, 1993*b*). I have also studied trains of mental imagery (Shanon, 1989*b*) and some patterns pertaining to dreams (Shanon and Eifermann, 1984).

With all this, it should be clear why I as a cognitive psychologist (as distinct from an individual person with his own curiosities and affinities) find the study of Ayahuasca—as well as other non-ordinary states of consciousness—so pertinent. As noted in the prologue, I encountered Ayahuasca as a traveller, by chance and without having had any prior special interest in either psychedelics or in Amerindian cultures. On the basis of my personal experience with the brew and my subsequent reading of the anthropological literature, it occurred to me that the visions and other non-ordinary experiential phenomena that Ayahuasca induces present a new, uncharted natural cognitive domain.[3] Since the number of natural

[3] The characterization of the Ayahuasca experience as a natural cognitive domain should be qualified with respect to the second criterion. While this experience does not depend on prior concentrated deliberation or active involvement on the part of the subject, it does not occur spontaneously without the ingestion of a foreign substance.

domains is very small, this makes the Ayahuasca experience of paramount interest for the student of mind.

Lastly, I will comment on the issue of biological reductionism. Reductionism is the thesis by which phenomena pertaining to what are referred to as higher-level domains can be explained in terms pertaining to lower-level ones; correspondingly, higher-level scientific theories can be formulated by terms pertaining to lower-level theories. In particular, by the reductionist perspective, the facts of psychology can, at least in principle, be fully explained in terms of neurophysiology. My talking of Ayahuasca, a particular plant concoction containing well-defined biochemical molecules, may give the impression that my theoretical stance is reductive. Specifically, the impression may be that my position is that whatever happens under the influence of Ayahuasca is a direct result of chemicals that one ingests and the pharmacological influence these have on one's brain neurophysiology. My underscoring of the specificity of some facets of the Ayahuasca experience may reinforce this impression. Appreciating this, I wish to emphasize that my overall psychological outlook is categorically *non*-reductive. For reasons which are totally independent from my study of Ayahuasca (for extended discussion see Putnam, 1973; Fodor, 1975, as well as Shanon, 1993*a*), I totally reject the possibility that biological accounts—detailed as they might be—can offer viable psychological explanations. Obviously, without a brain, nervous system, and body physiology, we human beings could not accomplish all that we do as cognitive agents. This trivial technical truth, however, should not be confused with theoretical cognitive-psychological claims. Like any theory, cognitive theory is concerned with the definition of lawful patterns in its domain of enquiry. The lawful patterns of cognition, I believe, are defined in terms of experience which is laden with meaning, not neurophysiological processes or brain events. The situation is analogous to that encountered in music. Admittedly, without a piano, piano music cannot come into existence. However, if one is to understand whatever is pertinent to the understanding of a piano sonata, it is senseless to study only the physics of the piano chords and their acoustics. Rather, one would make use of musically meaningful terms, such as those developed in the theories of melodic progression and musical harmony. *Mutatis mutandis*, the same applies to cognition and, by implication, to the cognitive study of the Ayahuasca experience. More on this topic will be said in the last two chapters of the book.

The Cognitive-Psychological Study of Ayahuasca

As noted, in this book I set myself to present a cognitive-psychological study of the Ayahuasca experience. On the one hand, I propose to study the experience as an experience—from within, based on the encounters with the brew that I, as well as other people, had. On the other hand, I am determined to keep the investigation within a scientific framework. In particular, both the questions I raise and the theoretical means by which I try to address them are grounded in contemporary cognitive-psychological thinking.

What, then, are the cognitive-psychological questions that the Ayahuasca experience raises? In essence, the cognitive-psychological enterprise consists of an attempt to describe the special phenomenology presented by Ayahuasca and to model it in terms of the current empirical and theoretical knowledge of the workings of the human mind. The cognitive-psychological questions of interest may be divided into several main clusters.

First are *phenomenological questions of the first order*. These are concerned with the systematic characterization of the phenomenology of the Ayahuasca experience. Essentially, these questions all pertain to one arch-question, namely—what is being experienced with Ayahuasca? What kinds of visions does Ayahuasca induce? What are the contents of these visions? What other kinds of experience are generated by the brew? These other kinds include non-visual perceptual effects, ideas and insights, and emotional and bodily effects. Note that some of these questions pertain to form whereas others pertain to content. Loosely speaking, the distinction between these two kinds of questions is analogous to the linguistic one between syntax and semantics—syntax pertains to the forms that linguistic expression can have, semantics to their contents. In the present context, both questions of form and of content are structural in that they are concerned with the space of possibilities, not with the actual occurrence and progression in time.

The *phenomenological issues of the second order* are concerned with lawful patterns revealed by relations between the elementary phenomena pertaining to the questions of the first order. Is there an order in what one experiences? Are there regularities in the progression of the visions and other experiences that Ayahuasca induces? Can distinct stages be defined? What are the patterns associated with moves between stages of visions? Also to be investigated are the progressions of experiences across sessions and their change over the course of long-term usage of Ayahuasca.

Third are the questions of *dynamics*. These are concerned with *how* the Ayahuasca experience unfolds in time. Closely related to the dynamic questions are the *contextual* ones. How are the various facets of the intoxication affected by the context in which one is situated—by the place, the social milieu, the interpersonal relationship at hand, the ritual being employed? Likewise to be checked are the contributions of one's attitude, emotional reactions, and past experience with the brew. In the framework of the study of altered states of consciousness, these two sub-clusters of questions are usually referred to by the labels 'setting' and 'set', respectively (see Leary, Metzner, and Alpert, 1969; Zinberg, 1984).

On the basis of all the foregoing types of questions and the analyses that they direct, we can take a more global, and more abstract, perspective and turn to the *theoretical* questions. In particular, we can examine whether the various facets of the phenomenology of the Ayahuasca experience can be characterized as the manifestations of a small set of psychological factors or not. With this, a general explanation of the various phenomena being observed should be attempted. Note that given the radical phenomenological approach I advocate, in the present context explanation should be understood in a strict psychological sense, not in any reductive computational or physiological one.

A last cluster of cognitive-psychological questions pertains to *interpersonal comparisons*. These examine similarities and differences between the experiences of different individuals. Are there commonalities in the experiences of different people? In particular, are there commonalities in the experiences of people from different backgrounds, cultural or otherwise? And what, as the case might be, are the differences between them?

Taken together, all the questions indicated thus far mark the potential contribution of a modern cognitive-psychological analysis to the study of Ayahuasca. However, as noted, the bringing together of Ayahuasca research and cognitive psychology defines a two-way interaction. Not only can a cognitive-psychological analysis make a crucial contribution to the study of Ayahuasca, the converse is also the case—the study of Ayahuasca may have implications of import to our general understanding of the workings of the human mind. Ayahuasca (along with other mind-altering substances) expands the horizons of psychology and reveals new, hitherto unknown territories of the mind. Thus, the study of Ayahuasca presents new data pertaining to human consciousness, and thus new issues for investigation, new ways to look at things, new questions, and perhaps even new answers.

In this regard, it is pertinent to cite observations made at the beginning of the twentieth century by one of the founders of modern scientific psychology, William James in his *Varieties of Religious Experience* (1929: 378–9):

> Our normal waking consciousness, rational consciousness as we call it, is but one special type of consciousness, whilst all about it, parted from it by the filmiest of screens, there lie potential forms of consciousness entirely different. We may go through life without suspecting their existence; but apply the requisite stimulus and at a touch they are there in all their completeness, definite types of mentality which probably somewhere have their field of application and adaptation. No account of the universe in its totality can be final which leaves these other forms of consciousness quite disregarded.

These lines are famous and they are quoted often, but not so often cited are the two sentences that introduce them:

> Some years ago I myself made some observations on [the effects of] nitrous oxide intoxication. . . . One conclusion was forced upon my mind at that time, and my impression of its truth has ever since remained unshaken. It is that . . .

And there follow the lines that I have cited above; for more information the reader is referred to James (1882).

The novelist-philosopher Aldous Huxley (1971) made similar observations following his personal experience with another psychoactive substance, mescaline. The mescaline experience led Huxley to write two essays, *The Doors of Perception* and *Heaven and Hell*. The following quotes are from the opening pages of the latter:

> Like the earth of a hundred years ago, our mind still has its darkest Africas, its unmapped Borneos and Amazonian basins. In relation to the fauna of these regions we are not yet zoologists, we are mere naturalists and collectors of the specimen. . . . Like the giraffe and the duck-billed platypus, the creatures inhabiting these remoter regions of the mind are exceedingly improbable. Nevertheless they exist, they are facts of observation; and as such, they cannot be ignored by anyone who is honestly trying to understand the world in which we live. (p. 71)

> A man consists of what I may call an Old World of personal consciousness and, beyond a dividing sea, a series of New Worlds—the not too distant Virginias and Carolinas of the personal subconscious and the vegetative soul; the Far West of the collective unconscious . . . and, across another, vaster ocean, at the antipodes of everyday consciousness, the world of Visionary Experience. (p. 72)

Likewise, when exploring the geography of the mind, claims and suppositions based only on ordinary states of consciousness (notably, the states of normal wakefulness and of dreaming) are not sufficient. Any general, comprehensive theory of cognition has to encompass both the ordinary and the non-ordinary facets of the mind. Thus, the new phenomena revealed by Ayahuasca (along with other psychoactive agents) have significant ramifications for psychology at large.

While the present investigation is focally concerned with cognition, the bringing together of the study of Ayahuasca and psychological research also has bearings on issues that extend beyond the cognitive-psychological domain proper. Specifically, the cognitive-psychological investigation may shed new light on phenomena within the domains of the other disciplines that have been involved with the study of Ayahuasca. In particular, the new line of research undertaken here may present new accounts of issues that have been of concern to anthropologists, amongst them those pertaining to music, mythology, shamanism, and healing.

Last but not least, I would like to note that Ayahuasca also raises intriguing *philosophical* questions. Some of these are closely related to the study of cognition—questions regarding the nature of mind and the relationship between it and the world. Other pertinent philosophical questions have to do with the human predicament, the nature of religion, and the study of culture. There are also fundamental questions—or rather, puzzles—pertaining to ontology and metaphysics.

One methodological comment before I continue. Throughout this discussion I shall be talking only of the phenomenology of the Ayahuasca experience. However, by no means do I intend to confer here any special status on Ayahuasca as compared to any other psychoactive agent. The discussion is confined to the Ayahuasca experience simply because this is the only domain of its kind of which I have good personal knowledge and which I have studied systematically. While many indigenous persons and some Western investigators have claimed that Ayahuasca is indeed special in comparison to other psychoactive substances (see, for instance, the comment by Schultes cited in the previous chapter), I shall say nothing on the substantive comparison between Ayahuasca and other substances, be they natural or synthetic. Methodologically, however, I would like to note that everything said here is, in fact, naturally extendible to all other domains of the so-called altered states of consciousness,[4] whether they are induced by means of psychoactive substances or not (as in various mystical states). Indeed, this text may be regarded as a general call for the cognitive-psychological (as contrasted with the medical, physiological, psychoanalytical, or clinical-psychological) study of the non-ordinary facets of human consciousness. In this vein, the methodological and conceptual distinctions introduced here may be applied to the study of altered states of consciousness in general, whatever the agent or method that induces them.

A More General (and also Personal) Perspective

These introductory programmatic comments would not be complete without noting the limitations of the enterprise to be undertaken in this book. Even before we start, I wish to emphasize the mysterious nature of the Ayahuasca experience. My goal here is to attempt a psychological investigation and, at times, I shall also tackle some philosophical questions. Yet, I have no illusions—the scientific study of Ayahuasca is bound to be limited. In fact, it seems to me that ultimately, no psychological (or any other scientific) investigation can fully unravel the secrets that Ayahuasca presents. Indeed, I am inclined to say that in various respects Ayahuasca brings us to the boundaries not only of science but also of the entire Western world-view and its philosophies.

Furthermore, my dealing with Ayahuasca in strict cognitive terms must by no means belittle the great personal—psychological, spiritual, metaphysical, and religious—significance that this brew imparts on those who partake of it, myself included. Fellow drinkers of this brew (notably, people who served as informants to me) have repeatedly pointed out to me that the visual experiences that are a prime focus of my empirical investigation are not at all the most important facet of the Ayahuasca experience. By far more significant, they said, are the psychological self-understanding, personal growth, and spiritual experience that the brew induces. As a person, I tend to concur. Here, however, let me just say that in this context I am doing the best I can as a Western student of cognition. One thing that the present investigation endeavours to show is that non-ordinary experiences such as those induced by Ayahuasca can and should be examined from a modern cognitive-psychological perspective. Many have claimed that the experiences usually referred to as ones of altered states of consciousness are ineffable, and that there is nothing objective that can be said about them. This entire project of research underscores a definite disagreement with this categorical claim. In fact, there is much that can be said about these non-ordinary experiences, and given

[4] Some use the term 'alternating states of consciousness'. Since to my mind continuous dynamical change and alternation between states is a key feature of the system of human consciousness, I prefer the term 'non-ordinary states of consciousness'.

sufficient empirical data and proper theoretical means, there is much that modern academic treatment can contribute to their understanding. (Obviously, there is also much that these very experiences can contribute to science and the Western world-view in which it is embedded.) Admittedly, there are facets of the Ayahuasca experience that objective discourse may fail to capture or convey; perhaps these facets are even the more important ones that Ayahuasca presents. Yet, none of this implies that nothing at all can be said on the subject matter at hand, and that academic investigation, such as the one undertaken here, is not feasible or important.

Let me add a more personal word. Numerous times during my Ayahuasca journeys people asked me whether my interest was professional or personal. In my own life, the line dividing the professional from the personal is almost non-existent. Obviously, the extensive, and not always easy, involvement in the Ayahuasca quest has had a deep impact on me as a person. Indeed, this quest has been one of the most meaningful enterprises I have undertaken in my life, and it has had a very significant effect on my self-understanding, life perspective, view of the world, and opinion on various matters. While my explicit aim in this work is to attempt a scientific investigation of the Ayahuasca experience, undoubtedly the investigation (like, for that matter, any intellectual endeavour) is not unaffected by my own personal and idiosyncratic perspective on things. While, in various manners and in various degrees, this will be evident throughout this book, I have done my best to ensure that it does not hinder the main thrust of the work, which is, as repeatedly noted, scientific.

Bearing in mind all foregoing comments, let me suggest a golden guideline. On the one hand, one should beware of mystification of the Ayahuasca experience. Indigenous shamans have appreciated this even though they had no knowledge of or concern with scientific work. As one Kamsá shaman told Taussig (1987: 455) 'Ayahuasca is the worst of liars.' The methods of Western science are perhaps the best practical tools at our disposal to guard us against the pitfalls of uncritical mystification. But on the other hand, throughout one's dealing with the non-ordinary phenomenology of Ayahuasca, one should always maintain the awe and humility that we—both as human beings and as scientists—should hold in the face of the mysteries that mind and nature present to us. It would be advisable for all of us Western researchers to bear in mind the words of that scientifically rigorous, non-sentimental Harvard patrician of natural science, Professor Schultes, namely, that Ayahuasca is offered to us as one of the 'plants of the gods' (Schultes and Hofmann, 1979). Whatever one's beliefs, opinions, and persuasions, it is advisable, I think, to always appreciate that this is, indeed, a *bebida sagrada*, a sacred drink.

3

Methodology and General Structure

> Like a geometer who sets himself
> To square the circle, and is unable to think
> Of the formula he needs to solve the problem,
> So I was faced with this new vision
>
> Dante

My study of Ayahuasca is based both on my personal experiences with the brew and on data I have gathered from other people. In this chapter I present information pertaining to the data and the procedures of their collection and analysis. Subsequently, the general orientation of my work is defined and the overall structure of this book is drawn.

My Own Data

All told, I have actively participated in more than 130 Ayahuasca sessions held in different settings. These span a period of ten years, beginning with my first visit to Brazil in 1991. Since then, I have returned to South America at least once a year. In particular, I shall mention a journey of more than a year (1993–5) which included an extended stay in the Amazonian regions of Brazil, Peru, and Colombia. On other occasions I have also been to the Amazonian regions of Ecuador. In all these visits, participation in Ayahuasca sessions and interviewing people about issues pertaining to the brew were central. I have also partaken of Ayahuasca in private settings outside South America. Except for the first few, all my sessions were summarized in writing immediately after they had ended.

I should note that, in all cases, I participated in sessions as a full-fledged participant, not as a researcher. In particular, when with traditional healers, I joined in not as someone who was investigating their practices, but rather as one who came to learn from them. Some of the people I was with knew that back home my profession is that of a university professor, but my interaction with them was not in that capacity. I might add that during sessions I never took pictures or video recordings. On occasion, though, I made audio recordings of sessions with my walkman.

My data were subject to a host of analyses, many of which were quantitative. Having to draw a line somewhere, I have decided to include in the quantitative analysis only those 67 sessions that took place before 1996, the date when I began writing this book. Henceforth, the data pertaining to these sessions will be referred to as the 'core corpus'. Details regarding both the core corpus and subsequent sessions in which I participated are given below. The non-quantitative aspects of the investigation take into consideration also sessions other than those of the core corpus.

I have drunk Ayahuasca in the following settings: the various rituals of the Church of Santo Daime, the meetings of the União do Vegetal, the Barquinha, healing sessions conducted by indigenous or mestizo medicine men, and privately—either together with a small number of persons or alone. After the closure of the core corpus I also participated in sessions conducted at the Takiwasi rehabilitation centre and in ones especially organized for foreign tourists in the Peruvian Amazonian region. General background information regarding the groups and settings just noted was presented in Ch. 1; following are further details regarding my own involvement in these and other contexts.

The Church of Santo Daime

Of the 67 sessions comprising the core corpus, 20 were *Daime* sessions. I have participated in all five types of sessions described in Ch. 1—regular concentration sessions, festive sessions, healing sessions, masses for the dead and, once, a preparation ritual. These sessions were held in different communities throughout Brazil. Later, I also participated in several sessions held in *Daime* communities in the Netherlands. In all these sessions, my status was that of a visitor.

União do Vegetal

The core corpus includes eleven UdV sessions. These took place in different communities throughout Brazil. Except for one preparation session, all the sessions in which I participated were either regular sessions or especially convoked sessions essentially similar to them. Here, too, my status was that of a visitor.

Barquinha

The core corpus includes only one Barquinha session. It was a festive session held by the congregation of Dona Chica Gabriel in Rio Branco in the Brazilian state of Acre. Later I participated in several other sessions with this group. Twice, I have participated in the meetings of other groups that bring together the traditions of the Umbanda and the *Daime*.

Traditional Healers

Sixteen sessions in the core corpus were held with traditional healers. They were conducted in various regions of the Peruvian Amazon—in the towns of Pucallpa, Iquitos, and Puerto Maldonado, as well as in villages in their vicinity. The healers were members of different indigenous groups—Shipibo, Yagua, Cocama, and Lamas as well as persons of mixed race. Later I participated in sessions conducted by *ayahuasqueros* from the Chalahuita, Secoya, and Shuar tribes. All told, I have been with thirteen healers; on a couple of occasions I have assisted in the healing work myself.

In addition, I twice drank Ayahuasca with Indians of the Inga tribe, in the Sibundoy valley and by the Rio Putumayo in southern Colombia. In both cases the session was convened especially for me—the first was communal and the second was an individual session in which a cleansing ritual took place. Both sessions are part of the core corpus.

Private and Other Settings

In Brazil, 12 times I have partaken of Ayahuasca in private settings with persons I call 'independent drinkers'. These are men and women who have in the past belonged to one of the institutionalized groups mentioned above but then decided to leave them. Further information about these independent drinkers is given below.

I have also drunk Ayahuasca alone. The core corpus includes 5 such sessions. In these I played classical music as well as *Daime* and Ayahuasca recordings I made in South America.

As noted above, after the core corpus was completed I participated in sessions conducted at the Takiwasi rehabilitation centre in Tarapoto. I have also attended sessions administered especially for foreign tourists and seekers of Amazonian shamanism.

In all places in which I have partaken of Ayahuasca, the constituent plants were those indicated in the previous chapter—*Banisteriopsis caapi* and *Psychotria viridis*. In several sessions held in indigenous contexts some other extra ingredients were used, but I do not have exact information in this regard. To the best of my knowledge, in two sessions I had alone with Peruvian *curanderos* the Ayahuasca brew also contained the juice of *toé* (*Brugmansia suaveolens*) and once, in a private session in Brazil, it contained *paricá* (*Virola* spp.)[1].

Data Collected from Other People

My second source of data were interviews in which I asked people about their Ayahuasca experience. Conducting such interviews, I should note, is not a simple matter. In many contexts, there is an explicit taboo against discussing the contents of Ayahuasca visions, and even when this is not the case the contents in question are highly personal and often people are reluctant to share them with others, especially strangers. Indeed, some anthropologists (see, for instance, Deltgen, 1993) have noted that they felt reports furnished to them by their informants did not reveal the entire story these individuals could tell about their Ayahuasca experiences. I would take the liberty of saying that in many respects discussing one's Ayahuasca visions is rather similar to discussing one's sexual life. And then,

[1] For further information on *paricá*, see Schultes (1954) and Wassén and Holmstedt (1963).

just as in the case of sex so also in conjunction with Ayahuasca—the best chance to get true and complete accounts of what happens is to actively share in the experience of one's interlocutor. Many of the interviews I have conducted were made possible because of such an active joint participation. The persons interviewed were ones I had met in sessions that I myself had participated in, and to many of them I have reciprocated by telling about some of my own experiences.

The interviews I conducted were of two kinds—non-structured and structured. All interviews were administered by me personally in an individual face-to-face fashion and with the only ones present being the informant and myself. All told, 178 persons were interviewed, 122 males and 56 females. Of these, 46 were queried by means of the structured interview and 128 were queried in a non-structured manner; to these should be added 4 persons for whom I have a full record of all the Ayahuasca experiences that they have had. Of the informants, 16 were indigenous or persons of mixed race, 106 were residents of urban regions of South America, and 56 were foreigners (that is, persons residing outside South America). All told, my estimate is that the total number of Ayahuasca sessions probed in these interviews is of the order of 2,500.

In the non-structured interviews informants were asked about their first Ayahuasca experience, the most remarkable experience they have ever had with the brew, and/or of the details of one particular session (the latter was normally a session in which I and the person in question had just participated). The informants included indigenous and non-indigenous persons, medicine men and masters of Ayahuasca ceremonies, people with long-time experience with the brew and some who had taken it for the first time; amongst the latter were both residents of South America and foreign travellers who knew nothing about Ayahuasca before they had partaken of it the first time. The interviews were conducted in many different locales in Brazil and Peru as well as outside South America.

The structured interviews were centred on a questionnaire that addressed informants' long-term experience with the brew tapping different aspects of the Ayahuasca experience. In addition to queries regarding the contents of their visions, the informants were presented with queries regarding non-visual perceptual effects, ideas, reflections and insights, moments of special psychological and spiritual significance, bodily effects and bodily transformations, and effects having to do with consciousness and the self.

In terms of their experience, the informants who participated in the structured interviews pertained to four groups. The first group is one which I call 'independent drinkers'. These were 18 individuals who partook of Ayahuasca regularly but who, at the time of the interview, did not belong to any institutionalized group. All were residents of Brazil, most of them from Rio de Janeiro, and almost all were middle class; 15 were males and 3 were females. For all of them, Ayahuasca was a most central facet of their lives, and all were partaking of it regularly at the time of the interview. All had at least four years of experience with the brew and all had consumed it at least 40 times; many had done so many more times than that.

I should note that my characterization of these people as a group pertains to the design of my research; while some of these persons know each other, most do not and by no means do they constitute a group in any social or interpersonal sense. The interview lasted for at least one hour; in some cases it extended up to six hours.

The second and third groups of informants presented with the structured questionnaire were members of the UdV and the Church of Santo Daime. These two groups consisted of 14 and 7 persons, respectively.

The fourth group consisted of persons who were not affiliated with any institutionalized group but, unlike the independent drinkers, had only small or moderate experience with Ayahuasca. This group consisted of 10 non-indigenous, urban persons. Of them, 4 furnished—by means of personal diaries and written records—complete accounts for all the Ayahusca sessions in which they had participated. The other 6 were queried by means of the structured questionnaire.

A final comment on the solidity of the corpus of data discussed in this book is in order here.[2] My assessment is that this corpus is solid in the sense that, by and large, it covers all the main types of phenomena that people may experience in the course of the Ayahuasca inebriation. Indeed, whenever the opportunity presents itself, I still continue to question and interrogate people about their experiences with Ayahuasca. At this stage, it is extremely rare for me to hear of phenomena of kinds not already encountered in my existing corpus.

More General Comments

Having defined the corpus of data, let me mark the special nature of the research enterprise reported in this book. First and foremost, this is both the first empirical cognitive-psychological investigation of the Ayahuasca experience ever conducted as well as the first and only theoretical cognitive treatment of this topic. Indeed, apart from the clinical-psychological testing administered in the framework of the medically oriented *Hoasca Project* conducted by Grob and his associates (Grob *et al.*, 1996; McKenna, Callaway, and Grob, 1998; and Callaway *et al.*, 1999), this is the only systematic psychological study of Ayahuasca that has ever been undertaken.[3]

To the best of my understanding, this is also the largest survey of data on the phenomenology of the Ayahuasca experience that has ever been collected and reported in the literature. In particular, this statement also applies to the specific topic of Ayahuasca visions. In fact, some aspects of the Ayahuasca experience examined here—notably, the ideations entertained under the intoxication and various aspects of the alterations in consciousness the brew induces—were seldom, or even never, discussed in the literature.

[2] The reader should bear in mind the difference between the present use of the two terms—'corpus' (or corpus of data) and 'core corpus'. The first term refers to all the data I have collected; the second refers to the set consisting of my first 67 Ayahuasca sessions.

[3] As indicated in Ch. 1 n.3, when the writing of this book was completed, Metzner (1999) was published. This book presents 24 verbatim reports of Ayahuasca sessions, mostly by North Americans with little or no experience with the brew, but it presents hardly any further psychological analysis.

This investigation is also the most systematic ever conducted. This holds true not only for the entire corpus as a whole, but also separately with respect to each of its subsets. Specifically, the data pertaining to my own personal experiences is the largest and most systematic account of the Ayahuasca experiences of any one person in the literature, the data collected in the non-structured interviews is the largest of its kind ever reported, and the data collected in the structured interviews is the first such systematic analysis ever undertaken. Furthermore, this is the first systematic study in the scientific literature that is based on a sizeable corpus of data collected firsthand. Many of the anthropologists who have investigated Ayahuasca and written about it have partaken of the brew only a couple of times and have had just rudimentary experiences with it (amongst others, this is true of Reichel-Dolmatoff, universally regarded as the leading anthropologist to have studied Ayahuasca); some have not had any firsthand experience at all (for further comments in this regard see Harner, 1973*b*, especially p. 16; Mabit, 1988; Davis, 1996; and Narby, 1998).[4]

The data reported here are special in still another respect— their heterogeneity. Normally, when an anthropologist studies Ayahuasca he or she focuses on one particular community. This is natural for, as noted earlier, the anthropologist's prime focus is on society and culture. In contrast, being interested in the psychological dimension, I have made a point of studying Ayahuasca in many different settings. These differ *inter alia* in the nature of the community in which the session was held (e.g. indigenous or not, religious or not), the kind of session (e.g. religious ritual, healing session, celebration, or one without any ritual), and locale. More specifically, all these variations apply to the various contexts in which I myself have experimented with Ayahuasca. In technical terms, I have served as my own control across these different variations.

Let me add a comment with respect to the study of the Brazilian syncretic groups, the Santo Daime Church and the União do Vegetal. Practically all the studies of these groups in the literature were conducted, either exclusively or in part, by investigators—be they anthropologists or natural scientists—who are members of these groups and/or sponsored by them. It will be noted that not always is this affiliation explicitly acknowledged. My work is fully independent. Furthermore, it is the only one in existence that examines all these groups, as well as other contexts of Ayahuasca use, in unison.

While the goal of this work is psychological, in the light of the foregoing observations, the research reported here is significant and pioneering from an anthropological point of view as well. In general, there is a rather common conception of 'Ayahuasca visions' or of 'the visions Amerindians have'. But actually, in the anthropological literature hardly any systematic survey of Ayahuasca visions has

[4] As this is being written, several doctoral dissertations regarding Ayahuasca are being worked on; some of these may already be published when this book comes out. Unlike their predecessors, these young anthropologists do have serious firsthand experience with Ayahuasca; in particular, I am referring to the works by Lagrou (1998) and Groisman (1999).

ever been conducted. Reichel-Dolmatoff's classic studies are based mostly on the reports of only one—admittedly, very experienced, most perceptive, and superbly articulate—informant. Practically all other anthropological works on the subject involve summary reports of the experiences of various non-individually identified members of a given ethnic group (see, for instance, Harner, 1973*c*). Normally, there is no indication of how many people were interviewed and no specification of the quantitative distribution of data amongst informants. Thus, the present enterprise may also be regarded as a foundation for more systematic anthropological studies of the Ayahuasca experience in its native context.

Last but not least, let me underline the pioneering theoretical nature of this work. While its orientation is phenomenological, it is not merely descriptive. As I am charting the various aspects of the Ayahuasca experience I also attempt to lay the foundations by which this special facet of human psychology may be conceptualized from a theoretical point of view. This endeavour is significant not only with respect to Ayahuasca but for the study of so-called altered states of consciousness in general. As cognitive-psychological studies of other psychoactive substances are very scant,[5] the present work could serve as a model for subsequent similar studies of other substances and provide a theoretical basis for further enquiries of the non-ordinary facets of human cognition.

With these last remarks, let me now turn from the variation across people and contexts to the variations between plants and substances and comment on the uniqueness (or non-uniqueness) of Ayahuasca. Is Ayahuasca special? Are the effects to be surveyed here specific to this particular psychoactive substance? How does Ayahuasca compare to other such substances? Without exception, all the groups—both indigenous and otherwise—that use Ayahuasca stress the uniqueness of this potion (see, for instance, the general assessment made by Schultes, 1982, cited in Ch. 1). Many indigenous persons have told me that only one other plant is considered to be more powerful—the datura (or rather *Brugmansia*, *floripundio*, and *toé*, in Latin, Spanish, and Quechua, respectively; the Amazonian indigenous term is *huanto*). However, unlike Ayahuasca, datura is seriously dangerous—its consumption can result in madness, irreversible physiological damage, and even death. Furthermore, whereas Ayahuasca intoxication normally lasts between six and eight hours, the effects of datura may last for a couple of days. Moreover, whereas with Ayahuasca drinkers maintain various degrees of self-consciousness and control, with datura this is usually not the case. Indeed, typical to datura, but not to Ayahuasca, is a radical confusion between reality and hallucination. Finally, in contrast to what is the case with Ayahuasca,

[5] Indeed, most were conducted in conjunction with LSD in the 1950s and 1960s, that is, prior to the advent of modern cognitive science. In particular, the reader is referred to the most interesting monograph by Masters and Houston (1966), which contains many firsthand reports of experiences with LSD and peyote, as well as to the anthologies edited by West (1962) and Siegel and West (1975).

with datura it is never clear in what state its user will return to normal life, if at all.[6]

But of course, the empirical question remains: What is the difference between the phenomenology of Ayahuasca and that induced by other hallucinogenic plants and substances? In order to answer this question one would have to conduct a systematic comparative study of Ayahuasca and the various other plants and substances. If a first-person 'within subject' perspective is to be taken, the same person or persons should experiment with all the plants and substances investigated and compare their experiences with them. Technically, this is truly a gigantic task, which I, certainly, could not undertake. A possible way to proceed is to have studies analogous to the present one, each conducted by a different investigator, followed by a meta-analysis; but that is beyond the scope of the present investigation. Here, I attempt to present a systematic analysis of the phenomenology of the Ayahuasca experience only, which I hope will serve as a basis for future studies that would extend the cognitive investigation to other plants and psychoactive substances.

The Systematic Charting of the Phenomenology of the Ayahuasca Experience

I state again, the primary goal of this work is to present a comprehensive survey of the phenomenology of the Ayahuasca experience. By and large, the survey undertaken here is qualitative and typological, not quantitative and statistical—it is an attempt systematically to chart the various phenomena that Ayahuasca may induce and to establish order in them. In this endeavour, my principal goal is to follow Aldous Huxley's lead and to chart the geography of regions of the psyche—the antipodes of mind, in Huxley's terminology—that in our ordinary state of being are hidden and which Ayahuasca reveals. This endeavour is tantamount to answering the question 'What can happen under the influence of Ayahuasca?' or 'What kind of phenomena does the Ayahuasca experience consist of?'. In essence, I regard this question to be completely analogous to that of the linguist who is asking 'What is natural language?' or 'What kinds of structures do human languages consist of?'. In both cases, the enterprise couples the descriptive and the analytic. First and foremost, it involves a detailed and systematic typology of the phenomenology of the cognitive domain at hand. This being achieved, the typology further aims at the discernment of internal patterns and regularities as well as lawful relationships and dependencies. Together, these are meant to define a coherent picture furnishing an answer to the principal questions by which the enterprise is defined.

The cognitive effects that Ayahuasca induces are many and multifarious. It goes without saying that in each single Ayahuasca session only a very small fraction of

[6] For general information on this subject the reader is referred to Reinburg (1921), Harner (1962), and Furst (1976). For a review of the use of datura by the Indians of North America, see Baker (1994). Especially interesting is Chango (1984), which is an account by an indigenous Amazonian shaman.

them are manifest. Similarly, it is not the case that all drinkers of Ayahuasca ever experience all the effects that this brew can induce. Essentially, the present enterprise is an attempt to construct a coherent unified picture out of the observations made by many persons in a great number of sessions taking place in different locations and in various contexts of use. Again, the situation is completely analogous to that which obtains in geography—the map that appears in the atlas is the cumulative product of records furnished by many surveyors, each measuring a particular region at a particular moment in time. In both cases, the final product requires a large number of observations coupled by analytical work grounded in factual knowledge of kindred phenomena (in our case, the study of human cognition in general and the phenomenon of consciousness in particular) and in theoretical conceptualizations thereof as they are made available by the scientific and intellectual state of the art of the day. This definition of the project of enquiry is crucial. Narrations of personal odysseys, adventurous though they may be, can be thought-provoking and most inspiring, but they constitute a categorically different enterprise, one that is outside the realm of science. As for compendia of accounts of the experiences of individuals who have experimented with Ayahuasca—these are valuable bodies of data but surely, at best (when the number of informants is sufficiently large, their backgrounds and affiliations varied, and at least some of them are reasonably experienced) they constitute only the first, and most basic, stage of the investigation. Curious tourists (or tourists to be) may be fascinated by colourful presentations of photos and slides. The cartographer—be it s/he who is charting the geography of the planet or that of the mind—will appreciate that it is in the passage from these to the drawing of the unified map that all the real professional work actually lies.

Before launching upon the charting of the phenomenology of the Ayahuasca experience (which I begin to do in the next chapter), by way of offering the reader an overall orientation, I will present a general overview of the various facets of the phenomenological domain at hand. The different items indicated in this general scheme define the topics of discussion in the chapters that follow.

First there is the *overall atmosphere* that Ayahuasca generates. I am referring to effects that may be very subtle, and much less dramatic than some others—notably visionary ones—for which the brew is famous. Yet, at the same time, these effects may be the most radical. Together, they amount to the definition of 'another reality', the primary characteristics of which are beauty, enchantment, deep meaningfulness, and sanctity.

Second to be noted are *affective* effects. These may be both positive and negative, heavenly as well as hellish. The affective facets of the Ayahuasca experience may be both intrapsychic and interpersonal. The most prevalent intrapsychic effects are bliss, elation, and awe on the one hand, and horror and immense fear on the other hand. The interpersonal facets of the experience are dominated by love.

Third are *bodily* effects. These include effects (like the notorious vomiting that Ayahuasca often induces) that are directly attributable to physiology and which are of no special cognitive import; these are not within the focus of my interest here. More relevant, albeit still rather peripheral, are kinaesthetic feelings whereby the sensation of one's own body changes. Bodily effects that are of clear cognitive significance are those associated with metamorphoses; that is, with the feeling that one's identity changes. Also to be noted are the curative effects of Ayahuasca. These bring together changes in body and soul. Very often people feel that the Ayahuasca inebriation is a curative experience, as a result of which people are healed. This applies both to body and spirit.

Fourth to be noted are the *perceptual* effects; these define some of the most prominent facets of the Ayahuasca inebriation. They are encountered in all perceptual modalities—the visual, the auditory, the olfactory, the tactile, and the gustatory (this list, note, is in descending order of both prominence and frequency). The visual effects are the most salient ones and they can be further divided into ones that happen with the eyes closed and ones that happen with the eyes open. There are also synaesthetic effects, in which the division between the different sensory-perceptual modalities is crossed. Special mention is deserved for visions in which light is central; these are often coupled with powerful spiritual experiences.

In the literature, Ayahuasca visualizations are universally analysed in terms of the different content items that may be seen in them; for instance, it is noted that serpents or felines are seen. This, however, is not the only dimension by which these visualizations may be analysed. Fundamental to the study of Ayahuasca visions is the typological one examining the various *structural types* or *forms* in which visualizations may appear. Furthermore, the study of the contents of visions is not confined to the *micro*-level of the identification of single *content items*. There is also a *macro*-level concerned with the *themes* of visions. And then, there are the various manners by which a drinker may be involved *vis-à-vis* the visions he or she has. Seeing is only one such manner of involvement. When visualizations are complex and the degree of a drinker's involvement with them high, drinkers may engage in active interaction with their visions and be involved with all sorts of performances associated with them.

Following the linguistic analogy already alluded to above, I would like to propose that the foregoing dimensions of analysis are similar to those encountered in the study of language. The structural typology may be said to define the syntax of Ayahuasca visualizations, the analysis of content items defines the lexicon of these visualizations, the thematic analysis pertains to semantics, and the interactional one to what may be regarded as the pragmatic of Ayahuasca visions. Finally, I shall note that when the visualizations are complex a narrative analysis, similar to that undertaken with written texts, may be applied as well.

Fifth are the *ideational* effects. Very commonly, people report that Ayahuasca makes them think 'faster and better'. The ideas, reflections, and insights that the brew generates may be of a personal and psychological nature, intellectual ones pertaining to whatever domains are of interest to the individual in question, and general philosophical and metaphysical ones. The latter usually have to do with what may be characterized as the fundamental riddles of Existence—the ontology of the world, the nature of life, basic questions regarding the human predicament, the meaning of history and culture, and—last but surely not least—with mysteries pertaining to the Divine.

Sixth are *spiritual* effects. Ayahuasca often induces powerful religious and mystical experiences. In general, these are associated with strong noetic feelings (that is, experiences in which one feels that true knowledge is attained) as well as with reflections concerning ethics and the conduct of one's life. Not infrequently, experiences of this kind result in actual, and at times radical, changes in the praxis of people's lives.

Seventh are effects pertaining to *consciousness*.[7] These are manifested in transformations of one's identity, in modifications in the relationship between the self and the world, in alterations in the epistemic status of mental events, and in the calibration defining the overall functioning of the cognitive system. Especially to be noted are experiences of *metamorphosis* in which drinkers feel that their personal identity is undergoing *transformation* and that they are changing into another human being or an animal. Many people also report *paranormal* effects; of these, the most frequent are ones involving telepathy.

Linked to the effects pertaining to consciousness are ones having to do with *time*. It is common for Ayahuasca to induce significant modifications in the experience of time. When radical, these may result in drinkers feeling that they have freed themselves from the dominion of time and have reached the realm of the eternal.

Intertwined with both consciousness and time are various patterns pertaining to *meaning*. In general, Ayahuasca induces an overall ambience of great meaningfulness. Often, the brew generates the feeling that existing, yet heretofore unknown, aspects of meaning are revealed to one. At times these are associated with a seemingly independent, Platonic-like realm of eternal truths. Symptomatic of these effects is the overall metaphoricity and aesthetic sensitivity that the brew induces.

Finally, let me note that the effects of Ayahuasca are not confined to the internal subjective domain. This brew also induces behaviours and patterns of activity that are manifest in the external, public domain. As a consequence, the level of one's *performance* may be highly enhanced. Especially marked in this respect are singing, the playing of musical instruments, and social interaction with other persons.

General Structure of This Book

This book is divided into three parts. The first introductory part, which ends with this very section, consists of the laying down of foundations for the

[7] Strictly speaking, all the effects experienced with Ayahuasca pertain to consciousness. Here (and subsequently, in the chapters devoted to consciousness) I apply the epithet 'consciousness' in a narrower sense, focusing on those effects that specifically have to do with the overall structure of experience and the self.

cognitive-psychological investigation of Ayahuasca. In this and the previous two chapters I have defined the rationale and the goal directing such an investigation, summarized the relevant literature, and furnished the pertinent background and methodological information. The second part of the book, which begins with the next chapter, focuses on phenomenological questions of the first order; by and large, the discussion of these follows the general scheme drawn in the previous pages. Some items indicated in this scheme are examined in separate chapters, some are grouped together into one single chapter whereas others still—due to their richness—are the subject of several chapters. Together, these chapters compose my drawing of the map of the Ayahuasca experience. By and large, the discussion is qualitative. By way of reference and corroboration, some quantitative information is presented in an appendix. The third part of the book is devoted to the examination of phenomenological questions of the second order as well as several theoretical cognitive issues associated with them. The last two chapters of this book are more global and more abstract. In them, I take a broader perspective and attempt an integrated theoretical account of the state of mind that Ayahuasca induces. Finally, In the last chapter of the book, more speculative philosophical and general intellectual questions that the psychological study of Ayahuasca raises are discussed. The book ends with reflections of a more personal nature.

PART II

The Phenomenology of the Ayahuasca Experience

4

Atmosphere and General Effects

> How wonderful it would be if one could only be worthy of hearing the song of the grass. Each blade of grass sings out to God without any ulterior motive and without expecting any reward. It is most wonderful to hear its song and worship God in its midst.
>
> Rabbi Nachman from Bratzlav

I begin my survey of the Ayahuasca phenomenology with the consideration of the distinct atmosphere and the general effects that this brew induces. In addition to its analytical import, this discussion will also serve to provide the reader who has no firsthand acquaintance with Ayahuasca with a further general feel for the non-ordinary state of mind that it induces. In line with the cognitive orientation of this book, the focus will be on subjective, psychological effects. Except when they bear on the psychological, externally observable physiological effects (e.g. changes in blood pressure, body temperature, pupil size, tremor, and the like) are not discussed here. For good reviews of these, the reader is referred to Chiappe, Lemlij, and Millones (1985), Naranjo (1983), Ott (1993), Stafford (1993), and Spinella (2001); recent systematic investigations of such effects are reported in Grob *et al.* (1996), McKenna, Callaway, and Grob (1998), and Callaway *et al.* (1999). For especially insightful and extensive personal accounts of the general effects to be discussed here the reader is referred to Polari (1984), Taussig (1987), Meyerratken and Salem (1997), and Fericgla (1998).

In this chapter I present a characterization of the overall ambience of the Ayahuasca inebriation, draw a sketch of the general course of its progression, and specify several general subjective effects that the brew induces along with some common reactions that drinkers have to them. All the patterns described in this chapter may be regarded as a general envelope, or frame, in which the more specific phenomena surveyed in all subsequent chapters of this book take place.

I should emphasize that it is not the case that in all Ayahuasca sessions all these patterns are encountered. Rather, the picture presented here is a constructed summary based on my own and my informants' cumulative experience. Thus, this picture may be regarded as a prototype of an Ayahuasca session. As with prototypes in general, the patterns noted need not be manifested in each and every instantiation of the set in question (in our case, each session). Furthermore, I would like to point out that the common typical patterns notwithstanding, each

Ayahuasca session is different. As Taussig (1987: 406) poignantly states: 'There is this paradox, that in trying to depict the general one has to seize upon the singular. . . . There is no "average" yagé experience; that's the whole point.' My assessment is identical, and the same was told to me time and again by many other persons: No two Ayahuasca sessions are alike. Even very experienced drinkers fail to predict what might happen to them in a session, and the experience may surprise and stupefy even them.

Overall Characterization and General Course of the Inebriation

The very first sensation that Ayahuasca evokes is the gustatory one. It is notorious that Ayahuasca tastes 'awful'. Many even go as far as saying that the taste of Ayahuasca is the worst that they have ever come across in their entire lives. The brew is bitter and there is something pungent and biting to it. In Ayahuasca sessions, it is very common to see, just after the sipping of the brew, people producing facial expressions of revolt and disgust in a kind of automatic reflex reaction. It seems that they try, in vain, to get rid of the repugnant taste of the liquid they have just consumed. Typically one sees eyes shut, heads shake, faces contort. People cough, spit, swear, and in some contexts cross themselves and pray for mercy.[1]

In the doctrines of both the Santo Daime Church and the UdV, a basic distinction is made between two aspects of the Ayahuasca inebriation; the two are also associated with the two plants of which the brew is made. The first aspect, associated with the vine, is that of 'force' (or 'power') or drunkenness; the second, associated with the leaves, is that of 'light' and eventually visions. These two distinct aspects are further characterized as the male and female aspects of Ayahuasca, respectively (for similar characterizations in the indigenous Amerindian context, see Lagrou, 1998).

The visual effects constitute the most striking, and the most famous, aspects of the Ayahuasca inebriation. However, in this early chapter (which is devoted to the consideration of the general, non-specific, effects of the brew), the visual effects are mentioned only in passing; in later chapters these will be discussed at length.

As the physiological effects of the brew start having their impact, the person who has partaken of Ayahuasca generally feels subject to what is often described as an internal invasion. The sensation is of something heavy and viscous crawling through one's inner parts, pressing and eventually taking possession of one's entire body. Often, drinkers sense that things are burning inside their system. An association that many have is that of snakes engulfing them from within.

[1] All this notwithstanding, with experience some come to savour the taste of Ayahuasca. While, admittedly, this taste may not be very pleasant, it is, many reckon, rich in character. Further, with experience drinkers discover that not all Ayahuasca brews taste alike—some strike one as very powerful whereas others are more mellow, having a rich and even velvety texture.

When the 'force' strikes, usually around forty minutes after consumption of the brew, many are prone to vomit.[2] It is a vomit like no other—drinkers often feel that they are pouring out the depths of both their body and their soul. Several informants have told me that when vomiting in this manner they saw snakes coming out of their mouths; some described the snakes as fluorescent or luminous. The people of South America refer to the vomiting induced by Ayahuasca as an act of *purga*, that is—a purge. Indeed, the moment of vomiting is often one of a major transition—from a situation one can hardly stand to one of coming to terms with the Ayahuasca experience. Many informants recount that the first time they underwent this vomiting experience a helping guide, usually a female figure, appeared and gave them moral support. Often this encounter and the relief it produces are taken by the drinker to be of substantive spiritual significance and they may even result in some sort of personal transformation.

In general, all during the Ayahuasca inebriation one's body feels different from the normal. Typically, drinkers feel as if a cloud occupies the inner space of their heads and that something that they cannot control has a grip on their bodies. Difficulties in motor control and co-ordination are very common. All this, however, need not imply that the inebriation is only distressing and disagreeable. With experience, drinkers often learn to establish a dissociation whereby the ongoing unpleasant bodily effects are left aside and the full focus of attention is with the otherworldly mental effects they are subject to.

Usually, the harshest symptoms of the Ayahuasca inebriation occur during the first 90 minutes following the onset of the effect. During this time, visions can be very strong and the entire experience may be tough and even frightening. Often the feeling is that the drinker has little or no control over what is happening. Thus, the initial phase of the inebriation is likely to present drinkers with moments of intense struggle. At times, the person who partakes of Ayahuasca feels he or she is losing his or her senses and even going mad. Quite commonly, people feel that they are about to die. Furthermore, it often seems that what is happening is irreversible and that one will never return to one's normal self. With this, thoughts like 'Why, for heaven's sake, did I make the mistake of partaking of this drink?' often cross drinkers' minds. Naturally, all this is likely to generate great trepidation. With experience, however, the fear can be better managed and the Ayahuasca drinker learns to gain more control over the intoxication.

The initial difficult phase is usually followed by a period in which the impact of the brew is strong but more manageable. In general, this period lasts from about one and a half to two hours. It is at this stage of the inebriation that drinkers usually begin to come to terms with the Ayahuasca experience and even enjoy it. Indeed, people may find that this experience presents them with moments of exhilaration and great wonderment.

[2] I should note that, contrary to the impression one may get from the literature, vomiting is not universal. Actually, some individuals seldom vomit.

The last phase of the Ayahuasca inebriation is usually mellower and is often accompanied with sentiments of serenity, extreme peace of mind, and bliss. One may still have visions, but normally, as time goes on the preponderance of visions diminishes and their strength and intensity are reduced. During this phase the experience is that of great happiness, harmonious well-being, and spiritual uplifting. Generally, people also feel great love for their fellow men and women and deep affinity to nature and to all existence. These feelings may remain with one for a while even after the intoxication proper has ended. In this vein, Reichel-Dolmatoff (1991) reports that after the Ayahuasca inebriation the Indians feel 're-created'.

Typically, the intoxication lasts between four and six hours. Usually, drinkers' spirits remain uplifted for another day, during which time they feel as if after a thorough 'psychic bath'. People may also begin to appreciate the significant amount of energy, both physical and mental, that partaking of the brew has exerted on them. With this, the increased wakefulness characteristic of the first hours immediately following the end of the intoxication often frequently gives way to tiredness, even exhaustion.

While the scheme sketched above is typical, it is by no means universal. On many occasions the Ayahuasca inebriation does not follow the tri-phasic pattern I have outlined. Often, the course of the inebriation is that of a spike-function where the strength of the various effects diminishes and takes off again several times before it starts to taper off. Not infrequently, significant visions appear quite late in the course of the Ayahuasca session. The same holds for various adverse bodily sensations as well as for the attacks of vomiting.

When the impact of Ayahuasca is most powerful, drinkers tend to close their eyes. During this time, they also tend to turn their attention inwards and delve into the visionary world that is revealed to them. When the effect is reduced, the tendency is to open the eyes. Especially typical of this stage are visions of the type which, in later chapters, will be referred to as seeing-as and seeing-in. Towards the last phase of the inebriation, many tend to shift their attention outwards. For some this is an especially interesting period in which the focus of attention is turned to other people. When this happens, it is not unusual for people to feel that they are endowed with an exceptionally keen perception of other persons' minds and souls.[3] Many times I have noted that during the last phase of a session there are several individuals who intentionally stop focusing on the visions, open their eyes and turn their gaze to the external world. With this, the prime focus of interest is on the effect of the brew on the other people around and on the social interactions between them. Personally, I can attest that this can be a most rewarding experience.

By way of summary, here are two general characterizations of typical Ayahuasca sessions. Both were written by anthropologists on the basis of indigenous accounts—

[3] For a general discussion of this phenomenon in the context of altered states in general, the reader is referred to Masters and Houston (1966).

the first is taken from Gow (1988: 26) and is based on a description furnished by a shaman from the Rio Napo in Ecuador, the second is taken from Goldman (1979: 210–11) and summarizes the information given to him by Cubeo Indians. Except for a qualification made below, both capture succinctly and accurately the overall flavour of the Ayahuasca experience as I myself have come to know it.

As noted by many ethnographers, the effects of taking Ayahuasca follow a fairly stereotypical course. Some time after ingesting the drug, drinkers experience severe auditory and visual disorientation: they hear loud rushing sounds and see patterns of coloured light. This is accompanied by a feeling of intense nausea, often leading to vomiting and violent diarrhea. This phase is often extremely frightening, and many people have reported their fear of going mad during it. It is, however, followed by more complex hallucinations, which become clearer and clearer: drinkers see distant and exotic landscapes and people. These hallucinations continue for one or two hours, gradually fading to leave those who have taken the drug with a sense of elation and beauty.

At the beginning the vision becomes blurred, things begin to look white and one begins to lose the faculty of speech. The white visions turn into red. One Indian described it as a room spinning with red feathers. This passes and one begins to see people in the bright colouring of the jaguar. . . . The hall begins to assume a disturbing and fearful form, one becomes aware of violent people milling about, shouting, weeping, threatening to kill. One is seized with fear that he no longer has a home. The house posts and trees come alive and take the form of people. There is a strong sensation that an animal is biting one's buttocks, a feeling of the feet being tied. The earth spins and the ground rises to the head. There are moments of euphoria as well, when one hears music, the sound of people singing, and the sound of flowing water. The Cubeo do not take *Mihi* [Ayahuasca] for the pleasure of its hallucinations but for the intensity of the total experience, for the wide range of sensations. I spoke to no one who pretended to enjoy it.

I disagree with the claim made in the last sentence. When the initial hurdles are overcome, the Ayahuasca experience is perhaps one of the most exhilarating that a human being can have. I say this on the basis of both my own extended experience with the brew and following extensive interviews as well as more casual sharing with hundreds of people.

Atmosphere

Under the effect of Ayahuasca, as with other psychoactive agents (see, for instance, Watts, 1962), people feel that something very basic changes. It seems that the world is no longer the same. Even when drinkers do not have any visions in the strict sense of the word they usually discover that the world has altered in a very fundamental fashion. The overall sensation is one of *otherworldliness*. The feeling is that things are not as they used to be and one has the sense of entering into another, heretofore unknown, reality. Let me emphasize that this may be one's subjective feeling even when visually the world looks just as it always does, and even without one's experiencing any hallucinatory effects as such.

It seems to me that the best way to characterize the experiential phenomenon of otherworldliness is to say that the world is seen as if it were a painting. What I have in mind are strictly realistic paintings, or rather hyperrealistic ones. Objectively, paintings of this kind depict a state of affairs just as it is factually in the world, yet they are not photographs. Figuratively, the paintings depict the world as it is, and still they are clearly the product of the hand of one particular artist. Specifically, the paintings exhibit particular signatures, and it is the same with the world as seen under the effect of Ayahuasca. Suddenly the world acquires a specific signature and, indeed, it may seem to be like a work of art. With this, several other phenomenological features are noted.

First is *beautification*. Typically, under the Ayahuasca intoxication, colours shine and are perceived as brighter and more saturated. Distinctions in hue and shade seem to be richer and, overall, the visual field becomes invested with a dramatic flavour it normally does not have. Furthermore, people usually feel that there is an all-encompassing harmony in everything that appears before their eyes. With this, the world appears to be strikingly beautiful. Once again, let me emphasize that what I am talking about is the perception, with the eyes open, of the external, real world, not of visual hallucinations.

Similar amplificatory effects are also encountered in the other sensory modalities. These are especially marked with music. Under the Ayahuasca intoxication, (real) music generally sounds more beautiful and more expressive than it normally does. Often people discover in the musical material fine distinctions and subtle nuances that they have not perceived or appreciated beforehand and which render the musical material remarkably richer and deeper. With this, drinkers are prone to find music especially touching and evocative; very often, it is perceived as sublime.

The second feature is *meaningfulness*. Under the Ayahuasca inebriation, things seem to be ingrained with meaning. One discovers that things in the world are not as they normally seem to be—'just like that'. Under the intoxication, it dawns upon one that there is sense and reason to it all. With this appreciation, insights are gained and new understandings are reached. The insights may pertain to the personal life of the individual or have to do with the world and its multifarious manifestations. They may be psychological or philosophical, intellectual or affective, spiritual or aesthetic. Thus, it is very common for drinkers to feel that they suddenly understand why things are as they are, to find deep, heretofore hidden, meanings in verbal expressions and in texts, to discover the true senses of their own lives. This often leads drinkers to theological meditations. Individuals who are less religiously inclined and more philosophically oriented may entertain ideations of a more metaphysical nature.

Along with beautification and meaningfulness, otherworldliness usually leads to *enchantment*. With Ayahuasca, things tend to appear as if under the effect of an all-encompassing spell. A common feeling is that one has entered a realm that is all magic. Often drinkers are under the impression that the world, this world in which one has lived for so long, is governed by invisible forces, energies, or beings. With this, they are prone to sense secretiveness all around. It suddenly appears that everything is engulfed in great mystery. Consequently, the world is perceived as an object of great marvel and utmost wonder.

Often coupled with enchantment is the appreciation of *powerful energy*. Under the intoxication, it seems that a tremendous force permeates and animates everything around. Over and over again, in different locales and contexts, I have heard people comment that this energy is the force that sustains all Creation. The powerful energy is also regarded as the source of all wisdom and knowledge, and the ultimate fountain of health and well-being. Typically, people feel a direct tie to this energy and come to appreciate that their very own livelihood comes into being and is nourished by it. Coupled with this is the recognition of the abundant bounty that impregnates all Existence. The feeling is that the world is rich with plenty and that in essence, it is so good and wonderful.

Both the investment of meaning and the appreciation of energy are likely to induce religious and spiritual sentiments. In particular, as perceived under the Ayahuasca intoxication, the world is shrouded in *sanctity*. Appreciating this, one feels that one is privy to what many traditional mystics have referred to as the *Tremendus Mysterium* (see Otto, 1957). With this appreciation, reality is apprehended as awesome and terrific. At the same time, once they have overcome their fear and when they open themselves up to the energies around them so as to be engulfed by the bounty, Ayahuasca drinkers usually feel that they are the recipients of utmost grace. With this, they are likely to be filled with deep gratitude. Indeed, in the course of Ayahuasca sessions, it is not uncommon to see people spontaneously utter words of thanks and blessing.

Also related to beautification is *eroticization*. The body of a person to whom I am attracted is seen by me in a radically different way from the way in which another person, who is not thus attracted, sees it. This is despite the fact that technically speaking my visual perception is not modified or distorted in any fashion. With Ayahuasca the entire world may acquire that quality of the body of the beloved. Drinkers often detect a sensuous, even sexual, flavour in whatever surrounds them. They may also feel that the world is the object for deeply meaningful intercourse. Many times it seemed to me that the branches of trees were soliciting caresses and love. On many occasions, I have seen people under the effect of Ayahuasca express their love to plants. One person told me that under the intoxication he passionately embraced the trunk of a tree; another person related an episode in which he lay down and stretched along the ground, immersing himself in a great union with Mother Earth. One of the independent drinkers described what she referred to as 'a feeling of oceanic eroticism'. As pointed out by Reichel-Dolmatoff (1991), in the indigenous Amerindian context of Ayahuasca use the erotic plays a major role. As further pointed out in Reichel-Dolmatoff (1971, 1975), the Tukano Indians of Colombia regard the Ayahuasca inebriation as a kind of cosmic coitus.

Lastly, let me mention an effect that impressed me greatly in my first, most powerful, Ayahuasca session—seeing the world as *primordial*. The feeling was that

I was seeing the world as it was on the first days of Creation, fresh as a piece of pottery just being taken out of the kiln. Later, I found this very same characterization in the anthropological literature (see, for instance, Reichel-Dolmatoff, 1975) and in the reports of my informants. Such a characterization is also encountered in Aldous Huxley's account of his experiences of mescaline which is cited at length at the end of this chapter.

Throughout the foregoing summary description, the non-ordinary character of the Ayahuasca has been highlighted. Yet, at the same time, it is very common for drinkers to report that under the intoxication they feel they are rediscovering a facet of their existence that is actually very basic, very much their own. It is as if life had estranged one from oneself and made one forget some very basic things pertaining to one's very essence. Time and again, drinkers say that the brew brings them 'back home'—to the true essence of their personality from which they have distanced themselves. Invariably, returning to this core is a delightfully comforting experience.

Subjective Effects and Personal Reactions

As indicated throughout the foregoing survey, in its various stages the Ayahuasca inebriation is, on the one hand, a tough and terrifying experience and, on the other hand, a wondrous one of marvel beyond description. Altogether, the extraordinary effects that Ayahuasca induces do not leave anyone indifferent. Various reactions that people usually have in response to these effects have already been noted above. *Grosso modo*, these effects may be divided into two main clusters—that of positive reactions and that of negative ones. By way of summary, in this section these reactions are presented in a more systematic fashion.

Before I go on let me note that both in the previous paragraph and in the discussion that follows the terms 'effect' and 'reaction' are used. Although in many contexts these terms are clearly distinct, here I use them pretty much interchangeably in a sense analogous to that employed in the phrase 'chemical reaction'. Generally, in contexts where the effect seems to involve a higher degree of choice and control the term 'reaction' may seem to be more appropriate; in those where this is not so (or less so) the term 'effect' is the more natural one to use.

The cluster of negative reactions is dominated by fear. Encountering otherworldliness and enchantment, drinkers are likely to question whether their normal standards of judgement still apply. They may even wonder whether they are losing their minds and going mad. Further, the appreciation of the great powerful energy that embraces the world often leads people drastically to confront their own smallness and weakness. Whether this appreciation is debilitating or instructive and psychologically constructive depends, of course, on the attitude of the person in question.

The positive reactions are more varied than the negative ones. The reaction that is most likely to be manifested first is that of wonder and marvel. This reaction is most clearly elicited by the visions. As noted above, these can indeed be unbelievably fantastic. I once sat beside a European young man the first time he experienced a vision with Ayahuasca. Over and over again, he uttered the phrases 'this is just so wonderful', 'how beautiful this all is', 'incredible, incredible'. The man repeated these phrases during the entire time he was having the visions. I fully understand this reaction. Second, I shall mention the sentiments of profound contentment and great joy. These often lead to bliss and elation and, as already indicated above, they are often coupled with a sense of deep gratitude. Third is a profound sense of well-being, both physical and mental. One major effect of the Ayahuasca experience is for people to feel both cleansed and healed. Time and again, I have heard people report that the brew connected (or reconnected) them to that energy which is the source of well-being and health and that as a consequence they were revitalized. Often, this is overtly manifest. Thus, in the later phases of Ayahuasca sessions drinkers usually appear to be remarkably relaxed—indeed, serene—and often they seem to be both rejuvenated and especially good-looking. I have experienced this many, many times and I corroborated the assessment when observing participants in sessions in which I myself have not partaken of the brew. The fourth effect to be noted is directly related to the third— the feeling that both one's mind and one's body are in a better functional condition. People feel that their minds are working faster and that their level of intelligence increases; many say they gain remarkable mental clarity. Likewise, people have a sense of heightened stamina and often they have the impression that they can surmount all obstacles and overcome all difficulties. With this, people who have partaken of Ayahuasca may reach the conclusion that there are no limits to what one can do. 'If only I had wished it, I could have done it all,' is a not uncommon thought. Last, but definitely not least, are the various spiritual and religious reactions that together may comprise a powerful mystical experience. Some of the strongest and the most meaningful experiences that Ayahuasca may induce are characterized as encounters with the Divine. Experiences of self-death and subsequent rebirth and salvation are also encountered. Often, these experiences have great impact on drinkers and they may lead to radical personal transformations.

There are also reactions of a more social character. As noted above, Ayahuasca generally induces great feelings of love and affection between people. During Ayahuasca sessions drinkers often feel that they understand their fellow human beings well and they often feel empathy and compassion towards them. Personally, I have experienced this many times. It also seemed to me that following Ayahuasca sessions I had a significantly better facility in interpersonal interchanges demanding tact and social finesse.

The feelings of love and affinity are not confined to human beings. Similar feelings towards both animals and plants are also very common. Once I partook of the brew in the vicinity of a zoo. Towards the end of the session I went out and stood in front of a cage of jaguars. Extremely engaged, I stood there for three-quarters of an hour and felt I fully comprehended the psychology and social interactions of the three beasts which I was keenly inspecting. I might add that the insights I gained on that occasion still remain with me now, several years afterwards. Many informants have reported to me that under the intoxication they not only gained a wonderful understanding of non-human creatures but could also communicate with them. Also common is the feeling that one can understand the life of plants and see how they grow and interact with their environment. Indeed, Ayahuasca induces a general feeling of a great closeness, even a tie, to nature at large. On the one hand, people come to regard the planet and all that exists on it as a living entity; on the other hand, they see themselves as part and parcel of that unified whole which is Life. Typically, this is coupled with a profound appreciation of the harmonious nature of the great matrix of Being.

I have mentioned increased psychological perceptiveness and an enhanced feeling of understanding other people. Along with the appreciation of meaning and secrecy, all these are symptoms of the sentiment that in the psychological and mystical literature is referred to as noetic (for discussion and references see Chs. 15 and 16). Under the effect of Ayahuasca, as in various other mystical experiences, not only do drinkers feel subject to powerful, non-standard emotional and affective states, but also that they gain special privileged access to knowledge. Indeed, many informants have reported to me that under the Ayahuasca intoxication they felt that their level of intelligence increased considerably and that they gained 'comprehension of everything'.

Ayahuasca may also induce slight adverse cognitive effects. Usually, during Ayahuasca sessions people do not talk. When they do, however, some problems of speech co-ordination may be exhibited. For instance, speakers may have difficulty in keeping track of different lines of thought that they express and some slips in their verbal output may be noted. A context in which this may be especially observed is that of UdV sessions, in which regulated verbal exchange takes place. Difficulties in the shifting of attention and the keeping of focus may also be encountered. Thus, on more than one occasion when summoned by a shaman to approach him for a curing treatment, it was only after two or three callings of my name that I realized that I was being asked to proceed. I have observed the very same phenomenon with other persons. Also encountered is perseveration, that is, the ongoing repetition of a given pattern of behaviour and difficulty in breaking out of it. Repetitive singing or excessive talking may be regarded as manifestations of this. Unsolicited, somewhat ungoverned, laughter is another possible adverse effect.

Performance

Under the intoxication, people often feel that they can perform better than they normally do. This feeling is pervasive and it encompasses both physical and intellectual achievements. As noted above, many report feeling that the brew makes them be able to do anything they would like to do. In general, of course, this feeling cannot be verified. However, at times the performances achieved under the intoxication are overt and can be observed publicly. Performances that are especially manifest are those pertaining to music—singing, playing instruments, and dancing. Objective inspection of these does attest to higher levels of performance. Also observed is an apparent relaxation of inhibition as a result of which drinkers may engage in some sort of acting out. More on the topic of performance will be said in Ch. 13, where the phenomenon of metamorphosis is discussed.

In this conjunction, let me comment further on the topic of motor control to which I have alluded earlier in this chapter. Under the intoxication, problems of motor control are especially manifest when drinkers attempt to change their bodily positions (as in standing up and sitting down) and with co-ordinated motor actions (such as walking or fine manual manipulation). Yet, at the same time, in the context of performance, it is possible to encounter under the intoxication levels of execution that are actually higher than the normal. I have discovered this myself in conjunction with dancing and kindred motions. In particular, let me mention swirling. For quite some time I was interested in Dervish dancing but never actually participated in any. Once when partaking of Ayahuasca by myself, while listening to Sufi music, I began to swirl in the Dervish manner. I did so for about an hour in a fashion which I could just not do in my ordinary state of consciousness. Yet, at the very same time, I had great difficulty in lifting myself up and, after my dancing, in returning to my original sitting position. The contrast between my agility during the swirling and the lack of it before and after were striking. It seems to me that the difference is due not to motor co-ordination alone but to the whole process of letting go and entering into a non-ordinary mode of action. Once I got into the swirling mode, so to speak, Ayahuasca allowed me to proceed and persevere.

Longer-Term Effects

The impact Ayahuasca may have on its partakers need not be confined to the time of the intoxication proper. Having experienced the extraordinary effects that the brew induces, many drinkers feel that they undergo deep personal changes. Indeed, it is very common to hear drinkers testify that having partaken of Ayahuasca they underwent major personal transformations and that their lives were no longer the same. The changes mentioned pertain to new psychological understanding and personal insights, modifications of belief systems, perspectives on life, and world-views, as well as religious and spiritual conversion. Not infrequently, these effects may result in actual, and at times radical, decisions and actions—becoming a member of a religious group using Ayahuasca is one of the most common examples.

Concluding Remarks

Summing it all up, let me mention one of the best characterizations of an Ayahuasca ceremony that I have heard. It was given to me in response to a question

I have presented to many people—why after so many years of experience, they continue to drink Ayahuasca. A friend who is a leading member of the Santo Daime Church replied that it is 'because the *Daime* ceremonies are "divine banquets" '. Indeed, *Daime* sessions, especially the festive ones, are, I vouch, magnificent celebrations of the wonder of life and creation. Many *Daime* hymns express this explicitly. By way of concluding this survey of the special atmosphere, feelings and reactions that Ayahuasca induces let me present several fragments of *Daime* hymns. These fragments were all received by the founder of the Santo Daime Church, Mestre Irineu Serra; the following are my own free translations from the Portuguese:

> The dwelling of my Father
> Is in the heart of the world,
> Where all love exists
> And there is a profound secret.
>
> This profound secret
> Is within all Humanity.
> If all will know themselves
> Here, inside the truth.
>
> I have taken this drink
> It has incredible power,
> It demonstrates to all of us
> Here in this Truth.
>
> I have climbed, I have climbed, I have climbed
> I have climbed with joy
> When reaching the Heights
> I encountered the Virgin Mary.
>
> I have climbed, I have climbed, I have climbed
> I have climbed with love
> I have encountered the Eternal Father
> And the Redeemer, Jesus Christ.
>
> *Daime* force, *Daime* light
> *Daime* love![4]
>
> *Daime* . . . the teacher of all teachers.

The reader will note that many of the effects surveyed throughout the foregoing discussion are described in these simple, non-analytical lines generated by a rubber-plantation worker in a state of high ecstasy.

[4] This is a play on words. The text means both 'The brew ('*Daime*') is force etc.' and 'Give me ("Dai-me", colloquial for "me-da") force etc.'.

To close this chapter I would like to quote at length from Aldous Huxley's *The Doors of Perception* (1971), a book written following the author's experimentation with mescaline. Huxley used a synthetic compound, but the same molecule is a principal ingredient of two major traditional Amerindian plants, both of them cacti—the North American peyote (*Lophophora williamsi*) and the Andean San Pedro (*Trichocereus pachanoi*). While I have had one or two experiences with each of these plants, neither personally nor scientifically am I in a position to make any statement regarding the experiences induced by these plants or to compare them to those induced by Ayahuasca. Informal conversations with various persons who have used these plants as well as the examination of the literature about the indigenous traditions associated with them, lead me to believe that in various respects these experiences are different from the Ayahuasca experience. Yet, as far as the general perception of the world is concerned, the description Huxley gives in conjunction with mescaline presents, I feel, a most sensitive and perceptive account of what we have examined here under the label of atmosphere and general effects. To my mind, the text to be quoted is impressive in the way it brings together intellectual insights and poetic talent. I cite at length because I find this text more pertinent than any similar account to be found in the literature on Ayahuasca. My appraisal is that this account fits especially well the early, pre-hallucinatory, stages of the Ayahuasca inebriation. Indeed, it is precisely because the report is concerned with the general ambient atmosphere, not with any visions or special effects, that it is so relevant to the discussion in this chapter. The parallels between the features reviewed above and those indicated by Huxley will, I hope, be readily appreciated:

> I saw no landscapes, no enormous spaces, no magical growth with metamorphosis of buildings, nothing remotely like a drama or a parable. The other world to which mescaline admitted me was not the world of visions; it existed out there, in what I could see with my eyes open. . . . I was seeing what Adam had seen on the morning of his creation—the miracle, moment by moment, of naked existence. . . . *Istigkeit* . . . 'Is-ness'. The Being of Platonic philosophy. . . . In their [the flowers'] living light I seemed to detect the quantitative equivalent of breathing . . . a repeated flow from beauty to heightened beauty, from deeper to ever deeper meaning. Words like Grace and Transfiguration came to my mind, and this of course was what, among other things, they stood for. [An essay by Suzuki] had been, when I read it . . . only a vaguely pregnant piece of nonsense. Now it was all as clear as day, as evident as Euclid. The books . . . with which my study walls were lined . . . glowed . . . with brighter colours, a profounder significance. . . . My mind was perceiving the world in terms of other than spatial categories. . . . [With mescaline] place and distance cease to be of much interest. The mind does its perceiving in terms of intensity of existence, profundity of significance, relationship within a pattern. . . . In this context, position and the three dimensions were beside the point. Not, of course, that the category of space had been abolished. . . . Space was still there; but it had lost its predominance. The mind was primarily concerned . . . with being and meaning. (pp. 16–19)

> Table, chair and desk came together in a composition that was like something by Braque or Juan Gris. . . . I was looking at my furniture, not as the utilitarian, who has to sit on chairs, to

write at desks and tables . . . but as the pure aesthete whose concern is only with forms and their relationships within the field of vision or the picture space. But as I looked, this purely aesthetic Cubist's-eye view gave place to what I can only describe as the sacramental vision of reality. I was . . . in a world where everything shone with the Inner Light, and was infinite in its significance. (p. 20)[5]

Except for the specific identity of the individual painters Huxley refers to, I would endorse every word in his text as describing what I have personally experienced with Ayahuasca. I may note that for me, the single occasion in which all this was most forceful was the very first time I partook of the brew. It is perhaps worthwhile to note that the experience described in the text just cited was a first time for Huxley too.

[5] Fragments skipped in this quotation will be cited later in Ch. 15.

5

Open-Eye Visualizations

Sometimes we see a cloud that's dragonish;
A vapour sometime like a bear or a lion,
A tower'd citadel, a pendent rock,
A forked mountain, or blue promontory
With trees upon't, that nod unto the world,
And mock our eyes with air.

Shakespeare, *Anthony and Cleopatra*

What, it will be question'd, when the sun rises, do you not see a round disk of fire somewhat like a guinea? O no, no, I see an innumerable company of the heavenly host crying, 'Holy, holy, holy is the Lord God Almighty.'

William Blake

Now that the reader has had a first feel of what the Ayahuasca experience is about, I will start to examine this experience more systematically.

While all the effects of Ayahuasca are extraordinary and all may leave a deep mark on those that experience them, in many respects the visual effects are the most salient. This, let me emphasize, by no means implies that a session in which a drinker has not experienced such effects cannot be a meaningful and/or powerful one. Yet, I would say that the primary language in which Ayahuasca expresses itself is the visual one. Undoubtedly, of all the effects of this brew, the visualizations are those that exhibit the richest phenomenology and they are also those that most readily lend themselves to phenomenological description and analysis. This is also reflected in the literature: of all effects in question, the visual ones are those that have been mostly reported on and discussed.

With practically no exception, when the question 'What does one see in Ayahuasca visions?' is discussed in the literature, the discussion involves a specification of the content items that appear in the visions. A typical statement would be: 'One sees jaguars or snakes or people one knows.' Logically, however, before examining the content of visualizations, we should ask what are the *structural types* or *forms* in which these visualizations may appear. It is with such a typological analysis that the present discussion of Ayahuasca visualizations begins. The contents of Ayahuasca visions will be considered in subsequent chapters.

An important terminological distinction is in place before we go on. I use the term 'visualization' to refer to all the visual hallucinatory effects that Ayahuasca induces. In contrast, as employed here, the term 'vision' refers only to those visualizations that are figurative and have semantic content; the term can also refer to the type I call 'visions of light'.

Ayahuasca-induced visualizations can occur both when the eyes are closed and when they are open. Admittedly, it is both easier and more common to experience visualizations with the eyes closed, but interesting visual phenomena also occur when the eyes are open. In fact, as we shall see below, some of the more powerful aspects of the Ayahuasca experience may occur precisely if and when one succeeds in having visions with one's eyes open. In this chapter I discuss the types of visualization experienced with open eyes; the types experienced with closed eyes are discussed in the next chapter.

Meaningfulness

I begin with a phenomenon already considered in the previous chapter. It does not involve visualizations as such, but to my mind it is intrinsically related to the visual phenomena described subsequently in this survey. I refer to meaningfulness. In the previous chapter we noted that Ayahuasca modifies the way people attribute meanings to things. Under the effect of the brew, one is prone to find in things and states of affairs meanings that are not associated with them under ordinary conditions. As we have noted, under the Ayahuasca intoxication there is a general tendency to confer great meaningfulness on whatever one perceives or encounters.

The enhanced conferral of meaning may pertain to a number of levels or dimensions. Globally salient is a dimension already mentioned in the previous chapter, namely, the aesthetic. One feature that distinguishes a work of art is that, unlike configurations and happenings in the world, it is composed. It is not a scene, but rather a *mise en scène*. With Ayahuasca, the world—the regular world of everyday life—seems to be precisely that: a work of art. As such, it appears to be composed, designed, directed (in the theatrical sense of the term). As noted in the previous chapter, most interesting comments on this are found in Huxley (1971); similar observations were also made by Michaux (1972) and Watts (1962).

Interpreting-as

The enhanced conferral of meaning may result in non-ordinary interpretations of things and states of affairs. I call this phenomenon *interpreting-as*, in analogy to the phenomenon of seeing-as, which will be discussed later. This phenomenon involves no visual hallucination as such—perceptually, the world is seen just as it normally is. However, the manner in which the world is interpreted changes radically and the person who has partaken of Ayahuasca feels that another reality presents itself to him or her. With this, one sees the world in a new, mysterious fashion. Castaneda (1971) refers to this phenomenon as 'real seeing'. It will be noted that many *Daime* hymns speak of the brew making people shift from ordinary perception which they characterize as illusory to the seeing of things as they truly are.

One striking instance of this phenomenon that I have experienced myself is the following one which occurred at a festive *Daime* session:

> The participants were dancing and chanting and then, suddenly, something dramatic happened. A woman lost her soul. Yes, I saw it. And apparently other people saw this as well. The dancing and chanting stopped. Assisted by another woman, the *madrihna* held the troubled woman. They sustained her, and at the same time they sang. Over a period that I would estimate as about twenty minutes, the woman's soul came back and was lost again. Eventually, the woman regained her soul and her normal self. She then appeared tranquil, clean and visibly younger than she had looked before the session started.

When recounting this story to friends who have never had any experience with Ayahuasca, I was repeatedly asked what it was that I had actually seen. No, I did not see anything white coming out of the woman's throat, nor did I see any ghost or spirit. Perceptually, there was just the woman in front of me. But the drama was evident, and I—along with the other participants in the session—saw it happen. I saw it with my interpretative eyes.

A second instance of this kind I would like to present also occurred in a *Daime* context. It was a festive session with a large number of participants:

> The intoxication was very strong and I was feeling that I was actually in the presence of angels. The epithet that crossed my mind was the Hebrew one—*mal'achei ha-sharet*, the angels of service, those in attendance of the Divine. There was no visual hallucination as such, but the feeling was very strong—the people around me were all angels praising the Lord.

Similar experiences were reported to me by my informants.

In still another *Daime* session the *madrinha* stepped aside and a man passed a vessel of incense back and forth in front of her. The smoke lifted up and it became perfectly clear to me: It was an act of cleansing, of protecting the woman from potential dangers that may be inflicted by evil spirits. There were no visual hallucinations as such, yet, I would not say that the act was merely symbolic. What I experienced was literally this—seeing the casting of a shield against evil powers. It all seemed to have a very serious and sombre allure, and manifestly, it was all invested with magic. If I were to define what made it all so mysterious I would say that it was the fact that on the one hand everything pertained to another reality, while yet at the very same time it was all real. Again, no hallucination as such was experienced—technically what I was seeing was real, and none the less it was all utterly non-ordinary, and enchanted.

Interpreting-as is often associated with an animistic outlook. In the course of a session, the Ayahuasca drinker may look at objects and feel that they embody hidden animae (I use this term expressly, in order to avoid the more natural term 'soul' which commits one to connotations that may not be meant here). This experience is extremely common. For instance, in his description of his first experience with Ayahuasca, Luna (Luna and Amaringo, 1993) notes that under the intoxication, the feathers on the *ayahuasquero*'s crown and the skins of animals hanging on the walls seemed to come alive (see also the description by Goldman, 1979, in Ch. 4). Like the other patterns surveyed in this section, the adoption of the animistic outlook may be regarded as an interim step between the overall, non-ordinary atmosphere that Ayahuasca induces and the visions it generates. The atmosphere is already non-ordinary, but visions are not yet experienced.

Non-ordinary interpretations may induce egocentric views whereby things in the world are seen as particularly related to the person under the intoxication. Inspecting other persons, the Ayahuasca drinker may feel that they specifically relate to him or her. When the attitudes attributed to others are negative, this can entail paranoiac-type feelings. The egocentric perspective may also be related to inanimate objects, in which case it is coupled with animism. A very common phenomenon—one experienced personally and by many of my informants—is that associated with the perception of faces in photographs. The halls in which both the Church of Santo Daime and the UdV hold their sessions are adorned with photographs of the founders of these sects. Under the intoxication, drinkers often feel that these photographs are invested with life and that, in addition, they orient themselves—for instance, in their looking or smiles—to the drinker and the other participants in the session.

Above, a contrast between the literal and the symbolic was made. This contrast deserves some further clarification. Surely, in a simple, uncontroversial sense the passing of the incense mentioned above is a symbolic act of cleansing. The cognitive level pertinent to our discussion here is, however, different. What I am talking about here is the perspective by which people under the effect of Ayahuasca see the world. In this particular example what was being 'seen' would have fully made sense for an external observer, but this is not the criterion by which the cognitive status of the example is to be judged. Rather, this status is to be based on the quality of experience, from the subjective point of view of the person under the intoxication. From an external point of view, the act is 'only' symbolic—that is, constituted by two levels, that of the act and that of what is meant by it (in classic semantic terminology—the sign and what is signified by it). However, experientially, the act involves one level only. Thus, the act is not symbolic. Like a smiling face it does not represent meaning, it presents it. The meaning of the act is, in other words, transparent in the act itself.

By way of further clarification, let me compare the viewing of the cleansing ritual recounted here to the standard perception of a person picking flowers. When I see a person picking flowers I see an act that has a certain well-defined meaning, but no symbols or symbolic interpretation are invoked. The meaning is perceived to be in the very act that is being seen.[1] Exactly so here. Admittedly, unlike the act

[1] This idea is consonant with an insight central in the non-orthodox school of ecological psychology founded by James Gibson (1966*a*, 1979) to which I am very sympathetic (for general discussions, see Shaw and Bransford, 1977; Michaels and Carello, 1981; Turvey, Shaw, Reed, and Mace, 1981; E. Reed, 1991). Gibson and his followers adopt a radical realism according to which information and meaning are out there in the world. They maintain that patterns of information that are meaningful to cognitive

of flower picking, the act of passing incense is one to which society attributes another meaning, thus making it a symbolic act. I knew of this cultural attribution and I am not denying that the knowledge played a role in the non-ordinary perception of the world that I am describing. What is significant in the episode considered here is that the symbolic became the literal, the real. In the other instances of this type that I have cited, the source for the non-standard perception might not have been related to prior general cultural knowledge but might rather have been situation-specific; this difference in source is granted. Yet, *qua* experiences, the various instances surveyed are not categorically different.

Concluding this discussion of the contrast between the literal and the symbolic, let me recount an answer I once heard given, under the intoxication, by a Master of the UdV to a disciple. The disciple asked whether a story the Master had told (one of the things Masters of the UdV do during sessions is tell 'stories') was to be interpreted literally or metaphorically. 'Both literally and metaphorically,' was the Master's answer. (For further, independent, criticism of the common distinction between the literary and the metaphorical, see Shanon, 1992.)

Another pattern of interpreting-as is one I shall characterize as seeing the particular as generic, or rather, seeing the generic in the particular. I have experienced this on a number of occasions. The first, which for me was very striking, occurred during the daytime. It was in a village and I, intoxicated, was sitting on a small verandah overlooking the meadows. A farmer (a real one) was passing by, and I saw *The Farmer*, the universal prototype of all farmers. Again, as in the previous example, the standard perception and the non-ordinary one are related. After all, I saw The Farmer, not The Fisherman or The King. Yet, while normally I would have seen just a farmer, this time I saw *The* Farmer. While semantically linked, experientially these two perceptions are totally different. I have heard accounts of the very same phenomenon from my informants and I shall return to it in Chs. 8 and 15.

Related to the phenomenon of interpreting-as is that of enhanced or non-ordinary psychological understanding. Under the effect of Ayahuasca, people often feel that they see other persons and they understand much about them. In the prologue, I recounted how, the very first time I partook of Ayahuasca, I watched people I had never seen before and I felt that I understood very well the meanings associated with their actions and interactions—the intentions, the affects, the thoughts. In a word, all the 'psychology' involved in these people's behaviour. But no, I should correct myself. The feeling was not merely that I understood the covert intentions of these people whose names I did not even know, but that I *saw* those intentions. I saw the people, I saw hands touching, I saw faces smiling, and I saw what was in their minds and hearts. The detection and determination of their patients' maladies by *ayahuasqueros* are, I suspect, akin to

agents are perceived directly in the environment; these are referred to as 'affordances' (see Gibson, 1982). Disciples of Gibson have extended the notion of affordance to the perception of patterns of social interaction; see, for instance, Baron and Zebrovitz-McArthur (1983); Heft (1989); and Costall (1995).

this. Routinely, traditional Amazonian healers say that Ayahuasca enables them to see the insides of patients' bodies (at times, the analogy to X-rays is made; see, for instance, Gebhart-Sayer, 1986); more on this will be said in Ch. 15 in conjunction with the noetic aspects of the Ayahuasca experience.

The feeling of enhanced understanding may be coupled with sentiments of great empathy and even identification. All these may be felt not only towards human beings but also towards animals, plants, and even inanimate objects. One informant participating in a private session told me of an experience he had while lying supine on the floor and looking up observing the lamp hanging down from the ceiling. 'I have seen this lamp many times before' he recounted 'but only now did I appreciate that it was so lonely. Down in the room there was so many things, but up there on the ceiling the lamp was all by itself. I sensed the lamp's loneliness and in order to alleviate it engaged in kind of silent mental intercourse with it.' Experiences of this kind, note, may be related to the various patterns of animistic thinking mentioned above; they may also be regarded as complements to what I have referred to as the egocentric perspective.

Seeing-in

The patterns noted at the end of the two previous sections lead to what is perceptually the simplest type of hallucination one can achieve with open eyes. I am referring to a phenomenon that is common even in everyday life—the seeing of figures in things. Ordinarily, this phenomenon is commonly encountered in people's detecting figures in clouds. We have all engaged in such detections and, I gather, we have all enjoyed doing so. Here I refer to this phenomenon as 'seeing-in'. With Ayahuasca, seeing-in is enhanced, intensified, and expanded, and the likelihood of it happening is great.

As a rule, the phenomenon consists in the seeing of figures or other items in a visual array in the real world. Usually, as in the day-to-day seeing of figures in clouds, some features of the real visual array may be incorporated within the hallucinated figure. At times, however, there are no shared features at all and the hallucinated figure is seen upon a completely smooth surface (in the psychiatric literature this phenomenon is referred to as *pareidolia*; see Scharfetter, 1980). What has impressed me very much in Ayahuasca induced seeing-in is the richness of the hallucinated figures and the complexity of their features. When examining the details of these figures, again and again I was stupefied. On the one hand, these details fitted so well with the figure of my imagination while, on the other hand, they all seemed to be generated from the texture of the real array actually in front of my eyes. Later, when the intoxication had ended, I would approach the array and inspect it closely and I would not be able to fathom how the imagined figure arose from it.

On the basis of both my own experiences and my interviews with other people, I can make the following generalizations. The most common substrates for seeing-in are ground terrain, walls, and tree foliage. The items most commonly seen are faces of either human or phantasmagoric beings; entire figures may be seen as well. At times, the faces and the figures are seen as works of art—notably masks, pieces of sculpture, or bas-reliefs. Ornamentations and mandala-like patterns may be seen as well. Of the dozens of images of this type I have seen, hardly any pertained to any other category of content. Usually, the images pop up. One looks at a real world visual array and, lo and behold, there is an image in it. Once the image has been seen, it usually stays there as long as the intoxication lasts. Many times I have turned my gaze from the hallucinated image and then returned my attention to the spot in the visual array where I had seen that image. This could be after a moment or after a longer interval. Invariably, when I would return my gaze, the image would be there. Exactly the same image would again be seen at the very spot it had been seen earlier. This experience was also reported to me by my informants.

Several patterns of seeing-in may be noted. The simplest is one in which isolated single figures are seen; it is exemplified by the following two episodes. The first took place during the course of a session taking place in the midst of the Amazonian forest:

> Feeling that the strong effect of Ayahuasca was over, I sat down on a bench looking outwards at the terrain in front of me. In the ground I detected a face I could interpret as that of a Mayan deity. I turned away, and when I returned my gaze to the initial point I still saw the same face. Further, when I looked at a different spot on the ground, I detected the contours of another face, this time that of what seemed to me to be a Cambodian deity. Again, I turned away and when I returned the face was still there.

The second episode occurred in a Barquinha session:

> I observed a (real) icon of the Virgin Mary with the baby Jesus in her arms. There appeared to be a third figure, a man, standing behind Mary. I was able to tell myself that the figure might have been generated by the folds in the canvas, yet the man's figure was so real that I could not be sure of this. I pointed this out to the person sitting next to me, and he confirmed seeing the same thing.

A somewhat more complex pattern is the following one that I experienced as I was observing a wicker shopping bag with an interlacing pattern of tan and black. I saw two alternating figures: at certain moments the wicker bag looked to me like a jaguar, and at others like the face of my niece. Later, when the intoxication was over, I approached the bag and examined it closely. The pattern was the simplest one possible. There was nothing in it that I could see as generating either the figure of a jaguar or that of a human face.

Usually—as is the case of all the examples described above—what are seen are single, isolated items, but I have also experienced seeing figures of multiple items. The following two episodes are examples of this. The first took place during a private Ayahuasca session held in the afternoon (that is, when there was light outside):

I was sitting in a garden in front of a cement wall of rough, uneven texture. And, lo and behold, an ancient Persian or Assyrian relief was there on the wall. The wall was large, and the relief extended all along it. And again, I had my eyes open and it was evident that the wall was a wall, a simple wall of cement. Yet, in it was the grand relief.

The second episode happened during a quite common *Daime* session, a special mass commemorating the passing away of the founder of the church, Mestre Irineu. The session took place one Sunday during the daytime in a very beautiful locale in the countryside of the Brazilian state of Minas Gerais. The building of the church was specially designed by a loving architect. The construction was that of a cone with a circle of open windows as its base. All around, uninterrupted by any wall or partition were hills and meadows. On my right side, the mountain was carved, so as to accommodate the path along the building. I looked at the furrowed rocks and they were all lions—a long row of lions that were watching us, the participants, in the holy ritual. Intermittently I would move my gaze inside, to the inside of the church, and whenever I moved it back to the rock lions, there they were, watching the service which was held in the hall.

In all the examples presented above the figure seen under the intoxication was static. My experience is that this is generally the case. Only once did I see (in the present sense of seeing-in) a figure that moved, and even then the movement was very limited. It happened during a session of the UdV. It was at night but, as is normally the case in UdV sessions, the hall was well illuminated:

On the upper section of the wall in front of me I saw a human figure. It was a handsome young man who was looking forward and ahead, as if observing the eternal. Every now and then, the man placed his hand on his forehead, as if to facilitate the scrutiny. Afterwards, the figure of the man was successively replaced by half a dozen other figures, all of individual human faces. All the figures appeared at the same spot on the wall, which was perfectly smooth and painted monochromatically. The figures were there all throughout the period of my intoxication. At each given moment only one figure out of the set was seen. Occasionally, a figure was replaced by another but over the entire period all the figures seen were items of the same set, and they all appeared in the same identical place.

After the session I checked—physically there was no special feature on that particular spot on the wall.

Seeing-as

Very related, and at times even coextensive with seeing-in, is a phenomenon to which I refer, following Wittgenstein (cf. Wittgenstein, 1953, 1980), albeit in a completely different context, as 'seeing-as'. The phenomenon is, I gather, familiar to everyone. One is presented with an item or scene and sees it as something else. The example Wittgenstein gives is that of alternately seeing the same line drawing as either a duck or a rabbit. Another, more famous example of this is the drawing of a human figure that can be seen either as an ugly old witch or as a pretty young woman. The visual input is identical, but somehow, sometimes you see it as one, and at other times as the other. It all depends on how you look at it. Notice, however, that how you look at it is actually not completely subject to your own will.

Seeing-as is very akin to seeing-in and at times it is practically impossible to say whether a hallucination pertains to the one type or the other. Still, in this summary survey I have decided to present these two as separate types. This I do in order to highlight phenomenological distinctions and better mark the variations that the hallucinatory experience can take. Appreciating the fuzziness between the two types, let me spell out the main features that differentiate them.

In seeing-in one sees an image in or on a given surface or visual array. In seeing-as one sees an image in lieu of an object in the real world. In seeing-in, the imagined item occupies a part (in general, a very small part) of the visual array upon which it appears. In seeing-as the image normally coincides in its contours with the real world object (although, as will be noted below, at times deviations from these contours are experienced). These differences are reflected in the verbal expressions that will be used to describe the experiences at hand. I see the figure of a person *in* a cloud, I see a face *in* the moon, but I see the face of a human being *as* the face of the devil. In Belshazar's vision in the biblical book of Daniel a written message appeared *on* the wall; in the book of Judges, a watchman is warned not to see 'the shadow of the mountains *as if* they were people' (9: 36).

As a first example, let me recount an experience having to do with a cloud. It was during a private session in which only I and one other person participated. I was sitting in front of the open window watching the sky an hour before sunset. 'Look at this cloud,' I called to my friend. Normally I would avoid talking during a session, but I just could not resist. The view was so wonderful, so grandiose that I had to share it. The cloud in front of me was a great bird, hovering with its immense wings spread open, protecting the universe. And my friend saw it all—the head which was slightly turned to the left, the body, the wings, the protection.

On another occasion, I was sitting in a *Daime* concentration session, and here, in front of me, were two Mayan faces, one looking at the other. When I first inspected what I saw, I realized that the faces were actually the shadows of the two shoes that the person sitting on the bench in front of me was wearing. The appreciation of this fact, however, did not affect my visual sensation. Quite frequently, the person moved his feet. When this occurred, the shadows moved as well, and an interaction, which seemed natural and meaningful throughout, was generated between the faces. No rational reflection could stop this perceptual hallucination.

In the simpler manifestations of seeing-as the overall contour of the figure that is being perceived matches that of the real figure that generates it. In more complex instances this need not be the case and deviations from the figural contours are noted. In general, these are minor—they are like the differences between the contours of a dressed person relative to the contours of his or her naked figure. There are also more *elaborate* cases in which many details not actually associated with the real object of vision are added. An example is one that I experienced late at night and in the dark when sitting in front of some potted plants. What I saw was an old Indian. Both during the hallucinatory stage and later I checked what was actually in front of me. The latter inspection revealed that only the base of the pot matched the lines of the figure I had seen. The greater part of the figure did not match any lines perceived in the external world.

A special subcase is that of seeing the faces of real persons as those of beings other than these persons actually are. The most striking episode of this kind I have experienced took place in the context of an Amazonian healing session. One of the patients, one whom I also treated myself, was a retarded girl, about 9 years old, who behaved, and was treated like, a child of 2. Apparently, she was having an epileptic fit, and she kicked and grunted violently. I approached her and saw the face of an ugly old woman. The sight, let me add, was appalling. Another episode occurred during a UdV session. I looked at the presiding Master and his head seemed to be half human and half bird with a richly coloured beak protruding forward. In still another UdV session, the Master appeared to me as a mythological, Promethean figure. On a fourth occasion, this time in a private session in which only one person other than myself participated, I looked at my companion and his face was that of an enchanted, somewhat devilish creature. Overall, all the contours of the face seen were that of my friend, yet the lines of the face—notably, those of the eyes and the mouth—were exaggerated. Actually, the distortion was akin to that of an actor when made up. Upon seeing the actor, the spectator sees the theatrical or cinematographical character, but at the same time he or she recognizes the real-life identity of the actor who is playing the role in question. Many informants reported similar experiences to me.

Whereas in the first and last of these examples the overall contour of the visualization matched that of the face of the real person, in the second and third the visualized contours extended those of the real face. For these latter cases the proper simile is not of make-up, but rather that of a mask. A case very similar to those described here is also encountered in a report cited in Taussig (1987: 325) in which the Ayahuasca drinker saw the shaman administering the Ayahuasca session as the devil.

Both experientially and conceptually, the phenomenon of seeing-as is most intriguing. On the one hand, the item seen is what it is in the real world (a plant, a cloud, a patch of ground), and on the other hand, at the same time, it is a visionary object that is completely different. How the two square together is experientially most perplexing and deserves further conceptual analysis.

Superposition

With further elaboration, the role of the sensory visual input diminishes, and the hallucinatory creation gains more weight. In *superposition*, there is little, or even no, overlap between the figure in the real world and that visualized under the effect of Ayahuasca. Thus, on one occasion in the Amazonian forest, I was looking at the trees, and dead people appeared to be hanging from them. The real figure (the trees) and the visualized one (the people) were related, but not by means of any overlapping of lines. In other words, the relationship was primarily *semantic*. Other instances of this kind I have experienced were seeing an (imaginary) jaguar resting on the branch of a (real) tree and an (imaginary) cow standing on a (real) truck.

The semantic relationship exhibited by the foregoing examples is, essentially, one of *collage*. More complex still are cases in which the superposition involves a meaningful, seemingly *causal*, relationship. Once, in an indigenous healing session in which I participated, the healer was blessing a patient and a handful of gems came out of his mouth as he did so. In addition to its being semantically rich, this superposition is *dynamic* in the sense that it is part and parcel of an on-going event taking place concurrently both in the real world and in the imaginary one. Several years later, when the writing of this book was almost finished, an identical episode was recounted to me by a woman describing her first experience with Ayahuasca.

By way of further example, I shall mention two other cases of superpositions which are visually complex, semantically elaborate, and dynamic. Twice, during sessions of UdV, I was gazing through the open doors. The bush that was seen through the door was full of beings. I might say that the beings I saw were 'tree-people', for their figures were all made of the leaves of the bush outside. What I saw was so rich: there were many figures, and their faces were distinct and expressive. At times it seemed that the faces expressed attitudes; for example, when the Master of the session spoke, the figures appeared to display interest and curiosity. Furthermore, it appeared that they were extending their heads towards the door so as to hear better, to be closer to what was going on inside the hall.

Scenes

Complex superposition may develop into full-fledged scenes. Visions of this kind constitute some of the most powerful of all Ayahuasca experiences. I experienced this the second time I partook of Ayahuasca. As recounted in the Prologue, in my vision I saw lots of animals—both ordinary and phantasmagoric—roaming the (real) forest in front of me.

There are also cases in which what is seen with open eyes is totally imaginary, and is not at all dependent on actual perceptual material. An example is my third Ayahuasca session, also recounted in the Prologue, in which with my eyes open I saw an enchanted city of gems and gold in front of me; together, they all composed a vivid scene, which was both semantically rich and dynamic. It was as if a screen had been raised and another world made its appearance to me. I shall note that whereas in the previous vision, that of the enchanted city, the hallucinatory material was superimposed upon the perception of the real world, in that discussed here the hallucination in its entirety consisted of constituents that were imaginary. In this case, there was no relationship between the content or form of the vision and the real scenery before my eyes.

In the last two examples, what was seen in the vision was like a scene in a film or a piece of theatre. In front, there was the vision, but when I turned my gaze sideways, I would see the ordinary world as it actually was. This is exactly like the situation in the cinema hall, when one turns one's head and, in the dark, sees the other spectators in their seats. More radical is the case in which the hallucination is all-encompassing. This happened with the most impressive open-eye visualization ever reported to me unfolded as the drinker was walking from one side of the hall in which the UdV session was held to the other, exiting to the yard, running down a staircase and heading to the bathroom:

> All during this walk I saw myself passing through a celestial palace of incredible beauty. There were opulent chambers, arcades, and ornate staircases. Noblemen wearing pointed, steeple-like mitres sewn with precious stones roamed around. It was explained to me that this was the palace for the affairs of the universe and that God was residing over there, in another pavilion. All during this time I had my eyes open and was not having any difficulty in pursuing my walk. All along I was simultaneously on two levels of existence. With my gaze forward and upwards I was fully immersed there in the celestial palace, which was experienced as totally real. When I lowered my head I saw the hall of the session as it was. The two realms were distinctly separated and there was no confusion between them. Nor did I experience any fear or apprehension. It was all extraordinarily fantastic; indeed, the most beautiful thing I have ever seen in my life.

The drinker further reported that he was fully immersed in another reality, and all along, with his eyes open, he was acting in it. The topic of immersion will be examined focally in Ch. 7, when patterns of action and interaction are discussed. Here let me underscore the remarkable nature of this vision from a structural point of view. Extraordinary as they were, all the details of the vision joined together and composed into one coherent scene. Furthermore, while they were all phantasmagoric, these details and their contours fitted ordinary reality. Thus, the contents of the vision did not counter the trajectory the drinker was following in the real world. In particular, no obstacle was visualized that barred the drinker's actual advancement, nor were false passages seen that led him astray in his actual walk. Examined in terms of the distinctions made throughout this chapter, it appears that this vision is the product of a wonderful interplay of all the different elements indicated in our survey: seeing-in, seeing-as, superposition, elaboration, richness of content, semantic complexity, and dynamic involvement. And keep in mind: all this was achieved on the spur of the moment, on-line, so to speak. The drinker's action was coherent both in itself and as far as the ongoing relationship with the ordinary world was concerned. Indeed, no external observer could detect anything unusual or wrong in the drinker's behaviour: it would seem that he was crossing the hall and going out to the bathroom in a totally normal fashion.

Veritable Hallucinations

The phenomenon to be reported here happened to me only twice; both times it turned out that the Ayahuasca brew also contained *toé*. What happened in these two episodes was almost identical. The first episode took place when I was alone with a healer in the upper Peruvian Amazon. At one point the healer retired to the other room to rest. Then, in the darkness:

I saw a man. I felt unwell and asked the man for help. He looked at me and may have smiled, but he did not offer any help. Indeed, he did not interact with me at all. I stretched out my hand and tried to grasp his. I did not succeed in doing this, and the man, strangely enough, did not make any effort to help me. In the morning I asked the healer who the other person was. He responded that there had not been anyone present besides us two. It must have been a vision, he said. It took me one full day to realize and accept that this was indeed the case.

The second episode took place in the inner yard of a house in which the *ayahuasquero*'s entire extended family resided. That the Ayahuasca brew also contained some *toé* was told to me only after the facts:

In the beginning, the session proceeded in a manner that I recognized to be totally standard and rather uneventful [this episode took place when I already had quite extensive experience with Ayahuasca]. At a certain point the healer retired to sleep and I remained in the yard stretching myself out on a long wooden log and half sleeping for, I gather, the last two hours of the night. When I opened my eyes it was the beginning of day and I discovered that I was having great difficulty in pulling myself up from my prostrate position. Eventually, I managed—only to see, at the corner of the yard, behind a door, a young man. He was a Caucasian, looked like a hippy, and had what seemed to me an overall suspicious presence. Around that time, the house began to fill with movement. The members of the family were getting up and preparing for the day. Children were criss-crossing the courtyard, and their mother, the healer's daughter, was helping them get ready for the day. Whenever this woman passed, the suspicious-looking man drew back and hid in a small space between the door and the wall. In that position he could not be seen by anyone but me. He stared at me, and it was clear he was indicating that I should not say anything about his presence. I did my best to avoid his look. I was also preparing to leave and go back to the village hospice where I resided. As I was doing this, however, I realized that one of my bags was gone. I had three items with me when I arrived, I reckoned—two shoulder bags that contained clothes and other personal effects and one smaller blue bag with all my guide books, diaries and phone booklets. My legal travel documents and money were tucked underneath my clothes on my body, but the material in the blue bag was my actual connecting line to the world, and without the information contained in it I could not go on with my journey nor establish contact with the many persons I was proposing to meet in the next stops of my planned itinerary throughout South America. In addition to these three items, I had a plastic bag that contained a flashlight, a bottle of mineral water, and a roll of toilet paper. So, here I was—getting up only to realize that one of my shoulder bags was gone. I looked around, the hippy man was gone too. Evidently, he had stolen my bag. I went outside intending to alert someone, but all the people present were too busy with their preparations for the new day. I tried to tell them that there was a suspicious person in the house, but no one would pay any attention to me. I stepped back into the inner yard, and I saw that now, the second of my large bags had disappeared. In my distress, I sat down on the log. And then I saw him again—that man. He was looking at me half maliciously, half in a friendly manner. It was clear to me that he was attempting to make a pact with me. If I collaborated, my belongings would be given back to me. At any rate, I knew that I should beware and not tell anyone of his presence. By now I was determined to get some help. I got up and looked for the owner of the house. I stepped out of the house and tried to explain what was going on, but, again, no one would pay attention to me. When I got back to the inner yard, only the plastic bag was there. I could have managed without my clothes and various travelling items, but with my blue bag gone I was completely lost. What happened later is not of relevance here. Let me just point out that in reality, all my personal possessions had been left securely in the lodging place where I was staying; when I got to the healer's house, I only brought along the plastic bag that contained those items that I deemed to be of potential use for me during the session.[2]

While some of the visions reported earlier in this book were much more powerful visually, these two episodes were the only times when I confused a vision with reality. For this reason I am referring to it under the perhaps not quite adequate label of 'veritable' hallucination.

Light

And then there is light. Many times, invariably towards the end of sessions and when I was stepping outside into the natural surroundings, there were lines and webs of light that interlaced everything. In time I came to learn that this experience is very common. Indeed, of the many people I have interviewed, only very few have not seen these patterns. The reader not having firsthand familiarity with Ayahuasca may, justifiably, regard these patterns as simple. Admittedly, in a strict geometric sense they are. However, invariably, drinkers associate deep meaning with these patterns. Upon seeing them, practically everyone I have discussed this matter with reported that they felt they were seeing the energy that permeates the world, makes it 'tick' and enables it to continue. To my question as to what she had gained most from her Ayahuasca experiences, one informant answered 'realizing that God exists'. When I asked her what made her reach this realization, she

[2] I have described this episode at length, for *toé* intoxications are very rarely described in the literature. I shall further emphasize that this experience was totally different from those usually induced by Ayahuasca. First, it lacked any fantastic allure. With Ayahuasca I have very seldom seen 'regular' persons—the people I have normally seen were royalty, historical figures, people of other lands and other times. In the episode just described, the person I saw was a man of our time that I could have met on the street. Second, the hallucination was completely incorporated within the perception of the external world around and there was no separation or distinction between them whatsoever. Indeed, and this is a third point to be noted—I took the hallucination to be real, never suspecting that it was not. With Ayahuasca this was never the case for me. On the basis of further checks with persons who had experienced *toé*, I tend to conclude that the special effects noted here reflect a categorical difference between this substance and Ayahuasca. With *toé*, one does confuse the real and the non-real; with Ayahuasca one feels that one is presented with other realities, but these are not confused with the ordinary one.

answered 'I went outside to the trees and I saw lines of light, like spider webs, connecting everything. These lines are the Divine presence without which the world could not be.' For a good pictorial depiction of this phenomenon the reader is referred to the drawings made by the Brazilian painter Alexandre Segrégio of his own Ayahuasca experiences (see Weiskopf, 1995). I shall also note that the phenomenon under consideration here is described by Castaneda (1972: 298). Castaneda's story is not related to Ayahuasca, but given the literary qualities of this story, I find it instructive to cite it:

> Suddenly I felt that my body had been struck and then it became enveloped by something that kindled me. I became aware then that the sun was shining on me. I could vaguely distinguish a distant range of mountains towards the west. The sun was almost over the horizon. I was looking directly into it and then I saw the 'lines of the world'. I actually perceived the most extraordinary profusion of fluorescent white lines which crisscrossed everything around me. For a moment I thought that I was perhaps experiencing sunlight as it was being refracted by my eyelashes. I blinked and looked again. The lines were constant and were superimposed on or were coming through everything in the surroundings. I turned around and examined an extraordinarily new world. The lines were visible and steady even if I looked away from the sun.

Other forms of light seen with open eyes are auras and halos around people. Usually, but not always, these are of white or gold. The different colours are usually associated with a general impression of the person concerned. A person who is manifestly uplifted by the intoxication will be seen by others as having a bright aura. Persons exhibiting problems or sickness were, by some informants, seen emanating red and black halos. I also heard reports of light descending upon participants in Ayahuasca sessions. The observers interpreted this as an act of help in which force is lent so as to sustain people in the difficulties they were confronting during the session.

Given the paramount significance of light in the Ayahuasca experience, I shall discuss it further, in a separate chapter—Ch. 17.

A Comment on Fuzziness

Readers have noticed, I am sure, that the demarcation between the types and distinctions made throughout the foregoing presentation is often fuzzy. Most notably, the boundaries between seeing-in and seeing-as are blurred and at times the division between the two may not be clear-cut. Likewise, fuzziness may be noted between interpreting-as and seeing-as. The same is true for the finer distinctions introduced in this chapter. Rather than defining well-demarcated subcases that are clearly separated from one another, these chart the dimensions and values upon which the structural types encountered in Ayahuasca visions may vary. This fuzziness is not a methodological matter pertaining to difficulty in labelling and categorization. Rather, it is a substantive feature that is intrinsic to human cognition and in fact is not specific to Ayahuasca visions. Indeed, I have encountered similar patterns of categorical fuzziness in my study of thought sequences mentioned in Ch. 2 (see also Shanon, 1989*a*). My general theoretical stance is that fuzziness of this kind is a fundamental property of human cognition. Yet, the fact that the variation between types and subtypes is fuzzy does not mean that from a conceptual point of view the distinctions in question are blurred. Types and subtypes may gradually blend into one another, but this does not imply that, on the abstract theoretical level, the conceptual distinctions are ill-defined. Appreciating this, my general strategy in cognitive phenomenological research is to accentuate the distinctions on the theoretical level while appreciating the rich, fuzzy, and often subtle gradation exhibited in the actual phenomenological manifestations at hand.

On a number of occasions I had the chance to experience the fuzziness directly and to see how, gradually, images of one type blended into those of another. Here is one such case:

> I was sitting inspecting the forest surrounding the hut in which a healing session was taking place. The forest looked enchanted. It seemed that figures were about to step out of it, but none did. My gaze rested on a hedge of bushes that lined the backyard. There was magic to these bushes. It seemed that something was concealed there—a secret, another existence. Two branches attracted my particular attention. They looked as if they had been sculpted, as if they pertained to a fairy tale. For a few moments I saw in them two elegant cranes standing one in front of the other as part of a wonderfully cast grille. Yet, all the time it was also clear to me that what I was seeing were bushes in the yard.

At this juncture, let me mention observations made by Klüver, one of the first scientists to study mescaline. Klüver (1928) describes a phenomenon he calls *presque vu*, almost seeing. This is the experience of feeling that one is just a small step from seeing a figure or an image. Most of the overall contour is there and one knows what the image is, yet it is not fully perceived as such. Drawing on my own experience with Ayahuasca I would mention in this regard experiences of feeling that the trees in the forest were invested with spirits and that of observing, but not quite seeing, that there were serpents in the wooden pillars of the building in which the session was taking place. I might add that this perceptual phenomenon of *presque vu* may perhaps be related to the phenomenon of extra meaningfulness with which this survey began. Specifically, the latter might be regarded as a precursor to the former. I would further venture that the animistic perspective by which one feels that objects, notably trees, are inhabited by souls or soul-like entities is also a manifestation of this phenomenon.

Two Final Observations

In closing, let me mention two patterns—both are, in a fashion, negative. The first is not-seeing, the second is one that has been observed with other psychoactive substances but is not encountered with Ayahuasca.

By the term *not-seeing* I refer to the phenomenon whereby one's eyes are open, one looks, yet one does not see what is in front of one. One case of this kind that has been reported to me is the following. The drinker looked at her legs—the trousers and shoes were there, but not her body. The perception was that the pieces of clothing had gained an independent existence. In particular, it appeared that the shoes were dancing on their own, and that inside them there was nothing. Another person reported looking at a mirror and seeing his reflection disappear. I myself had the experience of looking at the thatched roof of the hut in which the session was held and instead seeing the open skies or the high heavens. I have heard of this very same experience from several other people.

Second to be noted are sensory-perceptual effects that are *not* encountered with Ayahuasca. After extensive interrogation of many people, I tend to conclude that Ayahuasca does not induce perceptual distortions. I am referring to effects typically associated with LSD and which, in the culture at large, are often labelled as 'psychedelic'. With Ayahuasca the visual field is not fragmented into separate geometric pieces that gain independence from each other, the visual array does not become fluid, nor do contours twist and become distorted. Rather, the impression is that a secret world, hidden within the normal overt one, gradually reveals itself. Yes, there are perceptual effects, but they are not 'crazy', 'psychedelic' ones, but rather envelopes, like the skins of an onion, that are peeled so as to present senses and meanings that in ordinary states of consciousness are not perceived.

6

A Structural Typology of Ayahuasca Visualizations

> The secret is: Close your eye and roll your eyeball. Those colours that shine and glow will be revealed. Permission to see is granted only with eyes concealed.
>
> *The Zohar*

> In the upward movement the horses that draw the chariot change into birds, into swans.
>
> *The Rig Veda*

In this chapter I present a systematic typology of the structural types that Ayahuasca visualizations may take. Paradigmatically, the types surveyed are experienced when the eyes are closed. However, all the structural distinctions introduced here, as well as those pertaining to interaction, semantics, and narration which are introduced in the next chapter, can also apply to visualizations experienced with the eyes open. As indicated in the previous chapter, full-fledged visions are less likely to be experienced when the eyes are open; yet, this does happen. In fact, when the intoxication is strong the closing of the eyes may make no difference and the same visualization may be experienced both when the eyes are closed and when they are open. Thus, again, while the paradigmatic cases in which the distinctions in the present typology apply are most likely to be encountered with closed eyes, in principle, these distinctions apply both to closed-eye and open-eye visualizations. In contrast, the patterns surveyed in the previous chapter are specifically associated with visualizations experienced when the eyes are open.

An important qualification should be made before we begin. Naturally, the different types will be introduced in a particular order. In several respects this order corresponds to the order in which types of visualizations usually appear in the course of a session. Yet, this need not necessarily be the case. Conceptually, the question of types and the question of order are distinct. In this chapter, we shall be dealing with the former, not the latter. Admittedly, the progression traced here does reveal various patterns of development. At this stage of the analysis, however, I shall keep the issue of development in abeyance; I shall turn to a focal examination of it later, in Ch. 18.

In the following structural typology six main categories are noted: visualizations without any semantic content, visualizations consisting of primitive figurative elements, images, scenes, virtual reality, and visions of light. An early, partial analysis of the topics presented in this chapter and the next appears in Shanon (2002*b*).

Visualizations without Semantic Content

In the literature, it is common to characterize the first stages in an Ayahuasca session as containing a variety of visual elements devoid of content. Reichel-Dolmatoff (1978*b*, 1991) describes these as small, brilliant, star-shaped or flower-shaped elements that flicker in repetitive kaleidoscopic patterns. These luminous patterns alternate between bilateral symmetry, grid patterns, and zigzagging or undulating lines. Also encountered, he reports, are well-defined geometric motifs, multicoloured concentric circles, and endless clusters of brilliant dots. To my mind, Reichel-Dolmatoff lumps together several types that do not constitute one homogeneous group. Some finer distinctions seem to be needed; these are the subject of the following discussion.

Bursts, Puffs, and Splashes

First among the visualizations without content are what I shall refer to as bursts, puffs, and splashes of light. These may be white, yellowish, or in colour.[1] *Bursts* are local, discrete foci of light. They are akin to the patterns one sees when one's eyeballs are pressed. In the psychophysiological literature these are known as phosphenes (see Oster, 1970; Siegel and Jarvik, 1975; Horowitz, 1978; Tyler, 1978; Glicksohn, 1986/7; B. Jacobs, 1987; and Lewis-Williams and Dowson, 1988). When the bursts gain momentum, they turn into *puffs*. *Splashes* are generated when the foci of light begin to move; with this, the visualized elements become elongated and endowed with spatial extension. The splashes are like nebulous sprays, and they lack precise boundaries.

In the earliest stages of the Ayahuasca intoxication the elements of light appear sporadically, they do not extend across the entire inner visual field, and they do not form well-defined patterns. Later, the various elements of light multiply and extend throughout the inner visual field. Together, they may form undulating clusters and concentric circles, and may generate a rhythmic, pulsating presence.

Repetitive Non-Figurative Elements

None of the elements of light indicated above has distinct boundaries. For me, the simplest patterns exhibiting such boundaries are usually ones in which the same element appears many times. Examples are colourful round elements arranged in concentric circles, mango-shaped items ordered along a large arc, star-like forms extended as if on an invisible sphere throughout the inner visual field, and

[1] The term 'colour' here, and throughout the present discussion, denotes colours which are other than the white and yellow of natural light.

rainbows. Typically, the atomic elements are well formed and well defined. Each has an inner structure with distinct parts and various colours, and within a pattern, like beads in a chain, all elements are identical. In their totality, the elements define a pattern that extends throughout the entire inner visual field though it need not necessarily fill it up.

Geometric Designs and Patterns

More structured are full-fledged geometric designs and patterns. The geometric patterns are always composed of well-formed designs. Unlike the repetitive elements described in the previous subsection, these designs are not isolated atoms but rather they comprise the basic parts of a geometric whole. Paradigmatically, the contrast is like that between a bead in a chain on the one hand, and a square in a grid or a basic pattern in a carpet on the other. Typically, the designs are of wonderful fluorescent colours.

The geometric patterns may be two- or three-dimensional. In the former case they are like arabesques; these compose tapestries that entirely cover the inner visual field. Unlike the two-dimensional geometric patterns, the three-dimensional ones usually define structures positioned in space; hence, they need not be fully coextensive with the inner visual field. Often the patterns are like multicellular honeycombs whose cells are usually pentagonal or hexagonal. The total construction may be linear-polyhedral or oval-circular; it may be static, or it may be pulsating or vibrating. At times, the geometric patterns may seem to defy ordinary real-world Euclidean geometry; some persons that I interviewed made reference to higher orders of spatial dimensionality.

A characteristic worthy of special mention is that of symmetry. Reichel-Dolmatoff (1975, 1990) notes that at an early stage of the intoxication the patterns manifest bilateral symmetry and that at a later stage this symmetry is broken. My experience confirms this.

Finally, let me note that the patterns surveyed above are reminiscent of those defined by Klüver (1928, 1966), in his study of mescaline. Klüver distinguishes between four kinds of what he calls form-constants, that is forms that appear in practically all mescaline visions. The first type comprises gratings, lattices, honeycombs, and the like. Second are cobwebs. Typical of the third type are tunnels, funnels, and cones. The fourth type consists of spirals. As with Ayahuasca, Klüver notes that the forms are frequently repeated. For subsequent research employing Klüver's distinctions see Horowitz (1975) and Siegel and Jarvik (1975) as well as the review by Siegel (1977); in Siegel and Jarvik (1975) several pertinent artistic renderings of hallucinations are presented.

Primitive Figurative Elements

The simplest figurative elements, those to which I shall refer as primitive, are characterized by a lack of either independence or permanence, or both.

Rapid Figural Transformations

The nebulous splashes may gain form and generate a multitude of rapidly transforming figures. I would liken the transformation of the figures to sequences of photographs taken in time-lapse filming, such as the progression of stages of a growing flower and the transformation of shape a cloud undergoes.

A terminological point is in place here. In the literature, the rapidly transforming figures are usually characterized as *kaleidoscopic* (see, for instance, Reichel-Dolmatoff, 1975, 1978*b*). I take issue with the use of this simile. On analysis, the images of a kaleidoscope exhibit two features: first, a multitude of identical elements; second, rapid transformation. The patterns described here exhibit the second feature, but not the first; for this reason, I prefer not to call them kaleidoscopic but rather characterize them as rapid figural transformations. The term 'kaleidoscopic' will be used to refer to a type that will be defined in the next subsection.

Designs with figures

Another type is that of primitive figurative elements embedded in the geometric designs or emerging from them. Reichel-Dolmatoff (1960, 1990) characterizes these as constituting the second stage of visioning. He further notes that in this stage symmetry breaks down, the overall geometric aspect of these visualizations disappears, and figurative pictorial images emerge. Specifically, large blobs of colour are seen and from them emerge different shapes looking like people, animals, or all sorts of unfamiliar creatures.

As described in the Prologue, my very first Ayahuasca-induced visualization was of lizards popping out of arabesques. The lizard images were repetitive and embedded within the geometric pattern. Patterns that are very similar to this are seen in some Escher drawings. The figures indicated here differ from those described in the previous subsection in several respects. The figures emerging from the geometric designs always consist of many tokens of the same type,[2] they have well-defined boundaries, and they exhibit relative temporal permanence. In contrast, the figures emerging from the splashes are single tokens of one type, they are not always well defined, and they change rapidly.

In time, the figures embedded in the geometric design may gain independence. As this occurs, the design may become part of a larger picture composed of both geometric forms, typically three-dimensional, and figurative elements. Furthermore, the design may serve as a kind of scaffolding along which the figures climb. One such scene that especially stands out in my memory is that of a huge pulsating greyish-purple sphere around which all sorts of beings were dancing.

[2] Tokens are the actual occurrences of items pertaining to a type; for instance, in the word *David* there are two tokens of the type *d* and one token of the type *a*.

Images

The primitive figurative elements surveyed in the previous section lack either independence or permanence, or both. The simple images to be described in this section are independent and relatively permanent. Yet, in various respects they are still constrained and do not constitute full-fledged, stable, and independent visions.

Kaleidoscopic Images

The arabesque may change from a purely geometric pattern into a multitude of figures. This results in fast-moving kaleidoscopic images that usually consist of many items of the same kind. Patterns of this sort which I have seen several times are lines of semi-naked dancing women and of flowers. My experience has been that while in movement, images of this type are not subject to rapid transformations. Further, these images are independent in that they are neither part of a geometric design nor a nebulous pattern. Figuratively, the images are well defined. Yet they exhibit two features that differentiate them from standard pictorial images. First, they are always comprised of multitudes of the same basic figural element. Second, while not subject to fast transformation, they always appear to be in motion.

Presentations of Single Objects

Presentations of single objects consist of images seen in isolation, without their being placed in a setting. These images are single rather than multiple, they are stable, and, unlike the images of the cartoons or the animated movies, they are three-dimensional. The paradigmatic case of such images are faces.

Serial Images

Images of single objects may also appear in succession. In some of these cases, the different images may be interrelated. Two types of relationship may be noted. The first is that of *variations on a topic*. These consist of several images all of which pertain to one common topic. For example, a European with no previous experience with Ayahuasca reported seeing so many mosaics that eventually, she said, it was just too much. Some of the mosaics could be associated to various cultures in history, others seemed to be of types this individual had never seen or imagined. All were exceedingly beautiful. Several other persons told me of serials that concerned autobiographical themes. The second type of serial image is that of *serial transformation*. In these, one image of a single object changes into another which, in turn, may change into still other single images. For instance, one informant told me of a vision in which he saw a liana (an Amazonian vine, as *Banisteriopsis caapi* is), changing into a snake and then into a woman. Another informant saw a snake's tongue undulating and thus turning into a woman who in turn changed into a dancing fairy. Interestingly, exactly the same sequences appears in indigenous myths related to Ayahuasca (see Gow, 1988; Lagrou, 1998).

Scenes

In the present context, a *scene* is an entire, progressively developing visualized scenario that the drinker sees. In the following discussion, I distinguish a subset of scenes which I label 'full-fledged'. These consist of full-scope, contentful visions that present veritable narratives as in a film or in the theatre. In addition to full-fledged scenes, I present two other types of scene—snapshots and glimpses. These lack some features of full-fledged scenes and may be regarded as *proto-scenes*.

Snapshots

Snapshots are single shots that present themselves before the person under the intoxication. There is movement in these shots, but no further development beyond that. It is as if just a single shot in a video clip was made. Among the snapshots that I have seen are views of forests and riversides, scenes of animals, views of ancient and futuristic cities, and all sorts of human social activities.

Glimpses

Glimpses are similar to the snapshots, except that they seem more distant. It is as if a hole was opened and a remote scene is perceived as when peeping through a keyhole or a periscope.[3] Unlike snapshots, glimpses are not single shots: the things one sees change and develop. Characteristically, the glimpses seem distant not only in their texture but also in time. The feeling is that the scene has happened in the past, or will occur in the future, and the drinker has the privilege of transcending the confines of time and watching it. The glimpses often appear in advanced stages of intoxication, as a prelude to further stages of visioning.

Full-Fledged Scenes

As noted, full-fledged scenes consist of clear visualizations that are seen in their totality and that occupy the viewer's entire visual space. They extend continuously in time and present narratives exhibiting rich semantic content and well-defined thematic structure. As such, they are similar to the scenarios seen in a film or in the theatre. It is as if a curtain is lifted and the viewer sees something that is going on, happening before his or her eyes. In contradiction to what is perceived with snapshots and glimpses, the scenario is full and is taking place right there in front of one. Even though chronologically the time of the scene may be distant (for instance, the scene depicts a ritual in ancient Egypt), the feeling is that the viewer sees it as if at the time of its occurrence.

[3] Interestingly, in one of the very first firsthand reports of Ayahuasca by a European person, Reinburg (1921) notes that when he was visioning it seemed to him that he was looking through a little hole pierced in a card.

Grand Scenes

Of the full-fledged scenes, let me single out what I am referring to as 'grand scenes'.[4] These are characterized by their remarkable visual richness and the complexity of their content and narrative structure. In many cases, the visual scope of these scenes is large and often wide panoramas are seen. Typically, grand scenes also have special psychological and/or spiritual impact. Usually these scenes are accompanied by elaborate ideation and the drinker feels that he or she is the recipient of important, meaningful teachings. Admittedly, this characterization is impressionistic. Yet it seems to me that phenomenologically it is very clear: When a vision is grand, there is no question about it, it is grand. Applying again the cinematographic jargon, I would say that the difference between a scene and a grand scene is analogous to that between a video clip and a veritable cinematographical or theatrical masterpiece. Since many of the most impressive of Ayahuasca visions are grand scenes, a relatively large number of instances of this subcategory are reported throughout this book.

Virtual Reality

Throughout the foregoing discussion, all scenes were examined from a primarily visual perspective. After all, what we are dealing with here are visualizations and visual experiences. Indeed, first and foremost the Ayahuasca experience is a visual one, and scenes are visions. Yet there comes a point where scenes are so powerful that even though visual experiences are involved, the characterization as visions does not do full justice to them, and that of alternative states of being is more apt. In these experiences the Ayahuasca drinker is transposed to another realm of existence, one which he or she feels to be very real. Sensed as real, the context in which the drinker finds him- or herself to be situated in is, *inter alia*, seen. However, in its totality the experience transcends the visual and is felt to be a reality in which the drinker is immersed. For this reason I refer to it as a 'virtual reality'.

Let me clarify this by reference to the experience of reality *par excellence*, namely, our mundane, day-to-day being in the world. I am in my office, sitting with my assistant at the computer, examining the text on the screen. Of course, I see my assistant, and the computer, and the room with its walls and furniture and all sorts of objects around as well as, through the open window, a glimpse of the landscape outside. However, I do not describe the situation as *seeing* my assistant, the computer, and the room. Rather, the description is just as written in the second sentence of this paragraph: I *am* in the room, with my assistant, by the computer.

While Ayahuasca scenes in which virtual reality is experienced are unmistakably most powerful, the distinction between them and the visions involving immersion

[4] Throughout this book, I also refer to these as 'grand visions'.

(see the next chapter) can be fuzzy; for examples, see the instance of the celestial palace as well as those presented in conjunction with immersion and second-order visions in the next chapter. At times, the experience vacillates between one that is primarily visual and one in which the visual is, as ordinary reality, just one facet of one's being-in-the-world. A scene may begin as one of the former kind, gain strength and reach the characteristics of the latter, and then it may perhaps dissipate and turn into an experience that is again primarily visual. What characterizes very powerful experiences of virtual reality is that they involve no progressive process of immersion. As described in detail in the next chapter, immersion in a scene is often the end product of a gradual process of increased involvement in the vision one sees. This process may be comprised of stages such as entrance, penetration, and a shift from passivity to ever-higher levels of activity and interaction. In contrast, in full-fledged cases of virtual reality, the immersion is immediate and it involves no gradual process. After all, this is the case with ordinary reality. We do not enter the real world as we enter into a swimming-pool or a river. Being-in-the-world is not a product of a process nor does it involve any decision or choice. We are there, *tout court*. As Heidegger (1962) pointed out, we are thrown, or born, into the world. Indeed, in the philosophy of Heidegger, as well as other phenomenological philosophers (notably Merleau-Ponty, 1962), being-in-the-world is a pivotal concept.

Curiously, these most powerful of Ayahuasca visions actually are those most similar to dreams. Most Ayahuasca visions are not like dreams: in them, the primary experience is one in which things—usually, most magnificent—unfurl before the drinker's eyes. Ordinary dreams are not like that. In general, the dreamer is not an inspector but rather a principal actor, a hero. In a separate investigation, I have examined dream reports—mine, those of informants, as well as ones described in the literature, notably Freud (1900/1953) and Jung (1976). As a rule, dream reports begin with first-person reference (i.e. using expressions with 'I' as a subject) and they involve statements such as 'I was in such-and-such a place', 'I thought', 'I did', and the like. As I said, in ordinary dreams, the dreamer is the main hero. Not so in most Ayahuasca visions, nor in other experiences normally characterized as visions (e.g. those reported in the classics of the mystical literature). These usually open like that most powerful of all visions reported in the Bible[5] (Ezekiel 1:1): 'Now it came to pass in the thirtieth year, in the fourth month, in the fifth day of the month, as I was among the captives by the river of Chebar, that the heavens were opened and I saw visions of God.' And there follows the grand vision of the theophany.

Yet, those Ayahuasca experiences classified as virtual realities are different from dreams. Independent comparative analysis I have conducted (see the Appendix) reveals that the distributions—both qualitative and numerical—of the content

[5] Being Jewish, I use the term 'Bible' as referring only to the Old Testament. The comments made here, however, equally apply to the visions described in the Apocalypse of St John in the New Testament.

items that appear in dreams and in Ayahuasca visions (including those of the virtual reality type) are very different and there is almost no overlap between them. In particular, most dreams involve people the dreamer actually knows and the happenings in them relate in one way or another to the dreamer's concerns, desires, wishes, and conflicts in real life. In contrast, Ayahuasca visions—both those characterized as visual scenes *and* those of the virtual reality type—present phantasmagoric and/or otherworldly scenarios that have nothing to do with the drinker's life.

All this notwithstanding, a key difference between dreams and Ayahuasca visions (again, including those of the virtual reality type) should be borne in mind. In Ayahuasca visions one is fully awake and conscious; in dreams one is not. In response to this last statement, readers are likely to bring up the case of lucid dreams. Such dreams, which have been the focus of increasing scientific attention in recent years (see LaBerge, 1985; Gackenbach and LaBerge, 1988), are characterized by their dreamers being aware that they are dreaming. Yet surely, in these, subjects are *not* awake. In Ayahuasca visions subjects are. One experiences a vision and at the same time can interact with one's fellow participants in the Ayahuasca session, manipulate objects in the real world, move about, sing. Of course, nothing of the sort is possible with dreams, even if they are of the so-called lucid variety.

Much more can be said in regard to the comparison between Ayahuasca visions and dreams, but this should be left for another discussion. The only point I would like to convey here is that most Ayahuasca visions are quite different from dreams and that, curiously, the most powerful Ayahuasca visions—those characterized here as pertaining to the virtual reality type—are those which are most similar to dreams. And yet—this is the main point of the foregoing commentary—they are none the less so very different.

Visions of Light

Last to be mentioned are scenes in which light is the central constituent. These should not be confused with bursts, puffs, and splashes, which are merely light and colour stimulations and do not, in the structural typology employed here, constitute scenes. In particular, they are not associated with any substantive content or theme. Not so the visions of light noted here. They are full-fledged scenes having relative permanence and temporal extension, and in which light is the key element. Usually, visions of this kind especially impress one with their grandeur and meaningfulness. Because of their importance, I shall discuss them separately in Ch. 17.

Summary

Throughout the foregoing survey comments involving comparative evaluations of visualizations and marking ordered relationships between types of visions were made. In particular, some visualizations were characterized as stronger or more powerful, or as more complex or more impressive than others. Reaching the end of this typological survey, I will summarize the dimensions in terms of which the Ayahuasca visualizations can be evaluated and compared. As is apparent throughout the foregoing survey, since the various types can be ordered, at least partially, along these dimensions, these dimensions may be regarded as defining progressions[6] between the types. However, in line with what has been said at the very outset of this chapter, I must emphasize once again: these progressions are not to be taken in the temporal sense—the order to be noted here is qualitative and defines the relative strength of visualizations, not the actual temporal sequencing in which visions unfold. While the actual sequential progression of visualizations throughout a given session often parallels the relative structural order, this does not necessarily have to be the case. More on this will be said in Ch. 18 in which the temporal progression of visions in a session is discussed.

Here, then, is a summary of the lines of progression we have noted, which may be regarded as the various specifications of the relative strength of visualizations and together they provide an explication of the notion of 'visual strength' in the Ayahuasca context:[7]

1. A progression towards the *figurative*. The earlier, and weakest types, in the typology are non-figurative whereas the more powerful ones are figurative.[8]
2. A progression towards *well-definedness* and *well-formedness*. As visualizations become more powerful they gain visual acuity and distinctiveness and they become progressively more well defined and well formed.
3. A progression towards *stability*. As the visualizations gain strength, they gain stability and permanence. For instance, kaleidoscopic images turn into stable figures, and rather than transform into other scenes, scenes are maintained and exhibit inner development.
4. A progression towards *globality*. As visualizations become more powerful they encompass more material. Thus, single figures turn into proto-scenes, and these, in turn, into full-fledged scenes.
5. A progression towards *contentual richness*. This is a direct corollary of the three previous progressions.
6. A progression towards extended *scope*. As visualizations gain strength one may feel as though one were moving from a home screening to a cinematoscopic one.

[6] Note that this word, employed here in its standard sense, should not be confused with the technical term 'progression' introduced in the next chapter where it denotes a particular pattern in the semantics of visions.

[7] 'Visual strength' should not be confused with the 'force' mentioned in Ch. 4. Force is an overall characterization of the impact of the intoxication, and pertains to one's general state of being; it primarily manifests itself in somatic feelings and in alterations of awareness and consciousness. By contrast, the visual strength discussed here has to do specifically with the visual mode. In fact, if the traditional terminology for the characterization of the two major components of the Ayahuasca intoxication is to be used, 'visual strength' is to be associated primarily with 'light', not 'force'.

[8] Visions of light excepted.

7. A progression towards *sensed reality*. The more powerful visions are, the more they are experienced as real. Grand scenes are typically experienced as revelations in which other, independent realities are seen. Those experiences characterized as virtual realities are, indeed, experiences in which the Ayahuasca drinker feels that he is transported to other, non-ordinary realities.

8. A progression towards psychological *significance*. Often, visions that strike people as powerful are also characterized by having special psychological significance. In them, drinkers find meanings that are specifically related to their lives. Such visions may be coupled with transformative experiences and their impact on drinkers may be long-lasting.

9. A progression towards *spiritual* impact. Many of the most powerful Ayahuasca visions are associated with meanings that transcend the life of the individual and that lead on to higher spiritual realms. Visions of supreme light especially exhibit this quality.

The Style of Ayahuasca Visions

I conclude the typological survey with what I refer to as the style of Ayahuasca visions. This issue has intrigued me greatly. Since explaining what I mean by this phrase is not self-evident, I shall introduce it the very way I have defined it to my informants when asking about it:

> You know, when you have a painting, it is defined not only by its content but also by its style. Different painters may draw the same figure or place; their paintings will be 'realistic' yet they will each have the definite signature of their individual makers. Specifically, observers looking at the paintings would recognize that they depict the same content, however, they would also appreciate that the paintings are the product of different artists. Indeed, if they are familiar with other works of these artists, the observers could easily identify the paintings as being of this or that artist. This they would do on the basis of the paintings of each artist having a particular 'style'. My question to you is—what is the style of Ayahuasca? In other words, can you identify certain features as the defining characteristics of the style of Ayahuasca visions?

What then is the style of Ayahuasca visions? I list five stylistic features that I find especially characteristic. As they are associated with different types of visions, they may be taken as defining five different styles that the visions may have.

The first style was commented upon spontaneously by many people I have interviewed. Even without being asked about the style of the Ayahuasca visualizations, informants mentioned that what they saw resembled cartoons and animated movies as well as images similar to those encountered in pop art. Quite a few indicated that the visions reminded them of Disney-like designs. Usually, the images in question are described as two-dimensional, static, and having well-defined boundaries and homogeneous clear colouring. Often they are seen in sequence, as in comics strips. I too have experienced such visions several times.

Related to these observations is the characterization of some visions as kitsch or slapstick. Some informants explicitly expressed discomfort or embarrassment with this. Some thought this reflects their own taste in the arts, which is apparently not sufficiently cultivated. Yet, it appears that the phenomenon is not idiosyncratic: some visions are indeed of this style.

Second is the element indicated above in conjunction with powerful geometric designs—marked lines demarcating the boundaries between small colour elements in the manner of the metal divisions in stained glass windows. Such marked dividing lines can also be found in visions with semantic content. Visually, such visions will be composed of the same fluorescent colour elements of which the powerful geometric designs are, but like *vitrages*, they would define figurative elements, notably magnificent architectural complexes. Similar observations were made by both Harner (1980) and Reichel-Dolmatoff (1990) with regard to their own firsthand experiences with Ayahuasca; interestingly, they are also encountered in reports of subjects to whom harmaline was administered in an experimental-clinical setting (see Naranjo, 1973*b*).

The third style will be denoted by the term 'expanses'. I am referring to visions of wide expanses of open landscapes of either land or sea and to panoramic visions of the entire planet, the solar system, or the cosmos. In these visions one does not see the lines noted in the previous paragraph. Rather, the scenes have a realistic air and the overall atmosphere they induce is of eternal, profoundly meaningful serenity. Reichel-Dolmatoff (1975, 1978*b*) marks this as characteristic of what he defines as the third stage of the Ayahuasca intoxication.

The fourth stylistic characteristic is 'enchantment'. I am referring to the quality that is especially salient in paintings such as those by the French painter Henri Rousseau. It seems to me that one special feature in these paintings is the secretive light. An example from my own visions is a scene of a forest with the moon shining over the trees in a special bluish light. Reichel-Dolmatoff (1990) makes the same observation with regard to his own visions.

Fifthly, there are visions which may be characterized as having a Baroque flair. These typically depict scenes that may be characterized as fairy tales. For actual examples of such scenes the reader is referred to the drawings of Ayahuasca visions by the Brazilian artist Ademir Braga de Oliveira shown in Meyerratken and Salem (1997).

Finally, let me comment on what may be referred to as the general ambience that the visions exude. Many informants told me that when having visions, they felt that they were 'coming back home'. Even though the visions were phantasmagoric and 'out of this world', there was something in them to which informants felt very much connected. Furthermore, scenes in the visions were often characterized as 'inviting'. In addition, several informants said that beings in the visions told them they had been expecting them and/or waiting for them.

One Final Remark

In closing, it should be noted that it is not the case that all drinkers of Ayahuasca experience all the types described in the foregoing survey. Over the years, I have discovered that many persons do not go beyond the level of puffs, bursts, splashes of light, and non-figurative (that is, lacking semantics) geometric displays. My assessment is that only a small minority of those who partake of Ayahuasca experience full-fledged visions. I shall further add that often, individuals who do not undergo interesting experiences the first or second time they partake of Ayahuasca do not consume it any further. This is only understandable given the uncomfortable physical sensations that the brew induces. Comments made throughout this book with regard to the relative frequency of various visual effects should be read with this qualification in mind. Specifically, except for those informants interviewed about their first Ayahuasca experience, all the persons I have interviewed were ones that decided to partake of the brew more than once. Moreover, all the individuals to whom the structured questionnaire was administered had even more extended experience. Consequently it is very likely that these individuals have had more frequent and more powerful visions than the population at large. Thus, it may be the case that all statements with regard to the relative frequency of the various effects surveyed in this book pertain not to the general population, but to persons who have had some involvement with Ayahuasca and who are prone to be affected by it.

7

Interaction and Narration

> Prepare your inner thoughts to depict God and His highest angels. Depict them in your heart as if they were human beings sitting or standing around you. You are in their midst, like a messenger whom the King and His servants wish to send on a mission. You are ready to hear the words of the message.
>
> Abraham Abulafia (a medieval Jewish kabbalist)

In this chapter several further aspects by which Ayahuasca visions may be characterized are examined. Most of these pertain to two dimensions—the *interaction* between the drinker and his or her visions, and the *narrative* structure of the visions. Towards the end I also discuss the *epistemic status* of visions and make some conceptual remarks regarding the notion of *hallucination*.

The considerations discussed in this chapter mark an important shift in our analysis. The typology outlined in the previous chapter followed, in the main, a visual-perceptual perspective. Specifically, visualizations were analysed in terms of form and figure, colour and brilliance, acuity and distinctiveness, well-formedness and permanence. In this chapter, semantic and pragmatic considerations are introduced into the discussion. These, note, apply only to full-fledged scenes and they are especially salient when visions are powerful. Together, they introduce further refinements of the typology sketched in the previous chapter.

Interaction

Throughout the entire foregoing discussion it was implied that the only attitude or relationship drinkers have *vis-à-vis* the visualizations they experience is that of seeing or looking-at. With all visualizations that are not full-fledged scenes this is, indeed, the only kind of attitude possible. With full-fledged scenes, however, drinkers need not remain passive towards what they see. Indeed, the more involved a drinker is with the visions he or she has, the more powerful the experience tends to be. In this conjunction, it is pertinent to cite a statement made by Amaringo (Luna and Amaringo, 1993: 27):

> It is only when the person begins to hear and see as if he/she were inside the scene, not as something presented to him, that he is able to discover many things. There is nothing that he is not able to find out. I saw how the world was created, how everything is full with life, how great spirits intervene in every aspect of nature and make the universe expand. I was like a tourist, always asking the spirits what is this and that, asking them to take me from one place to the other, demanding explanation for everything.

Two dimensions in the relationship between Ayahuasca drinkers and their visions may be noted. The first pertains to the *stance* the drinker takes with respect to the vision; the second pertains to the drinker's *dynamic interaction* with the scene presented in the vision. The variations of stance are the following: The drinker may be *outside* the scene of the vision, *statically inside* the scene, or *in movement within* the scene. As for the variations of interaction, four cases will be noted. The first and simplest case is that of *no interaction*. Second, there is *passive interaction*: the drinker is passive but figures in the scene act towards him or her. Third are *active interactions* with figures in the scene. Finally, there are scenes in which the drinker has *control* over what is happening. The difference between these various cases of interaction may be sharpened by using the analogy of the film or the theatre. In the first case the drinker is merely a viewer or a spectator. The second case does not occur with films but may happen with certain kinds of theatre in which the audience is approached by the actors. In the third case the drinker functions like an actor. In the fourth he or she functions not only as an actor but also as the director.

Together, the different cases along the two dimensions generate a rich variety of types. These will be introduced in the order defined by the second dimension, that of interaction.

No Interaction

The minimal case is that of being in front of a scene without interacting with it. This amounts to *seeing from a distance*. This is what is always the case with both snaps and glimpses, and also with many scenes.

The next case to be noted is that in which the drinker steps inside the scene but remains static in one place. This constitutes what I shall refer to as 'witnessing'. The feeling is that one is there, but there is no interaction between him or her and what is happening in the scene. A situation I have experienced several times (an example of which is presented later in this chapter) is that of visiting a palace: I was invited by the monarch to observe how he ruled. I was standing in a front corner, neither affecting what was going on nor being affected by it, just watching. I will note that witnessing confers a feeling of protection. Thus, on two occasions I witnessed scenes in Nazi concentration camps; the experience was horrific, but I knew that nothing would happen to me.

When one steps further into the vision, one may find oneself in the midst of a scene without actively interacting with it. This results in *simple immersion*. Among the cases of this kind that I have experienced were finding myself in the midst of a beehive and being among jaguars in a forest.

Movement within a scene *without interaction* is akin to walking through a museum in which one is not allowed to touch the display cases, let alone the exhibits displayed in them. Instances of this kind are common. From my own experience I can cite visiting mansions, walking through gardens and forests, and travelling in space. Variants of this type may involve a minimal interaction with some elements of the vision. A typical case is that in which a vehicle of translocation is used. An example is a vision in which I found myself riding a horse. Riding, I passed through a landscape, continuously going forward.

The examples noted above may be referred to as cases of 'roaming about'—in these, the drinker's movement is simple: it consists merely of changing one's whereabouts in a given space. Other cases of movement, however, may exhibit directionality and be more structured. Thus, the drinker may follow a demarcated itinerary or pass a specific boundary or obstacle. Some instances of these define scenes of *passage*, that is, scenes in which the drinker moves from one well-defined domain to another. Common examples are those of crossing a bridge and of entering through a door or a gate. Passage can involve more than mere spatial translocation. The different places one passes through may define a structurally progressive series. The places in question may, for instance, be ordered in terms of difficulty or danger, significance or importance, and level of spiritual height. Even if in themselves the various places one passes through do not define an ordered sequence, the very passage from one place to another and the continuous sustaining of one's forward movement may itself constitute a non-trivial feat. Indeed, the progressive passage is often perceived as an accumulative accomplishment. A pertinent example will be given below.

Passive Interaction

In passive interaction with the drinker outside the scene, figures from the scene approach and act towards one. When the level of inebriation begins to increase, drinkers of Ayahuasca often feel that they are being enticed or lured by creatures or entities that attempt to touch them. Alternatively, the creatures or entities may be perceived as protecting the drinker or supervising his or her visionary voyage. Examples from my own experience are the following: a line of poor, sick people coming towards me, animals of the forest endowing me with their good energies, and good fairies protecting me as I was being increasingly immersed in the visionary realms. An example of passive interaction while statically within a scene is that of a magic operation; I experienced this on four different occasions. Here is the story of the first:

> I was lying down and a colourful weblike drape or mantle was stretched over me, as a spell. One person was in charge of the operation and others were watching, making sure that nothing bad would happen to me.

An example of passive interaction with movement about the scene is the following vision reported to me by an informant recounting his first powerful Ayahuasca experience:

> I was walking along a path in a forest accompanied by a guide. Below, alongside the path were people, apparently evil creatures, who were calling me, trying to lure me down to join them. I was tempted to do so, but did not. I continued along the path and eventually reached a wonderful place.

Active Interaction

Being outside a scene while interacting with it usually results in *affinity*. The drinker is outside, but has a strong affinity with whatever takes place in the scene. For instance, in one of my very first visions—that of the forest scene recounted in the Prologue—I was sitting on a terrace in front of the forest enclave. As I saw it, the forest was full of animals. I was here and the animals were there, yet, at the same time, I felt I was dancing with them.

Active interaction coupled with statically being inside a scene is a case for which it is difficult to find examples. The reason is conceptual and not at all specific to Ayahuasca: the case in question involves two seemingly contradictory attributes—being active and being static. The following rare example of mine was experienced with the eyes closed:

> Both in the real world and in the visionary one I was sitting in a meditating posture. Invisible yet very present beings instructed me how to position my hands and fingers in the manner of the *mudras* of yoga. Images of the various hand and finger configurations were shown to me, and I adopted each one of them in sequence. Doing so, I directly felt the differential corporeal and mental effects of each.

Roaming dynamically within a scene and acting in it results in *full immersion*. Here, the Ayahuasca drinker is not a mere traveller but a full-fledged participant in the scene. In the extreme, immersion into the world of the vision can be totally absorbing and manifest the same kind of totality that we normally experience in our being and acting in the real world. With this, the visionary experience turns into one of virtual reality. Thus, one informant recounted that he 'experienced himself walking through a landscape encountering all sorts of evil creatures. There was a spade in his hand and he was fighting these creatures. Another informant described a scene in which he was surrounded by Indians who then brought him to a cave. He wondered what to do. Then he saw a ladder. He climbed it and reached the Light.' Still another informant, a South American woman engaged in the theatre, described a scene in which 'she was roaming through the forest. She then encountered four jaguars. Getting nearer, one of the animals especially attracted her attention. She approached further and here, the two—the woman and the jaguar—were closely inspecting one another. And then, the woman climbed on the big cat and together they travelled throughout the forest.' Whose initiative was it, I asked her—yours or the jaguar's? My informant reflected, reimagined the situation and answered: 'It was both of us, simultaneously. We looked at one another, and it was clear that this is what should happen—me riding on the jaguar. Thus it came about.' In this example the shift from roaming about to action and immersion is especially apparent.

But there are cases that are even more powerful. In these, the Ayahuasca drinker assumes a new identity and acting upon it, he becomes, indeed, an actor, as in the theatre. Paradigmatically, this occurs when the Ayahuasca drinker undergoes identity transformations and changes into another person or an animal. Transformations of this kind are the subject of our discussion in Ch. 13. Of the various examples presented in that chapter let me just mention two. In one the drinker, a woman, found herself in the company of indigenous men, she herself turned into a man and joined in the activities of her companions—she danced with them in a masculine manner, interacted with them, and was initiated by them. In the second case the drinker turned into an eagle and flew high above in the skies inspecting the landscape below with extraordinary visual acuity.

Second-Order Visions

As a special case of immersion I would like to mention an experience I had only once, and never heard reported by any of my informants; it can be characterized as a vision within a vision, that is, a vision pertaining to a level distinct from that of the primary level of the visionary experience. Following the terminology commonly employed in mathematics and logic, I refer to this kind of vision as one of a second order. I shall not recount the entire story of the vision in which I had this experience—its plot is long and elaborate and most of it would not add anything new to the present structural typology; only that aspect of the vision that is special is recounted here:

> In a palace, I met a person whom I found impressive. This individual's attire was distinguished and his demeanour exuded sophistication and charm. Conversing with him, I readily detected a very quick and sharp mind coupled with a well-polished tongue. Manifestly, this was a man of the world. I was intrigued yet at the same time suspicious. My heart told me that this encounter is fraught with danger. The man invited me to join him in an adventure that, so I understood, would involve a radical change in my state of consciousness. Reflecting whether to accept the invitation or not, I found myself having a vision of the hypothetical state of affairs in which I saw myself being in the context of that altered state. In that vision I saw myself being amongst black pumas and undergoing a (hallucinatory in terms of the hypothetical scenario I was entertaining in my mind) transformation into a puma myself. When this happened—so did the vision within the vision proceed—my companion betrayed me and let the pumas (which, like the treacherous man himself, were real as far as the hypothetical scenario was concerned) devour me, the intoxicated human being who, in his state of hallucination, was under the illusion that he is a puma as well. On the basis of that experience, I declined the invitation that had been extended to me.

I cited this episode for the vision within a vision it presents. At the same time, the special character of the vision of the first order should be borne in mind. As far as the latter is concerned, total immersion was experienced. Having experienced a vision (that of my involvement with the pumas), the scene defining my state of being while having that vision (i.e. my stay in the palace and my encounter with the intriguing man) seemed totally real to me. But, obviously, it was not. That scene constituted a hallucination in which I was fully immersed engaging myself in extended interaction with the figures of that scene, experiencing it as being on a par with any other interpersonal interaction I would have in the course of my everyday, normal life.

The experience just cited reveals the complexity that the world(s) of hallucination may exhibit. This brings to mind two other domains. The first is lucid dreaming (see Green, 1968; LaBerge, 1985; and Gackenbach and LaBerge, 1988). In some instances of lucid dreaming one may have a dream within a dream (see, for instance, Green and McCreery, 1994). The second domain is the philosophical paradigm known as *possible worlds semantics* (see Montague, 1974; Barwise and Perry, 1983; as well as Hughes and Creswell, 1996). One aspect being investigated in this paradigm is the relationship between different worlds—the actual real one and various hypothetical or potential worlds that are associated with it (and also defining various relationships amongst themselves). One's memories of the past and one's wishes and plans for the future are examples of worlds defined as possible, but not actual, relative to the world of one's present.

Control

By and large, the degree of control an Ayahuasca drinker may exercise over a scene while being outside it is limited. In general, the most one can attempt to do is try to stop scenes that one does not wish to see. For instance, one may open one's eyes, shift one's gaze, interact with other individuals who are participating in the session, and so forth. However, when the intoxication strikes strong and the drinker is immersed in a vision very often he or she ceases to be cognizant of the fact that there is the option of opening the eyes or engaging in any other distracting activity.

Yet, some people do manage at times to exercise control over their visions. For instance, one informant told me of a case in which he placed himself within a drop of water on a table in the centre of a room. From this position he could observe all that was going on around him. He was also able to decide how much time was to be allotted to each view of the surroundings.

The most impressive type of control is a dynamic interaction in which one actually determines the ongoing development of the scene. Here the drinker may be likened not only to an actor in the scene but also to its director, as in a play or a film. This happened to me only once:

> There were seven men in front of me, each holding a black panther on a leash. At one instant the panthers were freed and moved towards me. I knew I had to act fast. And then it happened: a brook appeared between me and the panthers and I placed beautiful water lilies in it. The panthers were attracted to the flowers and they came to the river and drank. Having done so, they forgot about me and they all turned back and went away.

After this session, I felt a deep satisfaction; it seemed to me that I had advanced a grade in the school of Ayahuasca. More on control and related issues will be said in Chs. 20 and 21, in which dynamical aspects of the Ayahuasca experience are discussed.

Particular Types of Interaction

There are four particular types of interaction which, because of the special subjective impact often associated with them, I would like to discuss separately in some further detail. I refer to them as 'receiving', 'progression', 'voyages', and 'flying'.

Receiving

Receiving is precisely this—an experience in which drinkers experience that they receive something in their visions. The minimal kind of reception does not involve any specific item or content—it is the reception of stamina, well-being, and health. This kind of experience is very common.

A more specific kind is that in which knowledge is received. In the traditional Amerindian context, the ingestion of sacred plants in general, and of Ayahuasca in particular, is considered to be 'the only path to knowledge' (Chaumeil, 1983: 33). I discuss the experience of gaining knowledge by means of Ayahuasca at further length later, in Ch. 15 (further references to the anthropological literature are furnished there as well); here, I confine myself to a few comments on the interactional aspects of the experience of receiving knowledge. Typically, the person under the intoxication feels that a voice is addressing her and passing information or instructing her as to what to do or how to behave. In the previous sentence I say 'feels' and not 'hears' because often the experience is precisely that—that another agent is talking to one and that one understands what is being said, yet this communication is conducted silently. The code being deciphered is not phonological-linguistic but rather ideational, and the perception is not auditory but, in a fashion, telepathic.

In the indigenous context especially salient is the reception of instructions on healing (see Payaguaje, 1983; Luna, 1984*a*, 1986*b*; Baer, 1992; Langdon, 1992; as well as Luna and Amaringo, 1993). As indicated earlier in this book, Ayahuasca is said to instruct the shaman with respect to both diagnosis and treatment. The instruction is often attributed to a correspondence with supernatural beings, spirits, and various guiding and helping entities. Indeed, all the traditional *ayahuasqueros* with whom I conversed singled such instruction out as a major facet of their engagement with the brew. For these individuals, healing is, of course, a major personal and professional concern. Persons with other special interests may experience receiving information pertaining to these interests. One of my informants, an artist, reported receiving specific designs showing her how to proceed in her work.

Also paradigmatic is the receiving of music. As noted in earlier chapters, Ayahuasca often makes its partakers sing. More than that, the brew makes some people sing melodies they have never heard before, in other words, compose. Indeed, all the hymns of the Santo Daime Church are of this type: they are songs that prominent persons of the community have sung under the effect of the *Daime*. These songs are said to be 'received'. The musical reception may consist of a mere action, whereby one feels the energy working through one's system, so to speak, and one simply begins to sing. Alternatively, the reception may be linked to a vision in which one hears music and one is instructed to sing accordingly. For related discussion in the anthropological literature see Gebhart-Sayer (1986), Bellier (1986), and Luna (1992).

Last but not least is the experience in which the Ayahuasca drinker finds him- or herself being presented with a specific object. This case is by far less common than all other cases of receiving mentioned above. To me it happened only twice and I have heard it reported only by half a dozen informants. In all cases, the items reported as being given to the drinker were either kingly regalia or objects invested with special powers of magic and/or healing; examples are given in Ch. 11 in conjunction with tactile hallucinations.[1]

Progressions

As indicated throughout the foregoing discussion, the degree of immersion and movement within the scene of one's vision is a central parameter that distinguishes between scenes and marks the differential strength that they exhibit. The continuous maintenance of one's movement may involve merely moving about the scene or it may follow a predetermined structured order. Structured progressions are, of course, common in the religious mystical literature. Two paradigmatic cases will be noted. The first is the ascent to the heavens. In Western culture the classical case is that of Dante's *Divine Comedy*. In particular, the third book of this great work describes an entire structured hierarchy in which each heaven is higher than the one that precedes it in the sequence. The highest level in this hierarchy is that of the Supreme Light. Even more to the point perhaps is the *Book of Enoch*, one of a corpus of Jewish apocryphal mystical writings from the beginning of the Christian era whose subject matter are ascents to the heavens and encounters with the Divine (see Odeberg, 1973). Significantly, this literature is known, in Hebrew, as *Safrut ha-Heichalot*, the literature of the palaces. Hierarchies of the heavens are also encountered in various non-Western mythologies, notably that of the Aztecs, a culture in which psychoactive plants played a pivotal role. The second paradigmatic case is the heroic odyssey. In it, the hero passes through a sequence of struggles and battles until he conquers and wins. Obviously, this type of story is prominent in both the epic literature and in legends.

Progressions of these kinds are also encounterd with Ayahuasca. Here are two examples. The first is mine, and it presents a case of failure to sustain the progression of movement. The second was recounted to me by an Amazonian shaman as being one of the two major accomplishments he has had in his entire lifelong experience with Ayahuasca. Still another example of a progression is the vision of the shaman Arévalo Valera cited in Ch. 1.

My own story unfolds thus:

[1] It is interesting to compare the phenomenon of reception in the context of Ayahuasca to analogous experiences reported in the mystical literature. In particular, primarily on the basis of medieval Christian mystical reports, Underhill (1955) analyses the phenomenon of receiving knowledge. Her characterization of this phenomenon is extremely similar to that observed with Ayahuasca.

I entered a landscape. It was a meadow and there was a bridge in front of me. I crossed, found a trail, and went forward. At one turn of the trail I met a bearded old man sitting on a beautifully adorned chair. The man wore a splendid white robe full of rich embroidery, some of it golden, held a sceptre in his hand and his countenance was wise and benevolent. With his hand, he indicated that I should follow the trail to the left and go on. This I did until I met another old man, very much like the first, and he too indicated that I should go on. Soon I reached a magnificent mansion. Upon my entry, I was welcomed by a host of beautiful young maidens. They were clad in white, delicate lace and they invited me into a hall in which a sumptuous meal was set. I interacted with the various people present and then the vision faded. It took me a while to realize that I had forgotten to go on and follow the trail. Sadly I comprehended that I had been diverted and that I had missed the true goal of my journey.

And here is the shaman's vision:

In my vision, I entered a most beautiful mansion. It was Incaic or Mexican. I passed through the gate and there was a hall. I went on walking and found myself in front of a door. I entered it and there was a second hall. There was another door, and I passed through it too, thus entering into still another hall. I passed hall after hall until I arrived at the tenth. The hall was all gold—the floor, the walls, the ceiling. Bearded old men clad in white sat in two lines along the walls. As I entered they applauded saying 'Congratulations! You have made it!' In front was their leader. He was holding a great tobacco pipe. I advanced and when I reached him, he invited me to sit down. He passed the sacred pipe to me and blessed me. I felt immensely gratified and honoured and so emotionally moved that tears of happiness poured from my eyes.

When I heard this report, I thought of the High Priest on the holiest day of the Jewish year, *Yom Kippur*, the Day of Atonement. It was on that day only that the Priest, and only he, would enter the Holy of Holies of the Temple. It was a foreboding experience and in order to ensure that the Priest would come back a long ribbon would be tied to his leg. Indeed, the people outside were said to be waiting with anxious suspense which was relieved only when the Priest reappeared before them (for pertinent primary texts, see Jacobs, 1990).

Voyages

Voyages are scenes in which the Ayahuasca drinker experiences him- or herself travelling. By their very nature, all these scenes involve full immersion. I employ the term 'voyage' to refer to kinds of travel that cannot be achieved in ordinary reality. In particular, let me distinguish between *planetary*, *cosmic*, and *heavenly* voyages. In these, the drinker travels to the far reaches of the Solar system, to regions of the cosmos that extend even further, and to the supernatural, heavenly realms respectively.

Flying

Flying is a specific kind of voyaging, in which the drinker also experiences him- or herself transforming into a bird. Various distinctions, grades, and levels of flying can be noted. I discuss them in detail in Ch. 13, that dealing with transformation of identity and metamorphosis.

Semantic and Narrational Considerations

Like those pertaining to interaction, the semantic and narrational considerations apply only to full-fledged scenes. I should emphasize that while the terms 'semantics' and 'narration' are employed here, the topic of interest is not the content of visions, but rather the structured types in which they are manifested. Semantics and narrative complexity define a dimension that adds further richness and variation to the range of these types. Four topics pertaining to this dimension will be considered—serial thematic variation, narratives, grand scenes, and ideation.

Serial Thematic Variations

Serial thematic variations are series of visions experienced in sequence that together constitute variations on one common theme. The thematic content of visions is a topic to be discussed at length later, in Ch. 9. Here my concern is structural. Thus, my present use of the term 'thematic variations' pertains not to the contents as such but rather to the structural relationship between several short visions that appear in sequence and constitute one whole. This use of the term 'thematic variations' is completely analogous to the standard use in classical music, where 'variations on a theme' denotes a specific musical structure. The first time I experienced this type of vision was on the only occasion I partook of Ayahuasca in the midst of the virgin Amazonian forest:

> I saw a series of six visions presenting monarchs in their throne halls. The most spectacular of these were the first two scenes, which depicted ancient Egyptian pharaohs. In all cases I was invited to step in, stay in the corner, and witness the monarchs as they ruled. I was given the chance to observe the challenges and difficulties that absolute power presents. I appreciated the potential pitfalls as well as the grandeur associated with such power.

On another occasion, the theme of the series was the dancing woman. At least eight scenes passed consecutively before my eyes in which a woman or a group of women danced. These included a prima ballerina in a ballet performance, a hypermodern discotheque scene, a *pas de deux* of metal figurines, a parade of lascivious dames of ill-repute, and a very formal dance of a group of aristocratic ladies. Still another series concerned the animals of the night. Even though the hall in which the session took place was highly illuminated (the occasion was a festive *Daime* session), whenever I closed my eyes there was darkness—darkness of the night, one which, when one accommodates to it, actually turns out to be very far from being pitch black. The different visions that appeared all pertained to one common theme—the life of nocturnal animals. In each vision, a different species appeared: jaguars, jackals, several kinds of birds, insects, and organisms smaller than insects.

In each case, I was shown how the animals in question behave. My eyes accommodated and I could see what the animals themselves saw. In effect, the entire set of visions was a very instructive course on animal behaviour. Also concerned with animal behaviour is the series reported to me by one informant when recounting his first experience with Ayahuasca (this was in the context of the UdV). The series consisted of scenes depicting the collective aspect of animal life. These scenes presented several biological species, beginning with insects and ending with humans.

Lastly, there is one of the most marvellous Ayahuasca sessions I ever had—it consisted of two interlacing thematic series. The first series presented various scenes depicting the glorification of the gods in different contexts—scenes of religious rituals in various ancient civilizations as well as scenes of adoration by supernatural beings and by animals. The second series presented various scenes of dance. The first series was the primary one, and between every two successive scenes pertaining to it there appeared, as an intermezzo, a scene of the second series. The entire sequence consisted of about seventeen scenes extended over a period of almost three hours.

Narratives

When scenes are sufficiently complex, veritable stories unfold. With this, a whole spectrum of new considerations for analysis presents itself: What is the inner structure of the story? What are its parts and what are the relationships between them? What contributes to the complexity of the plot? And overall, how good a story does the vision make? With the recent developments in the study of narratives and texts, contemporary cognitive psychology has much to contribute in this regard. Here only very preliminary observations will be made.

From the perspective of narrative, the minimal type is that of mere presentation of an action in a given place and time. It seems to me that most scenes are of this type. But, in general, a good story consists of more than one scene. Similarly, powerful Ayahuasca visions may consist of several scenes. The serial thematic variations discussed above may be regarded as an interim case. They consist of several scenes that are semantically related but, from a narrational point of view, they are not related to one another. The relationship is that of a serial transformation, not of a narrative composition.

The simplest kind of narrative linkage is that associated with *spatial translocation*. For instance, when one experiences oneself as flying one may move from one place to another; with this the things one sees and encounters will change, and a narrative sequence will evolve.

Another simple type of linkage is one that has already been mentioned above, namely, *passage*. This type, recall, consists of the drinker moving from one well-defined locale to another; consequently, he or she may discover new things, persons, and creatures and thus the narrative of the vision will evolve. The different places he or she passes through may define a structurally progressive series. They may, for instance, be ordered in terms of difficulty or danger or they may, as in the case of the various heavens, define a hierarchy. Some examples were noted above when progression was discussed.

Semantically more complex are scenes that are connected by a proper narrative dynamics. In these, the development of the plot is dictated by actions and events rather than by physical movement or transformation. Here, in principle, all the distinctions made—by linguists, psycholinguists, and students of literature—in the textual analysis of stories apply. I should note, however, that in general the narratives of Ayahuasca visions are not very complex. By and large, it seems to me that structurally they are less complex than the narratives of dreams. Overall, what makes Ayahuasca visions impressive is their magnificence, grandeur, supernaturalness, and the psychological and spiritual impact they have on their viewers. Only a relatively small number of the visions in my corpus of data define multi-unit, complex narrational structures.

Ideation

Also pertaining to semantics is ideation. Usually, a rich vision is one in which one learns something. This may be achieved through conversation with persons or beings one encounters in the vision, through non-verbal communication with such persons or beings, or through enhanced mentation and insight. The ideas, insights, and reflections induced by Ayahuasca will be the subject matter of Ch. 10. Here I confine myself to some general, structural comments.

First, the noetic quality of visions will be emphasized. In some visions one feels that one not only sees apparitions, as in the cinema, but that the vision consists of a discovery in which another reality is revealed. With this, one's knowledge is felt to be expanded. Indeed, the key feature of some visions is precisely this—the experience that one attains knowledge and discovers the truth.

Second, it is important to note that structurally the perceptual and the noetic aspects of visions are intrinsically related. Often, the reception of knowledge is the main act in the vision. At times, the image itself, like a parable, is the mechanism by which truth is revealed to one. Furthermore, in powerful visions the distinction between seeing and acquiring knowledge dissipates. This has been noted throughout the writings of Carlos Castaneda (in particular Castaneda, 1971) in which a different kind of 'seeing' is prominent. Such seeing is akin, I find, to Spinoza's third mode of knowledge, on which I shall comment further in Ch. 15 (see Spinoza, 1670/1989). Similar reports are also encountered in the mystical texts of various cultural and religious traditions (*inter alia*, the writings of Plato, Plotinus, Swedenborg, and William Blake); I shall return to these in Ch. 16.

Metaphorical Parables

Finally, I would like to mention a special type of visual narration, one which may be regarded as a visual metaphor. Essentially, visions of this type are not unlike parables in the Bible: an image is presented and the viewer draws a moral from it.

In the data I have collected, almost all instances of this type involve personal psychological insights with ramification on the drinker's conduct of his or her life. For instance, one woman told me that in her first Ayahuasca session she saw herself fully encased in a transparent plastic sheet. Whenever she moved, the cover moved with her. She realized that she had been leading her life separated from other people. Even though it seemed that she was in contact with other people, in essence she was insulated and had no direct contact with anyone. The realization made this person change her attitude *vis-à-vis* human interpersonal relations. Of the same type is a vision of a dilapidated building I have heard reported independently by two different individuals. Apparently, when originally constructed the building was a nice one, but now it was shabby and in disrepair. Upon inspecting their visions, the Ayahuasca drinkers understood that the building was themselves; this, in turn, made them appreciate that they should make some basic changes in their lives. Only once was a metaphorical parable reported to me in conjunction with ideations not related to the drinker's personal life. The informant saw a chain of pearls. This visualization, she reckoned, revealed to her how time is structured. With this, she gained insight to the nature of time and new metaphysical understanding.

Summary

The patterns surveyed in this chapter mark a couple of progressions that should be added to the list presented at the end of the previous chapter:

10. A progression towards higher degrees of *interaction*. The stronger the visions are, the more involved the drinker is in them and the more active and complex his or her interaction with the content of the visions is.

11. A progression towards *immersion*. The more powerful a vision is, the greater the degree of the drinker's immersion in it. In the case of virtual reality this experience is total.

12. A progression towards higher degrees of *narrative* complexity. Most visions do not even allow any narrative analysis. Only the more powerful visions exhibit sufficient narrational complexity, structure, and development.

13. A progression towards enhanced *ideation*. Visions that strike people as powerful are often associated with reflections and insights. The experiences in question are not only perceptual but also ideational. In these cases, the visions are often regarded as significant learning experiences. At times, people feel that the visions carry special messages to them and that they have a noetic quality, that is—they serve as vehicles for the acquisition of veridical knowledge.

A Comment on Open-Eye and Closed-Eye Visualizations

Closing the typological survey of the last three chapters, I would like to return to our starting point, the open-eye visualizations discussed in Ch. 5, and comment on their difference from closed-eye visualizations with which the discussion in Chs. 6

and 7 was by and large concerned. Which are more powerful—the visualizations with open eyes or with closed eyes? All drinkers of Ayahuasca will vouch that the likelihood of experiencing visualizations is greater when the eyes are closed. Therefore, the likelihood of experiencing powerful visualizations is greater as well. For this very reason, drinkers wishing to have visions purposely close their eyes. With the closing of the eyes, one frees oneself from sensory input that grounds one's perception in reality, and thus enhances the power of mentation that is not grounded in external reality (for an independent characterization of perception along a similar line of thought, see Llinás and Ribary, 1994). Yet, it is also true that in those cases in which people do have visions with their eyes open the visionary experience can be very powerful; in extreme cases, even more powerful than visions had with the eyes closed. This is so because what is seen with open eyes is normally taken to be real. If one's eyes are open and extraordinary scenes are seen, the impact of the experience and the sense of reality conferred on it may be stronger than that of comparable visions one might have with the eyes closed. The difference in question is even more striking when the drinker actually moves about in the world while under the intoxication. Such cases are exceedingly rare—in most contexts of Ayahuasca use there are strict regulations forbidding independent movements that are not part and parcel of the ritual. However, when this does happen, the effect can be very powerful indeed; an example is the vision of the celestial palace recounted in Ch. 5.

8

The Contents of Visions

> To those who are in intelligence, there appear gardens and paradises, full of trees and flowers of every kind; the trees there are planted in the most beautiful order, combined into arbors, through which are arched entrances, and around which are walks, all of such beauty that they cannot be described.
>
> The New Jerusalem was seen of pure gold, its gates of pearls, and its foundations of precious stones.
>
> Swedenborg

> In the court of heaven . . .
> Are many jewels so precious and beautiful
> That they are inconceivable out of that realm.
>
> Dante

I turn now to a consideration of the contents of the visions that Ayahuasca induces. The following survey is based on a systematic analysis of both my own experiences with the brew and those reported by my various informants. Whenever relevant, citations from the literature are presented as background material. This review, I should emphasize, is qualitative, not quantitative. In other words, my purpose here is to chart the space of various contents that usually appear in the visions. A numerical, quantitative analysis is presented in the Appendix.

Investigators of Ayahuasca in the indigenous context have noted that the visions induced by the brew reveal certain common elements. Thus, Der Marderosian *et al.* (1970: 11) observes that:

> In spite of the individual nature of the hallucinogenic experience, there is a high degree of similarity in the content and frequency of occurrence of particular hallucinations from individual to individual during any one night of drinking. Certain themes also recur every time they drink Ayahuasca. The most frequent of these are: (1) brightly coloured, large snakes, (2) jaguars and ocelots, (3) spirits, both of Ayahuasca and others, (4) large trees, often falling trees, (5) lakes, frequently filled with anacondas and alligators, (6) Cashinahua villages and those of other Indians, (7) traders and their goods, and (8) gardens.

Similarly, reviewing the anthropological literature, Harner (1973*c*) states that the most common items seen in the visions reported by indigenous persons are snakes, jaguars, demons and deities, cities, and landscapes. Also noted are visions having to do with the resolution of unsolved crimes, flights of the soul, and experiences of clairvoyance.

The following discussion sets itself to answer the question 'What kind of things do people see in Ayahuasca visions?' It is the first attempt to examine the contents of Ayahuasca visions systematically and on the basis of a sizeable corpus of empirical data. The survey is structured in terms of the following main categories: personal and autobiographical material, human beings, naturalistic and non-naturalistic animals, plants and botanical scenes, beings which are neither human nor animals (henceforth, 'beings'), cities, buildings, and architectural complexes, works of art, objects and artefacts, vehicles of transportation, symbols and scripts, places and landscapes, historical and mythological beings and scenes, scenes depicting creation and evolution, celestial bodies, celestial and heavenly scenes, divine and semi-divine beings, encounters with the Divine, scenes of light, Platonic ideas and mathematical objects, and episodes pertaining to death.[1] A detailed explanation of the categorization system adopted here is given in the Appendix.

1. *Personal and Autobiographical Material*

First, one can see one's own life. In particular, one can see snapshots and scenes of one's personal past. The most elaborate visions of this kind reported to me are serials in which drinkers inspected different scenes of their life and as a consequence had a psychological insight of personal import to them. In one such case, the Ayahuasca drinker saw snapshots, each of which was depicting a certain moment of her biography. The shots were not ordered chronologically, but rather juxtaposed thematically. The juxtaposition revealed some patterns in the drinker's personality and conduct of which she had not previously been aware.

A special type of biographical memory I would like to highlight is that of snapshots of seemingly insignificant moments of one's life. The experience consists of the Ayahuasca drinker seeing himself or herself in the midst of an episode the likelihood of he or she normally thinking about is very, very small. For example, I once saw myself engaged in a conversation with an elderly English lady I met on a bus ride while travelling through the island of Malta. The event took place about ten years before I had the vision in question. During the entire intervening period, I never had any recollection of this episode nor had I thought about or reflected upon it. Yet, inspecting it in my Ayahuasca induced vision, I realized that I was gaining new insights regarding my own self. Similar experiences were reported to me by my informants.[2]

[1] While all categories indicated here appear, in this order, in the survey that follows, some of these categories define section headings in the survey whereas others are associated only with subheadings. The reasons for this are strictly editorial—the division into sections and the grouping of categories were dictated by the relative length of the discussion of the different categories in question. This division allows, I think, a better grasp and an easier comprehensive appreciation of the large body of data I attempt to summarize in this chapter.

[2] Interestingly, similar experiences are encountered in advanced levels of Buddhist meditation (Sogyal Rinpoche, 1993).

2. *Human Beings*

Personal Acquaintances

In conjunction with the biographical material indicated in the previous section, one may see individuals whom one knows or has known in the course of one's life. These may be seen in isolation, as single apparitions, or in the context of memories of one's personal past. While, for me, seeing personal acquaintances is rare, for many others such visions are quite common. Significantly, the most common episodes in which personal acquaintances appear are ones having to do with the actual death of a member of the drinker's nuclear family. Especially moving was the experience one European informant had at the very beginning of her first Ayahuasca session. She found herself at her father's deathbed. For reasons beyond her control, in her real life this informant could not attend her father's funeral; indeed, the last time she saw him she was not at all aware that she would never see him again. All this happened almost forty years before the Ayahuasca session took place and it all remained an unresolved, painful wound that my informant carried in her heart throughout the years. The Ayahuasca experience finally gave this person the opportunity to bid farewell to her father. This was a wonderful, unexpected gift for her, she said, and she was greatly relieved.

Famous Persons

One can also see people one knows of but does not actually know, or has never known, personally. For me, the most common such figures were royalty. Of these I shall mention King Solomon and Zedekiah, the last king of First-Temple era Judaea, as well as several Egyptian pharaohs and European monarchs. Other specific historical figures I encountered in visions included Moses, the biblical hero Samson, Plato, the apostle Paul, Cortes, Freud, and the Nazi Josef Mengele. Historical figures that informants told me they had seen included biblical figures, popes, Leonardo da Vinci, Newton, and Einstein. On several occasions I have also seen royal persons whom I associated with various ancient civilizations but who were not identified by name. Once I saw what seemed to me to be a high priest of the Incas. Many informants also reported seeing royal and religious figures of various cultures. Not necessarily famous, yet of special presence and fairly common, are warriors and strong men of all sorts. Notably, these include knights and muscular black men.

Guides and Teachers

A special category of human beings often reported is that of guides, guardians, teachers, and other wise men and women. Non-human beings (notably, fairies and angels) can serve in these functions too. In the Amerindian tradition, the most prominent of these is the Ayahuasca *mama*, the mother of Ayahuasca (see Luna, 1984*a*); in the *Daime* context such a figure is referred to as the *Rainha da Floresta* (Queen of the Forest) and the Holy Virgin (the two are often regarded as being the same persona). The seeing of such figures is usually associated with the reception of knowledge. Most notably, shamans have told me that they determine how to cure a patient on the basis of information presented to them by wise persons that they encounter in their visions.

Once, in a private session I was listening to Verdi's *Requiem*, and the scenes that appeared before me seemed to be the ones described in Dante's *Divine Comedy*. Like Dante, I was escorted by a guide. He stood at the back, in a corner of the visual scene, without me being able to see him. This is in line with what I heard other people describe: most typically, the guides are not seen, they stand behind the Ayahuasca drinker and direct him or her, make him or her understand what is going on, and protect him or her.

Also seen are persons one appreciates to be wise but who do not serve as guides or teachers. In my visions, these included a pre-Columbian high priest, an old man in Burma, and a magician from an unidentified ancient civilization.

Individuals

An experientially distinct category is that of individuals who do not have a name but who do have a specific, well-defined identity. Often, the figures seen have a distinct presence and those who report seeing them feel an intimate link or identification with them. Invariably, in both indigenous and non-indigenous contexts, this phenomenon is attributed to the figures seen being reincarnations of the Ayahuasca drinker's past lives. Putting the interpretative speculations aside for now, I would like to underline the special force of the experience in question. The first time this happened to me was before I had heard of such experiences from other people, and it was clear to me, instantaneously and without any doubt, that the old man who appeared before me in the vision, a lonely shaman in the icy tundra of Siberia was, while still being himself, me. Other individuals I have seen and felt a strong identification with included a wise old man on a balcony in a Chinese town and a young Spanish aristocrat of the Renaissance. People in this category described to me by informants included Amazonian Indians and Incaic figures, women of the ancient Middle East, princely figures, monks and priests from various lands and cultures, a prostitute, and a beggar. More on this will be said in Ch. 13, where personal transformations are discussed.

Social Scenes

Very common are scenes in which many people appear. The scenes in my corpus of data fall into five categories—dances, pilgrimages and processions, rituals, gatherings, and street scenes.

The *dances* I have seen may be divided into two kinds. The first is of colourfully decorated women celebrating the Divine. Strikingly, some of the visions of this kind that I have seen are exactly like ones found in Amaringo's work, notably the vision depicted in Plate 18 of Luna and Amaringo (1993). The second kind is of orgiastic scenes with semi-clad women, often black. Invariably, the dance was erotic and lascivious but there was no malice in it. In my own visions, the women were usually looking halfway to the side; this occurs rather frequently in the visions of other people as well (see also the report of Kusel, 1965 which was cited in Ch. 1). As reported in Ch. 7, twice I saw a serial of scenes whose theme pertained to dancing.

The scenes of *pilgrimage and processions* consist of lines of people marching forward. For me, the lines of people always ran diagonally from left to right, with the participants being the members of some ancient culture climbing towards a sacred place. Most of the processions that I have seen were royal processions. These include a series of scenes depicting kings and queens appearing before their subjects. Once I saw a procession of deities. A special subtype is that of people coming towards the drinker. For example, once I saw people climbing towards me in a procession so that I might cure them. They were poor, Andean persons and I received them with an understanding nod. Similar visions were repeated to me by many informants.

The *rituals* I have seen were usually pagan. Some seemed to be very ancient, and several included human sacrifice. The *gatherings* included receptions, feasts in palaces and mansions, and scenes in eating places and bar-like establishments. The *street scenes* were usually related to visions of villages and cities. Examples are market scenes in medieval Europe and a grand scene depicting the lives of farmers and fishermen along the Nile in ancient Egypt.

3. *The Natural World*

Animals

An inspection of the literature reveals that the items most likely to be seen with Ayahuasca are animals (for a review see Harner, 1973*c*). Indeed, there is no report in the anthropological literature that does not mention animals. In the data I have collected this is the case as well—animals are the most common category of content;[3] this holds true both for my core corpus and for the data provided by my informants. By far, the animals most frequently seen are serpents, felines, and birds. This is even though, obviously, both my own personal and cultural background and those of my non-indigenous informants is so different from that of the Amerindians studied in the anthropological literature. Below are some further descriptions and comments regarding the various animals that appear in the data I have collected.

Before proceeding, it is pertinent, I think, to cite observations made some thirty years ago by C. Naranjo (1973*a*). In an experimental setting, Naranjo administered harmaline (one of the chemical constituents of Ayahuasca) to thirty-five Chileans with no knowledge about Ayahuasca. Especially salient in the visions these subjects had were serpents, crocodiles, felines, and birds of prey (see also Naranjo, 1973*b*).

[3] Except for human beings, which is a category too large and too heterogenous to be really meaningful.

The felines mentioned included tigers, leopards, and jaguars (but apparently, not lions). I find the similarity between these findings and those surveyed here striking.

Serpents and Other Reptiles

Traditionally, Ayahuasca is closely linked to serpents (see Dobkin de Rios, 1973; Harner, 1973*c*; Luna and Amaringo, 1993; and Lagrou, 1998). Serpents are also extremely common both in my own visions and in those of my informants. Indeed, in both my data and those of my informants, serpents are the most common animal and, in fact, the most common single content item reported. As described in Ch. 1, the very first figurative items I saw with Ayahuasca were lizards and in my first major Ayahuasca experience serpents featured prominently. Subsequently, I have seen serpents on many occasions. On several occasions, I have also seen crocodiles, both natural and mythological, and dinosaurs. About a third of the serpents I have seen were mythical or non-naturalistic in one way or another. Some of the serpents were adorned with flowers or shining scales. Phantasmagoric and serpents like those of mythology were also reported by my informants. Amongst these were gigantic serpents, serpents characterized as 'cosmic', winged serpents, and ones made of or emitting spewing fire. At times, serpents appear intertwined in pairs, with one serpent coiling around another; this motif is discussed at length in Narby (1998). My own experience is that, in general, serpents appeared either as the first figures in a session or whenever I was embarking upon a new stage in the session. Why serpents are so prominent in Ayahuasca visions is an intriguing topic that I shall discuss elsewhere. Here I will just note that the associations I had with serpents in my visions were of wisdom, enchantment, seduction, and healing. A grand vision in which I saw serpents in conjunction with disease and cure is reported in the next chapter.

Felines

In the literature, apart from serpents, the most common type of animal associated with Ayahuasca visions are felines—notably, jaguars and pumas.[4] This is also the case with respect to the data I have collected, both mine and those of my informants. Indeed, for all groups of my informants, as well as for myself, felines are by far the most common kind of mammal seen in the visions.[5] On the basis of firsthand experience—both my own and that of persons who participated with me in sessions—I can testify that the seeing of felines also occurs in urban contexts which are distinctly non-Amazonian.[6] Indeed, seeing jaguars and pumas was also

[4] As employed here, the term 'feline' encompasses jaguars, pumas—both dark and light (that is, black and yellowish-brown)—and tigers, but not, however, lions or domesticated cats. The exclusion of lions was dictated by the data—unlike the other big cats noted, these are quite rare in the visions.

[5] Numerically, birds are more common, but as noted above, they comprise a larger and more heterogeneous group, and conceptually (that is, in terms of the semantic hierarchy of animals) they are of a higher order than the felines. Thus, the high frequency of the felines is, indeed, remarkable.

[6] I say this in the light of a comment made in Walsh (1990) that felines are seen only when Ayahuasca is consumed in the Amazonian forest.

reported to me by first-timers partaking of Ayahuasca outside South America. Overall, I would characterize the jungle cat as a manifestation of the energy of life in its full and pristine form. Seeing a feline is tantamount to an encounter with this potent energy. Indeed, my data reveal that many visions of felines exhibit a strong interactive aspect.

Of the felines, the most common are jaguars and black pumas. The prevalence of black pumas is especially striking. In my structured interviews I have asked my informants whether they have seen 'jaguars'. Again and again, the response I have received was 'Yes, but more frequently I have seen pumas, black pumas.' This detail was given even though my query did not specify pumas as such. Some of the cats are non-ordinary. Personally, I have seen jaguars with shining or flower-shaped spots. Once I had a glimpse of what I conceived to be the primordial Jaguar. One informant reported having seen a tiger of fire.

The seeing of felines may be coupled with an interaction with them. For example, once in a vision, I found myself confronted with a puma. Watching it, I experienced a transfer of energy—the puma's energy was passing over to me. An identical experience was reported to me by one of my informants. As recounted in the previous chapter, still another informant reported a vision in which she engaged in a magic journey riding a jaguar.

Birds

Birds of all kinds constitute the one single category of content that appeared most frequently in my visions. However, as explained above, it should be borne in mind that this zoological group is far more heterogeneous than that of the serpents or the felines and that it comprises several well-defined species; dividing the group into these will, of course, result in lower values of frequency. I have also seen many birds that I could not identify or label. The most common birds are birds of prey, notably eagles and condors. Also common are waterbirds (especially, cranes and herons), pheasants, hummingbirds, and parrots. For both my informants and myself, white birds are especially common. Many of the birds are characterized as being enchanted; these are often blue-feathered. Also mentioned are golden and two-headed eagles. Lastly, birds are salient in experiences of transformation and of flight; these are discussed at length in Ch. 13.

Other Animals

Mammals I have seen on more than two occasions are horses, elephants, and bulls and cows. I have also seen a dog, a lamb, a deer, a camel, a monkey, a raccoon, a hippopotamus, and a pack of jackals. In addition, I have seen fish, bees, ants, butterflies, as well as various lower organisms. Animals reported to me by more than three informants were (in a descending order of frequency) fish, butterflies, horses, bulls and cows, monkeys, lions, dogs, bats, dolphins, crocodiles, frogs and toads, all sorts of insects, as well as animals characterized as evil beasts and as prehistorical. Of these, the first seven were indicated by eight or more informants.

Again, note that 'fish' is a heterogeneous category which conceptually is on a higher order than the others.

Phantasmagoric and Mythological Animals

Phantasmagoric and mythological animals are especially common in Ayahuasca visions; some were already mentioned in the sections concerning serpents, felines, and birds. As noted in Ch. 1, dragons appeared before me most powerfully, and totally unexpectedly, in my very first strong experience with Ayahuasca. Other non-naturalistic animals I have seen include golden bulls, winged horses, a winged elephant, a taurus, a phoenix, and what I would characterize as 'evil animals'. Enchanted animals reported by my informants included dragons, winged horses, and a winged lion. As indicated above, enchanted birds are quite common. Both for me and for my informants, many of the phantasmagoric animals are creatures that are half-human, half-animal; these are discussed below.

Flora

Botanical items are seen very often in Ayahuasca visions. These may be viewed either in isolation or in scenes, either in natural settings, phantasmagoric or mythological ones, as part of works of art or decorative designs.

Perhaps the most remarkable botanical item seen in isolation is the Ayahuasca vine itself. Interestingly, such visions and the insights associated with them often resemble ones encountered in the indigenous Amerindian lore. Thus, in his first and only experience with Ayahuasca, in a private session in Europe, one European informant was struck by the feeling that a plant being was in his body and that he had a strong, intimate relationship with it. This man felt that the plant was a being in its own right and that it was passing knowledge to him. This is in line with the indigenous conception of plants as teachers, a topic with which this subject was not familiar (see Luna, 1984*a*). Another informant recounted a vision in which a beautiful woman turned into a serpent which in turn transformed itself into the Ayahuasca vine. Such stories of metamorphosis are central in Amerindian myths on the origin of Ayahuasca. A vision I myself had of the discovery of the vine from which Ayahuasca is made is recounted below.

Half-way between isolated items and scenes are visions in which the drinker's visual field is completely filled with flowers. Flowers can also appear as ornaments or decorations; an example of my own is that of wall drawings in a classical (ancient) Roman style. I find it in place to add a clarification here. Quite likely, the label 'flower' may fail to capture the experiential qualities that some visions depicting flowers have. A vision may present 'only' flowers, all naturalistic, yet be extraordinary. Such, for instance, was the first vision of one experienced informant, a vision he characterized as one of the most powerful he has had. In it, waves of flowers appeared one after another. I once had a vision in which my visual field expanded and was all replenished with garlands and wreaths of flowers. Just that. And yet, the vision was grandiose and its impact exhilarating.

The botanical scenes I have encountered include views of forests, gardens, meadows, and savannas. Many of the gardens that I have seen were enchanted; in some of these, fountains, pools, and waterfalls were seen as well. One image that especially stands out in my mind is a very realistic panoramic view of a forest, similar to those in Rousseau paintings. There were palm trees lit by the bluish light of the moon under a very clear night sky. Another was a garden the likes of which I have never, ever seen. It was purple-blue, and large pheasant-like birds were strolling in it. Still another was what seemed to me to be the garden of God. I did not enter it, but only viewed it through the large metal grilles that surrounded it.

On two occasions I had visions consisting of glimpses of the Amazonian forest and its mysteries. In the first, I saw Indians hunting with the aid of Ayahuasca. They were running through the forest in the dark but there was a strong white light hovering in front of them and guiding them. In the second, I saw how the plants of which Ayahuasca is made were discovered. This vision was similar to the first: An Indian was marching through the forest, and light was hovering like a halo above the plants, marking them.

An item to be mentioned also in the section on landscapes below is that of open meadows and fields. Visions depicting these are extremely common and were reported also by many first-timers with no prior knowledge about Ayahuasca. Typically, the meadows or fields extend far into the horizon and seeing them is coupled with an ambience of infinitude and eternity. This often imparts subjective feelings of serenity, faith, and optimism.

4. *Mythological and Phantasmagoric Beings and Creatures*

Beings and creatures that are neither human nor naturalistic animals are common in Ayahuasca visions. Mythological and phantasmagoric animals were commented upon above; the other beings and creatures may be divided into several types.

First are *mythological* beings and creatures. The epithet 'mythological' is employed here to denote beings and scenes that are not part and parcel of ordinary reality, present or past, nor of realms interpreted as being heavenly or divine. The mythological beings and creatures at hand may or may not be drawn from the cultural heritage of the Ayahuasca drinkers. I myself have seen such creatures only on a few occasions. Mythological beings reported to me by many informants included little green men, gnomes and elves, fairies, and monsters of all sorts. Various creatures characterized as 'Ayahuasca beings' pertain to this type as well. Many times, the creatures seen are said to be playful and mischievous (but at the same time benevolent).

Second are *chimeras* (in the broad sense of the term) or *hybrid* creatures—that is, creatures which are half-human, half-animal. Of these, by far the most frequent are mermaids—creatures that are half-woman, half-fish. Also reported are hybrids of humans and felines, reptiles, birds, and canines. Interestingly, such creatures are also common in indigenous reports (see, for instance, Waisbard, 1958/9 as well as the paintings presented in Luna and Amaringo, 1993).

A special category is that of *creatures with multiple faces*. Usually, these creatures are encountered in heavenly scenes in which the drinker feels that some ultimate secrets are being revealed to him or her. These are reminiscent of the Divine beings described in the visions of the biblical prophet Ezekiel as well as in the Apocalypse of St John. Both I and several of my informants had such visions. Especially remarkable is a grand vision in which one informant saw a creature with a great number of faces, each pertaining to a different animal. This creature was conceived of by the informant as the embodiment of all Life.[7] The multiple faces may also appear sequentially. Thus, one informant saw a human face changing, in succession, to the face of a puma, a tiger, and a wolf.

Third are beings referred to as *extraterrestrials*. Often these are seen along with spaceships. For some individuals these are especially common.

Fourth are *angels* and other *celestial beings*. The term 'angels' refers to winged humanlike beings made of light. Indeed, several informants have explained to me that the wings consist precisely of this—powerful light. Other, very common, supernatural humanlike beings are transparent figures. These are usually perceived as beings made of exceedingly delicate white webs of energy. Often they are explicitly described as 'beings of light'. Several depictions of these are found in the Amaringo paintings (see e.g. Luna and Amaringo, 1993: Pl. 36).

Fifth are *semi-divine beings*, that is divine beings which are not God. The most prominent of these is the figure of a beautiful bearded man in the prime of life. The man is clad in a simple white garment whose margins are often finely embroidered. Invariably, the man is seen *en face* (relative to the viewer), he radiates good energy and is full of love. Many of my informants identified this person as Jesus Christ. Also very common is a benevolent female figure standardly taken to be the Virgin Mary. Conceivably, the great frequency with which these figures appear in visions of my informants may be attributed to the Christian and semi-Christian contexts in which Ayahuasca is often taken.[8,9] I should note, however, that on several occasions, I too have seen both the male and the female figures of the type just noted. Other beings of the same category that I have seen are the Buddha and various Hindu and pre-Columbian deities. With open eyes, in the sky, I once saw the figure of the mischievous Hindu divine being Ganesh. Deities and divine beings reported to me by informants included ones pertaining to ancient Egypt, India, and various pre-Columbian cultures.

[7] When this book was nearing completion I saw, in the Amazonian Museum in Quito, a Shuar sculpture representing the Supreme Law of Life and Existence—it depicted a woman with many faces.

[8] The interpretive processes involved in this attribution are discussed in Ch. 17.

[9] Regrettably, nowadays even indigenous Ayahuasca rituals are contaminated with Christian elements.

By way of example, let me present one grand vision in which a goddess characterized as the Great Mother appeared. This vision was seen by one independent drinker from Rio de Janeiro the third time she partook of Ayahuasca:

> She was the mother of all atoms and the matrix of all forms. All the atoms were dancing and the Mother was pure joy. She looked like an Egyptian Goddess who was covering, and protecting, all of creation with her body. 'Why do you look like an Egyptian?' I asked the Goddess. 'Actually, I do not have any form, but I appear as Egyptian because the Egyptians were the first to comprehend my secret,' she replied. Then the Goddess lifted up her arm and turned it so that the fabric unfurled. I saw that the cloth consisted of all the possible forms and I understood that the Great Mother is the creator of Everything and that what holds this entire immense Creation together was Cosmic Joy. In this process, suns and planets are created. What creates it all is Sophia, wisdom.

The informant added that only later she found, in a book, a depiction of the Egyptian Goddess Nut which was very similar to what she had seen in the vision.

Last to be noted are *demons*, monsters, and beings of death. A vision of mine that especially stands out in my mind is the following:

> The Angel of Death presented itself in front of me. I knew that if I did not hold on to my life energy, he would take me. I also knew that as long as I sustained an unwavering will to live he could do me no harm. In other words, this angel will take me only if I manifested and/or conveyed a weakening of the will to live. 'But I do wish to live!', I reflected. With this, I summoned all my vital energies and the menacing figure in front of me retreated.

5. *The Cultural World*

Ayahuasca belongs to the Amazonian rain forest, but my data reveal that, apart from beings and animals, the items that feature most frequently in the visions this brew produces pertain to culture: cities and architectural complexes, palaces and temples, and various works of art and precious objects—all are very common for both me and the different groups of my informants. Indeed, palaces and objects of art and magic are, for some, the items most commonly seen in the visions. A perusal of the anthropological literature reveals that items pertaining to culture are also common in visions reported by indigenous persons.

Buildings

Buildings appear very frequently in the visions; most of them are magnificent. Most notable are, on the one hand, palaces and other palatial complexes and, on the other hand, temples and religious sites; castles, mansions, and archeological ruins are also common. Often the palaces and temples seen are made out of gold, crystal, and precious stones. Here, for instance, are two descriptions taken from Polari (1984: 197–8, 264):

I found myself in the salon of a castle illuminated by torches. The salon was oval, slightly oblong, with semicircular doors, placed at regular intervals. At the same time, I saw an atrium situated in front of the principal wall, and the hall of the throne.

Thousands of tunnels, galleries, corridors, secret doors, staircases, inclined planes criss-crossed in a composition akin to one of Escher's.... These led to sumptuous palaces, lofty halls, sarcophagi, caverns or temples. In some of these there were sentinels appearing as medieval figures.

I myself have seen palaces and temples many times. As described in the previous chapter, the only time I consumed Ayahuasca in the midst of the virgin Amazonian forest, I saw a whole sequence of palatial and regal scenes. I hoped to get a glimpse of the mysteries of the Amazonian jungle, but instead I saw palaces and royalty from different cultures on both sides of the Atlantic. The most impressive of these were two visits to the throne hall of an Egyptian pharaoh. In real life, I have been to many of the famous archeological sites in Egypt, and I have also visited the Egyptian galleries of many museums, but what I saw in this and other visions surpassed them all. The Egyptian palaces revealed to me in my Ayahuasca visions appeared to be totally new and full of life, as if they had just been constructed.

Like Polari, on many occasions I saw corridors, one hall opening into another, marvellous wall-paintings, sculptures, and reliefs. Architectural details that especially impressed me included sculpted marble colonnades in the form of white elephants, staircases adorned with golden lions, and finely carved gilded wooden ceilings. Several times, I saw most beautiful painted tiles. In the reports of my informants mosaics appear frequently; an example was described in Ch. 6 when serial images were discussed.

The temples I have seen were both ones of ancient civilizations and ones of the more recent Christian world. One of the more impressive architectural visions I have had was of a grand cathedral. The edifice extended up to the heavens and its dominant colour was emerald. I have heard of very similar visions from my informants. Several informants told me that looking, with open eyes, at the hall or the *maloca* (Amazonian hut) in which the *Daime* or Ayahuasca session was taking place, the whole place transformed into a holy edifice. In more than one instance, the actual roof seemed to disappear and the visioned construction reached up to the sky. Also reported by many informants are pyramids of all sorts, most either Egyptian or pre-Columbian.

As exemplified in the above citations from Polari, basements, hidden passages, gates, and doors are all very common. These often have a secret, enchanting allure.

Cities

Entire cities may be seen as well. Here I focus on cities qua architectural complexes; cities as a type of place are considered later in this survey. The cities seen in visions are usually exotic and most fabulous. Discussing the Amaringo paintings Luna comments that: 'About one third of the visions presented contain . . . cities. Their architecture is either diffusely Eastern—Chinese, Arabian, Indian—or futuristic, or both. They may be located in the underwater world or on another planet' (Luna and Amaringo, 1993: 41; see also Harner, 1973*c*). My own experiences corroborate this general statement: Most of the cities I have seen seemed to belong to ancient civilizations whereas others were futuristic or magical, whose identity I could not determine. The reports of my informants reveal the same pattern. The most impressive city I have seen was that described in the Prologue whose buildings were made of gold and precious stones. A couple of years after having had that early vision I saw a similar description in Villoldo and Jendresen (1994). An informant who partook of Ayahuasca with the Cashinahua Indians told me of seeing a city with strange, circular architecture in a style she had never envisaged or heard of beforehand. A European partaking of the brew for the first time in a private, urban context and who had no prior knowledge about Ayahuasca at all reported seeing what he characterized as being the most bizarre architectures possible: 'The buildings did not have a single straight line to them, and some had animal features. They had mouths, yet they were buildings. Some of the buildings also moved.' Similarly, Luna describes seeing 'a Hieronynmus Bosch-like city with buildings of organic form, like huge plants or organs of a gigantic animal' (Luna and Amaringo, 1993: 41).

Works of Art

Works of art and artistic objects are very common both in my visions and in those of my informants. Items I have seen many times include pieces of sculpture, pieces of furniture, chandeliers, ceramics and articles of gold, masks, carpets, tapestries and rich fabrics, embroidery, jewellery, crowns and sceptres, as well as swords and shields. All these were also reported by my informants. The materials of which most of these items were made were gold, brass, gilded wood, crystal, precious stones, and fine textiles (most of which were white and/or finely embroidered).[10] Usually, the items seemed to be exceedingly precious and some seemed to be endowed with magical powers. Once I saw an entire exhibition of the works of arts and crafts of one specific culture. Manifestly, all the items I saw belonged to the same culture, one that I could not associate with any particular culture I had ever known. Similar experiences were reported to me by two other persons.

Artistic works are also frequently encountered in the buildings one sees. As described above, often these are embedded in the architectural structure itself. An example drawn from my own experience are elephants of white marble that served as pillars to the front hall of a palace. On several occasions I have seen ornate frescoes. In many cases, these were in the art-deco style; on one occasion they were of the Roman Pompeian style, and on another they were eighteenth-century Baroque. As mentioned above, several informants have reported seeing all sorts of mosaics. Also common are carved wooden figures, engravings, and ornamental

[10] Remarkably, a separate analysis I have conducted reveals similar patterns in the visions reported in the Apocalypse of St John. Such items are also common in visions reported in Jewish mystical writings of the period from first century BC to the first century AD.

reliefs. Some of these were seen in conjunction with vehicles—notably sculptured prows of boats (see also Harner, 1980).

Especially to be noted are items pertaining to religious rites, be they generic or associated with specific services or occasions. Those I have seen include the Caduceus, the Golden Calf, the biblical *choshen* (a divination plaque with twelve precious stones, used by the ancient Jewish High Priest), and various idols. As noted above, objects endowed with magical powers are also common.

Vehicles of Transportation

The most common category of artefacts is that of vehicles of transportation. These may be of land, sea, air, or space. Of those that I have seen, the most frequent ones were on the one hand, carriages, mostly royal, and on the other hand, boats and ships.[11] Twice I have seen what I recognized as space ships. Some individuals seem to see these quite often. On many occasions, even if the vehicle was technologically standard (e.g. a train or a car), it would be fabulous in its decor—adorned with gold or wonderfully painted. Once I saw a joyous elephant playfully riding a scooter. On several occasions I have seen visions in which wheels or other revolving parts were central. Once, I saw a grand vision in which futuristic technology was prominent. Such scenes have been reported to me by my informants as well.

Musical Instruments

Another special category of objects pertaining to culture is that of musical instruments. These are often associated with the hearing of music. By far, the most common instruments seen by me or reported by informants were trumpets.

Books, Scripts, and Symbols

Many people report seeing inscriptions of letters, numerals, or other signs. Both in my case and in that of my informants, on some occasions the characters seen were made of, or engraved in, gold or silver. Often these are in scripts or languages that the Ayahuasca drinker characterizes as ones he or she cannot decipher or understand. Some informants say that they do manage to decipher and understand messages in scripts and/or languages that actually are not familiar to them. I could not substantiate any of these claims (especially to be noted are cases when the language in question was Hebrew, my own native tongue) and I prefer to regard them as reports of genuine subjective experiences, not facts. Mandalas and flags are also very common.

A vision of special significance to me was the following:

I was seeing what was clear to me was the Torah of God. It was a scroll of red fire on ice. As I examined it further, serpents came out of the fire. As the image developed, the serpents

[11] I may note that a ship featured prominently in a grand vision seen by the founder of the Barquinha, hence the name of the sect (in Portuguese, a diminutive of the word for boat); see also pls. 20 and 48 in Luna and Amaringo (1993).

crawled within the boundaries of the scroll in horizontal lines of alternating courses—from left to right, and from right to left. And then I realized that letters were engraved on the serpents. But the serpents were moving, and this scrambled the text. It was fine, I thought, what really matters in the holy text are the single letters, not the prose.

A year later I found the following description taken from the Zohar, the principal book of the Kabbalah: 'In the beginning, two thousand years before the heaven and the earth, seven things were created: . . . The Torah written with black fire on white fire, and lying on the lap of God' (Ginzberg, 1909: i. 3). As reported in the Prologue, one of my first visions consisted of a triangle with a hand inscribed inside it. On other occasions, I saw the entire zodiac, astronomical-like figures, and the primordial letters of the Hebrew alphabet.

6. *Places and Landscapes*

Cities and Sites

Above I mentioned cities in terms of the architectural complexes in them. Here I refer to cities as a special category of place. The cities I saw were usually not familiar to me from real life. Only on a very few occasions have I seen a city that I had actually visited. Most of the cities I have seen in my visions belonged to ancient civilizations; most were of ancient Egypt and the civilizations of pre-Columbian Central and South America. Other civilizations that featured in my visions included ancient biblical Judaea, Assyria, ancient Persia, the Hellenic world and classical Rome, medieval Europe, and the Far East. On several occasions I have seen what seemed to me to be futuristic cities. These featured skyscrapers, hi-tech constructions, and many neon lights. All these kinds of cities have been reported by my informants as well.

Landscapes

In the present terminology, landscapes are places that seem to be real but do not have a particular geographical or historical identity. It will be noted that the landscapes may be seen either from outside or from within. The landscapes seen from the outside can be either glimpses or broad panoramas. The landscapes I have seen included open vistas of meadows and grassland, panoramic views of mountains and of the sea, as well as scenes of lakes and riverbanks. All these were also commonly reported by my informants. My informants' data also include scenes of waterfalls and of the desert. Significantly more rare are underwater scenes. I myself felt that I was going under water only once, but I was worried that I might drown and I did not let myself go in further. A special category is that of the views one sees while flying; I shall say more about these in Ch. 13. Finally, snapshots of exotic places may be seen. Ones encountered in my own data include small pastoral mountain towns, an island in the South Pacific, and the African savanna.

Other places

Lastly, there are places of entertainment. These seem to fall into two main subtypes. One is that of places such as bars and cabarets, often lascivious and somewhat lewd. The other may be characterized as ludic. In these, the atmosphere is gaily frivolous; they include amusement parks and circuses. The frequency of facilities pertaining to amusement parks in the visions is, it seems to me, disproportionately high. Especially noted are carousels and Ferris wheels. Interestingly, a merry-go-round is also mentioned by Naranjo (1973*a*) in his experimental study of harmaline.

A puzzling question that arises is whether one can see real places that one does not know or know of. Kensinger (1973) reports that there were some Shipibo Indians who claimed to have seen the town of Pucallpa although they had never been there (see also Morton, 1931; for further discussion and references the reader is referred to Ch. 16). Campa Indians I talked to said exactly the same thing—that Ayahuasca enables them to travel and see foreign places. A lower-class, simple resident of the far west of Brazil told me that he had visions in which he found himself in Russia and Japan. In real life, this man had never left the Amazon region. Further, he assured me that he had never seen any pictures of these places. I have no means to check the veracity of these reports. Writing this summary survey, I wish I could have returned to the Amazon and shown art books of various civilizations to residents who have reported to me having had visions of foreign places and civilizations. Unfortunately, I do not know when, if at all, it would ever be feasible for me to do this. More on the alleged paranormal powers of Ayahuasca is said in Ch. 16.

7. *History and Evolution, Religion and Myth*

Historical Scenes

Scenes from different historical periods are very common. My impression is that in all cases events were seen as being in the process of their happening in a particular place at a particular time. In general, it is not that with the outset of the Ayahuasca vision a special event begins and with its fading off it terminates. Rather, it is as if out there life goes on and the intoxicated person is presented for a moment with the opportunity to witness a scene of other times, and other places. In glimpses this fragment of time is very short, in full-fledged scenes it may be long. The historical scenes may depict major historical events or episodes of daily life. The most common instances of the former are wars, coronations and royal pageants, and episodes in the lives of famous historical figures. Images of the latter kind that stand out in my mind are a market scene in medieval Europe and a very colourful street scene in China.

Especially impressive scenes in which I found myself in the position of a witness were of the Holocaust of the Jewish people in the Second World War, various biblical scenes, and scenes of daily life and rituals in various ancient civilizations. On several occasions I have also experienced scenes of prehistoric people and of prehistoric animals roaming savannas.

Of special significance are *panoramic* historical visions. These depict several historical episodes which together present a moral regarding the human predicament. As noted in the Prologue, one of my very first visions included a panorama of human history: the injustice it manifests on the one hand, and the bounty of culture and art—notably, religious—it produces. One informant told me that in the first vision he had with Ayahuasca he witnessed human suffering throughout history. With this, he understood the forces of evil, pride, and greed that make empires fall, and he came to the realization that the only hope for humanity is spiritual.

Religious Rites

On several occasions I have seen prehistoric sacrifices of both animals and humans, shamanic rituals, pagan rites, religious services in various cultural contexts, as well as processions and pilgrimages (these latter have already been described above). Religious rites mentioned by my informants included ones in the Amazonian forest, in ancient biblical Israel, pre-Columbian America, and in pagan and medieval Europe. An experience I myself had, and which was reported to me in identical terms by several other persons, consisted of a non-ordinary perception of the very Ayahuasca session in which I, or my informants, were participating. In all cases, the session was experienced as a very primitive (in the historical, chronological sense of the term) ritual—often a dance—in which the drinker him- or herself was actively taking part. Often it was commented how extraordinary the strong group feeling associated with these rituals was; some reports also alluded to a summoning of the forces of nature. In all cases the experience was described as exceedingly powerful.

Evolution and Creation

Akin to the grand, panoramic historical scenes are scenes of biological evolution. A Brazilian with extensive and varied experience with Ayahuasca reported:

> I was looking into the mouth of a cauldron. Within the circle defined by the rim, a cinematic presentation of evolution unfolded. Outside of the boundaries of the rim, everything looked utterly normal. This was the most powerful of the very many Ayahuasca visions that I have had.

I myself once looked at a tree, and, with open eyes, saw a reptile. The reptile grew and changed shape, sequentially transforming into several animals. First there were prehistoric, extinct species and later, modern ones.

Some of the most spectacular visions that can be experienced with Ayahuasca are those interpreted as depicting the Creation of the universe. Usually, these are shrouded in mystery and in general are characterized as being ineffable. Also characteristic of these visions is the great impression of harmony and bounty that they impart. One such vision I experienced myself consisted of a stupendous emanation of light which was bountiful and full of love. From the light emerged a multitude of flowers and with them the entire world rejoiced in the wonder that is Life. A European woman with extensive Ayahuasca experience told me of a vision in which she felt she was joining the Demiurge, the God who was creating not one, but many universes. Similarly, Reichel-Dolmatoff (1975) reports that indigenous users of Ayahuasca say the brew enables them to visit the place of creation, witness the act of creation, participate in the creation story, and comprehend the moral concepts it contains. Three of my informants related the creation of the universe to geometric forms.

Myth

In the indigenous context, Ayahuasca visions are intimately related to myths. Indeed, the visionary state is often regarded as that in which myths are revealed. Mythological scenes reported to me by my non-indigenous informants were akin to Amerindian myths, to Celtic ones, as well as to fairy tales. These portray beings already mentioned: heroes and supernatural beings; enchanted animals and creatures which are half human, half animal; fairies and gnomes; spirits and demigods. A vision that especially impressed me was one in which it seemed to me that I was understanding how myths are actually created.[12]

8. *The Divine*

Celestial and Heavenly Scenes

Of special impact are what may be called celestial or heavenly scenes. In these, the heavens part and the Ayahuasca drinker witnesses scenes akin to those described in the book of Ezekiel (ch. 1 and 10), the book of Isaiah (ch. 6), and the Apocalypse of St John. In such scenes drinkers encounter realms full of light which emanate great bounty and bliss. Celestial and heavenly scenes I have witnessed included chariots with magnificent white horses, a ring of animals around an expanding sky, and an opening of the heavens which might be called the entrance to the Kingdom of God. Both I and half a dozen of my informants have had visions in which a ladder was reaching the heavens and angels were going up and/or down it. For obvious reasons, I refer to such visions as 'Jacob's ladder'.

Related to the above are scenes in paradise. In these, the Ayahuasca drinker may see enchanted gardens and orchards, fountains and brooks, angels and other benevolent beings. Perhaps more significant is the general ambience which is characterized as paradisiacal. Typically, it is described as one of serenity and bliss in which the supreme Good reigns.

[12] In conjunction with the origin of myth, see the reflections of the anthropologist-philosopher Lévi-Strauss (1964) and my comments on them in Ch. 23.

Encountering the Divine

The Indians say that Ayahuasca enables them to see God (see Reichel-Dolmatoff, 1975; Taussig, 1987; Payaguaje, 1983). I would rather use such phrases as 'encountering the Divine', 'being in the presence of God', or the traditional Jewish one—'being under the wings of the *Shekhina*'.[13] I shall discuss this experience further both in Ch. 9 and in Ch. 16.

Light

Last but definitely not least are visions of light. Because of their special subjective significance and varied phenomenological complexity, these visions will be discussed separately in Ch. 17.

9. *Other Categories of Special Import*

In the last section of this survey, I shall mention several content categories that stand apart. I place them together here only for editorial considerations.

Celestial Bodies and Planetary Voyages

Both I and my informants experienced visions in which the sun, the moon, and/or the stars featured centrally. An example is presented in the next chapter, where themes of visions are discussed. In this conjunction let me note that one of the most important hymns of the Santo Daime Church is called 'Sun, moon, stars'. Likewise, these three entities feature centrally in the decoration of the halls in which sessions of the UdV are held.

Often Ayahuasca carries one higher and higher above the planet Earth and far into the cosmos. On several occasions I felt I was flying upwards. On one of these voyages I saw, from above, the whole planet. Similar interplanetary voyages have been reported to me by my informants. Some indicated having reached a particular astronomical body, such as the sun or a specific planet. Such voyages are also depicted in some of the paintings of Ayahuasca visions in Luna and Amaringo (1993); see, for example, pls. 21 and 37.

Anatomy and Physiology

Traditional healers and medicine men are famous for their (alleged) ability to see, under the intoxication, the insides of patients' bodies (see, for instance, Gebhart-Sayer, 1986). The shamans I have interviewed expressly boasted of having this ability too. I myself have experienced visions of the inner parts of my own body on several occasions. On one such occasion I caught a glimpse of the internal parts of my upper leg, and on another I travelled inside my cranium. Travels within the body were also reported by several of my informants. The most common of these are ones in which the drinker was getting an inside view of his or her own brain and

[13] In Judaism, the *Shekhina*, presence, is the female aspect of the Divine.

ones in which individual cells in the body were said to be seen; informants also reported seeing their own DNA (for a related discussion, see Narby, 1998).

Platonic Ideas, Archetypes, and Mathematical Objects

One special feature of Ayahuasca visions is the encounter with what may be referred to as the world of Platonic-like Ideas.[14] There, one sees not the particular and contingent but rather the generic. My first experience of this kind was reported in Ch. 5: a villager was passing by at a distance and I saw him as *the* Farmer; it will be noted that this happened with my eyes open. Other visions of this kind that I have experienced depicted the archetypal Man and Woman and the primordial Jaguar; both visions were seen with my eyes closed. Several informants reported identical experiences to mine.

Abstract entities may be seen as well. One informant told me he had a grand vision of perfect geometric bodies. Another reported a scene in which he spontaneously came to the appreciation that the physical world is harmoniously governed by mathematical laws. Three informants reported grand visions in which the manifold of all forms was seen. Several informants, all with an academic education, explicitly commented that Ayahuasca brought them to the world of Platonic Ideas. More on this topic will be said in Ch. 15 and 23.

One of the most impressive visions ever recounted to me was that seen by a South American man of mixed race, a labourer with only three years of primary school education. This person has had extensive experience with Ayahuasca and the vision he described as the most meaningful he had ever had was comprised of mathematical formulae which he understood as conveying the basic laws of the Universe.[15]

Birth, Death, and Rebirth

Birth and death, the two momentous events in human life, are both encountered with Ayahuasca. I shall begin with death, for it is the more common and the more characteristic of Ayahuasca. In the indigenous context Ayahuasca is intimately related to death. After all, this is what the name Ayahuasca means—the vine of the spirits, or of the dead spirits. I have seen death scenes twice, both times with my eyes open. The first time was at a *Daime* concentration session. Suddenly all the (real) people around me seemed to be dead. I was horrified and did all I could—turning my head, shaking myself—to stop this vision. The second time was in the midst of the forest. There were dead persons suspended on and from the trees. As noted earlier in this survey, once I had an encounter (fortunately, not consummated) with the Angel of Death. With my eyes closed, on several occasions I saw prehistorical rituals in which human beings were sacrificed. Seeing blood was reported by several informants.

[14] In order to avoid confusion, throughout this book I use a capital for 'Idea' when referring to the Platonic entities, and lower case when referring to ideas in the normal sense of the term.

[15] This is not the same individual mentioned in the previous paragraph.

Reports of my informants included scenes of self-death, scenes in which the drinker saw him- or herself as a victim in a ritual of human sacrifice, horrible scenes of wars, funerary ceremonies, as well as scenes in which skeletons, spirits, and sarcophagi featured prominently. Reports of self-death furnished by my informants are presented in Chs. 9 and 13. As will be described in more detail there, experiences of self-death are often coupled with experiences of rebirth. Drinkers feel that they almost die and then experience themselves reborn while gaining salvation and enlightment.

My informants also reported visions of birth. Both were reported: visions in which a woman was giving birth, and a vision in which the drinker herself experienced delivering a baby.

Small Details

Lastly, I would like to mention several small details that seem to feature rather often in the visions. What characterizes them is precisely their being details. In terms of their content, the items to be noted pertain to different semantic categories and are not of special significance in their own right. What is special is that they are disproportionally recurrent in reports of Ayahuasca visions. I have noted them in my own visions, and then was struck by their appearance, unsolicited, embedded in reports furnished to me by other people; at times I have found them mentioned also in descriptions of visions cited in the literature. What I am referring to are specific visual details seemingly without any psychological or cultural significance. Indeed, it is precisely the apparent insignificance of these details that makes them important with regard to the issue of interpersonal commonalities in the Ayahuasca experience. True, it could be argued that some items that are especially common in visions have been heard of by people, or read about, prior to their personal experiences. This might be true for serpents and felines, Jesus Christ and the Virgin Mary, but not for seemingly intrinsically unimpressive items such as those to be indicated below. Likewise, some prominent items, such as angels and palaces, may be associated with archetypes that are deeply rooted in the human condition; but again, this cannot be said with respect to the small details indicated here.

The first small detail I would like to mention is *disembodied eyes*. These are eyes seen floating in the visual space without there being either a face or a body of which they are part. The eyes may be those of human beings, of felines, or without any particular identity. Often, a great multitude of such eyes is seen. These are reported very commonly. Notably, they are also encountered in the most spectacular vision reported in the Bible—the prophet Ezekiel's encounter with the Divine (see, in particular, Ezekiel 1: 18; for a discussion of the motif of disembodied eyes in the context of pre-Columbian Mexican culture, the reader is referred to Ott, 1986). Also commonly reported are *detached faces*, that is, faces without bodies; bodies without faces are also reported. Interestingly, heads without bodies are also mentioned by Karsten (1935) in his report of what indigenous Amerindians experience with Ayahuasca. Especially prevalent are faces with slightly slanted, somewhat mocking or sardonic *smiles and grimaces*. Several informants specifically said that the faces approached and teased them (see Harner, 1980). Another frequently reported item is *mouths of animals that open* and from which something—for example, flames or other animals—comes out. These are encountered in several of Amaringo's paintings too, as well as in the early report of Harner (1980) of his first experience with Ayahuasca. Also to be noted are persons with *tilted heads* and ones whose torsos are inclined; especially marked are women who raise one arm and brush it up against their ear (see, for instance, the report by Kusel in Ch. 1).

As already mentioned, scenes of people—usually groups of women—dancing are very frequently reported; a beautiful depiction of such a scene, one which very much resembles a vision I myself had, is found in pl. 18 of Amaringo's paintings (Luna and Amaringo, 1993).

A recurrent detail in my own visions that impressed me quite a bit was that of a *person looking forward*, scrutinizing vast distances or gazing into the open and unknown. Typically, the act was associated with reflective serenity and the atmosphere encompassing it had a touch of the eternal. On several occasions the person had the allure of a sage, on others it was a valiant young man who seemed to probe optimistically towards the future. Once it was a group of small children who were engaged in the act of looking; in this scene, the entire ambience was of fresh hope. On still another occasion it was a chimpanzee. It was positioned way up on some very high boulder looking in the manner described above but not cognizant of the meaning of the act. To me it seemed that the ape was peering into the future progress of evolution.

Concluding Remarks

Having focused on details, let me now take a broader perspective and comment on the global picture that emerges from the foregoing survey. In their totality, the data we have surveyed define the semantic space of the visions seen with Ayahuasca. Embracing an alternative world-view, these data may also be regarded as defining what traditional users refer to as 'the world of Ayahuasca'. The semantic space, or world, in question comprises four main domains. The first is the domain of *nature*. As noted throughout the foregoing discussion, animals—especially serpents, felines, and birds—are some of the most common items in the visions. Natural landscapes and scenes of forests and gardens are also very common. Also common are scenes of heavenly bodies and the far reaches of the cosmos. The second domain is that of *culture*. Its prime manifestations are magnificent cities, the majesty of the royal, various products of artistic creation, religion, and magic. Usually, what are seen in the visions are not contents pertaining to the drinker's own socio-cultural milieu but rather ones associated with ancient civilizations. The majority of the constructions, objects, and artefacts that appear in the visions are either precious or wonderfully ornate or both. Further, most of the buildings seen in the visions are palaces or temples, and many of the human beings are either kings and queens or religious figures and persons with spiritual prominence. Third is the domain of *fantasy*. It comprises enchanted and magical lands and it is populated by all sorts of creatures which are neither human beings nor naturalistic animals. As just indicated, the objects and scenes that appear in the visions are usually not mundane; often they are associated with mythology, fairy tales, and magic. Fourth is the domain of the *spiritual* and the *supernatural*. Ayahuasca visions often reveal before one celestial and heavenly realms. In these, divine and semi-divine beings often appear. The supernatural domain is usually associated with spiritual and metaphysical meanings. Related to this domain are the items and scenes pertaining to death.

In addition to the specific contents that are typical to them, Ayahuasca visions are also characterized by the extraordinary beauty that they manifest. Time and again, drinkers report that what they see in Ayahuasca visions surpasses in magnificence anything they have ever seen either in reality or in works of art. As further indicated above, objects and artefacts that appear in these visions are usually extremely rich and wonderfully ornate. Many of the landscapes are fantastic, and even when strictly speaking they are utterly naturalistic, the landscapes often emanate special qualities such as eternal presence, divine bounty, or pristine meaningfulness. And then, when the visions lead one upwards, to the realms of the planets and beyond, what one sees can be utterly stupefying. Not infrequently a point is reached when the marvel is such that one feels that what is presented before one is overwhelming and that its scrutiny is beyond the grasp of one's human cognitive faculties. In the literature, the word 'ineffable' is often used, marking the limitedness of language. At times, however, one also experiences the limitedness of one's mind and one's heart. Especially powerful visions may impress one as being totally beyond one's mental reach. On such occasions, one feels that what one sees is too wondrous to be grasped and one finds that one's mind simply cannot contain the marvels revealed to one. In extreme cases one may feel it is too painful to watch what the visions present—it is all just too sublime, above and beyond the realm of the human.

The special character of the semantic space of Ayahuasca visions may also be marked by way of contrast. Comparative and contrastive analyses that I have conducted indicate that the content items that appear in these visions are different from items often seen in dreams. As further described in the Appendix, I have compared the content items of my dreams to those seen in the corpus of my Ayahuasca visions. Very little overlap was found between the content items that appear in the two sets of data. None of the content items characteristic of the visions are encountered in the dreams nor are those that usually appear in the dreams found in the visions. In particular, in my dreams I have never seen tigers or serpents, palaces or scenes from ancient civilizations, royalty or other special beings. Never have I seen places that may be characterized as phantasmagoric or magical, nor have I ever seen historical or mythological scenes or ones having to do with the meaning of either the human predicament, Life and Creation, Nature, or the Cosmos. Furthermore, my dreams never had any religious or spiritual quality nor did they ever seem to convey a message or serve as a vehicle for instruction. Rather, by and large, the content items that appear in my dreams are commonplace. In them, I usually see persons who are significant in my life, scenes echoing events that happened to me in real life, images incorporating details—often insignificant, such as telephone numbers—that I have encountered during the course of the preceding day or two, scenes that reflect my personal conflicts, issues with which I am especially concerned, my fears, my wishes and desires. In general, these do not appear in the visions I have had with Ayahuasca.

Having said all this, let me also note that what may be seen in Ayahuasca visions is, in principle, unbounded—essentially, there is no limit to it. The unboundedness pertains to both the types of content seen and to the tokens (that is, specific instantiations) of these types. With respect to the types, let me cite what, on two different occasions, I have been told by two very experienced drinkers. Checking whether they had seen the various items on my structured questionnaire, these informants answered in the affirmative for all items queried. Doing this, they smiled and said 'Well, you see, I have seen everything.' With respect to the tokens, the unboundedness manifests itself in people never seeing the same vision twice. I say this both on the basis of my own experiences with Ayahuasca and the repeated observations of many other individuals; no one has ever reported a case that counters this generalization. The generalization holds even though there are many items that are especially common in Ayahuasca visions. In terms of their type, the items seen do exhibit various patterned regularities, but the tokens associated with these content types display unbounded variation. Thus, I have seen palaces and interior decorations thereof many, many times; yet, each vision of these was totally novel—not even twice did I see the same building or decoration. Coupled with the fantastic nature of the contents of the visions and their extraordinary magnificence, I find this state of affairs to be truly remarkable. In sum, all things imaginable and non-imaginable can be seen with Ayahuasca. One can see all the moments of one's life, all the people and places that one knows, Nature and the Cosmos in all their manifestations, human history and the different cultures that it has and has not produced, and scenes that lead one above the planet, to the far reaches of the cosmos, to the heavens. One can see the inner parts of one's body and the deeper strata of one's soul, one can encounter the infinite richness of myth and fantasy, meet fairies and dragons, angels and devils, taste the nectars of the Eternal, be washed by the bounty of the Supreme Good, witness the perennial light, encounter the Divine. But then, if everything and anything is possible what sense is there to list and enumerate the 'contents of Ayahuasca visions'? I shall give several answers to this question; these answers are not mutually exclusive.

The present work is based primarily on reports that might be characterized as pertaining to a first stage in one's Ayahuasca schooling. By now, I have partaken of

Ayahuasca more than 130 times and only very few of my informants have had a more extended experience with the brew. With long-term experience the quality of the visions may change. Some (but definitely not all) Ayahuasca users say that after about two years of regular exposure to the brew, they see fewer and fewer visions. In contrast, long-term drinkers report increases in experiences of great insight, spiritually uplifting ambiences of light, and special performances with the brew. I myself have noted that after about 90 sessions a new phase in my Ayahuasca experience seemed to begin. In my more recent sessions, some of the more frequent items I have seen in the earlier sessions no longer appeared; this is especially true of serpents and jaguars. On the other hand, all sorts of new items appeared. With this, the set of content items that I have seen only once greatly increases with time. How then do I square the two? I would say that the items reported here are the ones most likely to be seen in what I would characterize, indeed, as a first cycle (see Ch. 18) in the school of Ayahuasca.

Two years after these last lines were written, I met an indigenous shaman whom I asked, as I had done with so many other people, why serpents and jaguars are seen in Ayahuasca visions. He responded that these two are, indeed, very common but that this is especially the case in the early stages of drinkers' experience with Ayahuasca. 'At first Ayahuasca shows one the natural worlds, later it reveals other worlds to one,' he said (see also Richman, 1990/1).

Let me put this in other words. In principle, it does indeed seem to be the case that given sufficient exposure to the brew, anything and everything can be seen with Ayahuasca. However, not all things are seen with the same probability. The foregoing survey seems to me to present a fair depiction of the types of content that are most likely to be encountered in Ayahuasca visions. Indeed, as already noted in Ch. 3, it seems to me that, as it stands, the corpus of data I have collected is robust; correspondingly, my assessment is that the survey presented here (as well as all other general data summaries in this book) is quite stable. At this stage, whenever I hear or read new reports of people's experiences with the brew, I do not find items that I have not heard of before and which are not included in my survey. Indeed, over and over again I am struck by reports of visions that are very similar to ones I have experienced myself or which I have heard about from my informants.

I would like to close with a comparison with the domain of literature. The question 'What does one see in Ayahuasca visions?' can be compared to the question 'What can be written in literature?' or 'What can be depicted in art?' Some, especially traditional, forms of both literature and the visual arts can indeed be associated with a relatively well-defined set of content items. Fairy tales are most likely to describe kings and queens, princes and princesses, good fairies and gnomes, wicked witches, dragons, and magical objects. Greek mythology often talks of deities and various types of heroic figures. Similar lists of most frequent items may be defined for other literary genres. But there comes a point when the question is meaningless: one can write, or draw, anything one fancies. There might be items that are more likely to be written about or drawn, especially in the more realistic genres of art, but in principle, the human imagination is without limits—eventually, anything is possible.[16] The situation with Ayahuasca visions is, I think, very similar.

Two Final Examples

By way of conclusion, I present two additional examples. Unlike all other examples in this chapter, which consist of specific visualizations pertaining to specific content items, these examples are records of sessions. They cite or summarize the various visualizations experienced by one individual drinker in one setting. I bring them in order to give the reader a more direct feel of the visionary experience induced by Ayahuasca.

The first example consists of a 'real-time' verbatim report of what I saw in one session in which I partook of Ayahuasca by myself. I spoke aloud describing what I was seeing and notes were taken by the person who watched me. This session is not part of the core corpus and it is the only session of which I have such a recording. Overall, I would characterize this session as one of moderate strength. In it, there were no grand visions and most of the visualizations in it are snapshots and relatively simple scenes. Furthermore, in this session I had very few ideations and no special psychological insights or spiritual experiences whatsoever. Yet, I find this report to be especially valuable in portraying the general flavour of Ayahuasca visions. Manifest in it is a fairy tale-like ambience and an overall air of magnificence and enchantment. Also featuring in the report are several details that are characteristic of Ayahuasca visions in general—these include fire (note the various ways it is incorporated within the narrative of the vision), light-producing objects, carriages, and processions.[17] Also recurrent in the report are turning movements, upward movements, and looking forward far into the distance. One comment made by a person seen in this visual sequence is a good example of how ideas relate to Ayahuasca visualizations. The entire sequence lasted about forty minutes:

A golden crystal chalice.

Flowers. In the flowers there are birds and insects and the birds go up and up.

A wheel is turning and there is a rod that is turning round and round. From it, a fire ignites. An old man holds a taper and from it the fire climbs up and up.

A futuristic city.

A Chinese king is sitting and turning his parasol. Now he is in his study. In the background, birds are kissing one another.

A great hall—like an animated movie.

[16] Let me reiterate: this statement should not be confused with the one made throughout this chapter marking the special characteristics of the contents of Ayahuasca visons. While, *in principle*, there is no limitation on what *might be* seen in the visions *de facto*, some contents (notably, ones pertaining to everyday life) are less likely to be seen than others.

[17] As explained above, fire and light-producing objects are not mentioned in this chapter; they, along with various other items, are discussed in Ch. 17, which is devoted to light.

There is a code here—like that of Morse or the genetic code. The code is constituted by many, many dots, the density between which varies. All this is a language calling to be deciphered.

There is something that pushes up and up. It is like a mountain train. All the time it goes up and up.

A car from the 1920s. Delightfully magnificent. From it emerge light and flowers. Advancing with this light, we pass along gold-plated walls and come out through a staircase made out of gold and ivory. The steps go up and down and reach a theatre.

Up in the heavens there is a woman escorted by a man. In the woman's hand there is a torch that swirls. Lights come out of it in the form of flags and the flags turn into hats full of gems. The gems are sparkling.

A scene in Europe in the sixteenth or perhaps the eighteenth century. Knights are riding. They are mounted upon magic motorcycles full of colours and light. All is like a cartoon and enchanted. It is all part of big procession. There are also small dwarfs there. Two of them are holding a banner with the insignia of the sovereign.

An Indian is smoking a big pipe. Through an old telescope, a man is peering into the far reaches of the universe. A view of the planet Earth turning round and round.

Beautiful gardens like Versailles and the Tuilleries.

There are ballerinas there. Like a cabaret. Their thighs are exposed. One woman gets to the balustrade and is watching the audience.

The Indian is smiling. The message is that 'all of these are the expressions of the same source, a source of bounty and grace'.

In a King's reception hall. There are chalices full of wine. Long processions of carriages proceed further and further. Slowly, all the time, the horsemen are pushing forward. In the hall, the seats are made out of silver. There is a feast. A big pot is placed in the middle. A fruit salad is offered in goblets of finely polished, very clear glass. Slowly, the chef pours some sort of syrup or gooey topping. The sauce covers the fruit and then it ascends upwards.

An elephant lifts up its trunk high and looks far, far forward. Up there are birds and they are looking in my direction. There are flowers, and butterflies are flying from flower to flower. All are washed in the light of the sun.

Women are dancing. Carriages come one after the other and the wine flows. An officer approaches a carriage and salutes. The footman bows and opens the carriage's door. The Queen is stepping out.

The second example was provided by a young man who partook of Ayahuasca in private sessions conducted in Europe. This individual is not amongst the informants whose data are analysed in this book; his report was communicated to me just when the typeset of this monograph was being sent to the publisher. I present this report as an illustrative example of the experiences of a first-timer. The following is a slightly edited synopsis of what this person saw during his first two sessions with the brew. I shall note that while the report is rich in details, the intoxication experienced was not especially strong; by the present structural typology, all items seen would be characterized as single, simple images.

Animals. Those seen most frequently were serpents, felines, and birds. Some of the serpents were ornate, like Chinese dragons; the felines included tigers and black pumas; the birds included parrots, peacocks, and toucans. Also seen were a galloping horse, dragons, monsters of all sorts, and evil beasts; with some of the latter blood was associated.

Many human persons were seen. Amongst these were Indians and a sensuous Caribbean dancer. A person present in the session appeared to have the face of a gorilla with the beak of a bird.

Palaces and mansions. Amongst the buildings seen were skyscrapers and pyramids. Also seen were interior decorations of buildings. These were very exquisitely ornamented; many were gilded.

Cities. Many different ones were seen; some had futuristic architecture.

Landscapes. These included forests, open deserts, river scenes, and scenes under water. Associated with the latter were corals and 'tornadoes of fish'. Overall, the landscapes had an ambience of serenity and silence.

Especially frequent were disembodied eyes; many of these pertained to big cats. Other items noted: an Indian in a boat, an old woman turning white and transforming into a young girl, cars of the 1950s that were colourfully painted in a style which was 'rather kitsch', streams of gold.

The trees outside looked like goddesses.

Significantly, the items reported by this informant include all those that are typical of Ayahuasca visions. As such, this report, I find, is a good example supporting the cultural non-specificity of these visions. For further comparison, the reader is referred to the quantitative analyses in the Appendix.

9

The Themes of Visions

Hallelujah.
Praise the Lord, O you who are in heaven,
praise Him, O you who are in the heights.
Praise Him, all his angels
praise Him, all His hosts.
Praise Him, sun and moon,
praise Him, all bright stars.

Psalms 148: 1–3

Sun, Moon, Star
The earth, the wind and the sea
This is the light of the firmament
That which I have to love.

Mestre Raimundu Irineu Serra

In the previous chapter I analysed the contents of Ayahuasca visions in terms of the specific items that appear in them. In doing this, I have followed what is standardly done in the anthropological literature when the issue of content is addressed. In general, what are noted are the types of things that drinkers see under the intoxication. Such an analysis pertains to the *micro*-level, but there is also a *macro*-level. Whereas the micro-level is concerned with content elements, the macro-level is concerned with themes. A thematic analysis focuses not on the details, but on the more substantive global facets of their meaning. By way of clarifying this contrast, let me again use an analogy from the domain of literature. What are the contents of Shakespeare's tragedies? In order to answer this question, the scholar may subject the texts of the Shakespearean tragedies to a statistical analysis and count the items that appear in them. Items likely to appear high on such a count are kings and queens, military chiefs, palaces, daggers, and so on and so forth. But lists of such items will not give us an answer to the question 'What are the tragedies about?' The categories that are pertinent to this question might be, 'the devastating effects of power and greed' and 'parent–child relationships'. Whereas the first, content-item analysis is technical, the second, thematic one is substantive. Obviously, it is the second, not the first, which is the more meaningful one.

In the previous chapter I focused on the first kind of analysis, the technical one concerned with items of content. This was done for several reasons. The first is methodological: the domain of Ayahuasca visions is so different from that of ordinary perception and cognition that a detailed mapping of the terrain is a prerequisite to any further, more substantive topical analysis. The second reason is practical: many (and for many people, most) Ayahuasca visions do not in fact permit a thematic analysis—these visions are simply too scant, they consist of brief, isolated snaps which can be identified in terms of their denotation (i.e. this is a person) but with which no broader or deeper meaning can be associated. In terms of the present structural typology, only full-fledged scenes—and at times, serial images—allow for a substantive thematic analysis. Thirdly, the analysis of content by means of items is necessary as the basis for the study of cross-personal commonalities and possible universals. Indeed, with regard to these topics, a reversal of import is noted so that commonalities on the level of content items are more significant than those pertaining to themes. Specifically, that all people are concerned with issues such as happiness, health, and death is by far more expected—hence, less surprising and less remarkable—than that their visions contain items such as serpents, pumas, crystal palaces, and nymphs. Fourthly, specific, localized items of content are more amenable to systematic statistical analyses that are necessary for comparisons across individuals and groups. Indeed, such an analysis of content items is conducted in comparable investigations of dreams (see Hall and Van de Castle, 1966).

In line with what has been said above, the following analyses are based only on those visions which, in terms of the present structural typology, qualify as full-fledged scenes. In fact, visions imparting insight and learning often pertain to the special subset of scenes that I characterize as grand scenes. It will be appreciated that visions of this kind are not frequent, and some persons never experience them at all. However, when such visions are experienced, the themes associated with them seem to be shared by many persons. Indeed, the most common themes associated with the visions appear to pertain to a relatively small number of clusters.

While in the following analysis I focus on visual scenes, there is no principled reason for the thematic analysis to be confined to the visual modality. The present focus on the visual is grounded not in conceptual considerations, but in factual ones: practically all the major Ayahuasca events in my data—both drawn from my own personal experiences and those I have heard reported from my informants—pertain to the visual modality. Apparently, this reflects the general primacy of vision in human cognition. There are, however, many very meaningful Ayahuasca events that lack any sensorial-perceptual quality. These pertain to the topic of ideas, one with which we shall be concerned in the next chapter.

Indeed, in a certain sense, the thematic analysis of visions occupies an intermediate place: focusing on the visual, some of its aspects are related to the analysis of content items presented in the previous chapter; being thematic, some of its aspects are related to the analysis of ideas, reflections, and insights to be presented in the next chapter. For this reason, there might actually be repetition and overlap between the present thematic analysis and the two other analyses indicated—that of content items and that of ideas. While I fully acknowledge this, taking into account considerations of completeness and global coherence, I find it unavoidable. Let me also point out that given that the thematic analysis is concerned with more global, higher-order phenomena, the examples of visions I present in this chapter are less fragmentary and more complete; these will also serve by way of giving the reader further appreciation of the possible richness of the Ayahuasca experience.

1. *The Domain of the Human*

Psychological Understanding

The most basic themes of Ayahuasca visions are those concerned with people's understanding of their own personality and life. I shall comment on the phenomenon of psychological insight in the next chapter; here I focus on the visual aspects of the visions in which such insights are gained.

Indeed, the visual experience often plays a crucial role in the process of gaining psychological understanding. One sees oneself as a child, one relives conflictual situations, one sees people who are especially meaningful to one and gains a new perspective on them and on one's relationship with them. As already indicated in Ch. 6, metaphorical parables seem to be especially linked with personal psychological insights. Also instructive are serial visions in which different biographical scenes are brought together. The concatenation may be chronological, thematic, or seemingly unordered. In the thematic concatenation all the scenes experienced are concerned with one particular aspect of the subject's psychological make-up or interpersonal conduct. What may at first blush appear to be unordered concatenations may turn out to be especially revealing. The juxtaposition of scenes that seem unrelated may reveal to the Ayahuasca drinker connections and lawful patterns in his or her personality of which he or she had not been previously aware.

Visions may also involve instructions and guidelines as to how the drinker should conduct his or her life. Several informants told me about visions in which they were confronted with two possibilities between which they had to choose. In all cases, the choice was conceived as being between Good and Evil. For example, in one such vision, one informant saw a black panther on one side and a wise man on the other. He understood that he had to choose between the two, and that in his life he should follow the course that is good.

Other visions are concerned with the lives of other individuals and offer the drinker a better psychological understanding of other persons. These may be persons one has met and/or personally known during the course of one's life, famous historical figures, or types of personae (e.g. persons having particular professions or life persuasions). By way of illustration, let me present two episodes from my own experience; these exemplify the second and third kinds just noted, respectively.

The hero of the first episode was Hernan Cortes, the conqueror and destroyer of the Aztec empire. Even before I was exposed to Ayahuasca I was especially touched by the enormously tragic fate of the high pre-Columbian American civilizations. For years, I have felt that the only historical crime comparable to that inflicted on the Jews during the Second World War is the destruction of these civilizations by the Spaniards. Thus, the basic attitude I always had towards Cortes was one of reproach and contempt. And then, once I had a vision depicting (what to me seemed to be) the encounter of Cortes with Moctezuma, the Aztec emperor; these are the thoughts and insights that passed through my mind as I was inspecting the vision:

The Aztecs believed that the white man was a god, and Cortes knew that they believed this. Thus, he was determined to act like a god. With a small retinue of escorts, he entered the heart of a powerful, militarily well-trained, empire. At that stage of the encounter, the Aztecs could have easily annihilated Cortes and all his companions. Surely, had the Aztecs discovered that he was a mere human, they would have tortured and sacrificed him. But, to use the colloquial expression, Cortes 'played it cool'—he maintained the act and presented himself as if he were a supernatural being. The slightest faltering on his part and he would have met a cruel end, but he went on with the act. I inspected the vision, and I was full of admiration. No, I did not condone Cortes' conduct, and my ethical stance towards him did not change, but I felt I gained some understanding of this person and appreciated how unusual a character he was. To play God one has to feel, in a fashion, like God. Taking such an attitude is far from easy, for it imposes so many ramifications both on one's interpersonal interactions and on one's self-comportment. The only other person in the Western world who has done this was Jesus, and indeed, he paid with his life for the enterprise he engaged in.

The figure presented in the second episode had no name, he was an opera singer:

The man was at the end of a grand performance, acknowledging the applauding audience. He was standing at the front of the stage, bowing. In front of him was the large, wonderfully ornamented hall. The singer saw the five or six levels of balconies, he inspected the multitude of people, none of whom he could discern individually, and again he bowed. An enthralling sense of accomplishment engulfed his entire being. Alone he stood there in front of hundreds and hundreds of persons and they were all treating him as an idol. They were thanking him for the enrichment that he had brought into their lives. But at the same time, I realized as I was inspecting this grand vision, it was, in fact, the audience who sustained the singer. In essence, the singer was a frail human being who needed others so badly to confirm himself as a person. Actually, he was a shy individual. So shy, that the only way for him to overcome his shyness was to present himself alone in the open in front of the multitudes of strangers. The only way for him to escape the criticism of other persons of whom he was so afraid was to excel and win their praise.

Human Life

Under the heading 'human life' I include visions that deal with issues pertaining to the life of human beings, but not specifically to that of the drinker or any other particular individual. Pertaining to this class are the themes of birth, death, and rebirth, the masculine and the feminine, as well as the meaning and values of human life. More specific consideration of these themes follows.

Death and Birth As noted earlier, the theme of death is one of the most prominent in Ayahuasca visions. It covers self-death, death of specific individuals and death in general, wars and other major calamities, bloodshed and human sacrifice, spirits and reincarnation. Significantly, the very first time they had visions, quite a few of my informants saw family members who had passed away. In all cases, there was something unusual about the death—it was caused by an accident, the deceased was young, or (as exemplified by a vision cited in the previous chapter) the informant could not attend the funeral and regretted it throughout the many years that had since passed.

An especially important theme is that of self-death and rebirth. In visions of this kind the person who has partaken of Ayahuasca feels that he or she is going to die, or actually that he or she does undergo death. Either just before this happens, or perhaps when this does happen, this person experiences rebirth. The experience is usually exhilarating, and those who undergo it equate it with salvation. Following are two examples of grand visions whose theme is a process of death and rebirth that leads to salvation. The first was reported to me by a South American psychologist who partook of Ayahuasca only once, in a session conducted in a private home; the second was reported by a European woman partaking of the brew with a traditional healer in the Peruvian Amazon:

> The scene was that of a grand Aztec ceremony. Human beings were being sacrificed and I was one of those thus designated. I first considered resisting, but soon I realized that this would be of no avail. Thus, I altered my attitude and willingly let the priests in charge carry me to the top of the pyramid. There I was placed on the altar and the priests were above me, to take my heart. Without any coercion, I was offering myself. With this, my entire feeling changed radically. From a terrifying scenario of torture and death it all turned into a wonderful process of rebirth and salvation.

> The scene was that of a religious Egyptian ritual in the course of which I saw myself being sacrificed; a priest bent over me with a knife and cut my breast. At first I thought to resist, but then I gave up. Subsequently, I had an out-of-body experience in which I inspected myself from above, and saw my heart being taken out. In its place was the sun, which was giving light and life to everything. Realizing this, and in order not to blind the other people present, I fainted.

Interestingly, a very similar report is given in Wolf (1992), a book that my informants had not heard of. Finally, there are the experiences of being born and of giving birth; both were reported to me by informants.

The Masculine and the Feminine In many traditions and schools, both esoteric and not, the masculine and the feminine are regarded as the two types of energy involved both in nature and in human psychological life. This is especially marked in Jungian psychology, where the masculine and the feminine are considered to be two constituents in the psyche of every individual (the terms used are *animus* and *anima*; for a general discussion see Hillman, 1985). As indicated at the beginning of this book, Ayahuasca, with its two constituent plants, also has a masculine and a feminine facet. Several informants reported to me seeing the interplay between these two facets in their visions. Several informants (all women) indicated that this theme was a major one in their experiences with Ayahuasca. The contrast and interplay between the masculine and the feminine is also a major theme in indigenous Amerindian lore concerning Ayahuasca (see Ramírez and Pinzon Castaño, 1992).

The Meaning of Life and Human Values Many visions reveal to drinkers what they interpret as the meaning of human life. Lessons in morals and instructions about the proper conduct of life are also very common.

The Human Predicament

By the label 'the human predicament' I refer to themes pertaining to the life experience of human beings as members in a collective group, that of the species *Homo sapiens*. These include the themes of human suffering, human history and its meaning, the relationship between the human species and the natural world—both physical and biological—human culture, and the place of human existence in the grand framework of Being.

As recounted in the Prologue, I experienced a vision of this kind in my first powerful session with the brew, which took place in a *feitio*, the ritual in which the Daime is prepared. In that session, recall, I had a grand vision in which different scenes of human history unfolded before me. These included a scene of slaves in the Roman period, the conquest of America by the Spaniards, and the Jewish Holocaust in the Second World War. All these scenes depicted extreme cruelty and injustice. Yet, at the same time they all pointed to a different perspective of human existence, the spiritual. Together, they carried a message of hope.

Remarkably, two other individuals, both independent drinkers, recounted to me that their very first powerful Ayahuasca visions consisted of grand scenes concerned with the meaning of human history. The first informant saw a panoramic view of human suffering—misery, hunger, war, death. But then, he saw light and he experienced the mercy of the Divine. With this, he understood that despite all the terrible afflictions there is hope, and human beings can attain illumination and grace. The second informant saw a grand scene of war—it is recounted in the section on history in the next chapter. Note that the moral reached by the two informants in their two separate visions is almost identical.

A theme associated with some of the most powerful visions I have had was that of the human predicament as a dialectics between nature and culture. On the one hand, human beings are animals and their fate is bound with that of nature. On the other hand, human existence is embedded in culture. From the very first sessions I have had with Ayahuasca, the tension between these two poles was apparent. Later, I discovered that this theme is also pivotal in many indigenous stories related to Ayahuasca (in particular, see the story on the origin of Ayahuasca and its analysis in Reichel-Dolmatoff, 1975, 1990). On the one hand, the brew, being made of plants, epitomizes nature. On the other hand, Amerindian societies regard the brew as that source from which they have obtained their language and music, religious beliefs and myths, laws of marriage and societal regulation—in other words, their culture. Indeed, in Amerindian mythologies, the discovery of Ayahuasca is usually conceived of as a unique, momentous point in human history: before it people were privy to a paradisiacal existence in which all their needs were provided by nature and where amiable relationships, including mutual communication, with the animals were maintained. With the discovery of the brew, this organic and harmonious tie with nature was severed but at the same time culture was born.

Health

The topic of health is especially prominent in Ayahuasca visions; this is the case both in reports found in the anthropological literature and in the data I have collected myself. For this reason, I discuss it separately. Specific themes pertaining to this domain include the body, its mode of functioning, and the manner in which one should take care of it, health and disease, healing and self-healing. As indicated in previous chapters in this book, shamans repeatedly report that in taking Ayahuasca they are taught how to heal (see Siskind, 1973*a*; Payaguaje, 1983; Chaumeil, 1983; Luna, 1984*a*, 1986*b*; Gebhart-Sayer, 1986; and Langdon, 1992). Similar claims were made by several of my non-indigenous informants, especially those associated with the health and helping professions.

Following are two examples taken from my data; the first was reported to me by an urban Brazilian with extended experience with Ayahuasca, the second is my own. My informant's vision took place after he had (actually) been stung by two wasps. The man was in agony and was served with Ayahuasca. In this vision:

> I saw two beautiful women who sucked the wasp poison out of the two sides of my nose. They passed it to my mouth and told me to spit. I did this and was, in reality, cured.

My vision took place when I was afflicted with malaria:

> I saw two snakes wrapped around each other. Seeing that, I realized that health and sickness are the two facets of one complex. Specifically, I reflected that if one engages with the snakes in one direction one is afflicted with disease, whereas in the other direction health and well-being are gained. Indeed, both illness and good health are energy. The symptoms of the body—that is, the illness—are pertinent to the process of cure. First, they are signs that the body sends to one and thus, they have important informative value. Second, if they are channelled appropriately, they can direct the organism to a new state of well-being. The key to healing is rooted in grasping the energy of the illness and transforming it into the energy of healing. It further occurred to me that this is analogous to the taking, in physics, of potential energy and transforming it into kinetic energy.

With this realization, I made the most significant step towards being cured from my illness.

On another occasion in which I was not feeling well a helping being appeared and presented me with a vial containing a soothing balm. I have heard of similar reports from other persons. On still another occasion, I felt I was subject to a magic operation. Several figures surrounded me and a very fine veil was spread over me and I let myself be carried into the realms of enchantment. A very similar vision was recounted to me by one informant.

As indicated earlier, healing is a major theme in the visions of traditional *ayahuasqueros*. Under the intoxication, the *ayahuasquero* receives instruction how to treat his patients and power to heal is invested in him. I felt this once when I treated a patient myself:

> As is common in such practices, I placed my hands on my patient's head and blew on it. As I was doing this I saw myself at the head of a long line of forest animals. The animals formed a long triangular column and one to one were passing energy and support forward. Eventually it all converged in my hands and breath giving me the power to heal. (I might add that the treatment of this patient was successful.)

Related to healing are the topics of human anatomy and physiology. As indicated in the previous chapter, it is not uncommon for drinkers to experience travel inside their own bodies. Typically, such experiences reveal information about the structure and function of the body. Very interesting accounts in this regard are those presented in Reichel-Dolmatoff (1981) in which theories the Desana Indians espouse about brain structure and functional specialization are described. These theories are highly sophisticated. Central to them is the notion of lateralization—that is, the functional differentiation of the two cortical hemispheres. Details may be different, but this notion is, indeed, central in modern neuropsychology as well.

2. *Nature*

The various items pertaining to nature that are revealed in Ayahuasca visions have been surveyed in the previous chapter. Here, let me take a broader perspective and focus on the major themes that may be the subject of such visions.

Flora and Fauna

In the survey of content items in the previous chapter I noted that animals are extremely frequent in Ayahuasca visions. Thematically, visions in which animals feature centrally may be concerned with the nature of animal intelligence and animal life and they may bestow upon the subject new understanding of either the process of life or of animal behaviour. Visions that impressed me greatly included ones in which I appreciated the collective, non-individuated, identity of ants, one in which I found myself inside an elephant's trunk and sensed how it felt to be an elephant, and various visions in which I felt I was undergoing animal transformation; more on this will be said in Ch. 13. Related to this is the feeling—very common with Ayahuasca—that one can understand real animals and communicate with them.

The forest and its mysteries also feature as themes of visions. Especially salient amongst these are those about Ayahuasca and its constituent plants.

The Majesty of Nature

While not having a specific thematic content as such, many visions of nature impress viewers in their portrayal of its majesty. A common feature in these visions is the viewing of all of reality as a great, wonderfully harmonious symphony. Such visions can have a deep impact upon their viewers. Typically, these visions induce a feeling of awe that is coupled with a new, more humbling, perspective by which people better measure their own value and place in the universe. These usually induce sentiments of profound appreciation and gratitude for all that nature offers. Panoramic scenes of open landscapes are especially conducive in this regard. The viewer looks forward, to the distant horizon, and in a direct, non-mediated manner, senses how immense and eternal the world is. I have experienced this many times and have heard similar observations by many informants, including ones with no prior experience with Ayahuasca.

Experiences of the kind described here are often coupled with religious sentiments and ideations. Drinkers feel that the whole of Existence is God's wondrous doing and that it is all there by virtue of its being permanently sustained by the Divine. Furthermore, drinkers feel that in its totality nature recounts God's glory, and the bounty and harmony of nature are interpreted as indicating that the whole of Existence is the manifestation of God's benevolence and grace. It will be noted that such feelings and reflections often come about without drinkers' having any visions: They may occur when, under the intoxication, people inspect natural scenes in the real world around them. More on this will be said below in conjunction with visions pertaining to the Divine, as well as in the next chapter.

The Basic Forces of the Physical World

Several informants told me that with Ayahuasca they encountered the basic forces, the elements that reign over nature. With this, they gained insights as to how the physical world functions. A vision of my own that I can give by way of example is the following:

> I saw the sun and appreciated the support and protection it bestows upon the planet Earth and its inhabitants. At one point I ventured to move upwards only to realize that I should know my limits. The sun was the provider of life, but, at the same time I realized that a distance from it should be maintained. Very clearly I sensed that if the delicate balance is not respected, the nurturing sun turns into a terrible destroyer.

Very similar visions were reported to me by two of my informants.

On another occasion, I felt I was embodying the primitive forces that were involved in the primordial waters in the earliest days of life on our planet. My throat was completely dry. Clinically, I was probably dehydrated and (as further described in Ch. 13), out of my mouth came sounds that, I felt, were not mine. They seemed to emerge from the cracks at the bottom of my digestive system exhausting the faint traces of humidity that could be dredged out from their dryness. On several other occasions, I felt I was embodying the wind. One informant, an experienced member of the UdV, told me that the most impressive vision he had with Ayahuasca was one in which he saw the forces of Earth, both constructive and destructive. This vision made him appreciate that creation and destruction are two interlocking facets, both essential and necessary to the workings of the universe.

Anima Mundi *and the Life Force*

Even more common are visions that reveal what is felt[1] to be the *anima mundi*—the cosmic energy that permeates all Existence and sustains everything that is. As noted in earlier chapters, this is often associated with the seeing of webs of translucent fibres that embrace the whole of Existence.

The Mysteries of the Universe

In the indigenous Ayahuasca lore it is said that the brew takes those who partake of it to regions where they encounter beings that teach them the mysteries of the universe. I have experienced—or rather, almost experienced—this once. It happened when I was taking the brew in my home, with just one friend who did not partake of the brew but served as a guard to watch over me:

> I found myself engulfed in infinite blue. [Later I referred to it as 'the blue place'.] There were beings there. I did not see them but I had communication with them. They offered to reveal the mysteries of the universe to me. There was no question about it, they were benevolent and their offer was genuine and sincere. However, there was a condition involved with it—a payment on my part was to be made. I had to relinquish any further contact with this world. In other words, I would never return. I opened my eyes and I looked around. I saw my living room, my piano, my friend who was supposed to watch over me but who was tucked up in the large armchair sound asleep. I thought of my family and friends, my teaching and writing. I looked through the large window and saw the trees outside. I thought of my sanity. No, I did not want to lose all these! Nor, I reflected, did I wish to lose my regular self, the way I am, the way I think and feel. I sat up straight and spontaneously got my hands moving and energetically slapped my lap. Again and again I slapped so as to break myself free from the spell. Thus, I had forsaken the opportunity to learn the mysteries of the universe.

Afterwards I regretted my decision. Later, I reflected a lot on this episode and have drawn many lessons from it. I shall not dwell further on them here.

[1] Expressly, I say 'felt' and not 'interpreted' or 'taken to be'.

One of my most experienced informants told me that some of the most impressive Ayahuasca sessions he has had involved his being transferred to other realms which apparently pertained to a different dimension. In these realms, the laws of physics were different from those that hold true here in this world. He encountered there beings totally different from those inhabiting planet Earth. They were flat and they proposed that he stay with them and learn the mysteries of the universe. The requirement was, however, that he lose his materiality. My friend declined and opted to return to this world. Incidentally, this informant too mentioned a blue place in his report. The vision with the mathematical formulae reported by a South American man and mentioned in the previous chapter is another example of a vision whose theme was the mysteries of the universe.

Creation and Evolution

As indicated earlier in the book, some of the most spectacular visions induced by Ayahuasca are those having to do with the creation of the world and the evolution of life. Commentary and examples were given in the previous chapter.

3. *The World of Culture*

As noted in the previous chapter, various manifestations of culture appear very frequently in Ayahuasca visions. Usually, the visions are depictions without any particular theme or message associated with them. However, as detailed below, at times, the visions do convey well-defined themes.

The Splendour of Culture

While not being associated with specific themes or messages, grand visions in which cultural creations are depicted may invest their viewers with a feeling of great fascination. As such, they may invoke a theme that may be referred to as 'the splendour of culture'. Often, such visions reinforced in me the appreciation of how extraordinary the story of human history and civilization is, and they have imparted to me a deep admiration for the wonderful accomplishments of the human race. Throughout its history, *Homo sapiens* has spread war and oppression, and human beings have generated so much evil and injustice; yet, from the midst of this very race also came such geniuses as Mozart and Bach.

Royalty

As noted in the previous chapter, royalty features prominently in many Ayahuasca visions. Visually, such visions are impressive in their own right—typically, they exhibit great beauty, opulence, and magnificence. They may also be instructive with respect to the themes on which they focus. Ones that I have noted in my own visions include the nature of kingly government, the relationship between monarchs and their subjects, the difficulties involved in being a king, and the values of the system of monarchy. With respect to the last topic let me mention one vision I had of ancient Egypt. The vision was with open eyes:

> The *maloca* in which I was sitting opened up and instead of the Amazonian forest, the landscape that revealed itself to me was the shores of the Nile. There I saw people of various persuasions—farmers, fishermen, government officials; all were engaged in their daily work, and all, it seemed to me, were happy. I attributed this to each person having a well-defined place in the well-ordered social hierarchy. Rather than aspiring to move upwards in society, each person found gratification in doing his job in the best possible way. This gave people a feeling of basic, solid security, a sense of meaningful existence, comfort, and contentment. The condition for that, I saw, was the deep conviction shared by all that the entire system is based on justice and fairness.

A couple of years later, I read some books about ancient Egypt and discovered that this conception is in line with the Egyptian notion of *Ma'at*, the divine order in which justice, truth, and harmony are encompassed (see Pinch, 1994).[2]

Artistic and Cultural Creation

Parallel to the creation of the universe in the domain of nature is human creation in the various realms of the arts. As we have seen in the survey of content items, works of art appear very often in Ayahuasca visions. Here I note that there are also visions whose main topic is the nature of art and the process of artistic creation. Thus, several informants—notably those who are themselves artists—recounted to me visions in which famous historical figures, renowned for their contribution to human culture, were seen as they engaged in artistic or intellectual creation. I myself had a vision of Moses writing out the Law. In this vision, pure white light laced with emerald green hovered upon the Hebrew prophet as confidently his hand flowed smoothly without pause. Manifestly, it was an act of supreme joy. Literally, he was inspired.

Especially impressive are visions that may be likened to visits to galleries or museums. As already mentioned in the previous chapter, once I found myself shown an exhibition displaying the works of an entire culture. Various works of art and artefacts were displayed. They were all of a style that resembled nothing that I had ever seen before in my entire life. What was striking was that the different objects on display all defined one coherent style. Similar visions were reported to me by several other persons.

Music

As explained in Ch. 1 (and further discussed in Ch. 19), music is central in all contexts of Ayahuasca use. At times, music itself can be the theme of one's visions. In Ch. 11, I describe one of my first powerful experiences with Ayahuasca, in

[2] Interestingly, these three, as well as love, are the four key constituents of the doctrine of the Santo Daime Church. Apparently, this reflects the involvement of Mestre Irineu Serra (the founder of the Church of Santo Daime) in the esoteric school *Círculo Esotérico da Comunhão do Pensamento* (see Sena Araújo, 1999).

which I sang accompanied by a grand choir of angels. On other occasions, my visions were associated with new comprehensions of specific musical compositions and with insights regarding the phenomenon of music, the relationship between music and the human psyche, and the process of musical creation. Here is the story of one of these visions:

High above, in a superior realm, immersed in nebulous transparency, subtle energy was vibrating. It was fresh, delicate, pristine. Carried along the waves the energy was creating and in perfect tune to its movement and rhythm, a fair maiden (now as I am writing this, I think of the woman in Botticelli's *Primavera*) was circling round and round. She was clad in very fine chiffon and her smooth, gentle movements flowed with delightful calm. She was floating in thin air, her feet bare, and her arms spread in an open, yet not too wide, inviting embrace. It all exhibited perfect serenity, deep self-assurance, and gay benevolence. The fine waves of energy washed over the woman, and like pearls of sparkling dew they scintillated with mellow pastels of celestial light blue and faint pink and the whitish green of buds that are just about to bloom. Through her dance the energy magnified and relayed ahead further and further. Like rain, it showered down and watered all creation. As they passed, the undulations gained ever more substance and, lo and behold, turned into music. It was music so sublime that no mortal could hear. By now, I realized, they had well-defined shape and form, so that they were so less abstract than the primordial energy from which they had originated. Yet, surely, they were still more abstract than pictures, more abstract than any sign or expression, more abstract than any word would be. This is, I comprehended, why music can exercise this unique, enchanting [*encantamento* (enchantment), *canto* (song), *cantare* (sing), I associated in Portuguese] power that it has on the human psyche. Down there on Earth, I knew, there were men and women and children who, unbeknownst to themselves, were drawing their hearts, like the heads of flowers towards the sun, so that they could be nourished and lovingly nurtured by the music of the spheres.

On another occasion, I had a vision in which I saw how the music of the spheres governs the motion of the planets. In a private conversation, Pablo Amaringo told me of a similar vision he had had.

Mythology

As noted earlier, in the indigenous Amerindian cultural context, Ayahuasca and mythology are intimately linked. The Amazonian tribes have many mythological narratives whose theme is the origin of Ayahuasca. Further, Ayahuasca is often regarded as the vehicle by which myths—along with other facets of the cultures at hand—were created. Similarly, much of the teaching imparted in UdV sessions consists of the telling of myth-like stories. Personally, I too have seen what may be characterized as mythological narratives. Specifically, I have had visions of the creation of the world, ones about the relationship of human beings with both the snake and the jaguar, the creation of fire, the beginning of cultures. On several occasions I have witnessed events recounted in Amerindian myths. In particular, I have seen a jaguar and realized that in order to reach the divine I must first identify with the animal. Later I encountered this motif in Viveiros de Castro's book on the Ararte (Viveiros de Castro, 1992).

4. *The Supra-Human Realms and the Religious Dimension*

The Supra-Human

As we have seen in the content analysis in the previous chapter, Ayahuasca often leads its drinkers to realms that transcend ordinary human existence. In these, other worlds are revealed and non-human creatures as well as divine and semi-divine beings may be encountered. Seeing what they present, drinkers feel they are privileged to inspect worlds or dimensions that in general mortals are not privy to and that they gain an understanding of the ultimate meaning of all existence and of the way the universe is governed. Invariably, scenes of the supra-human realms are glorious and often they are associated with a feeling of ineffability.

The Divine

Visions whose theme is the Divine are quite common. Thematically, they may be divided into two main clusters. The first cluster comprises visions that reveal the relationship between the Divine and the universe. These encompass the relationship between the Divine and nature, on the one hand, and that between God and human beings, on the other hand. The most notable facets of the first cluster are those of creation and sustenance, bounty and vitality, grace, love, and justice. The second cluster is of visions focusing on the converse relationship, that between human beings and the Divine. The themes of these are faith, praise, and the recognition of the sacred. Often, such visions depict rites and rituals in different civilizations. The ones most commonly reported are ones pertaining to ancient Egypt, the pre-Columbian civilizations of America, and primitive, prehistoric times.

Especially powerful are visions characterized by those who experience them as encounters with the Divine. Such visions are also reported by indigenous persons. For the Shuar, for instance, a major reason to partake of Ayahuasca is to be in touch with *Arutam*, the energies of the ancestors and the force that sustains the world (see Fericgla, 1998; Mader, 1999). The following is an account of an encounter with God recounted by a Secoya shaman (Payaguaje, 1983: 80–1); the text is my own free translation from the Spanish:

The angels carried me up there to the house of God. He welcomes each person speaking in the language of the newcomer. God lives with his spouse—he has a room, that is half of the house, the other half is for his wife where she prepares food for her husband. This is Ñañë, the God who combats with thunder, there is no other God that can exist. . . . Afterwards, the angels say to me: 'Look, son, watch God.' Obviously, I do not approach near, for where God is one cannot come near, not as the angels and the people of the sky who live there. But I heard as God addressed me: 'You shall be a healer, you shall have the power to heal; for this, you should love people and do the good, never the bad.' Thus God spoke and then: 'Take this so that you will be able to cure whatever person.' And he gave me a small package of rocks of salt. The angels brought this, repeating: 'Do not approach.' But then they led me by the hand and I saw God sitting on a throne—this was not exactly a chair. His garments were white, resplendent—how shall I explain this?—the house looked all illuminated, not in the manner of the house of the white people, but in a fashion whereby each thing shone by itself. This God rose up, touched my arm and said: 'Spit.' I spat and the saliva sounded on the ground like music. Also his footsteps had marvellous sounds and I remained paralysed, admiring this God of the vision. Soon afterwards I came to hear: 'We are at our home, and this is the time for you to return to yours.'[3]

Like this indigenous man, three of my informants reported having seen 'The house of God'. All spoke of palaces of indescribable beauty which were made of crystals and/or of light.

Encounters with the Divine are usually accompanied by various insights regarding God and his relationships with both nature and humanity. Several informants underlined their appreciation that the Divine exhibits a great sense of humour. With this, things are seen in a perspective different from the regular human one which is conceived to be small and limited in comparison.

As noted above, often the experience of divine presence is associated with the appreciation of the bounty of life and the harmony of the universe. With this appreciation, drinkers often report that Ayahuasca makes them recognize that there is a force of energy that governs the world and everything in it. Many identify this force with God. A pertinent vision of mine is the following:

Animals were running in the forest. In an unmediated fashion, I felt myself seeing that the animals were actually moved by a fundamental life force. It was further apparent to me that this same force is also the fountain of life, the source of health, and the key to all wisdom. This, I reflected with conviction, is why this energy is the object of human prayer and longing, as well as gratitude and praise.

More on the religious and spiritual experiences associated with Ayahuasca will be said in Ch. 16.

Hallelujah

Especially powerful are visions that I choose to refer to by the Hebrew term *Hallelujah* (literally, let us praise the Lord). In these the entire cosmos is seen as one great symphony of praise and glory. On this, I would like to mention a vision, the first time I participated in a large festive Daime session. The vision consisted of a grand scene whose hero was Zedekiah, the last King of Judaea who, in 586 BC was

[3] Reading this report, I could not help comparing it to that found in the *First Book of Enoch* (Odeberg, 1973), an apocryphal text of the first century AD: 'And I observed and saw . . . a lofty throne—its appearance was like crystal and its wheel like the shining sun . . . It was difficult to look at it. And the Great Glory was sitting upon it—as for his gown, which was shining more brightly than the sun, it was whiter than snow. None of the angels was able to come in and see the face of the Excellent and Glorious One; and no one of the flesh can see him—the flaming fire was round about him, and a great fire stood before him. No one could come near unto him from among those that surrounded the tens of millions (that stood) before him . . . Until then I was prostrate on my face covered and trembling. And the Lord called me with his own mouth and said to me, "Come near to me, Enoch, and to my holy Word". And he lifted me up and brought me near to the gate, but I (continued) to look down with my face' (14: 19–25).

captured by Nebuchadnezzar, the Babylonian king who destroyed the first Temple of Jerusalem, and was exiled. Here is how the Bible recounts his story:

> [The Chaldeans] took the King and brought him up to the king of Babylon . . . and they gave judgment about him. And they slew the sons of Zedekiah before his eyes, and put out the eyes of Zedekiah, and bound him with fetters of brass, and carried him to Babylon.
>
> (II Kings 25: 6–7)

And this is what I saw:

> King Zedekiah was chained and unable to move. He was positioned in front of a large furnace. The fire was ablaze and one by one his sons were consigned to the flames. Then his eyes were plucked out. I was standing on the side, witnessing the scene. What could poor Zedekiah do? He could not help his children and could do nothing to change their awful lot. He could neither resist nor fight. He could, of course, curse and blaspheme but that would have done him no good. The only thing that he could do, really, was praise the Lord. This, I saw, is what he did. The blind man who had just lost both his kingdom and his sons was singing a great *Hallelujah*. With this, he was both gaining strength to go on living and maintaining his dignity. And as he was singing he also understood. Powerful as the Babylonian tyrant was, he was just a player in a play that was of a still much larger scope. For Nebuchadnezzar was not at the top of the pyramid—still above was the creator of the universe and the ruler of the world. Nebuchadnezzar was playing a role allotted to him and one day his fate too was sure to come.

Personally, this vision had a great impact on me and I consider the lesson it conveyed as one of the most important and long-lasting I have received with Ayahuasca. Another most powerful vision that I had pertaining to the theme of Hallelujah is that of the cosmic lottery recounted in Ch. 11.

Finally, it is interesting to note that the visions encountered with Ayahuasca are often reminiscent of ones reported in the mystical traditions of different cultures throughout the ages. In particular, let me mention the famous vision described at the beginning of the book of Ezekiel already mentioned in the previous chapter. In terms of both specific content items and general ambience, the biblical theophany is very similar to visions seen with Ayahuasca. Indeed, technically, from a structural, textual point of view, this biblical vision and grand visions experienced with Ayahuasca could be classified as belonging to the very same genre.

The Religious and Spiritual Dimension

As noted above, many Ayahuasca visions lead one to an appreciation of the Divine and of the sacred dimension of being. Very early on in my Ayahuasca quest, in fact, the very first time I asked another person what was the main thing he got out of Ayahuasca, I received the answer: 'This drink introduced me to the dimension of the sacred.' During the years that have passed, I have often reflected on the meaning Ayahuasca has had for me. There are many things I have learnt with the help of this brew, but if I were to choose one, I would give the same answer that my very first informant had given.

Visually, experiences that have deep spiritual effect vary. Some specifically depict scenes characterized as celestial or heavenly. In many of these, divine and semi-divine beings make their appearance. Given the fact that most of the Ayahuasca sessions in which I participated were related in one way or another to a community in which a Christian belief system is upheld, and since the great majority of my informants were indeed Christian, it is perhaps not surprising that seeing Jesus Christ and the Virgin Mary (who is most prominent in the cult and doctrine of the Santo Daime Church) is especially common in the corpus of data I have collected. However, as indicated in earlier chapters, many of the deities and divine beings seen in Ayahuasca visions pertain to other cultural traditions.

Yet, while many of the visions that have powerful religious impact display celestial and heavenly scenes, this need not be the case: spiritual uplifting and profound mystical experiences can also be generated by visions of scenes that are more mundane. As indicated above, panoramic scenes of open landscapes, be it of land or sea, are especially conducive for spiritual and religious sentiments. Visions depicting historical scenes and the human predicament may have similar impact. As also noted above, powerful spiritual and mystical experiences may be felt even without any visual or hallucinatory effects at all. More on this will be said in Ch. 16.

5. *Philosophy and Metaphysics*

Many individuals report visions whose themes are metaphysical or philosophical. Since such visions highlight the ideational aspect of the Ayahuasca experience, they will be discussed in the next chapter, which focuses on ideas, insights, and reflections.

6. *Ambience-Related Themes*

The special ambience that Ayahuasca induces is itself the ground for several key themes. First to be noted are those themes associated with the ambience of *enchantment*. Sensing this, people often become concerned with the topic of magic and witchcraft and with what may be referred to as the hidden meanings of things and the secrets of the universe.

Second to be noted are those pertaining to the ambience of *well-being*. These themes pertain to health and its maintenance, to sickness and healing, to the good energies of nature and the harmonious balance associated with them, and to the joy and harmony that embrace the whole of creation. The ambience of well-being may be associated with feelings of bliss and at times, euphoria and rapture.

Third are the themes associated with the ambience of *love*. Thematically, love may be associated with one's feelings towards a particular person (notably, family members from which one has been estranged), towards human beings in general, and towards living organisms at large. The theme of love can also be related to a general outlook on life and/or an approach to the conduct of one's behaviour. Other similar ambience-related themes include those of compassion, mercy, and gratitude. In the latter case, especially prevalent is that for the very fact that one is alive.

Finally, there is the ambience of *eternity*, when powerful Ayahuasca visions lead one to what seems to be the realm of the eternal. With this, one feels privy to the perspective attained *sub specie aeternitate*. The ambience of eternity is, of course, just one aspect of the more general topic of temporality. Other aspects of this topic will be discussed in Ch. 14, which is specifically devoted to time.

7. *Metamorphosis and Performance*

Strictly speaking, thematic analysis is concerned with content, that is, with semantics. However, taking a broader reading of the notion of theme and interpreting it as that which a vision (and, in general, a text) is about, we observe that the themes of some visions pertain to praxis. In other words, what is significant about these visions, and what essentially defines their meaning, is actions or performances. In these, the primary aspect of the vision consists in what is happening to the viewer and in what he or she is doing in its course. Adopting a linguistic jargon, I would say that their themes should be approached not from a semantic (i.e. pertaining to information and meaning) orientation but rather from a pragmatic (i.e. pertaining to action) one.[4]

Several types of action-based visions are noted. The first is that involving *metamorphosis*. Most notable are visions in which one feels that one's identity changes and one is transforming into another human being or into an animal, notably a feline or a bird. Such visions allow one to feel what it is to be an animal and to engage in behaviours pertaining to non-human organisms; the most notable of these is flying. Metamorphosis will be discussed at length in Ch. 13.

Related to visions of metamorphosis are visions in which the drinker finds him- or herself adopting a *role* different from his or her ordinary one (e.g. a king, a healer, a peasant) and executing a *performance* that normally he or she would not be able to produce. This topic too will be discussed in Ch. 13.

Especially meaningful to people are visions that involve personal *accomplishment*. An example of my own is that

> I was climbing a very high alpine mountain. I have never done any mountaineering in my life and the feat was quite difficult for me. I almost got to the top but could not carry on any further. Then a fairy came, gave me a push and I reached the summit. It was a most gratifying experience.

Akin to these are visions in which one feels that one is *passing a test*. The main themes of such visions are not necessarily those defined by the specific visual

[4] The shift from a semantic to a pragmatic orientation in the study of cognition in general and of meaning in particular is one of the main themes of my general work in cognition (see Shanon, 1988*a*, 1993*a*).

materials that they depict but rather the process they make the Ayahuasca drinker pass through. In general, the morals of such visions are discipline and humility, patience and perseverance, proper ethical conduct and personal responsibility. Through them, drinkers also experience receiving specific guidelines for overcoming the hurdles encountered over the course of the Ayahuasca experience and instructions regarding correct interactions with the other persons who participate with them in the session. An example is the scene of passage recounted to me by a Peruvian shaman, cited in Ch. 6; recall that this scene was characterized by this highly experienced person as one of the most powerful visions he had ever had.

Another kind is that of activities involving the special *energies* or *force* that Ayahuasca imparts to one. Typically, these visions are intertwined with overt behaviour involving things or persons in the world. First to be noted are visions involving reception. As noted in Ch. 7, these may involve a reception of energy, of knowledge, or of the ability to perform in a special manner. Paradigmatic in this regard is the act of healing, which was discussed earlier in this chapter. Other pertinent examples were given in Ch. 7 and others will be given in Ch. 13.

Lastly, there are visions involving what is perceived by those who experience it to be an act of *magic*. To me this happened only once; the episode, which involves a confrontation with a group of black panthers, was recounted in Ch. 7.

8. *The Nature of Ayahuasca*

Ayahuasca itself may be the theme of visions. Such visions may be concerned with the origin of the brew, its discovery, its use in various contexts, and its metaphysical significance. A vision I had about the indigenous discovery of the plants of which Ayahuasca is made was mentioned in the previous chapter and is recounted in Ch. 17 when scenes of light are discussed. More on this topic will be said in the next chapter in conjunction with the ideas and reflections associated with Ayahuasca visions.

10

Ideas, Insights, and Reflections

> All reality is one in substance, one in cause, one in origin . . . and every particle of reality is composed inseparably of the physical and the psychic. The object of philosophy, therefore, is to preserve unity in diversity, mind in matter and matter in mind . . . To rise to that highest knowledge of the universal unity is the intellectual equivalent of the love of God.
>
> Giordano Bruno

> There will be seen what we hold by faith
> It will not be demonstrated, but intuited
> In the manner in which man believes the primal truth.
>
> Dante

As already indicated in our discussion of the themes of Ayahuasca visions, perceptual and sensory phenomena are not the only experiences Ayahuasca induces. Thoughts and reflections are just as common, and surely no less important. In this chapter I present a general review of the types of ideas and insights entertained under the effect of the brew. Logically, I should note, this chapter is misplaced: the chapter that should have appeared at this point of our discussion is that on non-visual perceptual effects which, in turn, should have been followed by the chapters discussing other sensory and sensory-related effects; only later should effects pertaining to the ideational level be discussed. Yet, I have decided to digress from this logical order and to present the ideas, insights, and reflections encountered with Ayahuasca now. I do this because of the close contentual affinity between these and the themes discussed in the previous chapter.

Given that ideational material is often more private than the perceptual aspects of the visions, much, but not all, of the discussion is based on the mentations that I myself had under the intoxication. I should emphasize that the ideas and insights, thoughts and reflections are surveyed here not for any putative intrinsic value—philosophical, theological, ethical, or otherwise—that they might have. In the context of the present analysis, it is not validity of the specific contents of the thoughts being entertained which is of primary interest to us; after all, what we are dealing with are the speculative cogitations of a particular individual (or individuals) under intoxication. Like the visions, the ideas and reflections are presented as facets of the phenomenology of a specific cognitive state. In the context of the present discussion, what is pertinent is the fact that mentations of certain types are entertained. The systematic charting of the space of these types is a cognitive issue of significance irrespective of the veracity, informativeness, or interest of the specific contents at hand.

The reader should bear in mind the difference between 'ideas' and 'themes'. Ideas are mentations that are entertained during a session; themes are the topics of visionary scenes. While, often, the ideas entertained in conjunction with a vision are related to the theme of that vision, it is not mandatory that this be the case. On the one hand, elaborate thematic visions may not be accompanied by any ideation; on the other hand, intense ideations may be experienced without there being corresponding visions associated with them. Furthermore, while in many cases there is a rather direct correspondence between themes and ideas, often this is not so. Thus, despite the partial overlap between the analysis in this chapter and that in the previous one, for the greater part, the categories pertaining to the analysis of themes and those introduced in this chapter are not the same.

General Considerations

Overall, Ayahuasca makes people think and reflect. Many individuals with whom I conversed said that under the effect of Ayahuasca they find themselves thinking faster than normal and that they become more insightful. Many further say that the brew makes them more intelligent and that it bestows upon them special lucidity and mental clarity.[1] It seems to me that the inebriation also renders people more involved with deeper psychological analyses and with philosophical contemplation.

Naturally, the thoughts that will pass through a person's mind in the course of the inebriation will depend on the interests and concerns this person normally has. Yet, when a more general perspective is taken, some striking cross-personal similarities are noted. These pertain to the subject matters being reflected upon, the general types of contents that become significant, and the overall perspectives by which things are viewed. In fact, there are also commonalities and similarities in the specific contents that are being entertained. By and large, the ideas of a non-personal nature reported by my informants pertain to these main topical clusters: ontology and the structure and meaning of reality, nature and the phenomenon of life, the human predicament and the meaning of human life, the nature of knowledge, consciousness and the mind, ideas pertaining to human history and its meaning, the Divine and the relationship between God and both humankind and the world, ethical values and proper human conduct, and, last but not least, Ayahuasca and the very experiences it induces.

I would like to emphasize that neither in general nor for me as an individual are the topics to be presented here necessarily novel. Personally, I have reflected upon many (but by no means all) of these topics before. Yet, there is something special about the reflections entertained under the effect of Ayahuasca. In particular, three special features will be noted. First, Ayahuasca enhances the likelihood of one's

[1] For an early attestation of this, see Reinburg (1921: 28).

becoming engaged with particular kinds of topics. Second, the insights that Ayahuasca brings forth seem to cohere into one comprehensive, unified picture. Third, Ayahuasca makes one see or appreciate the answers to these questions in a distinct and special way: one does not learn about the answer, one sees it.

The last point deserves further emphasis. The distinction between visions and ideas is not as clear-cut as the terms may imply. In fact, very often the two are strongly intertwined. One sees animals, or one is transformed into an animal, and one understands what it is to be an animal. One sees visions of human history unfold before one's eyes and gains insights regarding the human predicament and the destiny of the species. One feels that one is encountering the Divine and undergoes a spiritual transformation in which one's attitudes towards theological matters change. Thus, the ideas to be reviewed here were not only entertained, they were experienced. Often, no verbalization was needed either for their conceptualization or for their comprehension. In general, the knowledge gained with Ayahuasca is direct. As Castaneda (1971) said, 'one sees'. But in fact, many mystics have observed this long before (and without the aid of any psychoactive substances); see, for instance, Blake (1794/1961) and Swedenborg (1854) as well as the theoretical writings of Underhill (1955) and Wilber (1990). I shall return to the special noetic characteristics of Ayahuasca in Ch. 15.

Following is a typology of the different types of topics entertained and reflected upon under the intoxication. The emphasis is on the issues and questions, not the specific contents of the ideas and reflections—these will be presented sporadically, only by way of concrete examples. The foregoing comments notwithstanding, I should not deride the value of the insights people (myself included) have gained with Ayahuasca. Long after their initial conception, I still find many of the ideas I have entertained in the course of the inebriation to be of value to me. Some I have subsequently made use of, either personally or intellectually. Indeed, I find quite a few of these to be of interest in their own right and I intend to consider them elsewhere.

1. *Personal Concerns and Self-Understanding*

Universally, Ayahuasca makes people reflect about their lives and leads them to what they feel is an enhanced psychological understanding of themselves. Four years after my first *Daime* experience, I reread a letter I wrote to a friend about it. I was quite surprised: in the letter, there was very little written about the visions I had. The main topic was self-understanding. I declared: 'At long last my psychoanalysis is terminated.' Indeed, at times the Ayahuasca experience consists of a powerful and profound self-analysis in which deep comprehension and meaningful insights are gained. Over the years, I have heard many say that in one session of Ayahuasca one can learn more than in years of psychoanalysis.

Self-understanding need not be merely analytical and abstract, it can also have concrete practical ramifications. In the various contexts of Ayahuasca use, people often partake of the brew in order to find answers to specific personal questions. When ingesting the brew they would define a question, hoping that during the session they would receive an answer to it. Often answers, or at least guiding insights and clarifications, are indeed found. As noted, at times the Ayahuasca experience induces veritable transformations in both attitude and behaviour. I know many individuals whose lives have drastically changed as a result of this experience. A person who is especially dear to my heart is a simple labourer from a very poor neighbourhood in Rio de Janeiro. I met him when he was in his mid-thirties, by all accounts a loving husband, a devoted father of two small boys, and a devout religious believer. The man, his wife, and several mutual friends all testified that before he encountered the brew, this person was 'in the dumps'. He drank and smoked, did not work, and showed no social responsibility. The first time he partook of Ayahuasca he experienced an upheaval. The experience was horrendous, and the message conveyed to him was clear—'You have to change the course of your life.' An angelic woman appeared to him and offered guidance and help. My friend came out of it a different person. Now, more than ten years later, and with extended experience with Ayahuasca behind him, the radical change is still in effect. This man, as well as his acquaintances, attribute this change to Ayahuasca. In this conjunction I note that the brew has successfully been employed in the rehabilitation of drug addicts (see the comments on Takiwasi in Ch. 1).

Other Persons and Interpersonal Interactions

As described in some detail in the previous chapter, Ayahuasca also induces reflections having to do with other people. As further indicated in Ch. 5, it is very common for people to feel that under the effect of the brew, their psychological perception and sensitivity increase and that they gain deeper understanding of other people. As noted earlier, this applies both to those present in the Ayahuasca session and those who are not.

2. *Philosophical Issues*

In the present context, the epithet 'philosophical' refers to issues that are standardly entertained by philosophers. Because of my personal inclinations and professional background I am especially interested in these issues; hence, it is not surprising that I have often been concerned with them. Yet, under the effect of Ayahuasca, these issues are commonly reflected upon even by individuals without any philosophical background. For instance, thirteen of the eighteen independent drinkers I interviewed reported having entertained thoughts regarding the ultimate nature and structure of reality.

Metaphysics

Overall, Ayahuasca induces a comprehensive metaphysical view of things. I would characterize it as idealistic monism with pantheistic overtones. By this view, reality is conceived as constituted by one, non-material substance which is identified as Cosmic Consciousness, the Godhead, the ground of all Being, or the Fountain of Life. Coupled with this is the assessment that all things are interconnected and that in their totality they constitute one harmonious whole. This, in turn, entails an experienced realization that there is sense and reason to all things and that reality is invested with deep, heretofore unappreciated, meaningfulness. By and large, it seems that the metaphysical perspective induced by Ayahuasca is most similar to views entertained in classical Hindu philosophy (see, for instance, Phillips, 1995)[2] as well as by Plato, Plotinus, and Hegel. Remarkably, this view is essentially the same as that characterized by Huxley as the 'perennial philosophy' (Huxley, 1944; see also James, 1882); similar observations were also made in the context of LSD (see Grof, 1972, 1998).

In one form or another, I have heard ideas that are consonant with this metaphysical view from many of my informants. Indeed, no one reported to me metaphysical ideas that countered this view. Of the specific metaphysical ideas reported by informants perhaps the most salient one is the appreciation that all reality is interconnected. Furthermore, it is extremely common for Ayahuasca drinkers to talk of their discovery of there existing a force that is the ground of all Being and that gives nourishment and sustenance to everything. Often, this force is characterized as embodying love and its permeation as cosmic joy (see also Watts, 1962). Typically, it is interpreted to be the Godhead or the *anima mundi* and is characterized as being the source and fountain of everything good—life, wisdom, health, as well as intellectual and artistic creation. Interestingly, in the reports of these metaphysical insights often the self-same verbal expressions are employed by different individuals. Thus, many informants have reported to me that the brew made them appreciate that 'everything is interconnected', 'all is one', 'everything is spirit', and 'all is consciousness'. Other recurring expressions are 'this world is an illusion', 'everything has meaning', 'the different levels and aspects of reality exhibit the same essential structure', and 'I and the world are united'.

My own firsthand experiences with Ayahuasca revealed similar patterns. Personally, I have come to ideas of the kind just noted in conjunction with seeing the 'web' I described in Chs. 5 and 8, that is, a matrix of translucent strings that seem to tie everything together. I have experienced this many times and have heard of the same experience from many of my informants. The description of the visual effect was invariably the same and many persons used the identical phrase—'a web'—to describe it. For instance, one of the independent drinkers told me that the most important teaching she has received from Ayahuasca was

[2] In particular, the following characteristics of the Brahman (the Absolute) are specified by Sriharsa, the 12th-century Vedantic philosopher: 1. Brahman is self and consciousness. 2. Brahman is world ground. 3. Brahman is transcendent of finite individuality. 4. Brahman is unitary, the coincidence of opposites, and omnipresent. 5. Brahman has non-dual self-awareness. 6. Brahman is the essence or finest part of everthing. 7. Brahman is the locus of value. 8. Brahman is mystically discoverable. 9. Brahman is beyond the power of thought uninformated by mystical awareness. Essentially, all these characteristics are encountered in the world-view intuited with Ayahuasca.

the appreciation that the Divine does indeed exist. Asking her how she had arrived at this conclusion she answered by presenting a description of the translucent web that interlinks everything and sustains all existence. The experience described in the citation from Castaneda (1972) in Ch. 5 is similar.

Ontology

While I have been concerned with philosophy for years, ontology has never been of special interest to me. In fact, my basic position was that psychologists should not bother with ontology at all, and that when engaged in cognitive research they should not be concerned with ontological questions. The Ayahuasca experience forced ontology on me. Often, the things I saw under the intoxication impressed me as being so real that the conclusion seemed to be unavoidable: truly existing other realities are being revealed. Believing that this is the case is very common with the drinkers of Ayahuasca. Both during the course of the intoxication and afterwards the question repeatedly forces itself: Does this really exist? Over and over again I have heard people express the same feelings of puzzlement and intrigue: Where do all the wondrous things revealed in the visions come from? What do they mean? The things seen with Ayahuasca often strike people as so different from anything they have seen or known that they cannot be the products of their own intellect. Universally, drinkers of Ayahuasca feel that these things are too fantastic to be merely the products of the imaginative power of their own mind. If a supernatural realm (the term usually employed is 'the astral') exists in any sense,[3] many further questions ensue. What is the relationship between this realm and the physical world? What is the relationship between it and the human mind, between it and the brain? Does the supernatural or transpersonal consist of only one realm or of many?

Epistemology

Like traditional mystical states (see Bucke, 1901/1991; James, 1929; and Stace, 1961), the Ayahuasca inebriation is associated with a strong noetic feeling: under the effect of the brew people often have the impression that they are gaining access to new knowledge. Inevitably, the question poses itself: What is the status of this knowledge and what is its veracity? Some reflections on this will be presented in the last, and more speculative, chapter of this book.

The Ayahuasca experience often made me think of the contrast between two types of knowledge: intuitive, direct knowledge and knowledge amenable to explicit, perhaps verbal, articulations. While the appreciation of this contrast was not new for me (in fact, I discuss it at length in my book on the philosophy of psychology; see Shanon, 1993*a*), with Ayahuasca I sensed the poignant tragedy it presents. Intuitive knowledge, I felt, allows one to know everything. However,

[3] I should emphasize that in this discussion I bring these considerations only as topics often entertained by people who partake of Ayahuasca. No stance with regard to the specific ideas or positions associated with these topics is implied.

knowledge gained by means of intuition cannot be recorded in an objective form that is transmissible to other people. In contrast, knowledge that can be verbalized allows objectivity and interpersonal transmission but, given that it is framed in specific codes of conceptualization, it is destined to be constrained and limited. In one UdV session in which I participated, the presiding Master was asked what intuition was. He answered 'Knowing things in that manner which ordinarily is encountered in perception.' I found this to be a better definition than any I have seen anywhere in the psychological literature. I still do.

I will note that personally, Ayahuasca brought me, for the first time in my life, to doubt the validity of the Western world-view. With this, again for the first time in my life, I began to wonder whether science as we normally understand it suffers from some fundamental limitations and that, in fact, it may hinder and inhibit us from understanding some crucial aspects of reality. I began seriously to entertain the possibility (still unproven) that there are other, complementary sources for knowledge, ones that do not employ the instruments and methodologies developed by modern science. A full discussion of this extends beyond the scope of this book (but see Tart, 1972; Roberts, 1983; Vaughan, 1983).

Specific Philosophical Issues

In addition to those indicated above, several more specific philosophical ideas that appear in the data I have collected should be singled out.

On several occasions during Ayahuasca sessions I felt that I was presented with answers to questions I have been concerned with for a long time. One such question is that of *determinism and free will*. I present this question first for it is a clear case in which intellectual understanding was gained through direct, visual (i.e. perceptual, as contrasted to merely conceptual) seeing:

> In front of me I saw the space of all possibilities, that is, all states of affairs that can possibly happen. They were lying in front of me there like objects in physical space. Choosing, I realized, is tantamount to the taking of a particular path in this space. It does not, however, consist in the generation of intrinsically new states of affairs. All possibilities are already there, I saw, but one has the option of choosing different paths amongst them, just as when travelling through a terrain in real space. Further, while travelling in the space of possibilities takes time, the possibilities themselves are there, given in an ever-present atemporal space. Thus, I concluded, there is no contradiction between determinism and free will. With this, for the first time I felt I understood the Jewish sages in the Mishna—'Everything is laid out in advance yet freedom of choice is given' (*Ethics of the Fathers*, 3: 19).

Later I found out that the insight gained in this vision is in line with the philosophical perspective presented by Spinoza in *The Ethics* (Spinoza, 1670/1989) and is especially akin to the ideas on both determinism and the nature of time in his *Short Treatise on God, Man and His Well-being* (Spinoza, 1852/1985; see in particular ch. 16).

Second is the *mind–body problem*. Sensing my own behaviour when intoxicated, it occurred to me that the mind and the body are to be viewed as coupled dancers.

Are the dancers alone or together? Of course both are the case. The dancers are two individuals that move *in tandem*; each one is an autonomous organism yet, as they say, it takes two to tango. Note that this perspective is an alternative to both the monistic identity *and* the dualistic positions: mind and body are not two, separate entities nor are they two facets of one basic whole; more on this will be said in the last chapter of this book.

As noted, Platonic-like ideas reoccur throughout my informants' reports. I, too, thought of Plato several times in the course of my journey with Ayahuasca. For me, this was not a simple, straightforward matter. My professional work in cognitive psychology follows a strong anti-Platonistic line (see Shanon, 1993*a*). One of the most important effects Ayahuasca has had on me is a serious reconsideration of the Platonistic world-view. Some reflections on this matter will be presented at the end of this book.

Of the philosophical ideas reported to me by other people, a Platonic one that especially impressed me was furnished by an independent drinker, a resident of Rio de Janeiro, with no formal training in philosophy. When I interviewed this person, he presented an entire metaphysical picture which he said came to him from Ayahuasca. It was a radical idealistic view. When probing him with respect to the origin and possible veracity of this view, the man told me: 'You are a professor so you think that you teach me, that you pass information to me. But this is not so. You only talk to me, and through this come up ideas and knowledge that are there, stored in my own mind. It is all there and, in effect, you teach me nothing.' Plato's dialogue *Meno*, of which this person had never heard, strikingly comes to mind.[4]

Still another metaphysical view that appears in the reports of many informants is one that may be subsumed under the general heading of *animism*. In Ch. 5 I discussed the experiential aspects of animism, here I consider the ideational aspects. Ideationally, the animistic world-view may take different forms. The more metaphysical one consists of the appreciation (discussed above) that physical reality is permeated by forces of life and perhaps also by what may be regarded as cosmic or divine intelligence. 'All is energy', 'everything breathes', 'it is all the manifestation of cosmic consciousness', 'God permeates everything', are specific expressions that several of my informants used in describing what the brew made them feel. A second variant of the animistic world-view is one that may be characterized as ecological. Essentially, it reflects the view associated with what is nowadays referred to as the Gaia hypothesis (Lovelock, 1979, 1988), by which the planet Earth is an organism with its own homeostatic and intelligent

[4] In this dialogue, there is a famous encounter between Socrates and a slave in which the topic is geometry. Through a series of interchanges, the slave comes to prove a geometric theorem. Plato's point is that nowhere during the encounter does Socrates explicitly teach geometry to the slave. Where does the slave draw his knowledge from? In order to account for the slave's manifest intellectual achievement, Plato proposes a nativistic theory according to which human beings come to this world equipped with knowledge.

properties. This recognition makes people further appreciate the link between humankind and nature. Still another form of the animistic view may be referred to as semantic. It consists of the appreciation that all Existence and all creation are imbued with meaning and that there is a 'great story' behind it all. Time and again, people reported to me that Ayahuasca allowed them to see what is usually hidden and to appreciate what is always there but to which we are normally blind. As noted in Ch. 15, identical assessments are encountered in the indigenous context of Ayahuasca use. Other forms of the animistic world-view are spiritualistic and involve beliefs in spirits, ghosts, and other paranormal beings and entities. In the context of all syncretic sects using Ayahuasca such beliefs are extremely common. Lastly, many Ayahuasca drinkers report that the experience of having insights with regard to the animistic world-view is often accompanied by powerful, visceral sensations. In other words, these insights are gained not only through people's minds but in a non-mediated way through their bodies.

A topic that I have reflected upon on many occasions and which was also reported to me by several informants is *time*. As will be described in detail in Ch. 14, Ayahuasca introduces its partakers to a mode of being in which time seems to be of less, or even no, import. This naturally leads to reflections regarding the nature of time and temporality. A major conclusion that I have drawn from these experiences was the following tripartite distinction between three domains. First is the domain governed by both time and space. This is the domain of the physical world. In it, both temporal and spatial statements apply. For example, it is sensible to specify both the time and duration of a physical phenomenon and its place and spatial extension. Second is the domain governed by time but not by space. This is the ordinary mental or psychological domain. Thus, it is sensible to specify the time and duration of a thought but not its place or length. That there are phenomena to which time applies but space does not coupled by the absence of the converse state of affairs (space applies but time does not) led Kant (1781/1953) to conclude that time has primacy over space. In the third domain neither space nor time apply. By definition, this domain is outside the province of the natural world. Personally I never believed in (nor even thought of) this third option before my experience with Ayahuasca. With Ayahuasca I experienced it and subsequently began to reflect upon it. This domain which is outside both space and time is that which in the context of both the Santo Daime Church and the UdV, as well as in various Western-European esoteric traditions, is called the 'astral'. More on this will be said in Chs. 14 and 15.

Many informants expressed an appreciation of the relationship and analogy between what in the traditional philosophical literature is referred to as the *microcosmos* (the level of the human) and the *macrocosmos* (the level of the cosmos). Along with this was an appreciation of the great harmony that all Existence manifests. These, as well as all other insights mentioned above, are not mere ideas that pass through people's minds during the intoxication, but rather realizations that with one's utmost being one feels are true.

Lastly, Ayahuasca made me appreciate that there is a level of existence that defies standard *logic*. Over several sessions, I have developed the following general principles or rules for thought by which the non-ordinary logic of the Ayahuasca experience may be described; these are introduced with mnemonic labels I have devised for them:

1. Conjunction: Both *p* and *not p* are true. In fact, in an ultimate sense, truth is the conjunction of statements that are normally regarded as contrary to one another.
2. Disjunction: Between *p* and *not p* there is always a third possibility. Wisdom consists of bearing this in mind and searching for that other option.
3. The Zero: always keep things fresh, without interpretation.
4. The One: when drawing an interpretation, try to adopt a comprehensive view taking into account the broadest picture possible.
5. Reflection: when considering a question, remember that it is all a mirror. Thus, look at things from a distance and reverse the question and the perspective from which you are approaching it.

3. *The Natural World*

In addition to reflections that may be characterized as philosophical, Ayahuasca induces ideations about phenomena that pertain to the province of science. At times people also feel that the brew reveals to them concrete information regarding specific phenomena pertaining to the physical and biological domains. In the two previous chapters I have extensively described both general themes and particular contents pertaining to visions of the natural world. Here I comment further on ideas and reflections about physical and biological issues. While most of these ideas are linked directly to visions, some were generated in the course of the Ayahuasca inebriation without any direct relationship to any specific visualization.

Cosmological and Physical Issues

It appears that Ayahuasca has the special effect of enhancing people's concern with questions regarding the origin of the universe and the forces and laws that govern the natural world. (Significantly, these also feature centrally in Amerindian myths associated with the brew.) Typically, the questions entertained are grand: Why was the world created? How? What maintains it? What is the world's eventual destiny? What is the meaning behind it all? All these are issues that I, along with my informants, have found myself reflecting upon during Ayahuasca sessions.[5] The reader will observe that there are two kinds of questions here. On the one hand, there are questions of structure and process; these are the ones the natural sciences

[5] In his many studies of LSD, Grof found subjects entertaining the same kinds of questions (see e.g. Grof, 1972).

are standardly concerned with. On the other hand, there are questions pertaining to meaning, significance, and purpose. Usually these are regarded as belonging to the province of metaphysics.

Many of the visions of the physical world are associated with reflections on the various forces of nature—the sun, the rain, the waters of the sea. Usually, the contemplation of these different forces invokes a deep appreciation of the bounty of nature and the essential dependence of life upon it. Also encountered are visions and reflections regarding the forces of death and destruction. One informant told me of a vision in which he saw a great geological upheaval; with this, he came to appreciate that the forces of nature are not only creative and life-maintaining but also destructive. I once had the insight that indeed, the world was created as an experiment. As far as the Creator is concerned, I surmised, it would be fine if it worked, but not a terrible tragedy if it didn't. Tough luck, as they say. Later I learnt that such an idea is found both in Hindu philosophy and in the Jewish Kabbalah.

Biological issues

Reflections regarding the nature of the origin of biological life and the forces that guide the living are very common. With Ayahuasca, people sense that they encounter the ultimate force that makes life possible. They also feel that they are very much a part of this force.

Especially prevalent are reflections having to do with animals. What does it mean to be an animal? What is animal consciousness? What is animal intelligence? One striking insight I had in this regard sprang into my mind during a vision in which I felt that I was experiencing what it was like to be an ant. I found myself in the midst of an ant colony and I appreciated the strength one gains by being part of a whole larger than oneself and the harmony of the entire matrix of collective existence. For me, being an ardent individualist, this was a very novel experience. I should say that I also felt it to be quite frightening. On another occasion, I found myself reflecting on how ants find their way around. The standard explanation, I gather, is that ants secrete chemicals that they can retrace and that other ants can follow. But this, I realized when under the intoxication, offers only a partial explanation. What makes the ant follow the secreted trail? The scientific analysis does not answer this question. It seemed to me that in order to solve this question we have to invoke the ultimate force of life. There is a force that leads the ants and determines the manner in which they follow the trail. Going one step further I ventured that maybe this force of life is the very same one presented by the Ayahuasca brew itself.

A topic that I found myself focally reflecting upon in my first powerful experience with Ayahuasca was that of biological evolution. It occurred to me that even though it evolves in time, life is governed by one story which in itself is not dependent on time. In other words, Darwinian theory specifies the mechanisms and processes of evolution, but it does not provide for a full explanation of the story of life; a description of a vision concerned with biological evolution has been given in Ch. 8.

As indicated in the previous chapter, Ayahuasca also induces thoughts regarding the functioning of living organisms and biological systems. Narby (1998) developed an entire theory regarding the coding and storage of genetic information following Ayahuasca-induced visions and insights concerning DNA. Somewhat similar reflections are also presented by Harner (1980) in conjunction with his own experience with Ayahuasca.

4. *Psychology and Human Life*

Theoretical Psychological Issues

In addition to personal psychological insights, with Ayahuasca people also entertain ideas regarding human psychology in general. Perhaps the most acute are questions regarding madness and sanity. Along with the feeling that one might be losing one's sanity, the general questions often pose themselves: What is the difference between sanity and madness? What is the value of the former? How are people to protect themselves from the latter?

The one most important psychological insight that I—as well as, apparently, many other persons—have had with Ayahuasca is what may be characterized as the appreciation of the limitations of the psychological. Throughout my life, my approach to human life was fundamentally psychological (as opposed, for instance, to spiritual, political, or ideological). With Ayahuasca, for the first time, it occurred to me that there may be aspects of human life that transcend the ordinary psychological categories. With this, I came to realize that considerations pertaining to one's individual life history, to personality structure, and to the individual's interests, wishes, and conflicts might not be sufficient by way of accounting for the dynamics of the psyche. Thus, the brew made me entertain the possibility that psychological theories of personality are often limited and misguided. In particular, grounding psychological life with what is (metaphorically) below—the libido, instinctual forces, desires, and the *un*conscious may be fundamentally wrong. Perhaps a more profound understanding of the human psyche would be attained by drawing psychology's look (metaphorically) upwards. This is precisely the basic tenet of humanistic and transpersonal psychologies; authors whose writings made special sense to me after my experiences with Ayahuasca included Assaglioli (1965), Maslow (1968), Frankl (1975), and Wilber (1983, 1990).

One specific topic I have been concerned with is sublimation. Heavily influenced by Freudian psychoanalysis, my academic training in psychology has led me to view sublimation negatively—a process having to do with the confrontation of menacing and/or conflictual psychic materials and their redirection and transformation. With Ayahuasca, however, I found myself looking at sublimation in a new way. The sublime, I appreciated, is a facet of reality, and it is a great privilege to experience it and take part in it. Rather than a defence mechanism, sublimation is a process that allows human beings to step forward and attain higher levels of being.

The Human Predicament

Very often Ayahuasca evokes ideas and reflections regarding human life that are of a more philosophical nature. As noted earlier, these concern what might be referred to as the human predicament. On the one hand are ideas and reflections having to do with the generalized story of the life of a human being. On the other hand are ones having to do with the collective phenomenon of the species *Homo sapiens* and its history. Pertaining to the first type are issues such as life and death, the meaning and purpose of life, psychological and spiritual transformation, faith and salvation, the masculine and feminine aspects of human personae, as well as various ethical and moral issues. Of the second type are matters such as human suffering, human history and its meaning, the dialectics between nature and culture, and the place of human existence in the grand framework of Being. Examples were presented in conjunction with the analysis of themes in the previous chapter.

Cognitive issues

Cognition is the field of my professional expertise, so it is only natural that many of the reflections I have had with Ayahuasca concerned cognitive issues; however, similar reflections were reported to me by many other informants, practically all of them with no education in psychology or cognition whatsoever. Amongst the questions entertained are: What is the nature of knowledge? What aspects of human knowledge are innate? How does memory function? What are intuition and instinct? How is creativity achieved? What is the nature of intelligence, both human and animal? What is consciousness? As a university professor who has been professionally involved with these questions for many years, I can ascertain that some of the insights I have had on these with Ayahuasca are ones that I (still) find interesting and novel. Indeed, I can unequivocally state that my experiences with Ayahuasca have greatly contributed to my understanding of consciousness and the functioning of the mind. I intend to elaborate on this elsewhere.

Of the issues noted, let me underline those having do to with the mind and the nature of consciousness. Several informants reported to me that Ayahuasca made them appreciate that consciousness is a cosmic phenomenon. Several mentioned the assessment that 'eventually everything is consciousness'. One person with no philosophical education presented to me an entire radical idealistic Berkeleian-like theory of mind.

Also common are reflections regarding memory. Several informants told me that under the intoxication they came to appreciate that memory is out there in the world and recollection consists in one's making contact with this independent, permanently existing store of knowledge. I had such an experience, too. Later I discovered that this option is encountered in the esoteric literature; it is referred to as the Akasha records (see e.g. Steiner, 1969).

5. *Medical and Health Issues*

In the indigenous contexts of Ayahuasca use questions having to do, on the one hand, with health and disease and, on the other hand, with treatment and cure are central. Traditional healers say that in their visions they encounter beings who instruct them how to go about treating their patients (see, for instance, Dobkin de Rios, 1973; Payaguaje, 1983; Arévalo Valera, 1986; Langdon, 1992; Luna and Amaringo, 1993). Several non-indigenous informants have reported similar experiences to me.

During Ayahuasca sessions one is likely to confront health questions in a most practical, concrete manner. Physically, the Ayahuasca experience may be very rough, and often it forces drinkers' primary attention onto their bodily well-being. Given the urgency of the situation, people have no choice but to listen to their body and try and take care of it. With this, they often gain insights regarding health, disease, and cure.

Let me add that in the Ayahuasca context health is conceived not merely as a practical matter having to do with the biological condition of an individual, but as a supra-individualistic affair conceived in terms of harmony and balances between cosmic forces. I have heard one UdV master say that the source of health is the fundamental cosmic energy. Another master referred to it as light. These ideas are, of course, not specific to Ayahuasca; for an excellent general exposition the reader is referred to Grossinger (1990).[6]

As noted repeatedly throughout this book, Ayahuasca is closely linked with death. On the one hand, during Ayahuasca sessions one often undergoes such difficult experiences that one physically feels that one is about to die. On the other hand, visions in which one sees various manifestations of death are very common. With this, Ayahuasca is prone to bring about reflections about death and dying. An example from my own experience is that concerning the Angel of Death recounted in Ch. 8.

An especially common topic associated with Ayahuasca is reincarnation. As noted at the beginning of this book, etymologically, 'Ayahuasca' means the vine of the spirits or, rather—the spirits of the dead. As explained in earlier chapters, this is due to the experience, rather common with Ayahuasca, of seeing persons from other places and other times with which one strongly identifies. This is indeed a very powerful experience. Practically all who have undergone it interpret it as the meeting with one's previous incarnations. Personally, I do not share this interpretation. More on this topic will be said in Ch. 13.

6. *Ethical and Existential Questions*

Good and Evil

One good friend, an anthropologist with extensive firsthand experience with Ayahuasca, informed me that a shaman told her, 'You know, Ayahuasca is not

[6] I am grateful to Dr Alvaro Braz who introduced me to this book.

only good and generous, it also has a touch of malice.' Indeed, with its exuberant bewitchment the Ayahuasca situation often evokes an overall ambience of evil. The brew may be holy and be sanctified, but at the same time enchanted and full of secretive, and even malicious forces. Overall, it seems that Ayahuasca leads people to the conclusion that the world contains both good and evil, that the two are intertwined, and that the ultimate reality is beyond good and evil. Several indigenous shamans and masters of the UdV told me that it is very important to encounter the evil and get acquainted with it. This enables one, when the situation presents itself, to confront evil and overcome it (see also Fericgla, 1998).

Good and evil are not only metaphysical categories. They also apply to the lives of individual people. On several occasions I have seen ethical dilemmas as they are manifested in the lives of specific historical figures: the decadent popes of the Renaissance, Cortes in the conquest of Mexico, and Mengele, the monstrous Nazi physician. In all these cases, the brew made me view these figures from a new perspective, one which I had never entertained before. There is no way for me (or anyone else, for that matter) to ascertain the veracity of the novel observations I had. What I can say, however, is that now, several years after the experiences in question, I still find them all insightful and interesting.

Finally, there are visions in which one feels one is encountering the Supreme Good. A major impression these visions had on me is the (Platonic) conclusion that ultimately, the ethical and the aesthetical as well as the true are the same. I have heard similar assessments made by many other people.

Values

Very often, the Ayahuasca intoxication is experienced as a lesson in morals. Reflection about certain values and a sense of commitment towards them seems to be especially salient. Those reported by many individuals include personal responsibility, justice, and love. Also common is the appreciation of the significance of faith and hope, patience, and humility.

Common is the appreciation that values—in particular, love and justice—are not confined to the province of human life but they also apply to existence at large and to the forces or beings that govern the universe. In this vein, perhaps, is the insight of one *Daime* hymn that describes the day of final judgement. It is noted there that the situation is not sad; rather, it is one of joy and harmony in which all—both the just and the unjust—receive precisely what they merit.

The Meaning of Life

As indicated both in Ch. 9 and earlier in this chapter, reflections regarding the meaning of human life—both on the level of individual life stories and on the level of the species at large—are very common. In particular, many drinkers report that Ayahuasca made them appreciate the reason we human beings are here on this planet. The most common reason indicated is for people to correct errors they had made in previous incarnations and to contribute to the overall welfare of the world.

An answer I myself have found with the brew is to praise, to sing *Hallelujah*. More on this is said below.

7. *The Social Realm*

Related to the ideas regarding the human predicament are ones pertaining to the social realm. These ideas have to do with topics that are standardly subsumed under the fields of anthropology, history, and the social sciences.

Culture

With Ayahuasca, questions pertaining to culture are paramount. Reading the anthropological literature one gets the feeling that by and large visions involve the forest and its creatures. But as noted in previous chapters, and as further revealed in the paintings of Amaringo (Luna and Amaringo, 1993), what pervades Ayahuasca visions are cultures and their expressions—cities, palaces and temples, artefacts and works of art, royalty and religious figures, rites and rituals. This is also reflected on the ideational level. Specific questions I have entertained in conjunction with visions I have had include: How do cultures rise? How do they fall? What are the forces that maintain them? How are cultures to be evaluated and judged? What is the meaning of human history and its progress? What are the relationships between different cultures and civilizations? As noted earlier, a consideration of these questions was the subject of one of my very first major visual experiences.

A specific family of cultural questions that has concerned me greatly is the contrast between the Western and the Indian cultures and ways of life. With this, the questions regarding the dialectics between nature and culture are brought up. On various occasions I found myself considering the fruits of culture and the price they demand and in contrast, the fruits of a direct link with nature and the difficulties of maintaining them. I have seen this from a general historical perspective as well as from a more specific one focusing on the Indians of the Amazon.

Remarkably, the interaction between nature and culture features centrally in Amerindian mythologies having to do with the origin of Ayahuasca. Ayahuasca is a natural product, one taken from the Amazonian rainforest. At the same time it is often depicted as the source of the cultural products of societies: music, art, religion, and the very identity of specific ethnic groups (see, for instance, Reichel-Dolmatoff, 1975, 1996).

History

With the brew, I often had reflections regarding human history. The most powerful vision pertaining to this topic that I had is the one described in the Prologue. Another example is one recounted to me by an informant describing his very first Ayahuasca vision:

I found myself in the midst of a sand desert. It was surrounded by rocks and trees. In the desert, there were warring armies engaged in combat. I saw all the cruelty and suffering and I ran amongst the warriors in desperation. Then light descended on the heads of the people and they all lifted their hands in prayer. The moral of it all was clear: the solution to all the miseries of the human race lies in faith and trust in that which is above us. Thus, then, all wars are futile and there is always hope.

Both I and one of my informants had other visions whose message was the futility of war. Visions of two historical events impressed me most; both are tragedies—the conquest of America by the Spanish and the Jewish Holocaust in the 20th century. Not surprisingly, the latter event is one that has concerned me often. Why did it happen? How could human beings be so cruel? How could God allow the killing of six million innocent people, of whom more than a million were children? As described in the Prologue, with Ayahuasca I gained what seemed to me to be insights regarding the possible significance of this horrendous event.

Politics

Lastly, there are questions pertaining to what may be labelled political science. Especially, I am referring to the issue of royalty. As noted in Ch. 8, royalty and scenes pertaining to them appeared very often in my visions. These visions made me reflect upon the nature of government. They led me to entertain the idea that monarchy might actually be a perfect system, but on one condition: the kings and queens have to maintain justice. As already indicated in the previous chapter, I have also seen the difficulties of being a king: the power games, the lure of riches, the temptation to usurp, the danger of overconfidence and vanity.

8. *Aesthetics*

The Ayahuasca experience is cardinally aesthetic. First, the brew highly enhances people's aesthetic sensitivity and appreciation. Second, the visions the brew induces are exceedingly beautiful and having them is a most powerful aesthetic experience. Third, as already indicated, splendid works of art are prominent in the visions. Fourth, the Ayahuasca experience is intimately related to music and the intoxication often makes people sing and play on levels much higher than their performances exhibit under normal circumstances. Fifth, with Ayahuasca, people often feel that the aesthetic is the prime measure of things.

The last two points are specifically related to the ideational level and I would like to dwell on them, beginning with the last. Under the Ayahuasca intoxication, people feel that an entirely new world of sublime magnificence is revealed to them. In it, the aesthetic reigns supreme and the common impression is that it is the embodiment of eternal truths and the manifestation of the Supreme Good. Again, this is a Platonistic perspective. In his dialogue *Phaedrus*, Plato (1914) expresses the view that Beauty is unique in that it is directly manifest to the senses. This is in contrast to the other forms (in the Platonic sense), which are comprehended only through the exercise of the faculties of reasoning, which are always imperfect. Furthermore, Beauty is unique in that it has the power to reveal to us humans the existence of an ideal world beyond the world of sense. This, in turn, is a prime source for happiness. Again, these feelings and insights are very common under the Ayahuasca intoxication.

Turning to music, I shall note that on several occasions, the musical experience generated in me philosophical ideas that were very much along Pythagorean lines. A pertinent example is the vision on the music of the spheres recounted in the previous chapter.

Lastly, Ayahuasca is intimately related to artistic creation. The anthropologist Reichel-Dolmatoff (1975, 1978*a*, 1978*b*) claims that one main reason indigenous people partake of Ayahuasca is to gain inspiration for their artistic work. In his works, he argues that the geometric designs that appear in Tukano art are direct reflections of the designs seen with Ayahuasca. Similarly, the designs encountered in the ceramics, textiles, and house decorations of the Shipibo-Conibo are directly linked to the geometric figures seen in the visions (see Gebhart-Sayer, 1985; Gow, 1988) and so too in the case of the Cashinahua (Guss, 1989; Lagrou, 1998). Personally, during many sessions, I felt I understood that the source that great creators gained their inspiration from was the cosmic source of plenty. It was all there—the artist 'only' had to be sensitive so as to be in touch with this source, to grasp what it offers, and then, of course, to know how to translate it, or rather—how to create from it, to crystallize what he or she saw or heard into the particular medium that he or she masters. Several informants—notably ones who are artists themselves—have recounted similar reflections to me; a pertinent vision of my own was reported in the previous chapter.

9. *Linguistic and Semantic Issues*

Personally, I am most interested in natural language in general and in its semantics in particular. I have been formally trained as a linguist and for many years the primary focus of my work was in the field of psycholinguistics. Therefore, it is not surprising that I have had many reflections regarding the issue of meaning. Yet, it appears that with Ayahuasca the concern with meaning is quite common. Masters of the UdV often speak of word-games. For instance, I have heard it said that the word 'Deus', God, means '*De um*' (in Portuguese, of one), and that the word 'Jerusalem' means '*Jesus alem*' (in Portuguese, Jesus is above). Interestingly, such word-games are also very common in the Jewish Kabbalistic literature. In *Daime* sessions one standardly speaks of the two possible readings of texts: the simple, direct meaning and the second, esoteric one. Again, such a view is fundamental in Kabbalistic thought. More on this will be said in Ch. 15, where phenomena pertaining to meaning are discussed.

10. *Esotericism and the Paranormal*

The Ayahuasca experience is miraculous. Amerindian legends associated with the brew typically contain elements of the supernatural and the non-ordinary. The doctrine of the Santo Daime Church proclaims that the brew makes one enter another reality, the astral. The UdV characterizes the brew as *cha misterioso*, mysterious tea. Indeed, when one partakes of Ayahuasca one feels an all-engulfing enchantment. With this, people are prone to entertain ideas and reflections of an esoteric nature. Time and again, informants have reported to me that Ayahuasca made them appreciate the existence of a hidden reality to which human beings are normally blind. With this, people claimed, hidden forces were revealed and hidden meanings recognized. With this, questions regarding the paranormal and the possibility of miracles become especially pertinent. Indeed, having undergone the Ayahuasca experience, most people—including ones with higher levels of Western education—tend to believe that all sorts of paranormal phenomena are actual. More on this subject will be said in Ch.16.

11. *Theology, Religion, and Spirituality*

As noted, a key facet of the Ayahuasca experience is the encounter with the Divine. It is only natural that such a powerful and unusual experience would generate thoughts and reflections. On the one hand, there are reflections regarding the nature of God and his relationship with the world and with human beings. On the other hand, there are thoughts regarding human beings and their relationship with the Divine. Among these are the following: What (or who) is God? What is his relationship to the universe and to the world of human beings? What is the relationship between God and the self? What is faith and what is the relationship between it and knowledge? How should human beings address God? How should I do it? Should one change one's life having had experiences of encountering the Divine?

Of the various ideas pertaining to God encountered in my data, I single out the following three. The first two have to do with the nature of God. Several informants remarked that with Ayahuasca they discovered that God was beyond good or evil, or that he encompasses both good and evil. For some persons, because of their Christian persuasion, this was a revolutionary, and even disturbing, discovery. As indicated earlier, a couple of informants further noted that one prime attribute of the Divine is having a sense of humour. With this insight, human life is put in a very different perspective from the standard one. Some may conclude that the value of human life is dwarfed, others draw strength from it. The third idea is about the meaning of life, or rather—the reason why we human beings are here on this planet. Reflections on this topic are very common with Ayahuasca. As noted above, in my own personal case, the brew made me believe that we are all here in order to praise, to sing *Hallelujah*. Specifically, each of us is an instrument in the great symphony of life and the bottom line is that what we are to do in life is to find our own tune and perfect its singing. What the tune is and how to sing it is the very meaning of each person's life. Pertinent examples were given in the previous chapter; a more general discussion is presented in Ch. 16.

12. *The Nature of Ayahuasca*

Given that the Ayahuasca experience is so special and extraordinary, it is only natural to have reflections regarding the drink itself. What does it actually do? Where does its power come from? What is the meaning and value of the entire Ayahuasca experience? Indeed, reflections about these questions are very common.

A speculation I had and which I heard from several other people is that Ayahuasca is the ultimate source of knowledge and the presentation of all wisdom; that, in fact, it is the Tree of Knowledge (see McKenna, 1992). For related reflections, see Shanon (2000*a*) in which I present a hypothesis regarding the role of DMT enthenogens in early Israelite religion.

The reflections regarding the nature of Ayahuasca are not only theoretical; often they also have concrete, practical ramifications. Indeed, a major topic of the Ayahuasca experience is one's very handling of this experience which, as noted, can be very harsh. The harshness is concrete and immediate and one has no choice but to deal with it. With experience, drinkers of Ayahuasca feel that they gain better understanding both of the effects of the brew and of the ways to handle them. As one informant told me, a major topic of the Ayahuasca school is how to master the art of drinking Ayahuasca.

Concluding Remarks

Taken together, the ideas and insights that Ayahuasca induces seem to cohere into one consistent, unified picture—one that may be characterized as dynamic idealistic monism. Specifically, people partaking of the brew feel that there is an aspect or level of reality that is non-material and that this defines the essence or foundation of all Existence. Furthermore, it is appreciated that all things are interconnected, and that in their totality they constitute one harmonious whole. With this, drinkers appreciate that there is sense and reason to everything and that all reality is invested with great, heretofore unappreciated, meaningfulness. This world-view is also inherently spiritual and it often induces profound religious feelings. Consequently, people at times radically change both their perspective on life and their actual conduct.

As attested to by the foregoing survey, Ayahuasca often directs those who partake of it towards topics and lines of thinking that in their ordinary state of consciousness they are less likely to be concerned with. Yet, at the same time, people very often feel that the insights the brew induces touch the very core of their being. Many feel that these insights have to do with truths that are basic and in fact, self-evident.

The foregoing observations notwithstanding, it should be borne in mind that, in principle, the space of ideas that can be entertained with Ayahuasca is unbounded. There are indeed topics that are more likely to be entertained under the intoxication; these define the nucleus of common ideas associated with the brew. However, the ideas entertained under the intoxication may also reflect the personal-intellectual and/or personal-professional interests of individual drinkers. The range such interests can cover is, of course, vast—in essence, it spans the entire gamut of human experience, knowledge and concerns. Obviously, there is neither way nor sense to enumerate them. Indeed, it seems that there is great latitude on the part of each individual and that essentially one can use the enhanced cognitive state induced by Ayahuasca and direct it in any course of interest he or she may have. One of the most experienced of my informants described this as follows:

> The wisdom offered by Ayahuasca is infinite, and what one can learn from it is without limits. It all depends on the individual: if one wants to cure other people, one will learn about healing; if one is interested in philosophy, one will learn about that; if one would like to be a thief, one can learn about that too. It is in this sense that the brew confers what may be characterized as 'Divine knowledge'.

Idiosyncratic interests and concerns may affect the way visions and other experiences are interpreted. As alluded to in Ch. 8, one may see a fair young maiden and, in line with one's background, ideology, and other personal biases, may interpret the vision to be the Virgin Mary, the Queen of the Forest, the Mother of Ayahuasca, a guardian spirit, or perhaps just a maiden. Current reports by North Americans experimenting with Ayahuasca tend to be especially impregnated with flowery elaborations, usually with New Age interpretative biases. It is striking, I find, to compare the reports that have recently been burgeoning on the Internet and in the popular and semi-popular press with the reports of the early (up until the 1970s, say) European explorers. The earlier reports are by far more 'neutral'—invariably they are confined to strict descriptions of visualizations and they are devoid of personal and ideological elaborations. The more recent ones, I find, tend to be extremely tainted by beliefs, ideologies (notably of the New Age type), and expectations, perhaps also wishful thinking; this is especially apparent in the reports collected by Metzner (1999). Theoretically, the issue of interpretation and its contribution to the Ayahuasca experience is a complex topic that I shall discuss in Chs. 15 and 22. Here let me just say that in the surveys in this book I have done my best to keep, as much as possible, to the style of the earlier travellers to the realms of Ayahuasca.

11

Non-Visual Perceptions

> You can shout loudly in a 'small still voice'. You can scream without anyone hearing you shouting with this soundless 'small still voice'. Anyone can do this. Just imagine the sound of such a scream in your mind. Depict the shout in your imagination exactly as it would sound. Keep this up until you are literally screaming with this soundless 'small still voice'. This is actually a scream and not mere imagination. Just as some vessels bring the sound from your lungs to your lips, others bring it to the brain. You can draw the sound through these nerves, literally bringing it into your head. When you do this, you are actually shouting inside your brain.
>
> Rabbi Nachman of Bratzlav

> And it came to pass on the third day in the morning that there were thunders and lightnings and a thick cloud upon the mount and the voice of the trumpet exceeding loud; so that all the people that was in the camp trembled.... And all the people saw the thunderings, and the lightnings, and the noise of the trumpet, and the mountain smoking.
>
> *Exodus* 19: 16; 20: 18

While visual phenomena are indeed the most salient ones experienced with Ayahuasca, they are by no means the only ones that are encountered. In this chapter I review various perceptual phenomena pertaining to modalities other than the visual. Since the phenomenology of non-visual hallucinations is less rich than that of the visual ones, I discuss all non-visual perceptual effects in one chapter. In the sections that follow, I present all the observations I can make for each of the non-visual perceptual modalities. Subsequently, at the end of this chapter, I shall summarize all the patterns that have been reviewed and consider them from a unified and more general point of view.

The Auditory Modality

Sensory Effects

I shall begin with overall changes in the manner auditory stimuli are perceived. Strictly speaking, the non-ordinary perceptions in question are not hallucinations—they do not involve perceiving without there being actual stimuli in the real world nor do they consist of a false identification of such stimuli. Yet, the way these stimuli are perceived is, indeed, different from the ordinary. As in vision, where (real) colours are intensified and become more saturated and more brilliant, so too in the auditory modality. Overall, one's audition becomes subtler and more acute, and auditory stimuli sound fuller and stronger. With this, one may sense that one detects sounds that are well below the normal auditory threshold.

Another kind of alteration in the way real stimuli are heard is that of dislocation of sound. In other words, (real) sounds are heard as coming from directions other than they actually do. Thus, in several *Daime* sessions, I heard the music played by the musicians at the centre of the hall in front of me as if it were coming from just above my head as well. On other occasions, the music emitted from a speaker at one corner of the room sounded as though it were coming from all around. Identical experiences were reported to me by many other persons. The following indigenous description (cited in Baldran and Bareiro-Sanguier, 1980: 93) can, I think, be read as describing a similar perceptual effect: 'When a man drinks the yahé, when he is intoxicated with yahé, one thing arrives here, another there, a sound here, another there, all over there are noises, sounds, colours. There, there, there.'

There are also specific auditory effects that drinkers of Ayahuasca commonly report. One effect, usually described as most annoying, is the hearing of a continuous buzzing sound inside the ears. Another effect is that described by many as 'the sound of running water'. This very phrase has been articulated by several of my informants and is also used in Harner (1973*a*); curiously, 'the noise of great water' is also mentioned in the most spectacular vision described in the Bible, the Divine vision opening the book of Ezekiel (1: 24).

Structural Effects

One of the effects of Ayahuasca is to induce non-standard parsing of perceptual stimuli. Parsing is a term I borrow from linguistics where it refers to the way a sentence is factored into constituent phrases. The same sequence of words may be factored, or parsed, in two or more different ways, thus giving rise to two or more different semantic readings. For example, the sentence 'They are flying planes' can be parsed with 'flying' being read as an active, transitive verb, and 'planes' its direct object or, alternatively, with 'flying' being a predicate modifying the noun 'planes'. In the first case, the sentence is interpreted with its subject being 'they' (i.e. some persons who are operating planes); in the second, the sentence is interpreted with 'planes' as the subject. In the present context parsing refers to the different structurings of perceptual stimuli.

My own experience occurred in the course of a *preparo*, an Ayahuasca preparation ceremony in the context of the UdV. The ceremony lasted continuously for two days in the course of which there were two drinking sessions. Under the intoxication, I was listening to the sounds of the pounding of wood produced by a group of men to the side. Realistically, the banging was not co-ordinated, and each man hit the wood in his own irregular rhythm. However, I, under the effect of the brew, heard a magical composition. I did not hear any sound that was not actually emitted by the men who were pounding. However, it seemed to me that the sounds were miraculously structured and that together they generated a composition. No, there was nothing in the temporal spacing of the sounds that was altered. Yet, they appeared to me to be wonderfully patterned. What I heard in the pounding was a marvellous composition in the style of contemporary music (the name of the composer Ligeti came to my mind). I should say that, normally, avant-garde classical music is not my favourite style. Similar experiences were reported to me by two informants. Listening to the series of clicks made by the buttons of a machine, the first informant felt that the sounds constituted a musical piece. The second informant heard a verbal conversation as if it were composed, like music.

General Semantic and Interpretative Effects

An effect that borders between the sensory and the interpretative is that of beautification. With Ayahuasca, music sounds immensely beautiful. One's perception of minute aspects of the music is heightened and it often sounds—literally—heavenly. In this regard, I will mention the hymns of the Santo Daime Church. Musically, these are simple songs akin to folk songs of the Brazilian north-east. Under the intoxication, however, they genuinely sound like the most beautiful music there is. This feeling, I should note, remains with one even when one is no longer under the effect of the brew.

As in the visual modality, there is an enhanced conferral of meaningfulness. This is especially marked with music. In Ch. 8, I recounted a vision I had in conjunction with Verdi's *Requiem*. I have listened to this piece many times, but in this session the experience was like never before. It seemed to me that I was listening to this work precisely in the same manner that Verdi himself had heard it in his mind's ear when he composed it.

Often, Ayahuasca also induces a sense of heightened understanding. Once, in a private session I directed, a Mozart concerto was played. It was clear to me that I understood what Mozart was doing in the composition, what he wished to convey, how he decided to do it. I understood the harmonies, the developments and the modulations, the acrobatics performed with the lines of composition, and their resolutions. Indeed, often I felt I could anticipate what the composer was saying. In the same session there was also a person present with practically no knowledge of classical music and who had never heard that Mozart piece before. This person's experience was very much the same as mine. The phenomenon at hand seems to be analogous to that normally experienced in language. Listening to a person speaking or a lecturer delivering a talk, an involved interlocutor or member of an audience can assimilate the line of presentation or reasoning so as to be able, in real time, to predict what the speaker or lecturer is about to say. The first time I experienced such a mode of understanding with music was with Ayahuasca.

Hearing-as

The altered sensation of sound often induces non-standard interpretations of the source of the sound and its nature. For instance, once it rained and I heard this as lots of commotion taking place on the roof of the hut in which the Ayahuasca session was taking place; I thought animals were running around up there and I felt rather apprehensive. On another occasion, what I eventually appreciated as being a far-away motorcycle sounded as if a great, ill-comprehended, cataclysmic event was approaching. All these are examples of a phenomenon I shall call 'hearing-as'. It is fully analogous to 'seeing-as' in the visual modality: Just as one sees a figure as something else, one hears a sound as something else.

Personally, the auditory episode that puzzled me most was one that took place during an intimate *Daime* festive session in which only about a dozen persons participated. At one point in the session, I spontaneously moved towards the centre table. Space was made for me and I stood up at the head of the table and sang. Among the musicians on the other side of the table was a person playing the clarinet. What I heard him play were traditional Jewish Hassidic melodies. The clarinetist was a Brazilian who could know nothing of this kind of music, nor did he know me or anything about me. Unfortunately, I will never know how the music sounded outside the hallucinatory effect.

Another musical hallucination that impressed me very much occurred in an interval of a *Daime* session when I was lying in the meadow grass. I found that a mosquito was sitting on the tip of my nose. I was about to brush it off, but then, impressed by the fact that the mosquito was sitting there precisely on the tip of my nose, I changed my mind and decided I might as well let it be and observe. I did just this, and thus I heard the mosquito sing Mozart. Having been frequently disturbed by mosquitoes both in my home country, Israel, and even more so during my travells throughout the Amazonian regions, I often wondered what justification there might be for the existence of these annoying creatures. After all, their only contribution was, I surmised, negative. And here, I found out, here was an answer. What generally sounds irritating and monotonous is, in fact, on another scale, wonderfully musical. In this conjunction let me cite a report in the anthropological literature indicating that the Cashinahua say that during Ayahuasca sessions insects sing enchanted music (Chiappe, Lemlij, and Millones, 1985).[1]

Full-Fledged Hallucinations

A full-fledged auditory hallucination is one that is not evoked by any stimulus in the real world. One reported both in the anthropological literature and by inform-

[1] Some scientific findings in animal behaviour may be relevant here. The sound patterns emitted by insects (as well as by birds) exhibit subtleties that human beings cannot detect. Temporally, these patterns are extremely fine and they can be appreciated only with highly magnified graphic recordings of these patterns; in the case of birdsong the magnification speed may be fifty times slower (hence the corresponding amplitudes fifty times more extended) than normal (see Greenewalt, 1968; Cohen, 1983).

ants I have interviewed is that of angelic music performed in the heavens. I experienced such music in one of my first encounters with Ayahuasca. The session was held with indigenous people in the Putumayo region in southern Colombia. I was offered a large dosage of Ayahuasca and the effect came quickly:

> Immediately I felt I was thrown up into a realm high above the planet and found myself in the midst of what seemed to be a cosmic lottery. My understanding was that my entire existence, both physical and mental, was at stake: if I played it right, I would be saved, if not—I would lose everything and perish. In order to redeem myself, I began to sing. There was no reflection involved in this decision, nor any recourse to past experience with Ayahuasca (as would be the case today). For six or eight hours I continuously sang the praises of God. The words that spontaneously came out of my mouth were (in Spanish) *Gloria a Dios* (in English, Glory to God); the melody was being composed as it was being sung. As I was singing, I found myself to be surrounded by an immense choir of angels—I was taking the leading role and they were accompanying me. The music was exceedingly beautiful. Every now and then the choir was joined by guest groups that came in, performed, and then left. One such group that especially impressed me was one of Black, very sensuous, players. The music they played was very different from that sung by the angels, but it all fitted very well and the ultimate meaning and purpose of it all was one and the same—*Hallelujah*, that is, the singing of God's praises.

I have heard reports of angelic singing from several other persons; some describe it as the most beautiful music they have ever heard.

The Music of Silence

Lastly, there is an effect I shall call 'the music of silence'. I have experienced it only once. One person to whom I recounted this episode told me he recognized this effect from his own personal experience with Ayahuasca:

> The scene was celestial. It was a white place which I interpreted to be up in the High Heavens, where the Divine is praised. Acoustically, there was absolute silence there. Yet, I knew it all was full of what was, in essence, music. The ambience was most serene and I felt this was the music of the angels.

I shall note that cognitively, the effect at hand may be likened to the vision of darkness described later, in Ch. 17, where visions of light are surveyed.

The Auditory v. the Visual

Throughout the foregoing survey I have underlined analogies between the visual and the auditory modalities. Let me also note some differences between the two. As in real life, so with Ayahuasca: the visual effects are those that are deemed primary and more significant. As demonstrated throughout the discussion in this chapter and in those that precede it, they are richer and more complex than all other sensory effects; they are also the more salient and all-encompassing. It is not for nothing that the visions are the first effects of Ayahuasca that come to mind and that universally they are considered as the most paradigmatic manifestations of the Ayahuasca experience. Yet, at the same time, it seems to me that it is not uncommon for auditory hallucinations to evoke more fear than the visual ones. A consideration of this difference highlights several further aspects by which the two modalities differ.

First, it shall be noted that the possibility of checking the identity of stimuli one may suspect one is hallucinating is different in the two modalities. In the case of visual perceptions, the hallucinating person may know where the stimulus is, hence he or she can check and verify its identity. In particular, he or she may attempt to interact with that stimulus, for instance, attempt to touch it (in the case of an inanimate object) or speak to it or engage it in mutual action (in the case of a human being). In contrast, in the auditory modality the location of the stimulus may be unknown, and consequently verification may be more difficult. In particular, mere further scrutiny might give no more information than one already had. Indeed, my experience is that verification of the identity of an auditory stimulus succeeds only when a concurrent visual check is possible. Thus, what is really determined is not the nature of the sound itself, but the nature of the source, or producer, of the sound. In the visual case the inspection and the solution are both visual.

Related to the foregoing observations is the concreteness of the visual as opposed to the non-concreteness of the auditory. Objects that are seen are concrete and palpable, while hallucinations of them are not. Thus, if one touches an object that one experiences as seen and that object disappears one can conclude that it is a hallucination. In contrast, neither sounds nor their hallucinations are concrete, hence nothing like a probing touch is possible. With the possibility of verification being limited, uncertainty increases, and with it one's fear.

Thirdly, by extension of the above, it is generally the case that the range of interaction with visual stimuli is larger than it is with auditory ones. Furthermore, if one interacts with a visual stimulus and finds no response, one might conclude that it is hallucinatory. The auditory analogue does not hold. For instance, one might conclude that the source of the sound is inanimate or that it is an animate being that does not wish to interact with one. Given that there are fewer possibilities of verification with auditory stimuli, they have the potential of evoking more fear.

Last but not least, eyes can be closed, ears cannot. Or, more precisely, unlike ears, eyes can be opened. Usually, the easiest way to stop a vision causing discomfort or fear is to open one's eyes. However, this method is not foolproof—when the intoxication is very strong, opening the eyes does not change very much what one sees.

The Tactile Modality

As in vision and audition, so in the tactile modality— amplification is encountered, and one also becomes more sensitive to minute variations.

Coupled with the phantasmagoric imaginative interpretation that Ayahuasca induces, the overall heightened sensory sensitivity may lead to a phenomenon that may be characterized as *feeling-as*. Thus, on one occasion, I was touched by a fly and felt I was being pushed hard. It seems to me that indigenous reports of strong sensations that animals are 'biting one's buttocks' may be interpreted in the same manner. I myself have often experienced similar, out of proportion sensations when other people touched me during a session.

Tactile effects that are not directly related to stimuli in the external world are rare. In the corpus of data I have collected there are only ten instances of this phenomenon. Most consist of a feeling that the drinker has during the vision that someone (in all cases unseen) is touching him or her. One informant reported a tactile sensation associated with an item seen in the vision: he saw plants with thorns and felt that he was pricked by them.

The most special tactile hallucination I have experienced occurred at the end of the series of visions on royalty described in Ch. 7. As the reader will recall, the session took place in the midst of the forest. Four hours after the session began, a storm seemed about to break out, and the person in charge decided that we should terminate the session even though I was still highly intoxicated. I asked that person to support my arm as I walked and we started to march through the trees towards a hut several metres away. Even though in reality what my feet were touching was uncultivated forest ground, to me it felt as an artificially constructed flat, horizontal surface. I looked down—what my feet were stepping on was a floor made of beautiful *azuleijos*, tiles of emerald green and gold. It was the exit of a palace, and I was strolling out of it.

A full-fledged tactile hallucination would be one in which a person feels that he or she is touching an object that, in fact, does not exist. The closest I have come to this was once, when the effect of the brew was very strong. I found myself in 'the world of Ayahuasca' and there were phantasmagoric beings who approached me and offered to cure me. *Inter alia*, they presented me with a magic medicine vial and they placed it in my hands. At that moment, however, the vial disappeared. Interestingly, all the cases of full-fledged tactile hallucinations reported to me by my informants were similar—they all involved the reception of an object. Since this phenomenon is rare I report here all instances of it that appear in my data. One informant saw herself in an indigenous ritual in which she was presented with a magic potion. Another informant had a vision in which the Virgin Mary appeared to him and gave him a candle, which he then returned back to her. The strongest effect was one in which an informant experienced receiving a wristband made of cloth (such bands are very common in South America). He saw and felt the band on his wrist, but within seconds, it disappeared.

The Olfactory Modality

With Ayahuasca, olfactory perception, too, intensifies. This is especially noted when, during the Ayahuasca session, incense is passed around or when tobacco smoke is inhaled. In this conjunction, I shall note that in many contexts, drinkers of Ayahuasca are advised not to use bodily cosmetics (e.g. aftershaves, eau de cologne, perfumed deodorants) before sessions. Due to increased sensitivity to odours, these may disturb other participants in a way that normally they do not.

I myself have never experienced what may be regarded as a full-fledged olfactory hallucination, but such experiences were reported to me by half a dozen informants—all consisted of either of two types. The first involves the diffuse smelling of perfumes that evidently are not in the real world. This experience need not be associated with a vision and the perfumes are usually characterized as wonderful. The second type consists of the smelling of the scent of flowers seen in a vision.

Lastly, as in the visual modality (see Ch. 17), iconic perception may last for longer than normal. An example is presented below in conjunction with intermodal effects.

The Gustatory Modality

One does not eat during Ayahuasca sessions. Indeed, physiologically it is nearly impossible to do so. Usually people do not drink either. On a few occasions, however, I drank a little water just towards the end of a session, and once I was offered warm, sweetened milk. In all these cases I took only a very small sip, yet the experience was most powerful. The liquid that came into my mouth seemed to be extremely delicious. Twice I was so certain that fruit juice had been added to the water that I asked about it, but there was none. I have heard similar reports from other people. In line with the terminology introduced above, these experiences may be characterized as instances of *tasting-as*.

On another occasion, I felt that I could sense the various ingredients in the water, for instance, the different minerals that were in it. This made me wonder whether a person who is sufficiently sensitive could accomplish a full analysis of the composition of water (or any other material) by merely tasting it.

Cases that could be characterized as full-fledged gustatory hallucinations were neither experienced by myself nor reported to me by any of my informants.

Proprioception

Under the Ayahuasca intoxication, proprioceptual effects are very much enhanced. Many people report heightened sensibilities of their bodies; these are usually coupled with unusually fine degrees of sensitivity and control. For instance, I often felt the connection between my vertebrae in a way I normally cannot. This, in turn, made me able to work on my posture and correct it; these experiences were especially meaningful for me for in the past I have suffered from back problems. Other parts of the body I have felt when inebriated included ones that normally are not subject to proprioception at all—notably, the inner organs and the brain. Other persons reported sensing, and also seeing, individual cells of their body. Such sensations and perceptions may, of course, already consist of full-fledged hallucinations.

Intermodal Effects

Most important in the Ayahuasca experience are intermodal effects. These are known in the psychological literature as synaesthesia. Synaesthesia is the phenomenon of cross-modal perception—colours are heard, tunes are seen. Normally, synaesthetic perception is encountered in children but with adults it is, at least in its pure form, quite rare (for further discussion see Marks, 1978; Cytowic, 1989; the anthology edited by Baron-Cohen and Harrison, 1997; as well as Shanon, 1982; for discussions that relate synaesthesia to altered states of consciousness, see Klüver 1928, 1966; Hunt, 1985; and Cytowic, 1989).

With Ayahuasca, synaesthesia is paramount; indeed, this is true of psychedelic experiences in general (see Ludwig, 1969). As noted throughout this book, music often guides one's visions. The tempo and rhythm of the music one hears is often reflected in one's visions. Notably, the music determines the rhythm and movement of figures in the visions as well as the rate of change between images. Thus, the circles of light one commonly sees when intoxicated often pulsate to the rhythm of the singing one hears and are co-ordinated with the movements of one's body in dancing. Many times during festive *Daime* sessions I have seen visions of women dancing. Invariably, the dance was in synchrony with the real music being played in the hall. Likewise, the chanting of *ayahuasqueros* is sensed as directing the course of the visions of those who listen to it and as determining their general flavour and colouring (see, for instance, the report by Kusel in Ch. 1).

As indicated, auditory-to-visual synaesthesias are extremely common with Ayahuasca. The following myth told by the Kamsa and Inga of southern Colombia and recounted by Ramírez de Jara and Pinzón Castaño (1986: 175) may be read as an attestation to these effects:

> At the beginning earth was obscurity. Men already existed, but they lacked intelligence. They took yagé and parted it in half. For the women it gave menstruation, for the men—the piece grew up and up until the sky. Little by little the shadows took on shape and form and the silhouettes got details. In the sky, they saw yagé entering an immense flower. It got fecundated and transformed into the sun. And from there came the people of the sun with their distinct music played on flutes and tambors. Each melody transformed into a distinct colour. When they arrived to the earth they gave it colour, and when the world was illuminated, played the symphony of colour and the music awakened the comprehension of people, creating intelligence and language. This is why yagé is used: with it the world is seen as it is, and the intelligence expands making everything clear and harmonious with the spirit of cure.

While the auditory-to-visual synaesthesias are by far the most common ones, other synaesthetic patterns occur as well. Especially to be noted is the olfactory-to-visual synaesthesia. I have encountered it several times when smelling incense or smoking tobacco. The smell, along with the air that passes through my nostrils and respiratory system, was perceived as waves of beautiful colours. Once during a session, I was presented with a special perfume that was put on the palms of my hands to savour. I brought my hands to my face and inhaled deeply. The sensation was visual and truly enchanting. On that occasion, I also observed that my olfactory sensitivity heightened significantly. Furthermore, I could smell the traces of the perfume for a long time after having been presented with it.

A case of tactile-to-visual synaesthesia occurred when I was pressing my temple and I saw a bird with a beak whose shape was congruent with the movement of my hand. I have not experienced a gustatory-to-visual synaesthesia, but I can envisage it—for instance, tasting a fruit and seeing a colour (see also Cytowic, 1993).

The reader will observe that the synaesthesias described above are from a non-visual modality to the visual one. Can there not be synaesthesias from vision to another perceptual modality or from one non-visual modality to another? The answer to this question is, I find, somewhat unclear. Inspection of my data suggests that, indeed, full-fledged synaesthesias are to the visual mode, not from it. Other intermodal effects may be fuzzy cases for which the applicability of the epithet 'synaesthesia' might not be very clear. For instance, auditory stimulation may (literally) move one, shatter one, induce physical sensations in the inner parts of one's body. Similarly, gustatory sensations may be amplified so as to be felt concretely as (or almost as) tactile ones. Should these be regarded as instances of synaesthesia? I am not certain, but rather than attempting to split hairs, let me leave the matter with these phenomenological observations and not push it from a conceptual, theoretical point of view. More on synaesthesia will be said in Ch. 20.

Summary and Conclusion

Having presented the various effects encountered with the different sensory modalities, let me consider them in unison and attempt a more abstract structural analysis of them. In particular, the following analysis highlights the commonality of perceptual effects across modalities.

The most basic effects are the strictly *sensory* ones. With Ayahuasca sensory stimuli are perceived to be stronger, ampler, and more intense. Further, one's perception becomes remarkably acute and one is aware of perceptual variations that normally one cannot detect. These effects are encountered in all perceptual modalities. With them, the external world is perceived to be both richer and more beautiful than it normally appears to be.

Coupled with the heightened perception is an extended perceptual memory. This is most pronounced in the visual modality. In particular, I am referring to what, in the cognitive-psychological literature, is called iconic memory: the ability to keep in one's memory a vivid, perceptual-like image of sensory stimuli. With Ayahuasca this ability is greatly enhanced. Typically, when drinkers of the brew gaze at the real surrounding environment and then close their eyes, the image of what they have seen, or part of it, remains in their mind's eye (for further discussion see Chs. 20 and 21).

The next cluster of effects are the *structural* ones. Phenomenologically, these need not be perceived as categorically different from the sensory ones. Yet conceptually, it is, I believe, worthwhile to distinguish between the two. The sensory effects pertain to what may be regarded as first-order sensory properties; the structural effects pertain to second (or higher) order ones. In the auditory modality we have noted structural alterations in both the spatial localization and temporal organization of stimuli.

The sensory and structural alterations lead smoothly to the *semantic* and *interpretative* ones. Confronted with non-ordinary qualities of stimuli, the drinker of Ayahuasca can be puzzled, perplexed, and perhaps also frightened; he or she may also be enchanted and feel enlightened. With this, the drinker is prone to wonderment and look for explanation. This in turn may result in unconventional interpretations of what is happening.

The non-standard interpretations vary in how lasting they are. They may be quite temporary and be dismissed with further inspection and reflection. For instance, as mentioned previously, a sound that had been thought to be an impending disaster is, upon further scrutiny, recognized to be an approaching motorcycle. When the non-standard interpretation is more durable we encounter the phenomenon of *perceiving-as*: seeing-as, hearing-as, and the like.

There are variations in the complexity of the phenomenon of *perceiving-as*. All the features associated with an item or a state of affairs may be ones that are actually there in the external world, and only their qualities—e.g. amplitude and intensity—may be modified so as to induce an altered interpretation and identification. Alternatively, features may be elaborated or embellished so that items become more complex. This generally results in a sensation that things in the world have transformed and in the perceiving of things as other than they actually are.

And finally, there are *full-fledged hallucinations* in which the person who has partaken of Ayahuasca perceives things that are not there in the real world at all. In the visual modality, that in which the effects of Ayahuasca are most varied, I have made a distinction between effects occurring with the eyes open and those occurring with the eyes closed. It seems to me that this distinction is applicable to the other modalities too. Specifically, the likelihood of perceiving what is not there in the external world increases when the eyes are closed not only for the visual modality but for the other perceptual modalities as well. In particular, I have heard 'non-existent' sounds only when my eyes were closed.

As is, I hope, appreciated from the foregoing analytical summary, the gradations between the different perceptual effects are subtle. This is true both structurally and dynamically. Structurally, the different effects encountered with Ayahuasca define an entire space of possibilities. My theoretical stance is that this is typical of natural cognitive domains in general (for analogous patterns in natural language and in thought sequences, see Shanon, 1989*c*). Likewise dynamically: usually, the movement from one effect to another that is stronger than it (for instance, from the mere sensory to the interpretative) is smooth. Experientially, the distinction between structurally distinct effects is often blurred. The person under the intoxication may perceive something as completely phantasmagoric, then reflect a moment or physically shake his or her head and appreciate that what is perceived is actually this or that. More on this will be said in Ch. 21, which is devoted to the dynamics of the Ayahuasca experience.

In the foregoing survey, I have set myself to highlight the general characteristics of the perceptual effects induced by Ayahuasca. This is crucial for any theoretical cognitive understanding of the phenomenology at hand. However, in closing this discussion, I would like to emphasize what is clearly manifested by the empirical observations made here—namely, that as far as the Ayahuasca experience is concerned, the various perceptual modalities are definitely not on the same par. Without any question, the effects encountered in the visual modality are more common, more varied, and more complex than those encountered in any of the other modalities. Next in line is the auditory modality. With all the other modalities, non-ordinary perceptions are by far less common and they present a significantly less rich phenomenology. Of these, the tactile ones seem to be slightly more common and varied whereas the gustatory ones are the rarest and most rudimentary.[2] By and large, this order seems to correspond to the relative functional preponderance of the different sensory modalities in ordinary cognition. In particular, vision, which is the most important of our sensory modalities, is also that with which, under the Ayahuasca intoxication, hallucinatory experiences are most common. Interestingly, the order of relative frequency just noted differs from that encountered in psychopathology. In psychosis, the auditory hallucinations are by far the most common ones (for a review, see Slade and Bentall, 1988). The foregoing, I should emphasize, are general, summary statements. They do not imply that a single non-visual effect may not be experienced as most powerful.

A Comment on Motor Co-ordination

To close the survey of the various perceptual effects encountered with Ayahuasca, I append a comment on the motor effects that correspond to them. Above, I have noted that the brew enhances proprioceptive sensibilities. Yet at the same time, under the intoxication, drinkers often exhibit difficulty in fine motor co-ordination (see also Chs. 4, 13). The conjunction of these two patterns may seem a contradiction, and indeed, there is something paradoxical about the effects in question. At the very same time, Ayahuasca may hinder motor co-ordination *and* facilitate motor performance. I myself have experienced this on many occasions. On the one hand, I had difficulty in lifting myself up and changing from a sitting to a

[2] It should also be remembered that the gustatory modality is the least employed with Ayahuasca; as noted, during the course of the inebriation, people refrain from eating.

standing position; on the other hand, once I was on my feet, I could dance in a manner by far superior to that which I would in my normal state of mind. These two seemingly contradictory effects may, in fact, be viewed as the symptoms of one underlying factor, namely, enhanced sensitivity. This enables drinkers both to be more sensitive and finer in their performance and also more fragile and vulnerable. By and large, it seems to me that the difficulties are encountered in the changing of position and in the execution of abrupt, isolated acts. In contrast, the enhanced performance seems to be associated with the execution of psychologically meaningful activities that extend in time. In other words, moving from one state to another may be difficult; however, once within a state, and swept along by the special energies that Ayahuasca confers, one can actually be more agile and display higher motor control than usual.

12

Consciousness I

There must be a fourth state beyond the waking, dreaming and dreamless states in which the absolute oneness of Brahman-Atman is what should be known.

The Upanishads

Ayahuasca is an agent that induces an altered state of consciousness; as such, it does precisely that—it modifies people's state of consciousness. In this chapter (and in Chs. 14 and 16) I examine these modifications and the effects associated with them. Before doing so, however, I will make some more general remarks regarding the nature of human consciousness.

General Preliminary Remarks

Consciousness may be dealt with from several different perspectives. Logically, the first perspective is the *substantive*, *ontological* one. It is concerned with the question of what, in the grand order of things, the phenomenon of consciousness is. Questions pertaining to the substantive perspective are: What does consciousness consist of? How does it come into being? Is consciousness a phenomenon that is part and parcel of the material world or does its existence indicate that reality is made up not only of matter, but also of other, non-material constituents? While these are all intriguing questions, we shall not deal with them here; they pertain to metaphysics and as such they are not within the framework of our cognitive-psychological investigation. In saying this, I am not denying the importance of these questions. In fact, I believe that in the context of altered states of consciousness in general and in that of Ayahuasca in particular, such questions gain special significance. It might even be that, ultimately, the cognitive understanding of the phenomenon of consciousness will itself require reference to metaphysical considerations (for such a view see, for instance, Goswami, Reed, and Goswami, 1995, and Close, 1997). Thus, in excluding the metaphysical questions, I am not advocating a principled negative stance in this regard. Rather, I am keeping to a certain framework by which the present work is defined, namely, a cognitive-psychological framework adopting a strict phenomenological line. Important as they are, the metaphysical questions are outside the scope of this framework (for an interesting defence of a similar position, the reader is referred to James, 1890/1950, especially chs. VI, IX, and X).

Even when the ontological questions are put aside, from a *structural* and *phenomenological* perspective the question 'What is consciousness?' still remains. The phenomenological psychologist does not ask how consciousness comes about or what is the relationship between consciousness and the brain. Rather, she accepts consciousness as given and she sets herself to determine what the structural features of this phenomenon are. From a structuralist, phenomenological perspective consciousness is, essentially, a cognitive system that defines human subjective experience. That we, human beings, have subjective experience is not a trivial matter. In principle, one could envisage a cognitive system lacking this quality. Furthermore, subjective experience could be defined in a number of different ways. Hypothetically, cognitive agents could have been built so that they feel, sense, and perceive the world differently than the members of the species *Homo sapiens* actually do. For instance, they could perceive colours differently, feel that their bodies are heavier or lighter, see things in the world as larger, more distinct, or more energetic, and experience the flow of time as faster or slower. Thus, from a structural point of view, consciousness can be defined as the set of parameters that specify the particular way human beings experience the world, both physical and mental. If the specifications of these parameters were to change, the phenomenology of experience would be altered. However, since we never get out of the confines of consciousness, it is extremely difficult for us to be aware of the structure of consciousness and be cognizant of these parameters.

By way of further clarifying this structural perspective, let me use an analogy. I wear glasses. Without them I cannot see. Usually, I pay no attention to the glasses. Indeed, I normally take them for granted. However, were the specifications of the lenses to be altered, the features of the world that I perceive would drastically change. For instance, were the lenses tinted pink, the colouring of all I see would be modified. With this, the existence of the glasses, their specifications, and their contribution to my vision would become apparent. Likewise with consciousness, only that in this case the situation is more radical. Glasses can be taken off, but consciousness is always with us, or rather—everything we experience is always grounded in consciousness. It is here that the fundamental cognitive-psychological importance of the study of altered states of consciousness is manifest. As noted, the basic goal of the structural analysis of consciousness is to identify the set of parameters that together define the nature of human subjective experience and determine the specifications pertaining to them. But, as in the case of the glasses, it is only when our state of consciousness changes that we can begin to appreciate these parameters and values.

Let me repeat. To my understanding, a structural study of consciousness is precisely this: the definition of parameters and the determination of their values. For this, what one basically has to do is address the question of what, in effect, consciousness is not. This question is not easy to tackle. Some of the key characteristics of consciousness are so ingrained in our psychological existence that they are taken for granted and we are usually blind to them. In order to define the structural parameters of consciousness and to determine their actual values, the student of mind has to consider the potential range of these parameters and their values. The great potential contribution of the study of non-ordinary states of consciousness to the scientific understanding of the mind lies precisely in their rendering the parameters of the cognitive system apparent and in their revealing the various possible values that these parameters may take.

Before proceeding let me introduce yet another analogy; it comes from the field of mathematics. In the algebraic context a term such as *a* indicates a certain value: it is *a* as contrasted with, say, *b*. But in fact, *a* is 1*a*, and furthermore, it is also 1*a* to the power of 1. Normally, the constant 1 and the power 1 are not indicated or even thought of. However, once a broader, polynomial perspective is taken then one realizes that indeed the single term *a* is actually 1*a* to the power of 1. Thus, a distinction is noted between two components of the mathematical expression. On the one hand is the term *a* as contrasted with other possible terms, such as *b* or *c*; on the other hand are the multiplication and power factors—in the example given here, both are equal to 1. Turning to the domain of consciousness, I suggest an analogous distinction between, on the one hand, the particular mental state that one experiences at any given moment (i.e. specific sensations, perceptions, ideations, and other mental states) and, on the other hand, the parameters that define consciousness as a cognitive system. As I conceive it, the subject matter of a theory of consciousness is the latter, not the former. Again, the cognitive impact of non-ordinary states of consciousness is that they reveal the rich range along which the parameters in question can vary. A theory of consciousness cannot be confined to one specific case, as important and prevalent though it might be. Such a theory, if it is to be scientific, must address itself to *all* possible variations of consciousness.

A third perspective for the study of consciousness is the *functional* one. Consciousness serves several cognitive functions. The first pertains to the definition of the *self*. First and foremost, consciousness defines the background of all mental material. All the mental materials that I experience are, by definition, mine. When a thought such as 'This is a beautiful day' passes through my mind, what is actually being entertained is something like the following: 'I [B.S.] am thinking that this is a beautiful day.' In general, it is as if all mentations are encompassed within quotation marks—inside is the content expression and outside is an indication of *agenthood*.[1]

An extension of the above is the sense of personal *identity* that consciousness confers. In essence, the identity of this agent is constituted by the totality of mental activity this agent is experiencing—his or her thoughts, sensations and feelings (for all these characteristics, see James, 1890/1950).

Personal identity also manifests *unity*, and this in two respects. On the one hand, every cognitive agent has one identity (not multiple identities); on the other hand,

[1] After this text was written, I discovered that some authors refer to this feature as 'agency'; see Graham and Stephens (1994), and Radden (1996, 1998).

the thoughts that constitute each agent's mentality are, in principle, all interconnected.

A person's individual identity is further defined by means of *contrast*. People have a sense of what constitutes their self and what does not. What is not me consists of both the physical world outside me and the community of social others with whom I interact. With these contrasts, a *differentiation* is made between the internal world and the external one. There is also a differentiation between different *mental states*—a differentiation between one's thoughts and one's perceptions, between one's memories and one's perceptions, between perceptions and dreams, between perceptions and hallucinations.

Another main function of consciousness is that it defines the *calibration* of experience of both the cognitive agent and the way he or she perceives the world. As a rule, in order for an attribute to apply, a normal standard must be assumed. Things are big or small, heavy or light, slow or fast, and so on, relative to certain standards. Such standards are a prerequisite both for the perceptions of the outside world and for one's own proprioceptions.

All functional aspects indicated above pertain, in one way or another, to the realm of inner experience. However, consciousness also faces the world outside. A fundamental (according to some, *the* fundamental) property of consciousness is *intentionality*, that is, directedness towards an object (see, for instance, Dennett, 1987; Searle, 1992). Brentano (1874/1973) singled this out as the key property of all mental states. Thus, when one perceives one perceives something, when one is thinking one is thinking of something. James (1890/1950) too includes the directedness towards an object as one of the key characteristics of consciousness.

Intertwined with intentionality is *connectedness* to the world and having *knowledge* of it. As pointed out by both Bergson (1944) and Merleau-Ponty (1962), it is by virtue of our experiencing it that we come to know the world in which we live. That this is the case is reflected in lexical morphology: The word 'consciousness' is a cognate of the word 'cognize', to know. Indeed, in the Romance languages, 'consciousness' and 'knowledge' are denoted by the same or by very similar words, for example, 'conscience' and 'connaisance' in French. The same is true in Hebrew, a totally unrelated language—*toda'a* (consciousness), *muda'ut* (awareness), and *yedi'a* (knowledge) are all from the same root. In Sanskrit, a language extremely rich in terms for mental phenomena and activities, such relationships are encountered as well—examples are the verbs *budh* and *jña**, both of which have derivatives denoting consciousness or awareness, on the one hand, and knowledge, on the other. Related to knowledge is the *conferral of reality*. As pointed out by Merleau-Ponty (1962), perceiving involves not only discerning specific information about the objects seen, but also taking these objects to be real. When I see a flower, not only do I detect that it is a particular object with a particular shape and colours; seeing the flower, I also take it to actually exist there in front of my eyes. Intertwined with knowledge, the conferral of reality is thus another key function of consciousness.

Finally, our being conscious affords our being conscious of ourselves. Not only are we conscious of the world, we human beings are also conscious of ourselves as cognitive agents that think and act in the world. Endowed with *self-consciousness* we can reflect on both our cognitive activities and their products. Moreover, we can be conscious of our very selfhood and identity.

A Typology of Non-Ordinary Effects of Consciousness

This section is a survey of the various non-ordinary patterns of consciousness that Ayahuasca induces. Bearing in mind the foregoing general discussion, the following discussion centres on a series of structural parameters of consciousness and the values they can take. Eleven such parameters are presented; they pertain to five clusters that correspond to the five main functions that consciousness serves—the self and its identity (items (1), (2), and (3) in the following survey), contrast and differentiation (items (4) and (5)), the calibration of experience (items (6), (7), and (8)), reflective self-consciousness (item (9)), and intentionality (item (10)), connectedness, knowledge, and the conferral of reality (item (11)); I discuss self-consciousness before intentionality, connectedness, knowledge, and the conferral of reality because the latter present some more general philosophical issues which I consider briefly at the end of this chapter.

1. Agenthood

As noted above, normally all the mental material that a person experiences and entertains is his or her own. In particular, this applies to the thoughts that pass through one's head. Prima facie it seems inconceivable that things could be different. Indeed, how could there be any other possibility? Yet, with Ayahuasca, people do at times experience thoughts as not being their own. Two phenomena will be noted.

The first phenomenon is a dissociation between the self and the mental material that one experiences. Content is passing through my mind, but I am not experiencing myself as being the source that generates it. Instead of being the generator of the thoughts that I am entertaining, I feel that I am a channel that receives them. The source of the mental material may or may not be given a particular identity.[2]

In the context of both the Santo Daime Church and the UdV such experiences are standardly characterized as cases of *receiving* (see Ch. 7). In particular, the hymns sung in the services of the Santo Daime are said to have been received, not composed. This, incidentally, is also true of the Barquinha. I have experienced such receiving once. After a *Daime* session, in the middle of the night, I went to the meadows and contemplated. My mouth opened and I sang an entire hymn. The hymn was in Portuguese and it consisted of four well-structured stanzas. Both the lyrics and the music were novel, and I still remember them.

[2] Similar phenomena are also encountered in psychopathology; see the works of Graham and Stephens, and Radden cited in n. 1.

The second phenomenon has to do with control. Experientially, the feeling is that one is no longer in full control of the thoughts one entertains. Rather, one feels that other people or agents are controlling one's thoughts. The converse may also be experienced, namely, that one feels one is controlling the thoughts of other people.

Together, the two phenomena lead to a state of affairs in which thoughts are experienced as being not private. Often, this is accompanied by a strong sense of telepathy. Reports of telepathy are extremely common with Ayahuasca. While I have heard some remarkable stories with regard to both telepathy and precognition, and even though I myself have experienced some very perplexing episodes that seemed telepathic, I reckon that it is extremely difficult to ascertain the validity of these reports in an objective manner; more on this will be said in Ch. 16.

2. Personal Identity

An elemental aspect of being conscious is having a definite personal identity—'I am me' is axiomatic. With Ayahuasca this may no longer be the case. On a couple of, admittedly rare, occasions, it was no longer clear to me who I was. One informant told me of a phase in a session in which she felt she was acquiring the identity of whatever she was looking at. For instance, she looked at the hand of a watch and felt she was becoming one. As will be noted at length in the next chapter, metamorphosis—that is, the transformation of personal identity—is a relatively common effect of Ayahuasca. In the Amerindian context, such transformations are a major feat of *ayahuasqueros*. The transformation of the shaman into a jaguar is especially significant (see Reichel-Dolmatoff, 1975; Lagrou, 1998). More on personal identity and its transformation will be said in the next chapter.

3. Unity

Normally, people feel that they have one cohesive identity, and that their mental life coheres into one unified whole. But perhaps consciousness may be divided and one may have one's standard identity but also another, new identity. In our culture, this phenomenon is associated with the psychopathological condition of multiple personality disorder. In the Amerindian context of Ayahuasca, the ability to maintain a double identity is regarded as a key characteristic of a competent shaman. Shamanistic flight is not just having an illusion that one's identity and location have changed. Rather, the experience consists of being in two realms—being oneself and at the same time being someone or something else: the shaman is here and at the same time he is there, up in the sky or in the heavens (for general discussion, see Eliade, 1964). On this, too, more will be said in the next chapter.

An account which I find extremely insightful is by Taussig (1987: 141) of his first experience with Ayahuasca. After describing his torso as being distorted, his limbs becoming detached and a feeling that at times his body did not belong to himself, he says:

I learn to use dissociation as an advantage[,] as a way of escaping from the horror. I am not the person got at; rather I am the disembodied face-presence calmly peering in and watching this other and unimportant me. I watch my other self, safely now. But then this second me, this objective and detached observer, succumbs too, and I have to dissociate into a third and then a fourth as the relation between my-selves breaks, creating an almost infinite series of fluttering mirrors of watching selves and feeling others.

Division can also be perceptual. A Brazilian with extensive involvement with Ayahuasca told me that once he experienced a split of his visual field. On one side, he had powerful visions; on the other side, he observed the real world around him, just as it was. Further, he could navigate between the two realms and readily shift between observing the visions, on the one hand, and acting in the real world and interacting with other (actually present) persons, on the other hand. I may note that while this never happened to me with Ayahuasca, I once had a similar experience with the San Pedro cactus: on one side of my visual space I saw daylight whereas on the other it was night (as it was objectively at the time). Another pertinent example is the division between the upper part of the visual field and the lower one mentioned in the grand vision of the Celestial Palace cited in Ch. 5.

4. Boundaries and the Differentiation of States

With Ayahuasca, the boundary between the inner and the outer reality may dissolve. One may feel that one's 'I' is blended with that of others, that one immerses oneself in the world and becomes united with it, that there is no neat distinction between one's internal mental world and one's perceptions of the external world.

Likewise, the contrast or differentiation between the various mental states may fade. Under the Ayahuasca intoxication, it is often difficult or even impossible to know whether one is perceiving or remembering, whether one is perceiving or imagining, whether one is thinking or perceiving the thoughts of others. On several occasions, I have found myself reflecting upon the events of the last day (that which ended with the Ayahuasca session in which I was participating) wondering whether they were real or not. I was asking myself whether people I have met and the interactions I had with them were not actually the orchestrated acts of an enchanted *mise-en-scène*. I have heard the exact same description from several other persons.

5. Individuation

Together, the unity of the self and its boundaries define a person's individuation. Normally, consciousness is assumed to be associated with well-defined individual identity. With Ayahuasca there are experiential phenomena that seem not to be in line with this seemingly fundamental tenet. There are experiences in which the notion of the individual self dissipates, and yet consciousness is maintained. The following citations from Polari (1984: 278–9) describe such experiences:

> In that other, parallel existence that I assume exists in some other part of the cosmos, the concept of individuality is not very clear. I was part of a larger body that congregated with a series of 'astral "I"s' pertaining to the same spiritual matrix that nourished various 'physical "I"s.'
>
> The passage gravitated towards increasingly more complex planes and I was able to reach increasingly elevated forms of consciousness. And the more elevated my consciousness was, the more extinguished was the notion of 'individuality' and it dissolved in a manifest ocean of divinity where the total and the all were just One.

In Polari's reports the non-individuation of consciousness is defined with respect to the cosmos or to some sort of super-consciousness, that is—consciousness that is not confined to the bounded selfhood of human beings. From an ontological point of view the postulation of such consciousness is, of course, very radical. Here, however, I am making no metaphysical claims. I am only presenting and analysing various patterns of human experience. Phenomenologically, it is a fact that with Ayahuasca—as well as in other non-ordinary states of consciousness—human beings feel that they are being connected to or integrated within some sort of super-consciousness. The ontological ramifications of such experiences are outside the scope of the present discussion (see also Ch. 16).

The non-individuation of consciousness may also be manifested in the blurring of the distinction between the individual and his or her fellow human beings. As a consequence, one may feel that one's identity is defined not individually but rather in group terms. Thus, strong identification with the other persons who participate in the Ayahuasca session is common. One clear manifestation of this is the communal singing in the rituals of the Santo Daime Church. Many times I have observed how sessions begin—the leading persons start to sing and the others in the hall readily join in, as if tied to them by hidden strings. Furthermore, the singing may be extremely co-ordinated, both with respect to tempo and rhythm and as far as immediate adjustments in tune are concerned. On such occasions, the group becomes a kind of a single organism that acts in a precise and highly concentrated fashion. Once I gave a cassette recording I had made of such singing for inspection to a musical laboratory equipped with high-tech measurement instruments. The experts were astonished at the perfect degree of synchrony between the people singing. In a direct, non-technical manner I have felt this many times as well. As recounted earlier, once I also had a vision that made the notion of group-consciousness even more apparent to me. In the vision I found myself in the midst of an ant colony. I felt the relationship between each ant, as a biological organism, and the colony as a whole. Consciousness was the property of the latter, not the former.

6. Calibration

Usually one does not even think about the fact that one has a certain sense of the size of one's body, its weight, one's posture in space, and so on. Under the effect of Ayahuasca all these may change. For instance, one may feel that one's body is larger, lighter or heavier, and the like. This naturally brings to mind some of the experiences that Carroll's Alice had in Wonderland. Indeed, this very analogy passed through the mind of one of my informants when she felt that the size of her body was changing and shrinking to the size of a match.

A number of times I have felt that my body was elevated above the ground. Two informants told me that they felt as tall as the trees in the Amazonian forest. In this conjunction, let me mention a Cashinahua myth that there is one kind of Ayahuasca that makes its drinkers light and thus enables them to fly out to heaven (see Lagrou, 1998). It is plausible, I think, that this myth originates in the change of the feel of body weight that the brew induces.

Several Western persons have told me that Ayahuasca made them feel that their limbs were not their own; one of them further reported seeing his limbs being torn apart. Still another informant described a terrifying experience in which she felt the entire cohesion of her body was lost. Interestingly, the experience of dismembered limbs is also mentioned by Karsten (1935) in his report of what indigenous Amerindians experience with Ayahuasca.

Of special significance is the change of the scaling of one's perceptual field. I am referring in particular to an expansion of the inner visual space. Close your eyes. What is the spatial extension of that which you see? With Ayahuasca the inner field of vision is significantly enlarged.

7. *The Locus of Consciousness*

A special aspect of calibration is the experienced locus of consciousness. Where is consciousness located? At first glance, this may appear to be a meaningless question—mental events do not have a 'place'. Yet, subjectively, people do feel that there is a place in their bodies where the centre of their awareness is located. Usually people locate this centre in their heads. Some people, and some cultures, situate the centre in the heart or in the stomach; for further discussion of this issue, the reader is referred to Jaynes (1976).

What does this definition of locus mean? Surely, it is not a symptom of the physiological fact that the organ responsible for thinking is the brain—until recent historical times most people did not know that that was the case. Besides, since the brain does not have proprioreceptors, human beings do not have any sensation of their brains. If consciousness is to be placed where there is a feeling of oneself, a feeling of energy, agitation, concern, it does indeed make more sense to locate consciousness in the heart or in the belly. Alternatively, the locus of consciousness may be defined in a quasi-geometric sense as the primary reference point of mental spatiality. By and large, the main point of spatial reference is at the middle point somewhere behind the eyes and between the ears. This makes sense as far as the phenomenology of perception is concerned and it is, I gather, not an accident that this point is related to the anatomical position of the brain.

Normally we locate our consciousness within our physical body. With Ayahuasca this may change. A phenomenon reported by several of my informants is that of the out-of-body experience, where the self dissociates itself from the body. One may even find oneself watching one's body from the outside. (For further discussion of this phenomenon in other contexts, see Green, 1973.)

8. Time

Another major aspect of calibration pertains to time. Just as we have normal feelings of the size and weight of our bodies, we also have a normal feeling of temporality. In particular, psychological time is felt to have a certain rate of flow. Often, Ayahuasca changes this flow: either it may appear that things change very rapidly or it may be felt that change is minimal, irrelevant, or even non-existent. The linearity of time may be disrupted as well. As a consequence of these alterations, one may feel that one has an existence that is not subject to temporality in the standard sense of the term. In the extreme case, there may be the feeling of having reached the realm of eternity.

Another aspect of psychological time is that we live in the present. We may have memories of the past and plans or speculations regarding the future, but our being is in the here and now. Ayahuasca may act as a kind of time machine whereby we may observe past and perhaps future events perceptually, as if they were taking place in the present.

Because temporality is of fundamental cognitive importance, I devote a separate chapter, Ch. 14, to its consideration.

9. Self-consciousness

A question I have often been asked is whether under the effect of Ayahuasca one's consciousness of oneself is maintained. I would say that by and large it is. Admittedly, as described above, there are moments in the Ayahuasca experience when it seems that one's individuated selfhood has totally dissipated. In such moments, self-consciousness might indeed be lost. Yet, it seems to me that self-consciousness can be maintained even when consciousness and selfhood undergo radical changes. One main lesson I have learnt with Ayahuasca (some will say, from Ayahuasca) is that radical as the effects of the brew are, I can always maintain a residue that is my own, normal self. Maintaining such a residue is, in fact, one major ability that, with accumulated experience with Ayahuasca, one learns to achieve. This residue serves as an internal compass that keeps track of what is going on and sustains a sound judgement. It holds things in check and balance, and it maintains contact with the ongoing reality around one. Again I want to emphasize—this residue of the normal self can be reached even when all other facets of one's consciousness are completely altered. High as one's soul may rise, one's normal self may still be able to maintain a distance and take cognizance of the special state in which one is and of what is happening in it. Several informants with extensive experience with Ayahuasca have described this very same state of affairs to me. Taussig's dissociative experience, cited above, also exemplifies the compartmentalization of self-consciousness and highlights the multi-layered character that it may take.

Yet, as indicated, on extreme occasions, there can be situations in which one's notion of one's own self is completely lost. I would relate this to what may be subjectively characterized as an enchantment, a kind of a spiritual seduction. Being absorbed by the marvels of Ayahuasca, one may lose track of everything else. One forgets where one is, what is happening in the real world around one, and who one is. By and large, it seems to me that losing oneself is, to a great extent, a matter of a decision. One may decide to lose touch, let go and immerse oneself in the non-ordinary realms that Ayahuasca presents one with. More on this will be said in the discussion of the dynamic aspects of the Ayahuasca experience, in Chs. 20 and 21.

10. Intentionality

As indicated above, intentionality is often taken as the most basic and requisite feature of consciousness. Yet, in non-ordinary situations, states of mind may be experienced in which this requirement seems not to be met. In these there is no object to which thought is directed and no content is being entertained by the mind. With respect to Ayahuasca, by and large, the experience is of the *kataphatic* type—that is, oriented towards imagistical filling, which is contrasted with *apophatic* type—that oriented towards empyting (see Forman, 1998). Yet, on occasion the latter type may also be encountered with Ayahuasca. The one occasion I experienced this is described in Ch. 17. As noted there, in retrospect I conceptualize the experience as one in which I was observing my own soul. I have also associated this experience with ones I have read about in the mystical literature. Indeed, in various mystical traditions a state of mind lacking intentionality is central (see Stace, 1961; Forman, 1990, 1998). Traditional mystics have referred to this state as 'the Void', 'the One', 'the infinite'. Modern scholars have employed the terms 'cosmic consciousness' (see Bucke, 1901/1991), 'oceanic consciousness' (see Parsons, 1999), and 'pure consciousness' (see Stace, 1961; Forman, 1990, 1998; Wilber 1996; as well as Grof 1998, 2000).

Given the fundamental—perhaps even definitional—status of intentionality, the question poses itself: is a state of pure consciousness possible? If intentionality is a defining feature of consciousness, then, in principle, one should object to the feasibility of states of consciousness in which this feature is not met. Indeed, several scholars have argued precisely that (see S. T. Katz, 1978). However, firsthand acounts from many mystical traditions do present attestations that people have experienced such states. As pointed out by Stace (1961) and by Forman (1990, 1998), such accounts constitute empirical data that cannot be dismissed. The Ayahuasca experience—along with experiences with other psychotropic substances—presents further phenomenological indications to the possibility of there being mental states without content or directedness to an object.

11. Connectedness, Knowledge, and the Conferral of Reality

When, in my preliminary analysis above, I characterized consciousness as affording connectedness with the world, and hence knowledge, I did so as a theoretician;

here I shall be doing this as a phenemenologist describing facts of experience. Under the effect of Ayahuasca both connectedness and its role in the gaining of knowledge are directly experienced. I have already noted that Ayahuasca often weakens the differentiation between the inner and the outer worlds. With this, drinkers often feel strongly connected with the world. The connectedness is manifested in the feeling that the boundary and division between one and the world dissolve and in ensuing sentiments of great empathy towards entities and beings other than oneself. These feelings may be coupled with the sensing of strong affinity, blending, identification, and even unification with what ordinarily is 'other'. The enhanced connectedness may be felt towards single, particular objects—both animate and inanimate—as well as towards the world at large; it may also be experienced *vis-à-vis* other human beings. Many have told me of such feelings in conjunction with other persons participating in the session, with photographs (notably, those of the leaders of either the Church of Santo Daime and the UdV), and with plants (see examples and discussion in Ch. 5).

Intertwined with knowledge is the conferral of reality upon the object known. Normally, human beings have a certain feel of the real. This feature is so fundamental to our perception of the world and our being in it that in general we are totally oblivious of it. There are situations, however, when our sense of reality is reduced—for instance, extreme fatigue and being in the dark. This also happens with Ayahuasca. The Ayahuasca inebriation, however, also presents cases in which the feel of reality is enhanced. Many drinkers report that what they see during the course of the intoxication seems to be 'more real than real'. Phenomenologically, this is related to the intensification of sensory qualities of both perceptions and visions, as well as to ideations entertained along with them. The feeling is so strong and compelling that it may be coupled by the assessment, very common with Ayahuasca, that what is seen and thought during the course of the intoxication defines the real, whereas the world that is normally perceived is actually an illusion. In many *Daime* hymns, this claim is explicitly made (see also Ch. 1).

As indicated, the experience of connectedness is often associated with powerful noetic feelings—that is, feelings that one is privy to true knowledge (see also Chs. 4 and 15). Indeed, the feeling is that the knowledge one gains is ultimate and veridical. It is achieved not by means of analysis and reflection, but rather by means of a direct contact, or even identification, with the objects to be known. This mode of knowing is very much in line with Spinoza's intuitive mode of knowing, which he characterized as the highest that may be attained (see Spinoza, 1670/1989). Also pertinent here is the notion of 'introception' proposed by Merrell-Wolff (1995). Unlike perception and conception, introception pertains to immediate knowledge that transcends the subject–object division as well as the distinction between knower and known. The theoretical analyses provided by both Spinoza and Merrell-Wolff fit perfectly with the phenomenological patterns experienced with Ayahuasca.

The issue of knowledge raises, of course, very difficult philosophical questions. Are the noetic feelings Ayahuasca induces justified? Is the knowledge involved indeed veridical? The treatment of these questions lies beyond the scope of this psychological work. Here, I merely note that I have presented these very questions many times to my informants. 'How do you know,' I asked 'that what Ayahuasca made you see is indeed true?' Common answers were 'I just felt it,' 'I experienced this as evident. That's it.' Analytically minded philosophers or scientists would dismiss such answers as unfounded and unprovable. In contrast, it seems to me that Spinoza would accept them and demand no further proof. Indeed, by its very nature intuition, or introception, allows no proof other than itself. By its very essence, this mode of knowledge is internal, and any putative proof—be it empirical or deductive—would be external, hence already outside the realm of knowledge at hand. Obviously, this does not close the philosophical discussion, but I shall leave the matter at that; other aspects pertaining to the experience of knowledge will be discussed in Ch. 15.

More General Ramifications

The primary focus of the discussion in this chapter is the phenomenological analysis of the special effects pertaining to consciousness that Ayahuasca induces. However, as explained at the beginning of this book, the study of these effects also has theoretical ramifications with regard to the cognitive study of consciousness in general. Here, I would like to spell some of these out.

First, the very existence of the special effects described here indicates that some of the most fundamental conceptualizations students of mind usually have of the phenomenology of human consciousness are neither general nor mandatory. The effects we have surveyed violate what are usually regarded as very basic features defining the structure of human consciousness. As indicated throughout the foregoing discussion, most of the features at hand have been put forward by James (1890/1950) and are usually considered to be paradigmatic. The special phenomenology of the Ayahuasca experience puts all these features in a new light. It shows that these features are neither general nor necessary. Admittedly, they do define that state of consciousness in which we find ourselves most of the time we are awake. Yet, there are other states of consciousness in which, to varying degrees, these features do not hold. Thus, rather than being universal defining features, the characteristics in question are to be viewed as specific values defined upon a set of parameters; in different states of consciousness, the values are different. Obviously, all these empirical findings are of fundamental import for any general theory of consciousness. Furthermore, it appears that many of the patterns described in this chapter are also encountered in psychopathology; reviews are presented in Nemiah (1989) and G. Reed (1972), further references are indicated below.

From a positive point of view, the special effects encountered with Ayahuasca indicate that our picture of human consciousness may have to be modified. The phenomena we have surveyed mark the possibility of there being states of consciousness other than those normally conceived in academic psychology. In particular, what I have in mind are states that are transpersonal and non-individuated; as noted, these are at times referred to as 'super-consciousness' or 'cosmic consciousness' (see Bucke 1901/1991; Stace, 1961). The existence of such states suggests that the scope of the psychological phenomenon of consciousness has to be expanded and that new states of consciousness should be defined. A proposal as to how to incorporate the non-ordinary patterns encountered with Ayahuasca into a general theory of consciousness is drawn in Ch. 16.

Second, the present discussion can be related to findings that I have considered in Ch. 10 when discussing the ideas and reflections that Ayahuasca induces. As noted in that chapter, Ayahuasca causes many drinkers to reflect upon consciousness and its nature. This is true also of individuals without any prior intellectual interest in this topic. Moreover, in general, the specific ideas that different drinkers entertain with regard to consciousness fall into one consistent picture. As indicated earlier, consciousness is conceived of as the basic constituent of reality and the ground of all Being. Many further say they experience, and consequently conceive of, consciousness as a supra-human and non-individuated phenomenon of which human consciousness is a derivative. Obviously, that different people have and share these ideas proves nothing. Yet, perhaps this has some bearing on the topic being entertained? In other words, perhaps the similarity of these insights does indicate something with regard to the nature of consciousness? I leave this as an open question (see also Ch. 16).

Third, let me turn from the descriptive level to the theoretical one. Ayahuasca has led me to question some theoretical tenets I had held as obvious and to consider new ways in which to view consciousness. In particular, I shall mention three issues. The first pertains to the relationship between consciousness and the individuated self. Before I encountered Ayahuasca it was obvious to me that either the individuated self is fully commensurate with consciousness, or that it is the progenitor of consciousness. With the special experiences that Ayahuasca presented to me, I began to wonder. Perhaps the individuation of cognitive agents is secondary to the more general phenomenon of consciousness? Perhaps the consciousness exhibited by individual cognitive agents is a derivative phenomenon, which, in turn, depends on a super-individual form of consciousness? If so, consciousness precedes individuated selves. By such a view—which can also be found in Indian philosophy—the latter are formed when the former crystallizes into specific, bounded structures. As suggested by Polari (see subsection 5 above), there might perhaps be one super-consciousness of which the various seemingly individuated cognitions partake. This super-consciousness may be either something cosmic—a kind of *anima mundi*—or it may pertain to a higher-level self, to which we are all connected. Here are some further observations that Polari (1984: 278, 282) makes on this matter:

Our consciousness . . . could be . . . directly related with . . . the totality of the Cosmic Energy [which, in turn, is] merely a provisory point in the comprehension of other, infinite totalities that conduct us to the veritable totality which is God. . . .[Hence] the discovery of the veritable 'I' of which the rational ego is just a pale caricature.

Similar notions of consciousness are, of course, encountered both in the traditional mystical literature and in the modern literature on psychedelics; see, for instance, Bucke (1901/1991), Stace (1961), and Tart (1969). In psychology, it is tactfully (yet explicitly) hinted at by James (1890/1950: 346) and pronounced in the non-orthodox school of transpersonal psychology; for further pertinent discussions the reader is referred to Assaglioli (1965), Maslow (1968, 1971), Wilber (1977, 1990), and Walsh (1993), as well as the anthologies edited by Walsh and Vaughan (1980) and Tart (1992).

The second theoretical question pertains to knowledge. Since Descartes (1641/1960), Western philosophy has been puzzled by how our knowing of the world can be proved to be veridical. As pointed out by Russell (1967), logically there seems to be no such proof. Yet, as further pointed out in Shanon (1983), factually, no sane human being will doubt the validity of this knowledge. How are we to square the philosopher's sceptical argument and the manifest psychological certainty? The way out involves a change in our basic conceptualization of what knowledge is. Often, knowledge is conceptualized in terms of a representation, or a picture, cognitive agents have of states of affairs in the world (see, for instance, Wittgenstein, 1922). The picture theory suffers from a serious problem: between the cognitive representation and the world there is always a gap that cannot be bridged. The question is how these representations tie with the world. If cognition, and with it the human faculty of knowledge, is to be accounted in representational terms, then cognition is bound to be confined to the realm of the internal. As Wittgenstein himself remarked in the conclusion of his book, 'The boundaries of my language are the boundaries of my world.' As I argued at length in Shanon (1993), for a psychological account of knowledge to be viable, it has to be founded on a conceptual framework that takes the tie between cognitive agents and the world in which they live as basic.[3] With Ayahuasca, the intimate tie of mind and world is experienced in a most concrete fashion. Under the intoxication, drinkers actually feel that our knowledge of the world is the result of a direct connectivity that ties cognitive agents with the world. Thus, the problem of the gap is circumvented; in fact, it does not even exist in the first place.

Thirdly, at least for me, Ayahuasca sheds a new light on the relationship between what are standardly referred to as the inner and the outer worlds. Normally, it is assumed that the two are distinct and separate. With Ayahuasca, this is often felt no longer to be the case (for similar observations with other substances, see Watts, 1962). Furthermore, it is common for drinkers to experience a deep, intimate connection between the two realms. The connection is such that there comes a point at which the very differentiation between the inner and the

[3] This stance is in line with the insightful observations of Gibson (1979) and his followers in the school of ecological psychology (for a review, see Michaels and Carello, 1981).

outer realms seems to be devoid of meaning. With this, one feels that the individuated consciousness of each person is part of a higher, more extended supra-personal consciousness. This ties in well with the experience of non-individuation described above and with the idea of cosmic consciousness to be discussed in Ch. 16. If the feelings at hand do, indeed, reflect an actual state of affairs, then radical changes in the conceptualization of consciousness are entailed. More on this will be said in the next chapter.

Taken together, the foregoing observations raise the question as to whether it is indeed possible to study consciousness only structurally and functionally while ignoring the substantive, ontological dimension. Before my encounter with Ayahuasca my position was categorically affirmative in this respect. Since then, I have come to question my erstwhile position and am open to other possibilities. An in-depth discussion of these, however, extends beyond the scope of this book.

13

Transformations

> The Divine influx will begin to prevail in you and will weaken your external and internal organs. Your entire body will begin to tremble, until you think that you are about to die. This is because your soul is separating itself from your body as a result of the great joy that you experience when you perceive and recognize these things.
>
> Abraham Abulafia (a medieval Jewish kabbalist)

> Then I became a bird whose body was of Oneness and whose wings were Everlastingness, and I continued to fly in the air of the Absolute until I passed into the sphere of Purification, and gazed upon the field of Eternity, and beheld there the tree of Oneness.
>
> Abu Yazid of Bistam (a Sufi mystic)

Under the effect of Ayahuasca, people not only see, hear, and think in a special manner, they may also experience transformations in which they feel that they are subject to metamorphosis and that their personal identity is altered. Indeed, in Amazonian shamanism such transformations are of paramount significance. Specifically, the transformation into an animal, notably a jaguar, and the ability to fly are usually taken to be the key features of shamanic competence (see Reichel-Dolmatoff, 1975 and Lagrou, 1998). This chapter focuses on various transformations and patterns of metamorphosis that I and my informants have experienced.

Before I begin, let me note that the transformations to be described here are not voluntary. It is not the case that the person under the intoxication decides to adopt another identity and consequently he or she undergoes metamorphosis. Rather, the metamorphosis just happens—and typically, both the very fact of the transformation and its specific identity surprise the individual who is subject to the effect. This having been said, I should note that whether willed, intended transformation is ever possible is an issue that remains open. While I myself have never experienced such an effect nor have I been told of it by any of my informants, logically I cannot preclude its theoretical possibility.

Identity Transformations

Being Another Person

The encountering of individuals with which the drinker feels a very strong identification was reported to me by many informants. The experience varies from one of special affinity with a person seen in a vision to that of a transformation in which the drinker feels that he or she actually assumes the identity of another person. Usually, the other persons at hand are from other places and other times. The following example was reported to me by a woman without much experience with Ayahuasca:

> I saw an enchanted mirror, like that in *Snow White and the Seven Dwarves*. I looked at the mirror, and in it saw the image of a woman. The woman looked very different from me [here the informant made a motion with her hands showing how the hair of the woman in the mirror was arranged], but it was very clear to me, I fully felt it—that woman was me.

I myself have never experienced a total change of personal identity, but once I did feel a very powerful identification with a king who was the key figure in a vision. It seemed to me that I could enter the skin of this person, so to speak, and be one with him. Several times I experienced powerful identifications with people who appeared in my visions. As indicated in Ch. 8, these included a Siberian shaman, a young Spanish aristocrat, and an old Chinese man.

Most prominent amongst the cases reported to me by my informants were ones in which drinkers felt themselves being transformed into kings, Incaic figures, shamans, and simple people of various past civilizations.

Three informants—two men and one woman—reported visions in which they experienced themselves as members of the opposite sex; in one of these the drinker felt he was transformed into his mother when she was giving birth to him. Here is the report of the woman, a non-indigenous Peruvian in her early thirties:

> A young Indian man, apparently the leader of his group, appeared in front of me and I felt a strong identification with him. At first, I felt resistance—after all, this person was a man, and I a woman. But then I went on and accepted full identification with the man. With this I became an Indian and was accepted as such by all other men around. They taught me to dance their dances, the men's dances, and also gave me a potion that heightened my stamina and well-being.

Also reported to me are cases of age change—that is, experiences in which drinkers felt they were younger or older than they actually were. In particular, one informant experienced himself being a baby. He reported he physically assumed the body posture of a baby and that he could not speak, only babble like an infant.

In practically all contexts of Ayahuasca use, the assumption of other personal identities and the strong affinity with persons that appear before one in visions are usually interpreted in terms of reincarnation. Personally, I do not believe in reincarnation. At the end of this chapter, I shall further comment on the phenomenon and my stance as to how to account for it.

Finally, let me note that personal metamorphosis may also involve people other than the drinker. Specifically, one may see a person from other times or from other lands and identify that person as an individual that one actually knows. As already exemplified in Ch. 5 in conjunction of the phenomenon of seeing-as, it is not uncommon for drinkers to see fellow participants in the Ayahuasca session as persons other than these individuals actually are. I, for instance, once saw the Amazonian farmer who was leading the *Daime* session as the patron of a Mediterranean—Italian or southern French—tavern of dubious repute. Several years later, I recounted my experience to the man. He smiled and said that he might very well have been one in 'another life'. Rather than the terminology of past lives and reincarnations, I would employ the analogy of casting. When choosing particular actors to impersonate particular roles, the director of a film or a play imagines how the actor in question would appear in the role assigned to him or her. Of course, the actor is neither Hamlet nor Lady Macbeth but he/she is one that would fit into these Shakespearean personae. Good as he or she might be as an actor, that individual might not befit a Lear or an Ophelia. Picking up the right actor for a role is certainly one of the skills that characterize accomplished directors. Matches of this kind, as the *mise-en-scène* as a whole, create all sort of scenarios that are not real (after all, they are fictional), yet this does not mean that 'anything goes'. So, too, with the phenomenon of seeing-as experienced with Ayahuasca.

Whereas this episode, as well those of seeing-as discussed in Ch. 5, occurred with open eyes, the identification of one individual as another can also occur with the eyes closed. Thus, one informant reported having a vision in which she saw a Spanish inquisitor and knew that this was—in the informant's terminology a reincarnation of—a friend of hers. The vision made the informant understand her friend's psychology from a new perspective, one which she found insightful.

Transforming into an Animal

Animal transformations are experienced not only by indigenous people. The anthropologist Luna reports that the first time he partook of Ayahuasca he felt himself changing into a serpent (see Luna and Amaringo, 1993: 11). Experiences of animal transformations are also encountered in the data I have collected. Most animal transformation reported by my informants were into felines or birds, notably eagles or condors. Also reported were transformations into a serpent, an ape, a dog, a dolphin, a frog, a spider, a butterfly, a cockroach, and a unicellular sea creature. Also to be noted is the case of one informant who reported experiencing herself being transformed into a nymph.

I myself have experienced transformation into an animal on several occasions. In all, the change was either into a feline or into a bird. When I first had these experiences, I did not know anything about their significance in the Amerindian context. As described earlier, I also had visions in which I experienced what it feels like being in the midst of an ant colony as well as being inside an elephant's trunk.

Let me describe my feline transformation in some further detail. I have experienced this kind of transformation three times. The first was during one of the first times I consumed Ayahuasca. My eyes were open and it seemed to me that my left hand was turning into a black paw. I felt as if my face protruded forward. I was frightened and did not let the experience develop further. Several years later, an identical experience was recounted to me by one of my informants, a European partaking of Ayahuasca for the first time. My other two feline transformations were part of a full-fledged vision. In both instances I found myself being a jaguar amongst jaguars. In both, the change was problematic for me. I was concerned that the other jaguars would discover that I was not a real member of their species, that I was an outsider, that I did not know how to act as they did. On the first of these two occasions I solved the problem by acting as a weak female jaguar that seemed to be ill and withdrew to the side so as to be separated from her community. The second time I remained with the pack but was constantly on guard lest they discover my ignorance in matters of group activity such as playing and hunting.

One of my informants, a European with little experience with Ayahuasca, partaking of the brew in a private session outside South America, reported the following:

I felt myself being transformed into a cheetah. I was still me but at the same time a cheetah. I moved and ran energetically and with great agility, just like a big cat. It all felt very real and I enjoyed it tremendously. At one point, in front of me, I saw a deer. It was there for me to hunt. However, even as a feline, I felt I had to ask the deer for forgiveness. Eventually, I could not bring myself to kill it.

Transformations into a bird are discussed below, in conjunction with flying.

Other Types of Transformation

Transformations into entities that are neither human nor animal are relatively rare. Apart from the transformation into a statue recounted below, the only experience of a kindred sort I have had was one in which I felt I changed into a *plant*:

All around me I saw the blossoming branches of a tree. The air was light and the atmosphere springlike. The white and magenta flowers were fresh and fragrant and a host of butterflies and other insects were attracted to them and hovered all around. Only upon further reflection did it occur to me that the tree itself was me.

Finding out that I was the provider of nourishment and good energy for so many living organisms was a wonderful, very gratifying experience for me.

Next to be noted are transformations into *inanimate* objects; these were reported to me by only a very small number of persons. Akin to personal transformation is the transformation into a statue. Both I and two of my informants experienced a transformation into the statue of an Egyptian pharaoh. My own story is thus:

I was sitting comfortably in my chair, with my feet hanging down freely. At first, I felt like relaxing, letting my spine rest leisurely and be fully supported by the back of the chair. But then the force of the intoxication hit me strongly and I pulled myself forward. I straightened my back and planted my legs vertically on the ground in front of me. I drew even further forward, so that only my buttocks were touching the chair. My position was such that my thighs were horizontal, my knees defined an exact right angle, and my feet were pressing firmly on the floor. On the one hand, energy was flowing from my hips downwards to the earth below; on the other hand, my entire torso was being lifted up. All of a sudden it dawned upon me: I was being transformed into a statue of a Pharaoh—a king of yellowish sandstone, erect and facing forward. My stature was somewhat larger than my own human size; with me was the bountiful energy of that which is the source of everything, before me was the eternal.

Another informant reported having been transformed into a stone statue of the Buddha.

A Peruvian shaman told me that he regarded a transformation into a grain of sand as the most remarkable feat in his entire lifelong experience with Ayahuasca. He explained that he underwent this transformation in order to make himself invisible. In a vision, enemies were after him and he had nowhere to flee. With the said transformation, he could not be detected and he was saved. Note that this episode presents a case of control in the sense explained in Ch. 7.

An impressive sequence of transformations to inanimate objects was recounted to me by a Brazilian artist who had some of the most powerful Ayahuasca visions I have ever heard of. The first is that in which the informant was experiencing herself as a drop of water in a fountain (the Trevi fountain in Rome). Then she looked at her watch and felt she was transforming into one of its hands. She also changed into a fig. In the end, my informant reflected later, she could transform into anything she looked at. Woody Allen's film *Zelig* readily comes to mind.[1]

The ultimate transformation of identity is that into *nothingness*. This is an experience in which one's identity as a particular individual is lost and one feels that one is becoming one with the universe. Understandably, the experience can be extremely frightening, but when accepted, it can be gratifying to the utmost. In the mystical literature of different traditions such an experience is central (see Bucke, 1901/1991 and Stace, 1961). I myself had only a fraction of a glimpse of this phenomenon, but I heard it reported by several informants. In addition, many have reported to me experiencing a strong *affinity with the universe* and having that 'oceanic feeling' in which one is immersed into the universe and where totality and nothingness converge. One informant told me of an experience in which he felt that he was turning into God.

Calibration

Associated with identity transformations are changes in the calibration of one's body. Thus, when undergoing animal metamorphosis one may feel that the very texture of one's body changes. This may be coupled with visualization. For instance, as mentioned above, I once felt and saw my palm and arm turning into the heavy and hairy paw of a puma. Radical changes in calibration are also experienced in conjunction with perceived self-death. In this context, the experiences of paralysis and of numbness were reported to me by several informants. Changes in the texture of the face and the skin were also reported.

[1] In this film, the hero, like a chameleon, adopts the characteristics and demeanour of whichever person he is interacting with.

Flying

One of the most salient transformations that Ayahuasca produces is that associated with flying. Personally, I have experienced flying on several occasions and found it to be a most exhilarating experience. When exposed to testimonies of flying, people lacking any firsthand, personal familiarity with the experience often react by asking, 'But do they really? Do they really fly?' But surely, this is the wrong way to look at it. Not only does the Ayahuasca flyer not lift his or her physical body off ground, also he or she is not supposed to lift, or wave with, his or her arms or hands either. The more experienced drinker, and the more respected one, would sit still and only his or her spirit would travel. Indeed, the more he or she kept still, the bolder and more extensive his or her flight experiences would be (for related discussion in other contexts, see Eliade, 1964).

Flying as a Bird

I have experienced flying as a bird on four occasions. Since these experiences define a progression—different stages in the acquisition of a skill, as it were—I shall describe them in full.

The first time I experienced flying was during a festive *Daime* session. As usual, the session lasted an entire night and consisted of the chanting of hymns while dancing in lines. Between the hymns there would be short breaks—a time for the lead singers to recollect the melody, for the musicians to tune up, for the dancers to have a little pause. For me, these breaks also presented an occasion to close my eyes and enter the visionary world. It was in one such moment that I found myself successively presented with several different pairs of wings. There were the wings of a vulture, the wings of a swallow, and the wings of two or three other birds whose exact ornithological identity I did not know. When I took possession of them, it felt as if the wings were attached to my shoulders, and that they became a new pair of limbs at my disposal. I used them—I lifted myself off the ground and flew. It was a short flight, and more than anything I was using the wings and feeling what it is like for a bird to fly. Having got the feel of it, I returned the wings, upon which I was given the next pair to try. The process repeated itself four or five times. During all that time I was standing still and erect. Indeed, it seemed that standing firm on the ground was a necessary condition for me to go on flying.

The next time I flew was when I drank Ayahuasca alone, in a hotel room in a small Amazonian port town. Sitting on the floor and listening to a song whose topic was a white dove, I experienced myself (in a vision) spreading my two arms to the sides, and starting to fly. Like a dove, I flew over the meadows. The flight was calm and steady. Now and then I would move up or down a bit, but overall the level of the flight and its speed were constant. And then, as the song was coming to a close, I found myself descending. My arms-wings were slowly folding, and eventually I landed. The difference between this experience of flying and the one I had had before is telling. The first time, I was presented with wings, I tried them out for a while and then I gave them back. But, of course, a bird does not fly in this fashion. Most significantly, when a flying bird comes to rest, it does not stop in one go. Rather, it changes the pattern of its flight, it flutters its wings, begins to descend, lands, and folds its wings. In my flight as a dove I did precisely this. And note: I first did this, and only afterwards did I realize the difference between the two kinds of flying experience.

On a third occasion, I flew as a seagull. I was sitting in my chair, towards the end of a UdV session, and there I was—a seagull in the air:

> I was high above a long stretch of coast, between sea and white cliffs. *La Manche* (The English Channel), I thought. I circled up and down, soared, and even performed acrobatics in the air. During this entire time I had views of the land and sea down below; the experience was both very realistic and wonderful.

My fourth experience of flying was of the same type. In it, I flew as a stork, crossing the ocean. My body was large, my wings long. The flight was a long, arduous feat that constantly demanded stamina and perseverance.

Several informants have reported experiences of flying to me. The most common of these was flying as an eagle. For example, a European taking Ayahuasca for the first time experienced himself as an eagle which flew up above the hut in which the Ayahuasca session took place. The man experienced himself high in the sky looking down at the people participating in the session. He discovered that his eyesight was very acute and that even though he was observing from a great distance he could detect the most minute details. The experience was, he said, exhilarating. Wolf (1992) reports a similar experience. Experiences of flying were also reported by C. Naranjo (1973*a*) in his study of the psychological effects of harmaline.

Other Types of Flying

When people speak of flying, they would normally think of flying as a bird. But flying as a bird is not the only kind of flying. Indeed, one may envision flying without one's human identity being transformed. Though I have experienced this while dreaming, with Ayahuasca I never experienced flying through space without being transformed into a bird. Yet, on two occasions I experienced something related—lifting myself up in the air. In both cases this happened to me when the effect of the Ayahuasca intoxication was subsiding and without any hallucinatory visual effect. One of these occurred towards the end of a session with a *curandero* in the Peruvian Amazon. I was standing, facing the meadows and looking at the full moon. I felt I was lifting up, to about two-thirds of the distance to the moon above me. When I felt I could no longer hold myself up, I stopped and landed. Several informants have reported similar experiences to me. As in my case, many of these flights terminated in mid-course. The informants attributed this to their not having sufficient strength and tenacity to go on.

Spatial Dislocation

Spatial dislocation consists in the feeling that one is at a place other than that in which one physically is. While not involving transformations of personal identity as such, spatial dislocation is none the less akin to identity transformations of the kinds discussed above. Just as one may be transformed into a bird and fly in space, one may also be transformed in the matrix of space-time and find oneself in other places, at other times.

In the indigenous context, Ayahuasca is famous for its power to translocate persons and take them to other places. One Campa Indian with whom I spoke characterized this as the prime feature of the brew: It makes you see distant places (for similar reports by indigenous persons cited in the anthropological literature, see Kensinger, 1973 and Harner, 1973*c*). Usually the experience involves what might be characterized as remote seeing. Some haze dissipates or a curtain is lifted up and one has the privilege of inspecting scenes of other locales, of other times. At times one may feel that one is actually entering the scene that is being envisioned; one might even feel that one is interacting with the beings encountered in the visionary realm.

The experiences of spatial dislocation and of identity transformation may be intimately related to one another. Indeed, the question does present itself—where actually is one when spatial dislocation is experienced? In milder cases, especially ones that involve no interaction, the answer seems to be clear: one is where the Ayahuasca session is taking place, but one has at the same time the ability to inspect other visionary places. On other occasions one may characterize one's place as the physical spot in which one is really situated (the place in which the Ayahuasca session actually takes place) but at the very same time feel that one's consciousness is over there, within the scene one is envisioning. For instance, in the powerful session reported in the Prologue I was sitting on a bench in front of the forest. I saw all sorts of animals and found myself dancing with them even though it was clear to me that, all the time, I remained in the same place, sitting on the bench. When the force of the experience is greater, bilocation, and dual existence, may be experienced. One may feel that one is actually in two places at the same time: at the actual spot in the real world, and within the scene being envisioned. In still more extreme cases one may lose touch with one's actual physical location and feel that one is situated solely in another place. This happened to me only once. I was in a *Daime* session in the Amazon and suddenly I found myself in my bedroom in Jerusalem. It was not that I saw myself in Jerusalem, I actually felt that I was there. I should say that the experience was quite frightening. It seems to me that the experience was so powerful precisely because it involved no extraordinary content. Indeed, the only remarkable thing about this vision was the very sensation of being in another place. A similar experience is recounted by the indigenous *ayahuasquero* Payaguaje (1983: 63) who observes that 'one is seated in one's hammock but, at the same time, is in another world'.

Likewise, in one of his session reports, Polari (1984) notes that while physically he was in the Amazonian forest he actually saw himself sitting in his study in his home in a town on the other side of the continent. Polari also recounts an experience he refers to as 'ubiquity', that is, being everywhere at the same time.

Lastly, let me mention a phenomenon that is somewhat different. It does not involve a vision, but the very experience of being elsewhere, or rather, of participating in an activity in another place, at another time. Stepping along the lines during *Daime* festive sessions both I and several of my informants (the reports are totally independent from one another) have experienced ourselves participating in a dance that had been performed long, long ago—in ancient and even prehistoric times. We felt that the activity in which we were taking part was carrying us to other places and other times. In a fashion, the dance transcended time—it was an eternal ritual in which both we, here and now, and people very distant from us, were taking part.

Death

As attested to by the very etymology of its name, Ayahuasca is traditionally associated with death and it is assumed to enable one to meet with the spirits of the dead. One special facet of the encounter with death is the experiencing of one's own death. Examples of self-death were presented earlier. I discuss this experience here too for it involves a radical change in one's state of being.

As pointed out by Reichel-Dolmatoff (1971), for the indigenous Amerindians 'to take Yagé [was] to die'. I have never experienced myself dying, but several of my informants told me of visions in which they found themselves undergoing death. In all cases, the agonizing experience culminated in profound transformation, experienced as salvation and/or rebirth. All informants characterized these experiences as the most powerful they have ever had with the brew. Two examples were given in Ch. 9; another one was reported to me by a European who partook of Ayahuasca with an indigenous healer in Peru:

> I experienced the tormenting loss of almost all of my body. From the toes up to the neck, one by one all the parts of my body were becoming cold and the flow of blood in them was stopping. The process was slow and terrifying. At several points in this process I was certain that this was the end and that I was dying, but a voice told me that my time had not come yet and I was ordered to persevere. I was on the brink of perishing and then I saw a grandiose vision. It depicted a myriad of people who emanated great light. This vision, I felt, conveyed some important lessons regarding human life. With this, I felt reborn.

A very similar experience of self-death is recounted by Harner (1980: 3).

Performance

All the changes considered above were, in one way or another, parts of visions. However, as already noted in Ch. 4, Ayahuasca also induces actual, overt behaviours which may be quite remarkable. In these, people often reach patterns and levels of performance that far surpass anything they would normally be able to do.

Singing and Other Musical Activities

Perhaps the most prevalent manifestation of heightened performance is that of singing. This is highlighted in many indigenous reports in which Ayahuasca is often characterized as a brew that makes people sing (for a general discussion, see Dobkin de Rios, 1984; Bellier, 1986; and Luna, 1992). All *Daime* sessions attest to this—in them, most people find themselves singing in a fashion they would not normally be able to. Here is a report by a Brazilian man of what happened to him the first time he partook of the *Daime* (Pellegrini, 1994: 46):

> Without knowing what to do, I decided to sing. I opened the hymn book and I found myself singing in a voice that for me was extremely high. The voice came out of the depth of my being, without stopping, giving me a sensation of great pleasure. I remember thinking in those moments, that this should be the sensation that opera singers have in the most overpowering moments of their arias.

Indeed, under the effect of Ayahuasca, people (myself included) sing better and in a much ampler voice than normal. I have observed this many times, including once when I myself was not under the effect of the brew. When the power of the brew is strong, the Ayahuasca drinker may feel that it is not he or she who is the generator of the song. The person feels, rather, that he or she serves as a vehicle for the music to convey itself through his or her voice. Typically, such singing is an expression of the deep gratitude that people feel when under the Ayahuasca intoxication and of their acknowledgement of the other forces whose effects the brew leads them to appreciate. Indeed, as indicated earlier, singing plays a pivotal role in all contexts of Ayahuasca use. Recordings I have of *Daime* sessions corroborate the foregoing assessment. In these recordings one clearly hears the transition between the normal (rather mediocre) singing of non-professional people to a powerful, well-coordinated chant. Similar observations are made by Reichel-Dolmatoff (1990) with respect to a Tukano Ayahuasca ceremony.

I have engaged in powerful singing many times. As demonstrated by tape recordings that I took on several occasions, the quality of my singing surpassed anything that I could have achieved normally. Many times I found myself producing high notes that were definitely way above the normal reach of my voice. To do this, I knew I had to sit up straight, breathe, and transform my entire body into a vocal apparatus. (Later, I showed what I did to a singing teacher and he indicated to me that, indeed, intuitively I had practised what a singer should do.) When I was quite experienced with this kind of singing, I would close my eyes and lift up my imagination. With this, I would find myself in the heavens immersed in light and energy. From here my music emanated. On some occasions, I produced sounds that emanated from the cranial and nasal cavities as well as chords in which two sounds were emitted simultaneously, through two different resonance cavities;

only much later, a musicologist explained to me how this is technically possible. Very often, the melodies I sang were invented (some would say, received) on the spot—they were never sung before either by me or by any other person.[2] And I could keep singing on and on for extended lengths of time. On some occasions, I sang for an entire day following the Ayahuasca session itself.

A couple of times, when taking Ayahuasca alone, I found myself emitting sounds that I felt were not mine at all—the grunting of jaguars, the whispering of snakes, the calls of birds of all sorts, the sounds of rushing water, and the blowing of the wind. The feeling was that it was not me who generated these sounds. Experientially, the sounds took over me. My body served as a channel through which, through forces that are beyond me, the sounds came into being. At times the production of the sounds was coupled with a vision. It was not a vision of the animal or being associated with the sound being emitted. Rather, it consisted of a voyage to other realms. It was the wondrous beauty of these realms and the enchanted energies that radiated from them that enabled the non-ordinary vocal performances that I was actually producing in this world. On other occasions, I was immersed in total darkness. There was nothing there, only the primordial energy from which, through my mouth, a breath of air was emerging.

Unusually high levels of performance are also exhibited in the playing of musical instruments. I have been to many concerts in my life, in which I have heard some of the best musicians in the world today. Yet, some of the most marvellous musical performances I have ever been privileged to listen to took place during *Daime* sessions. Ones I was most impressed with were moments in which a guitar player, whose normal role is that of accompaniment, took liberties and improvised freely. This happened in the short intervals between hymns, always towards the end of festive sessions, when the overall atmosphere was that of universal elation and spiritual triumph. Tape recordings I have made document some of these extraordinary musical performances. Once, at the end of a festive *Daime* session, I saw one man pick up a violin and play in a manner that amazed all those present. Yes, the man knew how to play the violin, but by his own avowal, he had never played in this manner before. The playing was most impressive and everyone was captivated: it seemed that the Muses had descended upon the man who was playing. On another occasion, at the end of a *Daime* session, a large Brazilian drum was passed to me. I have never touched such a drum before, but in the context I felt I had no choice but to play it. I fully identified with the drum and I played it with my entire body and being. Indeed, my feeling was that I was dancing with the drum, that I was in a special kind of intercourse with it. It was a wonderful experience for me.

Lastly, let me mention an experience that has had a concrete, long-lasting effect on me in my daily life. In an amateur fashion, I have been playing the piano since childhood. I have played only classical music, always from the score, never improvising, and very seldom with an audience. Once during a private Ayahuasca

[2] Incidentally, the song still remains with me, and even now I can reproduce it at will.

session, I saw the piano in front of me and decided to play. A score of a Bach prelude was there. I played the piece repeatedly and felt I was entering into a trance. Then, I left the score aside and began to improvise. I played for more than an hour, and the manner of my playing was different from anything I have ever experienced. It was executed in one unfaltering flow, constituting an ongoing narration that was composed as it was being executed. It appeared that my fingers just knew where to go. Throughout this act, my technical performance astounded me. At times, I felt that a force was upon me and that I was performing at its command. No, it was not that I was an automaton nor was I enslaved by any external agent, rather the playing was the expression of a wondrous co-operation between me, my very self, and forces I felt were superior to me. Another person was present and he listened to my playing. Later he told me that he felt the music was a story and he was very moved by it. 'It seemed that the Muses descended upon you,' he said. When the session ended, it occurred to me that I had had the most wonderful piano lesson of my life. Since then I have been free-playing without Ayahuasca. The quality of this playing is not like that under the intoxication, but it does exhibit some features that my piano playing never did before that session.

Related to music is dance. Again, the superior dance performance of people under the influence of Ayahuasca is regularly manifested in *Daime* sessions. In the festive sessions participants can dance for up to twelve hours. In terms of their steps, the *Daime* dances are actually very simple, but the extended duration of the dance and the co-ordination exhibited in it are far from ordinary.

In my personal life, I like dancing and I am good at it. Thus, during several freely regulated, private Ayahuasca sessions I have danced. On these occasions, my physical performance was better than it usually is. For instance, I found myself swirling in a Dervish-like manner for quite a long time. Twice, I spontaneously engaged in t'ai chi-like movement sequences that surely I could not perform in my ordinary state of consciousness. I have seen this happen once to another person, a European without any previous experience with Ayahuasca or t'ai chi.

Role Playing and Other Performances

The drumming episode recounted above is also an example of role playing. Extraordinarily enhanced performance may be regarded as a precursor of role playing, identity transformation being an intensified extension of it. It is not for nothing that many indigenous *ayahuasqueros* are also actors of a sort (see Narby, 1998; for a related discussion in the general context of shamanism, see Lévi-Strauss, 1964). In my regular life, I have never been involved with acting, but on several occasions, in sessions I myself was directing, I experienced the joy of acting and role-play. Twice I found myself reciting Amerindian stories recounting the origin of Ayahuasca. The stories I was telling were based on mythologies I had read in the anthropological literature, but when I was recounting them, these stories changed significantly. They became much more elaborate, intricate details were incorporated into them, and general human implications were associated with them. Experientially, I felt I was at the same time an indigenous storyteller and very much myself. Indeed, assuming the role of the indigenous shaman I actualized my role of teacher (a role that is central in my regular life) to the utmost.

Ayahuasca may also enhance motor balance and control. As recounted earlier, in private sessions, I have found myself engaged in dancing. For instance, I managed to swirl in a sufi, Dervish style for about an hour. I would not have been able to do this (indeed, I never did) in my ordinary state of consciousness. Also related to this are the t'ai chi movements and Dervish-style dance mentioned above.

An enhanced performance that I have noticed several times following Ayahuasca sessions involved my command of Portuguese. When I was still not fully conversant in this language, my articulation became more fluent than normally and my pronunciation became more like that of a native; several times I checked this with my interlocutors and found that they shared my appraisal.

Finally, on many occasions I discovered that under the intoxication my interaction with other people became smoother and more sophisticated. Often, I was quite astonished as to how apt were the words my mouth was uttering.

Typological Summary

Reviewing all the various patterns indicated in the foregoing discussion, let me attempt a typological summary.

The weakest pattern of identity transformation is actually not a transformation as such, rather, it may be regarded as a precursor of one. It consists of a strong identification with another being. One is oneself, the other is him/her/itself, yet one has the feeling that in some sense one and that other are the same. When the identification is strong a double identification may be felt: there is an assumption of another personal identity while at the same time one's own normal identity is kept. The relative weight of the normal self and the transformed one may vary. Especially to be noted is a balanced state in which one maintains an existence on two levels, or in two locales: One is both this and that, one is both here and over there. When the transformed identity outweighs the ordinary one, one might still, at the background of one's consciousness, keep cognizance of the fact that one is oneself, or that one is undergoing a transformation. In the extreme limit, all contact with the normal self is lost.

Thus, as described above, in my own animal transformations, I felt I was transformed into a jaguar, yet at the same time I was me. Using terms taken from the philosophy of language, I would say that I was a jaguar in the attributive, as contrasted with the predicative, sense. (In other words, the qualities of a jaguar applied to me, yet I was not one.) The different instances of flying I have recounted also mark a progression in the change of personal identity. As mentioned above, analogous patterns are also noted in conjunction with spatial dislocation. As with personal identity, the weakest case is that of a strong affinity with another place that is experienced while one is fully aware of being in the place where one factually is. Then, in complete analogy to experiences of double identity, there are experiences of bilocation. In the most powerful cases, the only location that one feels one is actually in is the visionary one. Often the two types of transformation are related, and the spatial dislocation may go hand in hand with heightened change in personal identity. *Mutatis mutandis* the same may be said with regard to role-play.

Metamorphoses also vary in the degree of involvement that drinkers experience in their new identity, position, or role. The more active a drinker is, the more powerful the metamorphosis is sensed to be. Above I have mentioned that I felt myself being a jaguar, but I remained passive or avoided interaction with other jaguars. In the experiences of flying I have reported, a progression of involvement occurred. The increased involvement manifests itself in several aspects—the length of time the assumed identity is maintained, the depth and intensity of one's immersion in it, the range of activities performed, their richness and fullness. With respect to the last aspect, let me recall my experience of flying as a seagull. In it, not only did I assume the bodily movements of a bird, I also found myself immersed in a landscape and, for an extended duration, I was situated in the world being, acting, and perceiving as a bird. In the episode recounted above of the woman who changed into an Indian man, the rich activity and the full immersion in the new situation were coupled with ongoing interaction with the other persons in the scene of the vision.

When one does enter the visionary realms, one can do so in stages. These stages differ both in the magnitude of perceived spatial distance that one has traversed and in the complexity of the activities in which one engages. With respect to the first aspect, the successive passing of well-defined stations can be noted. In particular, the drinker may find him- or herself going through gates and doors, crossing rivers and bridges, overcoming obstacles. The more of these he or she passes, the more distant ordinary reality is perceived to be. Often, one may even wonder whether one will ever be able to come back. Such an experience may be overwhelming even when no metamorphosis of personal identity is involved.

A Final Comment on Reincarnation

As indicated above, in practically all contexts of Ayahuasca use, the assumption of other personal identities and the strong affinity with persons that appear before one in visions are usually interpreted in terms of reincarnation. It is believed that Ayahuasca makes people see their previous lives. (In both the Santo Daime Church and the UdV, the expression usually used is the 'seeing of one's previous incarnations'). To me, reincarnation is foreign: I have never believed in it, it does not fit with my general world-view, and Ayahuasca did not present me with any personal experiences that would warrant a radical change of my position on this matter. I should make clear that I do appreciate the special flavour and the strength of the experiences that people interpret as a manifestation of another incarnation of their selves. I have had this experience myself and indeed found it to be most powerful. Yet, rather than viewing this phenomenon in terms of reincarnation, I would prefer to approach it from a cognitive-psychological point of view. As I shall argue at length in the last two chapters of this book, it seems to me that a key effect of Ayahuasca is to greatly enhance the mind's creative powers. The expressions of these powers are multifarious and they are especially manifest in imagination and role-play. Consequently, under the intoxication, the human mind can create new personae of its agent (towards the end of this book, I shall argue that with Ayahuasca, the mind can actually create new worlds). I can attest that the half a dozen persons I have encountered in my Ayahuasca visions and which I felt myself identifying with were all characters that I greatly sympathized with. Indeed, in a hypothetical sense, had I lived in other places at other times I might have been one of them. These possibilities are all with me in this life—they are the different facets of who I am here and now. Some of them are realized in my life in other articulations, some—unfortunately perhaps—have not reached realization. That what happens is not the 'seeing of previous incarnations' by no means belittles the significance and impact of the experience. After all, the seeing of possibilities may be very instructive—even revelatory. At times, it may also lead to concrete changes in the actual conduct of one's life. As indicated, more on this will be said in the last two chapters of this book.

I shall further add that reincarnation may also be conceived in non-personal, cosmological terms, and thereby be rendered—at least as I see it—intellectually more palatable. In particular, it may be regarded as a manifestation of the fact that ultimately all living beings—past, present, and future—are the different compositions made out of a finite pool of elementary material elements. Thus, viewed from a broad metaphysical perspective, the phenomenon of life is an ongoing flow of permutations in which these elements compose, decompose, and recompose and with this the various forms of life come into being. Such a conceptualization of reincarnation (which, of course, is akin to ideas entertained in Hinduism and Buddhism), however, extends beyond the specific topic of personal identity and its transformation discussed in this chapter and the common interpretations given to it in the different contexts of Ayahuasca use. At the same time, I shall note that under the intoxication, the grand cosmological picture is itself, rather frequently, experientially intuited; observations in this regard were made earlier, in Chs. 9 and 10.

Let me end with an example taken from the verbatim record transcribed by Allen Ginsberg of the ongoing ruminations articulated by the beatnik writer Jack Kerouac when under the intoxication, partaking of Ayahuasca brought back to New York by Ginbserg in 1960. This example is of interest because it is an early report of a Western person, one of the very first to consume Ayahuasca outside South America; in it, as in the traditional indigenous context, personal transformation and reincarnation are central:

I'm positive that in 1860 I was a great thief in London—My name was Robt Horton—I had troupe of Boys who were sent out by me to attract men of title & money—with whom I was not in love—I was in love w/ a girl a brunette, dancer in Soho whom I murdered—for which I was hanged— (Cited in Ginsberg 1977: 143.)

14

Time

This timeless moment

(The title of Laura A. Huxley's memoir of the life of her husband, Aldous)

Although all things have succession and progression in heaven as in the world, still the angels have no notion and idea of time and space.

The angels who speak with men, never speak by the natural ideas proper to man, all of which are from time, from space, from the material and from things analogous to them; but by spiritual ideas.

Swedenborg

Temporality is perhaps the most fundamental constituent of human cognition. Kant (1781/1953) characterized time as the primary formal condition of experience, both external (of the physical world) and internal (of the psychological one). Empirical phenomenological inspection reveals that, indeed, human cognitive behaviour is grounded in time: Everything psychological takes time. This is true even of performances—like sensory perception—that, on the face of it, are either immediate or static. In Shanon (1993*a*) I discuss this characteristic of human cognition at length and incorporate it within a theoretical framework in which situated action in the world and unfolding in time are conceived as the basic properties of the human cognitive system (see also Shanon, 1998*a*).[1] The theoretical orientation I endorse follows the footsteps of many thinkers before me, notably James (1890/1950), Bergson (1944, 1950), Heidegger (1962), and Merleau-Ponty (1962).

Given this status of time and temporality, it should come as no surprise that these also feature centrally in altered states of consciousness in general and in the Ayahuasca experience in particular. Indeed, it seems to me that one of the most basic effects of Ayahuasca is the induction of a new temporal mode for the mind. Specifically, the brew induces mental states that, experientially, defy the dominion of time. In the light of my basic stance regarding the intrinsic temporality of human cognition, this effect is of special theoretical interest to me. Hence, I decided to examine this parameter of consciousness separately and to devote a special chapter to it; an early version of this discussion appears in Shanon (2001).

[1] This is in contrast to the dominant theoretical frameworks in contemporary cognitive science that stipulate that the basic capabilities of mind are information processing and symbol manipulation.

Preliminary Observations

I begin with what is perhaps the most basic, and the most commonly reported, facet of the modified temporal experience encountered with Ayahuasca. Under the intoxication, people's perception of time's rate of flow often changes. As mentioned in the Prologue, I discovered this the very first time I partook of the brew. Being confronted, unprepared, with marvellous visualizations, I decided to be cautious and close my eyes for only a limited period of time. Feeling that so much had happened, I reopened my eyes only to discover—by looking at my watch—that barely two minutes had elapsed. The contrast between perceived time and real time was striking.

Once I had a vision whose focal theme was this very experience. The vision consisted of a very vivid scene in which I was the main actor:

> I was dancing. It was a mad dance, or rather, an almost mad one. As long as I continued to dance, I realized, time stopped. This, I felt, was due to the energy that I was investing: as long as the energy was there, an existence outside time was maintained. In the vision, there were two people observing me and standing guard. Their policy was to let me go on as much as possible, but also to make sure I did not collapse. I myself felt that as long as I was in control and maintaining my sanity, I could go on and on.

Eventually, the scene stopped when a person who was guarding the session (a real person this time) touched me gently on the shoulder, asking me to calm down. Objectively, this entire episode lasted a couple of minutes; what I experienced, however, was of practically unlimited duration.

By way of further attempting to convey the flavour of the experience of non-ordinary temporality and of timelessness let me cite an account I found in the literature about a year after the episode described above had taken place. The account is by a friend and a colleague of Aldous Huxley, Christopher Mayhew (1961: 294–5), who had one session in which he experimented with peyote:

> What happened to me between 12:30 and 4 o'clock on Friday, December 2, 1955? After brooding about it for several months, I still think my first, astonishing conviction was right—that on many occasions that afternoon I existed *outside time*.
>
> I don't mean this metaphorically, but literally. I mean that the essential part of me (the part that thinks to itself 'This is me') had an existence, quite conscious of itself, enjoying itself, reflecting on its strange experience, in a timeless order of reality outside the world as we know it.
>
> I count this experience . . . as the most astounding and thought-provoking of my life.

Struggling to make sense of this experience, Mayhew (ibid. 297) proposes that

> From my peculiar disembodied standpoint, all the events in my drawing-room between one-thirty[2] and four existed together at the same time.

[2] Above it is indicated that the episode began at '12:30'. Both citations are taken from the text as it is. While apparently there is an inconsistency here, I prefer not to alter the original.

And he presents an example (ibid.) that I find most illuminating:

> When we take off from an airport at night, we are aware of individual runway lights flashing past in succession. But when [we] look down a little later, we see them all existing together motionless. It is not self-contradictory to say that the lights flashed past in succession and also that they exist together motionless. Everything depends on the standpoint of the observer.

On the basis of my own experience, I endorse this characterization. I too felt that even though time stopped, the vision that I was seeing was full of motion.

In line with this are the following observations made by Watts (1962: 27) on the basis of his long-term experiences with several psychoactive substances (LSD, psilocybin, mescaline). Watts, too, characterizes modified temporality as the major aspect of the non-ordinary states of consciousness that he had experienced:

> To begin with, this world has a different kind of time. It is the time of biological rhythm, not of the clock and all that goes with the clock. There is no hurry. . . . the present is self-sufficient, but it is not a static present. It is a dancing present—the unfolding of a pattern which has no specific destination in the future but is simply its own point.

The foregoing observations will serve as a point of departure for us. Bearing them in mind, we shall now turn to a more systematic analytical discussion. In the next section I attempt a typology of the various possible characteristics of non-ordinary temporality. Naturally, the definition of non-ordinary temporality goes hand in hand with that of ordinary temporality. The discussion will be based, on the one hand, on the consideration of different manifestations of the experience of temporality and, on the other, on pointing out how these might cease to apply in non-ordinary states of mind such as those induced by Ayahuasca. Thus, while laying the foundations for the systematic study of the non-ordinary temporal experiences encountered with Ayahuasca, I shall also be outlining a framework for a phenomenological study of ordinary temporality. The reader will note that, again, this is in line with the basic orientation defined in Ch. 12, namely, that the study of consciousness involves not only the definition of what is actually the case, but also that which is potential, as well as that which theoretically can never be.

Temporality, Modified Temporality, and Atemporality

In order to characterize the various aspects of altered temporality we should first distinguish the parameters that define the ordinary human experience of temporality. Six such parameters shall be noted. First and foremost is the *passage* of time—time flows. Metaphorically, the master clock is always ticking. Second, the *rate* of time's flow is—to use the terminology introduced in Ch. 12—calibrated. In other words, there is a particular measure defining the speed that the flow of time is experienced to have. Theoretically, time could be experienced as flowing either slower or faster than it actually does, and this could be uniform or not. Third, time exhibits *order*; this is defined in terms of the sequential relations of 'before' and 'after'. Fourth, time exhibits *directionality*—time moves forward; experientially, this is manifested in the distinction between past and future: as time flows (or perhaps, flies) what but a moment ago was experienced as future is now experienced as present and in a moment will be experienced as past. Fifth, a temporal *metric* applies so that specific and well-determined values can be assigned to both temporal location and temporal intervals. Operationally these values define those statements that constitute answers to when-questions. Sixth, all temporal distinctions pertain to a particular *frame of reference*. The frames vary with respect to the singular point each person takes as the basis of reference or co-ordinate definition. Different agents or observers will be grounded in different reference points and thus have different temporal perspectives.

The six parameters define the various possible alterations in experienced temporality. First of these is the stopping of time—the experience whereby the flow of time seems to come to a halt. This is an extreme result of that aspect of modified temporality which corresponds to the second aspect of temporality noted above, namely, a change in the perceived rate of the flow of time. In other words, time may be experienced as passing faster or more slowly than normally it does; this change, note, need not be uniform. Third is a possible confusion in the ordering of time—in particular, the determination of the relationships of before and after may be disturbed. Likewise, the distinction between past and non-past may be blurred or become irrelevant. A possible corollary of the confusion of the ordering relations are confusions related to the distinction between memory, perception, and thought regarding the future. Further, the modifications in the ordering relations may result in a failure adequately to establish temporal locations and intervals. Lastly, non-ordinary frames of reference may be adopted. With respect to this last parameter, three possibilities will be noted. The first consists in a modification of the perceived distance of time. This modification entails the feeling that things are more or less distant than they are normally sensed to be. It should be noted that this experience need not involve a change in the determination of the specific temporal location at hand. The second possible effect consists in the adoption of another agent's frame of reference. Specifically, one may look at things past or things future as if they were in the present. The third effect consists in adopting an absolute frame of reference where time is no longer pertinent. In the philosophical tradition of the West this has been referred to as looking at things *sub specie aeternitate*—from the perspective of eternity.

In addition to the structural parameters I have introduced, the human experience of time may be inspected from a *semantic* point of view, one which is grounded in considerations pertaining to meaning and narration. As pointed out by James (1890/1950) and Bergson (1950) as well as by the psychologist Gibson (1975, 1979), whereas physical time is defined in terms of an abstract, universal matrix, psychological time is defined in terms of events. As such, human time is intrinsically semantic, or rather, as conceptualized in Shanon (1993*a*), intensional.[3] This being the case, an entire manifold of considerations pertaining to meaning and narration may apply. Thus, people's attitudes towards time may change; with this, the sense and meaning they attribute to temporal patterns will be altered. In the limit, two extreme states of affairs can be noted. In the first, time ceases to be relevant and temporality gives way to *pure semanticity*; in the second, neither time nor meaning are pertinent.

Taken together, the foregoing distinctions define various possible manifestations of non-ordinary temporalities. Here, I shall distinguish between two clusters—that of *modified* or *altered* temporalities and that of *non-* or *a-* temporality. Modified temporalities constitute states of mind in which temporality is experienced but whose specifications are different from those of ordinary temporality. Modified temporality may be manifested in changes with respect to one or more of the six parameters noted above. Non-or a-temporality consists of the experience that one is 'outside time'. With respect to this, two distinct cases will be noted. Either time may, so to speak, cease or a new, non-ordinary semantics of time may be assumed. The latter case includes states of affairs in which it is felt either that *any* temporal frame of reference may apply or *all* temporal frames of reference are applicable or even co-present. It may also be that all points of reference are lost. With these experiences one feels that one has reached the realms of the eternal. I discuss all these cases in further detail below.

The foregoing distinctions were general theoretical ones, spanning the space of all possible non-ordinary effects associated with the experience of time. Using these theoretical distinctions as my conceptual frame of reference, I shall now turn to the actual phenomenology of the Ayahuasca experience. With Ayahuasca, both modified temporality and atemporality can be experienced. Indeed, to one degree or another, all the possible effects surveyed above are encountered.

Modified Temporality

The various aspects of modified temporality encountered with Ayahuasca will be presented in three clusters. The first cluster is associated with the first and second of the six parameters indicated above, the second cluster with the third, fourth, and fifth parameters, and the third cluster with the sixth. The reason the presentation is by clusters and not by parameters is that phenomenologically the effects pertaining to the various parameters within one cluster are closely related to one another and, in fact, may even be intertwined.

The Passage of Time and Its Rate of Flow

The most common aspect of modified temporality encountered with Ayahuasca is that of changes in the rate of time's flow: time may be perceived as moving either

[3] Note that this term is written with an 's'; it means 'intrinsically invested with meaning'.

faster or slower than the clock would indicate. Usually the effect is that time moves faster. In other words, the person under the intoxication experiences lots of happenings and change, hence it seems to him or her that more time has passed than actually has.[4] In extreme cases, it may seem that time has altogether stopped or that time and temporal distinctions are no longer relevant. These effects relate to the experience of atemporality which will be discussed in the next section.

The relationship just indicated between an increased amount of things experienced to be happening and a change in time's rate of flow is in line with a basic theoretical tenet of the modern cognitive research of time perception. By it, people's estimation of temporal intervals is a direct function of the amount of information processed during the intervals in question. Under conditions of higher informational load people would tend to overestimate the amount of time that has passed (see Fraisse, 1964; Ornstein, 1969).

However, when extreme changes in the experience of time are considered, the conceptualizations in the psychological literature are not suitable. Specifically, this is the case with respect to the stopping of time and the experience of eternity. On these effects and their bearing on the psychological literature I shall elaborate in the next section, that dealing with atemporality.

Temporal Relations and Metrics

The second cluster of effects to be noted are modifications related to the determination and evaluation of temporal locations, intervals, and relations. Just as one may not be sure as to where one is, under the Ayahuasca intoxication, one may not be fully cognizant of when one is. Further, blurring or confusing of the ordering relations of before and after or those of past and future may be experienced. I should note that in themselves such confusions are rather rare. More common is the experience whereby temporal relations and the distinctions associated with them are simply deemed to be irrelevant. Furthermore, intertwined with

[4] Some would characterize this effect as one in which time moves more slowly, rather than faster, than it normally does. This, I gather, they would associate with phenomena such as are commonly experienced in a boring lecture. There, people intermittently inspect their watches, hoping that the greater part of the lecture period is already behind them. Unfortunately, they discover that, in fact, 'so little time has passed', and they conclude that 'time moves so slowly'. In their turn, when Ayahuasca drinkers consult their watches, they are surprised to find out that objectively very little time actually elapsed. For this short duration to contain the many experiences that they have undergone, the flow of time should have been greatly accelerated. The two cases are both similar and different. They are similar in that in both, the clock shows that 'only a short period of time has passed'. They are different in terms of the perspective of the temporal judgement and of that which serves as the basis for the comparison. In the case of the lecture, the comparison is within the experience, and is based on a sensation of the flow of time itself: being bored, people feel that 'time hardly moves' (note that the phrase is in the present tense, as it characterizes something as it is being experienced). By contrast, in the Ayahuasca case, the comparison is from without, after the experience, and its basis is the happenings that have been experienced, not a modified feel of the flow of time proper. In the visionary world, time passes neither faster nor slower; rather, things happen. *Post facto*, when the clock is consulted, people are astonished as to how much could have happened in such a short period of time. By way of resolving the puzzle, they conclude that within the visionary world in which they experienced themselves to be, time must have moved extremely fast (note that the phrase is in the perfect).

modifications along the parameter of directionality are indeterminacies with respect to mental states. Specifically, both I and my informants have experienced a blurring of the distinction between memory, perception, and thought regarding the future. Thus, one may find that one cannot ascertain whether mental contents one entertains are memories or reflections about possible states of affairs. I shall note that this confusion that directly bears on time is similar to other confusions and indeterminacies encountered with Ayahuasca that are not related to temporality. I refer to the confusion between thought and perception, the real and the non-real, between perception and hallucination (for further discussion, see Ch. 16).

Frames of Reference

Last to be noted is a phenomenon which primarily involves the relationship between two temporal frameworks, that of the Ayahuasca drinker and that of what he or she observes.

The most rudimentary effect pertaining to the parameter of frames of reference has to do with the perceived temporal distance of perceived events. Specifically, the feeling is that things are at a temporal distance different from the normal: what is seen is felt as closer or more distant than it is usually felt to be.[5] With Ayahuasca, the former effect (that is, distant events experienced as closer in time) seems to be more prevalent than the latter. As indicated earlier, this experience need not involve a change in the specific determination of temporal location as such. In other words, there need not be a change in the temporal location attributed to what is being seen, yet the perceived temporal distance of the scene may be felt to be altered. For example, a scene identified as taking place in ancient Egypt may be felt to be very close to one.

A more radical effect is that the drinker remains in his or her ordinary temporal frame of reference but sees what seems to be taking place in periods of time other than his or her 'now'. Essentially, the effect may be likened to a voyage in time. Significantly, this experience does not involve stillness. Whatever the drinker inspects are events that occur *in* time, and as such they exhibit order and sequentiality, change and movement. It is only that the drinker is situated in a temporal frame of reference other than that to which the perceived events pertain. Often, episodes are seen that are associated with ancient, enchanted, or futuristic civilizations. Typically the episodes portray scenes as they proceed in the midst of their progression. In contrast to scenes of a film or a theatre play that begin when the curtain lifts and where the narrative is viewer-oriented, here it seems that the personae in the scene lead their course of life independently of

[5] The reader should note that I am not saying 'than it usually is'. What we are discussing here is phenomenological experience, not physical states of affairs in the real world. In our ordinary state of consciousness too the distances—temporal as well as spatial—things have are felt to be of a certain magnitude. Experientially, the same physical distance could be felt to have a greater or a lesser phenomenological magnitude.

the viewer and with no consideration relating to him or her. The experience is that these people are there as they would have been in their place, in their time; what is not ordinary is that the viewer, now and here, has the privilege of seeing them. This phenomenon is curious, for it involves the intersection of two distinct points in time—that of the observer and that of the scene being viewed. I am here now and I am observing (not remembering, not hearing about, not reflecting about), and at times even acting in, a scene taking place at a different time, in a different place.

I find it pertinent to emphasize that the experience I have just described is quite different from that of memory. In remembering, one feels that one obtains information about the past; here, one feels that one is actually observing the past. Furthermore, in memory one feels that the information entertained reflects one's own past history and is defined by one's own perspective and point of view; in contrast, in the case discussed here the states of affairs being inspected seem to be totally independent of the observer. Thus, phenomenologically, the experience at hand is akin to perception: the person having the experience feels that he or she is privy to actually inspecting states of affairs and events as they are taking place in a point in time other than that in which he or she is situated. An experience that may give partial flavour to what I am describing is that encountered when viewing a scene in a *camera obscura* (that is, through a small and distant hole). This may sound trivial, but anyone who has seen this would surely attest to the fact that this mode of inspection induces a very special experiential effect. More on this will be said below, in the section on eternity.

The confluence of two distinct points in time is surely paradoxical. The observer is in the present, yet he or she is seeing episodes drawn from the past or from another, undetermined time. The paradox is both intrinsic to the experience and apparent when an attempt is made to conceptualize the experience theoretically. One way to accommodate the paradoxical state of affairs is to assume a domain in which the two points are indeed co-present; this domain is the non-temporal realm of eternity. The following reflections present an attempt to find order in this kind of experience. I shall note that the reflections are not presented here merely as speculations, let alone theoretical proposals, regarding metaphysics or ontology. Rather, they are an attempt to find internal order in the non-ordinary experiences surveyed here. In part, the reflections presented are grounded in experiences I and other individuals had with Ayahuasca.[6]

With the foregoing reflections, we embark on those aspects of the Ayahuasca experience having to do with atemporality and eternity.

[6] In the context of psychoanalysis it is often said that the free associations a person has in conjunction with his or her dream cannot be separated from the dream proper; indeed, it is argued that conceptually the separation is not tenable. Similarly, reflections entertained during the Ayahuasca experience, as well as entertained when drinkers attempt (or even struggle) to make sense of this experience, are often part and parcel of this very experience.

Atemporality

Being Outside Time

When the Ayahuasca inebriation is strong, drinkers may feel that they are altogether outside time. Significantly, this may obtain with or without a concurrent cessation of movement and change, that is, with or without an experience of stillness. What is encountered with Ayahuasca seems to defy the views prevailing in the cognitive-psychological study of time. Extrapolating from the models currently held in the field, the cognitive theoretician may deem that psychological atemporality necessarily entails an experience of stillness, but the actual inspection of the phenomenology of the Ayahuasca experience reveals that this is not the case. Specifically, being outside time without experiencing stillness is a pattern definitely encountered with Ayahuasca. An example is a scene of whose kind I have seen many times. In this scene women are engaged in a dance that is perceived as being in a realm totally different from the ordinary one. The dance involves movement and may even be associated with specific rhythmical patterns. Yet, the dance is not taking place *in* time. In turn, this special experience of non-temporality may have semantic ramifications whereby special meaning is attributed to the dance; for instance, it may be interpreted as divine or cosmic. In this regard, the reader is called to look back at the text cited from Watts (1962) at the outset of this chapter; in particular, I refer the reader to the phrase 'a dancing present'.

While atemporality need not necessarily entail the experience of stillness, there are instances in which the two effects are encountered *in tandem*. Furthermore, the cessation of movement and change may be coupled with a cessation or annihilation of the Self. This, I gather, is a key feature of the states of mind called *nirvana* and *samadhi* sought in Eastern meditative practices.

In the previous section I have considered changes in the perceived rate of time's flow in the light of contemporary cognitive-psychological thought. The extreme experience of atemporality, however, does not seem to be consistent with the current psychological conceptualization of time. As a rule, the models of time perception and time estimation in the psychological literature are linear: According to them, the change in the perception of time and its estimation define a direct linear function (in the mathematical sense of the term) applied upon a given continuous measure such as the amount of information processed (Treisman, 1963; Ornstein, 1969; Zakay, 1993; Zakay and Block, 1997). Admittedly, cognitive psychologists have never treated the experience of atemporality. However, extrapolating from the line of reasoning prevailing in the literature, the theoretically minded cognitive psychologist would venture, I gather, that atemporality comes about through an extreme modification in time's rate of flow. Specifically, as this flow becomes slower and slower eventually time comes to a halt and with this atemporality ensues. But it seems to me that with Ayahuasca just the opposite happens. The experience of atemporality seems to be associated with an increase, not decrease, in visual stimulation and a corresponding increase in the rate by which psychological time is experienced to flow. Indeed, the stimulation and flow may be so great that the person under the intoxication may feel inundated. Almost literally, he or she may feel sucked into the world of the vision. Typically, this happens with powerful abstract (that is, geometrical, non-figurative) patterns. A great swirl may be experienced, perhaps an immense vertigo, and then time no longer applies. It is as if one has moved so fast that one has managed to exit time and arrive at a realm outside it.[7]

In the light of these considerations, I would suggest that the stopping of time experienced under Ayahuasca should be viewed not in terms of speed but in terms of changes in temporal frames of reference and in an altered semantics of time. Essentially, the experience is not of time passing more slowly, but of time ceasing to be relevant. It is not that one moves more slowly in time, but rather one enters a frame of being which is, so to speak, outside the province of time. In the same manner, the experience of eternity that I shall be considering below should not be conceived as an infinity of temporal extensions but rather, as a state of mind in which time gains a different significance or perhaps is no longer pertinent. The following observations made by Huxley (1971: 20) in conjunction with his experience with mescaline are consonant with this view:

> ... Along with indifference to space there went an even completer indifference to time. 'There seems to be plenty of it', was all I would answer when the investigator asked me to say what I felt about time. Plenty of it, but exactly how much was entirely irrelevant. I could, of course, have looked at my watch; but my watch, I knew, was still, of an indefinite duration or alternatively of a perpetual present made of one continually changing apocalypse.

I shall note that similar observations are also encountered in various mystical traditions (see, for instance, de Cusa, 1960), and, further, that some of the most interesting manifestations of atemporality encountered with Ayahuasca involve modifications in what may be regarded as the semantics of temporality. At the heart of these experiences is the impression that ordinary measures and notions of time no longer apply. Thus, Polari (1984: 70, 66) observes:

> The *Daime*, simply, delivers us to all the Times. Overcoming the difficulties of our consciousness, which still has to establish differences between what was, what is and that which will be.... [In] this other dimension of time ... past, present and future are mere arbitrary, didactic representations.

It is to the consideration of the various aspects of the experience of atemporality that I turn next.

[7] The pattern observed here is analogous to that encountered in physics in conjunction with the special theory of relativity. By this theory, the higher the velocity of a moving body the slower is the passage of time in this body's frame of reference. Theoretically, when this velocity reaches the speed of light time stops. For further discussion of analogies between patterns in cognitive psychology and modern physics see Shanon (1991).

The Experience of Eternity

By and large, our discussion so far proceeded along a *via negativa*: the experiences of modified temporality and of atemporality were examined in relation to ordinary temporality and in the main, they were characterized in terms of what they are not. However, a complementary perspective may be taken by which an attempt is made to characterize the non-ordinary experiences of temporality in *positive* terms. Doing so, we embark on an examination of what is experienced as eternity. In this examination, I shall further expand on issues I have touched upon earlier in this chapter. In line with what has been said above, in the following discussion semantic considerations pertaining to meaning and interpretation will feature centrally.

Time and again, Ayahuasca drinkers report that under the intoxication they have experienced eternity. What does the experience of eternity consist in? As Polari points out in the text cited above, Ayahuasca brings people to a realm that is beyond past, present, and future. From the perspective of the eternal, everything that has ever happened, as well as everything that will ever happen, has an equal temporal status. In a certain sense, all happenings—past, present, and future—are there and one only has to look at them. This, I suspect, is what is meant when the term *sub specie aeternitate* is invoked: a perspective is taken by which all that will have happened at all times is co-present. In this limit situation, the temporal may, in a fashion, be reduced to the spatial. I understood this once in a vision—it was recounted in Ch. 10 in the section on philosophical ideas associated with Ayahuasca visions; there I also noted the similarity between the insight gained in the vision and Spinoza's view of both determinism and the nature of time. It seems to me that the insight gained in this vision is in line with the philosophical perspective presented by Spinoza in *The Ethics* (Spinoza, 1670/1989) and is especially akin to the ideas on both determinism and the nature of time in his *Short Treatise on God, Man and His Well-being* (Spinoza, 1852/1985; see in particular Ch. 16).

In conjunction with this, a useful analogy is to be found in Mayhew's (1961: 298) discussion of his non-ordinary experience of time: 'When we read something, we are aware of one word coming after another. But when we detach our minds from the sense of the words and look at the page as a whole, this impression fades, and we are aware that the words all exist together at the same time.' I would use the analogy somewhat differently. Specifically, I would focus not on the written words but on the content they express. The book is there—everything described in it is there at the same time, now, before the reader. The story that the book narrates progresses in a certain sequence in time. Yet, we the readers (especially if we have already read the book or heard the narrative) can open the book and read it in any order we wish. One can start by reading an episode that took place in the middle of the sequence, then turn to an episode towards the end of the book, and later to one at the very beginning.

Eternal co-presence may be associated with *omniscience*. The feeling that one becomes all-knowing is quite common with Ayahuasca. The two are also commonly encountered in mystical experiences reported in various religious and spiritual traditions (see Bucke, 1901/1991; James, 1929; Stace, 1961). What I would like to propose is that it is not an accident that the two experiences are encountered in non-ordinary states of mind, be they induced by psychoactive substances or other means (meditation, solitude, fasting, and the like). Rather, it seems to me that there is an intrinsic relationship between the two. Conceptually, the relationship between the noetic experience and the experience of atemporality can, again, be captured by means of the analogy of the book. A book takes time to read. It is also the case that if enacted, all the episodes in it would take time. Yet the information contained in the book is at once present in its entire totality. Indeed we may say that, non-standard as it is, the stopping of time is encountered in everyday life, in a fashion. Think of a body of information you know well: a novel, a scientific theory, a piece of music. I ask you to think what it means, to appreciate what it conveys, to assess your understanding of it. How much time did it take you to do so? A few moments, I presume. At any rate, much less than the time it would actually take to read the novel, to retrace the theory, to listen to the musical piece. One's appreciation is, in a way, instantaneous. And the more knowledgeable or competent one is, the stronger the effect. For example, Mozart was said to have been able to grasp an entire symphony in an instant (Ghiselin, 1952). But in truth, all of us know this feeling: when one knows something well, the grasping of this knowledge (as distinct from specific demonstrations of it) seems to require no time at all.

Typically, the experience of eternity is associated with a distinct ambience. Drinkers of Ayahuasca often report that the scenes appearing in their visions have an intrinsic 'quality of eternity'. This quality may be associated with several features. First is the *co-presence* of distinct moments of time noted above. Second is perceived *spatial extension* and *distance*. This is especially marked in visions of great expanses of open landscape. The spatial qualities of these scenes are readily translatable to temporal ones—the vast expanses and the distances of inspection generate an ambience in which time is no longer relevant. Third is *serenity*. When one comes to terms with the Ayahuasca state, the visions are often endowed with great serenity. It seems that everything proceeds in order, as it should, with no hurriedness and in an ambience exhibiting great mindfulness and dignity. Serenity may, in turn, be linked to silence. The real-life experience that comes closest to this is, I find, deep-water diving. This silence imparts a feeling of otherworldliness in which all things seem to stand still in a dimension that is outside time. This non-applicability of time is often sensed as a manifestation of eternity. It is also sensed as sanctity.

Lastly, let me note that the experience of eternity and its ambience are not confined to the dimension of temporality (or rather, atemporality). Central to the experience is a change of value and meaning. Inspected from the perspective of the eternal and the omniscient, all that is taking place here on this planet gains very different proportions. With this, the importance of things changes and the significance attributed to them is altered. One feature of this change is that one takes things less seriously and with more tolerance, forgiveness, and also a (benevolent) sense of humour. When describing this, several of my informants employed the phrases, 'The Divine sense of humour' and 'God's laughter'. Here, too, we note the import of semantic considerations in the analysis of altered temporality.

Atemporal Reality

As noted, Ayahuasca drinkers feel not only that their perception and interpretation of time change, but also that the non-ordinary temporality they feel subject to transports them to a domain that is outside time. With this, drinkers feel that the realms of the eternal are revealed to them.

The state of affairs characterized as eternal defies standard cognitive-psychological conceptualizations. By these, psychological activity may indeed be devoid of spatial location, but it cannot be without or outside time. While there is no sense in fixing a particular place as the locale in which mental events—thinking, conceiving, perceiving, remembering—take place, it is universally agreed that these events always happen at a particular time, and that they always have a certain duration (for a related discussion in the contemporary cognitive literature, see Pylyshyn, 1979). Thus, whereas it is meaningless to say where a particular thought (or feeling, or act of remembering) has occurred, it is an intrinsic feature of it that it began at a particular moment and ended at another. Mental events thus differ from events in the physical domain, which always have both a particular time and a particular place.

Ayahuasca presents experiences that do not fit this picture. Many drinkers report that the brew brings them to realms that are outside time. I have had such experiences too. The realms at hand are conceived as constituting other realities, ones that transcend the distinction between the internal and the external and in which time and temporal distinctions are no longer relevant. On the basis of both these experiences and the repeated attestations of my informants, I would propose to view the state of mind under the intoxication as the third in a set, the two first members of which are the physical domain and the ordinary mental one. The physical domain is defined in terms of both space and time. The ordinary mental domain is defined by time but space cannot be attributed to it. The non-ordinary domain experienced with Ayahuasca is attributed by neither time nor space. As further explained below, this domain is defined by the parameter of meaning.

Remarkably, characterizations similar to those made here are found in mystical texts in different cultures (for a review, see Huxley, 1944). Thus, consider the following comments taken from the writings of a contemporary Jewish Kabbalist, Ashlag (1956: 1); these comments are from the very beginning of his comprehensive book, *The Study of the Ten Spheres*:

> One has to remember that all Kabbalistic study is based on spiritual matters that occupy neither space nor time, and there is absolutely no lack or transformation associated with them. All the changes discussed here do not imply that one aspect is lacking and is receiving

another form, but that the change in question consists of the addition of form only and the first form does not move from its place, for lack and changing are of the material matters. And this is the difficulty for the beginners, for they perceive things in their material expressions within the scope of time and place, change and transformation, that authors employ only as signs for their upper sources.

In sum, states of affairs pertaining to the spiritual domain have neither time nor place. Some Jewish Kabbalists have referred to this domain as *nivdal* (in Hebrew), being apart (see Mallin, 1979).

Very similar ideas also appear in the writings of the eighteenth-century Swedish scientist-philosopher-mystic Swedenborg (1854: 83–5):

Although all things have succession and progression in heaven as in the world, still the angels have no notion and idea of time and space. . . . The angels who speak with men, never speak by the natural ideas proper to man, all of which are from time, from space, from the material and from things analogous to them; but by spiritual ideas, all of which are from states and their various changes. . . . But still angelic ideas, which are spiritual, when [the angels communicate] with men, are turned in a moment and of themselves into the natural ideas proper to man, corresponding altogether to spiritual ideas; that it is so, the angels do not know, nor men.

In both the indigenous Amerindian context and in the modern syncretic religious sects employing Ayahuasca, the brew is said to bring its drinkers to other, separate realities. In the doctrines of both the Church of Santo Daime and the UdV, these are referred to as 'the astral'. Usually, the otherworldly realms are associated with spirits and other supernatural beings (again, this is reflected by the very name Ayahuasca). I have had those experiences that are characterized as pertaining to the astral myself, and I think I know what the protagonists of spiritualism are referring to. Yet, I do not subscribe to the spiritualistic, supernaturalistic view. Instead, I would propose that the domain in question be characterized as the realm of meaning. This realm of meaning will be the topic of our discussion in the next chapter.

Concluding Remarks

Throughout the foregoing discussion, states of mind experienced as being outside the dominion of time have been considered. The very existence of such states challenges the general statement with which this chapter opened, namely, that human psychology is grounded in time. This also contrasts with my own basic view of cognition. Indeed, for me, the discovery of the atemporal was intellectually most unsettling. As noted above, following extensive research, both empirical and theoretical, I have come to maintain that human cognition is embedded in time and that all cognitive performance is achieved through action temporally unfolding in the real world. This, I have argued (see Shanon, 1993*a*), is not merely a technical constraint but rather it is an essential, most fundamental, feature of human cognition. Theoretically, one could perhaps envision intelligent systems that operate in a different fashion, but these would be radically different from the human one. Ayahuasca revealed a totally different picture to me: it presented me with a cognitive mode that, in a fashion, is not dependent on time. The question as to how to reconcile these new discoveries with my general view of cognition perplexed me quite a bit. Full discussion of this question is outside the scope of this book. Here, let me confine myself to the following brief suggestion.

It could be suggested that human beings have the ability to operate, and exist, in two different states. Metaphorically, these may be conceived in terms of the shifting of gears. The first state is the ordinary one, and it is fully grounded in time. The other, non-ordinary state consists in the freeing of the mind from the ordinary temporal constraints. That such freeing is possible is a major feat of the human psyche. The study of the dynamics of the shift between the two states is, I think, a cognitive-psychological topic of utmost significance. A theoretical framework that accounts for it will encompass both ordinary consciousness and non-ordinary consciousness and view them as specific cases obtained by means of variations in a common, general structure. Thus, the enterprise in question is, in essence, the development of what may be regarded as a general theory of consciousness.

A Biblical Aside

I would like to close with an episode taken from a very different context but which, I believe, is directly related to the discussion entertained in this chapter. I refer to the famous biblical episode of the burning bush:

> Moses kept the flock of Jethro . . . and he led the flock to the backside of the desert, and came to the mountain of God, even to Horeb. And the angel of the Lord appeared unto him in a flame of fire out of the midst of a bush: and he looked, and, behold, the bush burned with fire, and the bush was not consumed. And Moses said, I will now turn aside, and see this great sight, why the bush is not burnt. And when the Lord saw that he turned aside to see, God called unto him out of the midst of the bush, and said, Moses, Moses. And he said, Here am I. (Exodus 3: 1–2)

My way of reading this biblical text is based on the hypothesis that the origins of the Israelite religion, as those of many other world religions, are rooted in the use of psychoactive plants (this hypothesis is developed in Shanon, 1999; for a general presentation of the thesis linking the origin of religions to the use of entheogens, see La Barre, 1972, as well as Wasson, 1968). In defending this hypothesis, I have examined various episodes in the life of Moses as they are described in the Bible and marked the great similarity between them and experiences encountered with Ayahuasca. The experience cited here is the first; indeed, it is Moses' own first encounter with God. Why is the bush not burning? I would like to suggest that this is because Moses was under the influence of some psychoactive substance and that he had experienced the same phenomenon with which the discussion in this chapter opened, namely, a marked slowing of the rate of time's flow. In actual, physical terms nothing special was happening to the bush—it was aflame. The time Moses looked at the bush was very brief and not sufficient for it to be fully consumed by the flames. However, subjectively, because of the change in time's rate of flow, it seemed to Moses that much time had elapsed. In such a period of time, he reasoned, the bush should have already been consumed by the fire. It was not, and Moses was puzzled.

15

Meaning and Semantics

> Besides these two kinds of knowledge there is a third, as I shall show in what follows, which we shall call intuitive knowledge (*scientia intuitiva*). Now this kind of knowing proceeds from an adequate idea of the formal essence of certain attributes of God to the adequate knowledge of the essence of things.
>
> Eternity is the essence of God in so far as this involves necessary existence. Therefore to conceive things under the species of eternity is to conceive them in so far as they are conceived through the essence of God as real entities, or in so far as they involve existence through the essence of God.
>
> Spinoza

> Vision or imagination is a representation of what eternally exists, really and unchangeably. . . . Imagination is surrounded by the daughters of inspiration, who in the aggregate are call'd Jerusalem.
>
> William Blake

As summarized at the end of the previous chapter, in our discussion of temporality, meaning featured centrally. Meaning is also associated with other facets of the Ayahuasca experience that we have discussed. At the very outset of the phenomenological survey presented in this book, the overall enhanced meaningfulness induced by Ayahuasca was highlighted. As noted in Ch. 4, the general ambience that Ayahuasca generates is that things are richly invested with meaning. Furthermore, the inebriation makes people feel that things, even the most banal, do not just happen to be: they are there because they are the manifestations of ultimate truths and essences. Similar observations are also common with LSD and mescaline (see Ludwig, 1966; Huxley, 1971; Watts, 1962; and Michaux, 1972). Here, for instance, are comments made by Aldous Huxley (1969: 137) in conjunction with his mescaline experience; in these lines he describes what he experienced when listening to music by Bach:

> The tempo of the pieces did not change, nevertheless they went on for centuries, and they were a manifestation, on the plane of art, of perpetual creation, a demonstration of the necessity of death and the self-evidence of immortality, an expression of the essential all-rightness of the universe—for the music was far beyond tragedy, but included death and suffering with everything else in the divine impartiality which is the One, which is Love, which is Being.

In this chapter I examine in further detail some of the manifestations of what may be characterized as the realm of meaning associated with Ayahuasca. Pivotal to this discussion is the intimate association of meaning with atemporality indicated at the end of the previous chapter.

Several different topics pertaining to meaning will be considered. The first is the global experience of enhanced meaningfulness and several psychosemantic patterns that are ramifications of it. Second are phenomena that people usually associate with an independent realm of ultimate meaning. In particular, these pertain to the essence of things and to Platonic Ideas and entities. Third is the experience of noesis—that is, the feeling that one is gaining access to true knowledge. Lastly, theoretical ramifications—both psychological and philosophical—of these topics will be discussed. This sequence, note, marks a progression from what is within the province of individual psychology to patterns that seem to pertain to domains that transcend this province. With it, our discussion also develops into being progressively more theoretical.

Non-Standard Semantic Patterns

With Ayahuasca, the experience of enhanced meaningfulness is pervasive. This results in several non-standard semantic patterns. By and large, these exhibit insight and creativity and they impart to one a new and enriching view of things. However, as will be indicated towards the end of this section, at times people may go overboard, so to speak, and the same basic state of affairs may result in what I regard as infelicitous conferral of meaning.

Metaphoricity

Many phenomena pertaining to the Ayahuasca experience may be regarded as the products of affording the world an intense metaphoricity. A context in which this is readily observed is the linguistic one. Under the intoxication, verbal utterances and linguistic texts often gain what are felt to be deeper meanings. Interestingly, the affinity between the Ayahuasca experience and metaphoricity has been noted in the indigenous context. Sherzer (1986) marks the prevalent metaphoricity of Ayahuasca songs and notes that the Indians attribute this to the fact that metaphors allow direct communication with the spirits. In the Church of Santo Daime people often speak of the 'second reading'—an esoteric understanding of the hidden meanings of seemingly simple texts, such as those of the hymns sung during the rituals. Essentially, this consists in tuning into language with a pervasive metaphorical, and poetic, perspective. In Ch. 20, I shall further discuss the metaphoricity induced by Ayahuasca and shall relate it to more general theoretical cognitive considerations.

The Narrative Dimension

In the previous chapter, I have cited the contrast between physical and biological time made by Watts (1962) in conjunction with his description of non-ordinary temporality. Here, I would like to suggest another, more cognitive, distinction between two types of time—*sequential* and *narrative*. Sequential time is defined in terms of physical units and consists of the ongoing progression of moments one after the other. Narrative time is defined not by means of any objective measure but in terms of content. It consists in the temporality of meaningful events.

By way of clarification, let me turn once again to the domain of cinema. The scenes in a film each take some time. Yet the time the scene actually lasts (that is, the time that an external viewer will measure, by means of a clock, from the beginning of the scene to its end) and the time associated with the narrative that the scene depicts are not the same. The former is usually significantly shorter than the latter. An entire film may, for instance, depict the seventy or eighty years spanning a lifetime, yet in reality it would last only ninety minutes. What I would like to argue is that under the intoxication, the Ayahuasca drinker—like the person watching a film or a theatre play—is situated in a world of fiction. In this world, the story reigns. As long as the story is being narrated, the normal flow of time and its measurement do not apply. As in a film, there is no direct, linear relationship between the actual duration of events in the world and how long they are seen for. Furthermore, the depiction is not uniform in the sense that the ratio between the time events depicted in the film or the vision and the corresponding duration in the actual course of events may vary. Thus, the determinant factor will not be the sequence of time as such, but rather the semantics of the narration.

In passing, let me refer again to the comment made by Swedenborg (1854) regarding the angelic realm that I have cited in the previous chapter. In this realm, Swedenborg says, 'Progressions are not in time and space, but changes of states.' This seems to me similar to the present contrast between sequential and narrative time. The words and the theoretical contexts are different, but the basic observation is the same.

The Salience of the Verbal Medium

The enhanced investment of meaningfulness is also associated with a pattern that on first sight may seem to exhibit the very opposite, namely, a disregard of meaning. Under the Ayahuasca intoxication, people often focus on the contingent medium of words and, disregarding their meaning, they parse the phonological form in a new fashion so as to come up with new, at times ingenious, readings of the linguistic expression at hand. As a consequence, the initial disregard of the standard meaning of words results in the investment of extra meaningfulness to them. An example already mentioned in Ch. 10 is an answer I once heard a master of the UdV give in response to a question regarding the meaning of the word 'Jerusalem'. The answer was 'Jesus alem', which in Portuguese means 'Jesus is above'. This answer, of course, has no factual or etymological validity whatsoever. However, it is significant in that it points to a dynamic cognitive pattern that is common both with Ayahuasca and various mystical traditions. In particular, it is pervasive in the Jewish Kabbalah. I myself once had a vision in which such a verbal reparsing was central. I was escorted through an assembly of people and of physical objects. When I arrived to the end of the line, a wise old man told me, 'You see, this is the "Nature of things".' With this I had a novel insight: usually this expression is read as denoting the basic character of things and state of affairs in the world. Now, however, I understood that it carried a metaphysical-ontological statement, namely, that the ontology of nature is defined by things, in other words, the world is made of things. More on this will be said in Chs. 20 and 21.

Synchronicity and Syntony

Enhanced meaningfulness and non-standard semanticity may also apply to the temporal dimension. In particular, I would like to single out two phenomena which are especially common with Ayahuasca—*synchronicity* and *syntony*. Both consist in the interpretation of temporal relations as meaningful, and hence non-accidental.

Synchronicity (see Jung, 1951/1957 as well as Progoff, 1973; Peat, 1987) is manifested in the co-occurrence of two events that cannot be explained in the normal causal fashion; typically, this co-occurrence is highly improbable and it is associated with special meaningfulness. Indeed, coincidences feature very commonly in the stories of Ayahuasca drinkers (myself included). Whether these do or do not occur in reality is not important for us here. What is significant is the subjective cognitive feeling that synchronicity has taken place.

Syntony is manifested in the co-ordinated occurrence of two seemingly unrelated events. The paradigmatic cases of syntony involve rhythm. For example, more than once I saw large butterflies enter the open hall in which a *Daime* session was being held. It seemed to me that the flapping of the butterfly's wing was co-ordinated with the music played. At times, the feeling of syntony can be quite complex and it results in what may appear to be one harmonious and coherent composition. An example is an experience I once had when chanting in the context of a traditional healing session. The session was held in the forest, and whenever I reached the end of a stanza in my song, a (real) bird in the forest chirped (for another example of this, see the report by Kusel in Ch. 1).

Infelicitous Meaningfulness

In general, enhanced meaningfulness is associated with semantic richness, poetic flavour, and insight. Yet, this is not always beneficial. At times, the enhanced meaningfulness experienced with Ayahuasca may lead to conclusions that are manifestly exaggerated and faulty. In one of my first Ayahuasca sessions, I found myself sensing that all that had happened to me during the day before the session had been orchestrated. The people I met, those I joined, and those I avoided were all part of a scheme. There was sense, I felt, and order to it all, nothing was by chance. In a vision that followed I saw what seemed to me to be that which would have happened had I acted differently. The things I saw were not happy ones and I felt very fortunate to not have pursued a bad path. Once I observed a European man, a member of a 'shamanistic excursion' group travelling to the Amazon, just at the termination of his second Ayahuasca session. The man was overtly agitated. He told me that he suddenly realized that all that had happened to him, and to his companions in the group, was all a *scene montée* (in French, a mounted scene). Paranoiac feelings, which are not uncommon with Ayahuasca, can also be related to the enhanced meaningfulness induced by the brew.

Another case of what I would regard as infelicitous enhanced meaningfulness is that related to the phenomenon referred to in other contexts as 'speaking in tongues', that is—uttering sentences in a language the individual does not actually know (in the psycholinguistic literature this phenomenon is called glossolalia). Several informants told me of experiences in which they heard persons in the Ayahuasca session begin to speak in languages that were not known to them. In one particular case, I listened to a recording that was said to be Hebrew, my native tongue. It surely was not. In one of my first, and very powerful, experiences with Ayahuasca, I myself addressed one of my fellow participants in Hebrew. Of course, this Brazilian could understand nothing of what I was saying, but I had the impression that he could. More on this will be said in the discussion of paranormal effects in the next chapter.

In the same vein, I would address various parapsychological phenomena which are so often reported in conjunction with Ayahuasca. Under the intoxication, people very often feel that they can read the thoughts of others. They also feel that what other people do and say is specifically related to them. At times, these feelings exhibit a most remarkable insightful intuition, yet in many others they are false. As noted by Taussig (1987; see the citation in Ch. 2), Ayahuasca can reveal truths, but it can also lie. I have been repeatedly told the same by many of my fellow seekers in the realms of Ayahuasca. How to discern between the two is one very important skill that one has to develop in the course of one's long-term travelling with the brew.

Platonic Phenomena

Enhanced meaningfulness and deeper senses are naturally linked to what may be regarded as the essence of things. As already indicated in earlier chapters, with Ayahuasca people often feel that they gain access to the domain of essences and ultimate meanings. As further noted earlier, these domains are akin to ones that, in the Western philosophical tradition, are associated with Platonism. With these, too, temporality either receives a different status or becomes altogether irrelevant.

Essences and Design

In my first powerful Ayahuasca session (once again, this was the *Daime* preparation session described in the Prologue), a new (for me) insight forcefully crossed my mind, namely, that in explaining reality, two levels of explanation should be distinguished. On the one hand, there is the level of actual processes. These evolve in time and are normally accounted for by the sciences. This level and the accounts associated with it do not, however, explain the essence of things. These are—thus my reflection ran—determined on another level that is not in time. During that early session I found myself entertaining new (for me) ideas on theories in biology, psychology, and history. Evolutionary theories may account for how biological variation and change come about, I reflected, but there are aspects of the phenomenon of life that these theories assume and cannot account for. In particular, this holds for the notion of fit and the meaning of biological forms. In other words, the theories specify mechanisms by which life evolves but they do not explain the design that living forms seem to exhibit. Likewise, psychodynamic explanations give accounts of the genesis of certain psychological traits and personal behaviours, but above and beyond the contingent dynamical processes, there is what may be regarded as 'the story of a human being'. This is best characterized in semantic and narrative terms. As indicated at the end of the previous section, *mutatis mutandis*, these ideas may be applied to the story of peoples and cultures, that is, to history.

As described in the Prologue, later in the same session I suddenly had (what seemed to me) the insight that what I was perceiving was precisely that world of which the Jewish Kabbalah is speaking. Essentially, what the Kabbalah—notably, the analyses referring to the realm of the Spheres—is concerned with is the grand design of Being. The relationship between this Kabbalistic realm and that of the physical world may be succinctly illustrated by means of the following contrast made in *Sefer Yetzirah* (The *Book of Creation*; for an English translation, see Kaplan, 1997), one of the most ancient Kabbalist texts: it is not that there are ten spheres because man has ten fingers, but rather the converse, man has ten fingers because there are ten spheres. The reader will appreciate that the contrast made here is precisely that noted above in conjunction with my early Ayahuasca experience. This contrast may be further characterized as that between mechanism and design. Natural scientists are concerned with the mechanisms by which the world operates. But in addition, questions of design may be entertained. The mechanisms and their concrete manifestations may be many and varied, but taken in unison and regarded from a more abstract perspective, a unified order may be grasped. This specification is to be made in semantic terms (i.e. terms that pertain to meaning). In general, in the context of modern scientific thought, questions of design are regarded as teleological, hence outside the realm of science; there are, however, scientists and philosophers who think otherwise (Teilhard de Chardin is perhaps the most famous one; see, for example, Teilhard de Chardin, 1965).

Essentially, semantics transcends time. We can all appreciate this when reading literary texts of past generations, inspecting old photographs, or looking at works of art from ancient civilizations. The child or the young enamoured couples that appear in them are long, long dead, yet for the reader or the viewer they are forever children, adolescents, or young adults. Infancy, adolescence, and adulthood are not just moments in biographical chronology, they are also chapters in a story. It is the story of human life, that which we all enact in the course of our own individual lives. In the Jewish tradition, when a baby is born he or she is given the blessing to enter the bondage of the law (i.e. enter into societal obligations, thus becoming an adult, at the age of 12 (for females) or 13 (for males)), of marriage (traditionally at the age of 18), and to be engaged in 'good deeds' (throughout his or her life). All this will take time to happen and the particulars of the events are not known, but *grosso modo*, the overall outline of the story is already laid down. That dimension of temporal progression which, none the less, is not contingently dependent on time pertains to the domain of meaning.

With Ayahuasca, similar appreciations may be gained on a much larger time-scale—that of human history. In grand Ayahuasca visions (especially those I have referred to as 'panoramic') drinkers are often privy to a special perspective by which they observe such historical events with the detachment of distance (which does not mean lack of empathy). They observe periods of peace and periods of war, times of success and times of failure, Kings reign, exercise great power, and eventually die, entire civilizations rise and fall. With this, drinkers feel that they savour a little bit of the taste of the eternal and appreciate the perennial meanings of life and existence. As the Biblical sage Ecclesiastes (1: 9) said, 'There is nothing new under the sun.'

In line with all this is an insight reported in Polari (1984: 61). This leader of the Church of Santo Daime characterizes time as a major mystery revealed to him by the *Daime*: 'I understood that this truth is absolutely immune to any historicity. ... A Truth that is not relative nor relativized by anything, which entered and syntonized within a dimension of eternal permanency. ... Above and beyond a finite consciousness depending on matter and ending with it.' Discussing dreams, McKenna (1991) makes a rather similar observation. He contrasts dreams with history, suggesting that dreams offer an escape from history. Dreams, he argues, are eschatological. They last for no time and are outside history.

Interestingly, the association between atemporality and meaning is also encountered in the indigenous Amerindian context. Overing (1985*a*, 1985*b*) reports that in the cosmology of the Piaroa, a tribe from the Orinoco basin of Venezuela, in the world of the gods relationships between beings are defined in a manner that defies ordinary logical and temporal relationships. Overing explains this by noting that in this world relationships pertain to the world of meaning. Analysing the cosmology and metaphysics of the Piaroa, Overing also notes that the world of the shaman, and that of the gods, is outside time. Instead of being defined by linear temporal contiguity, this world is defined by relationships of meaning.

Platonic Experiences

Naturally, all this brings Platonism to mind. Platonism is the metaphysical view according to which there is an independent realm of reality that is constituted by abstract ideas (hence, this view is characterized as idealistic realism). The Platonic Ideas are entities by virtue of which things in the world, as well as the words of language and the concepts associated with them, gain their meaning. All existing horses, for instance, are manifestations of the eternal, independently existing, Idea of the Horse. The Ayahuasca experience exhibits various features that are Platonic. As described in earlier chapters, a salient experience is that of the perception of the generic—seeing the general in the particular. In Ch. 5, I have recounted having seen, with open eyes, one Amazonian farmer as *the* Farmer. On later occasions, in visions with closed eyes, I have seen what I interpreted to be the primordial Jaguar and the archetypical union of Man and Woman. As mentioned earlier, very similar experiences were reported to me by other people.

Also observed is the experience whereby one feels that one is encountering the abstract and seeing it as though it were concrete. This, I take it, is what one experiences when one sees the world of Ideas, that comprising the Platonic realm. I have had such an experience once, and have been told of similar experiences by three of my informants. One of them told me she saw what she characterized as 'the Platonic spheres'. These spheres, she recounted, embody all true knowledge—everything that is not affected by time and change. Intimately related to Platonic ideas is the world of mathematics. Once I had a vision of Plato's academy, in which I saw the great philosopher observing the perfect geometric bodies. Visions of these bodies were reported to me by two other informants. Still two others reported having encountered a deity who showed them all possible mathematical forms; one of these visions was recounted in Ch. 8.

As indicated in our discussion of the themes of Ayahuasca visions and the ideas and reflections associated with them, Platonic features are associated also with the overall world-view that seems to be associated with the Ayahuasca experience. As noted, this may be characterized as dynamic idealistic monism. Precisely this was noted by Huxley (1944, 1971). As indicated earlier, Huxley's observations were made on the basis of his experience with mescaline, yet given their wonderful insightfulness and eloquence let me quote from them once again; in the original text, the following sentences are embedded in the fragment quoted in Ch. 4:

> The Being of Platonistic philosophy—except that Plato seems to have made the enormous, the grotesque mistake of separating Being from becoming, and identifying it with the mathematical abstraction of the Idea. He could never, poor fellow, have seen a bunch of flowers shining with their own inner light and all but quivering under the pressure of the significance with which they were charged; could never have perceived that what rose and iris and carnation so intensely signified was nothing more, and nothing less, than what they were—a transience that was yet eternal life, a perpetual perishing that was at the same time pure Being, a bundle of minute, unique particulars in which, by some unspeakable and yet self-evident paradox, was to be seen the divine source of all existence. (1971: 17)

Huxley's criticism of Plato is, I think, astute. It is in line with the distinction made in the previous chapter between the two types of non-ordinary temporality. As pointed out, not all manifestations of non-ordinary temporality are associated with stillness. Thus, a paradoxical relationship holds between meaning and temporality. While it expresses itself in time, meaning in itself is above and beyond the temporal. Specifically, meaning, like the contents of a book, the gist of a musical piece, the essence of information, is there, in a non-temporal realm that is not dependent on any of its real manifestations in time and matter. Such ideas have also been presented by several contemporary thinkers in contexts totally unrelated to mysticism; in particular, I shall mention Popper's (1972) World Three, the world of cultural creations, and the non-orthodox philosophical approach to grammar proposed by J. Katz (1972). And last but not least, similar ideas have been presented on the basis of very different lines of reasoning in the writings of Jung (see Jung, 1969). Indeed, the Jungian archetypes—e.g., the Great Mother, the Sage, the Eternal Youth—may be regarded as personality patterns and narrative moments that recur in the Story of Human Life. More on these issues will be said in the final chapter of this book.

In this conjunction, let me mention again the study of the Piaroa conducted by Overing (1985*a*), who reports that according to the Piaroa there is a realm that is before time in which the special powers of the gods are guarded in boxes of crystal.[1] *Inter alia*, these powers include rationality and the words of songs. The knowledge pertaining to this realm is directly associated with hallucinatory experiences induced by psychoactive substances (not Ayahuasca).

Music and Mathematics, Aesthetics and the Sacred

Still another domain that may be associated with the Platonic realm is that of music. As we have repeatedly noted, in all contexts of Ayahuasca use music plays a pivotal role. A full discussion of music and Ayahuasca is beyond the scope of the present discussion. Here I would highlight only the relationship between music and the atemporal realm of meaning that is the focus of the present analysis.

The relationship between music, mathematics, and the world of eternal forms was, indeed, made by the Greeks themselves. It is at the heart of Pythagorean thought, where mathematics is regarded as the music of the spheres (see Plato, 1935; as well as Meyer-Baer, 1970; Stolba, 1990). Furthermore, in Platonic metaphysics the beautiful is equated with the true and the good. Significantly, as noted in Ch. 10, in *Phaedrus* (1914) Plato makes a specific relationship between aesthetics and noesis. He observes that there is a special class of intelligible objects whose apprehension does not require a mediation of our reasoning powers. They have the unique power of spontaneously revealing to the psyche the existence of an ideal world beyond the world of sense. The class of objects in question is that which we credit with the property of 'Beauty'. I may add that the intrinsic relationship between the realms of truth, aesthetics, and ethics is also central in the doctrine of the Santo Daime Church: the four key constituents of this doctrine are truth, harmony, justice, and love (see Sena Araújo, 1999).

The enhanced aesthetical sensitivity may be regarded as one manifestation of the overall enhanced meaningfulness that Ayahuasca induces. The same may be said with regard to sanctity. Tying the foregoing observations with those made in the

[1] Crystal, recall, is one of the most common materials of which objects seen in Ayahuasca visions are made.

previous chapters, a general hypothesis may be suggested. The closer one draws to the realm of the eternal, the more immersed one is in a domain governed not by procedural mechanical dynamics but rather by semantic considerations. In this realm, one invests more meaning in things. With this, both one's aesthetic and religious sensitivities are enhanced.

Noesis

Noesis is the experience of knowledge. As will be further noted in Ch. 16, noetic feelings are characteristic of mystical experiences in general (references are indicated there). With Ayahuasca, noetic feelings are common. As already indicated in several earlier chapters, people very often feel that Ayahuasca puts them in touch with the ultimate reality of being, and thereby reveals to them the deeper meaning of things.

The belief that Ayahuasca is a vehicle—in fact, the *prime* one—for the attainment of knowledge is prevalent throughout the indigenous cultures of the Amazon. In these cultures, knowledge of ultimate realities as well as all major cultural achievements are attributed to this brew. As indicated in Ch. 7, Ayahuasca—along with other psychotropic agents—is considered to be 'the only path to knowledge' (Chaumeil, 1983). The world revealed through the consumption of these agents is taken to be real whereas the ordinary 'real' world is often regarded as illusory. Citations regarding these attributions were presented in Ch. 1; here is another one made by the Tatuyo Indians and cited by Baldran and Bareiro-Sanguier (1980: 98): 'The yagé recounts the words of the ancestors.... Thus, the yagé is for teaching... For learning to dance, for learning to sing, for learning to speak. One is being taught to know how to speak, the other how to dance, and still another how to whistle.' For further discussion, see Reichel-Dolmatoff (1975); Langdon (1979*a*); Chaumeil (1983); Luna (1986b); Lagrou (1998).

It should be noted that similar ideas with regard to the noetic quality of Ayahuasca are encountered in the new syncretic groups employing the brew. In the context of both the Church of Santo Daime and the UdV it is maintained that what the brew reveals is the ultimate, true reality, to be contrasted with what we normally take to be real, which is actually an illusion.

Noetic experiences were reported to me by many informants, and I had them too. These were associated both with personal insights and with intellectual and spiritual ones. Examples were already given in the discussion of ideas, insights, and reflections in Ch. 10. Here I cite firsthand observations made by the anthropologist Deltgen (1978/9: 69) who, by his own avowal, was very sceptical about Ayahuasca and when he began his research not at all interested in experiencing it himself: 'I was "seeing" with my brain directly, not with my eyes. Or more adequately: I was seeing with my mind. And what I saw was of pure spirit by nature. I said to myself "gee, this is pure spirituality". And I understood those mystics who had written about "marriage of the soul with god".'

Here, let me note that when powerful, the experience of noesis induced by Ayahuasca is quite extreme. Many drinkers told me that under the intoxication, they felt they could know everything, reach everything. For some this is intertwined with the feeling, very common with Ayahuasca, that reality is one and everything is interconnected. As a consequence, every bit of reality reflects all reality.[2] Indeed, all that can be known is presented to one in one shot. It is merely up to one to get it. It is as if one is presented with the most sumptuous meal imaginable. How much one could eat ('grasp' is a better term here) will depend not on the offering, but on one's capacity. Likewise with the knowledge Ayahuasca presents. The reader will note that this characterization echoes one made in the previous chapter in conjunction with time; there I underlined the unbounded experience drinkers feel they may have when time comes to a halt. There and here the same state of affairs obtains: all can be known, all can be experienced—only the knower, being human, is limited.

Let me conclude this discussion with a description taken from a context quite different from that of Ayahuasca (or psychoactive agents, for that matter). These lines from Hesse's *Magister Ludi* (Hesse, 1969: 423) describe the experience I have been talking about here both accurately and beautifully, and despite their different context I find them most pertinent to our discussion:

> For a moment it seemed to him that the mind could grasp everything—the sure, soft course of the planets above, the life of man and animals, their bonds and hostilities, meetings and struggles, everything great and small along with the death locked within each living being. He saw or felt all this as a whole in a first shudder of premonition, and himself fitted into it, included within it as a part of the orderliness, governed by laws accessible to the mind.

The reader is advised to return to Ch. 1 and compare this text to the indigenous characterizations of *yagé* cited there.

The Issue of Interpretation

Related to the experience of noesis is the issue of interpretation. How much of what is seen with Ayahuasca is the product of drinkers' interpretation? Experientially, the visions usually impress those who have them as revealing other, independently existing, realities. Usually, what is seen is so foreign and unexpected that drinkers feel that they themselves contribute nothing to the contents of the visions. I too have had these feelings. Yet, in the course of my interviewing I have encountered several cases in which the contribution of interpretation was explicitly admitted. Thus, consider a report furnished to me by one of the most experienced of my independent drinkers. The report was of the first powerful vision that this person had with Ayahuasca; the main figure in the vision was Jesus Christ. On a subsequent occasion when this informant again described this vision to me he said 'Well, what actually can be said about the vision is only that the figure of a loving

[2] This, of course, brings to mind monistic metaphysical views, notably those of Leibniz (1714/1945).

young man, clad in white and radiating light was seen. The identification of this man as Jesus was perhaps mine, reflecting my cultural background and religious beliefs.' Should we therefore say that what was seen was a radiant young man clad in white and that the identification of this man as Jesus was merely the biased product of the drinker's interpretation?

Phrasing this question in general theoretical terms, should we say that what is seen in Ayahuasca visions is to be divided into two: that which is 'really' seen, and that which is the product of interpretation? While there might be instances—like, perhaps, the example just cited—where interpretation may be relegated to a separate, secondary process, I am reluctant to regard this as the paradigmatic, general case. Because of my previous work in both psychology and semantics, I have difficulty accepting the two-stage analysis dividing perception and interpretation. My general theoretical stance in cognition is that there is no demarcation line between 'raw' perception, on the one hand, and semantic, meaningful interpretation, on the other hand. Following the philosopher Merleau-Ponty (1962) and the psychologist Gibson (1979), I believe that it is impossible to draw a clear-cut line dividing between naked, interpretation-free sensory inputs and interpretative processes that are subsequently applied to them so as to render these inputs into meaningful percepts. In the spirit of Heidegger (1962), I maintain that cognition is always 'laden with meaning'. Applied to the example cited, this view implies that, from a cognitive-psychological point of view, if the figure seen was identified as being Jesus, then phenomenologically this is indeed who was seen.

Maintaining that interpretative processes are involved in perception implies that the experiencing agent contributes to what he or she sees. This is true of perceiving agents in general; this may be true of Ayahuasca drinkers as well. Yet, all this does not imply that what is seen in the visions is merely subjective. Or rather, Ayahuasca visioning need not be more subjective than ordinary perception is. Following the Kantian tradition, practically all modern students of mind adhere to the view that both our perception of the world and our conceptualizations of it are filtered by the apparatus of our cognitive system. We can never step outside this apparatus and know the world as it 'objectively' is. Kant distinguished between the cognitively determined *phenomena* and the unknowable things as they are in themselves, the *noumena*. I, along with other contemporary thinkers (as, for instance, those cited above) would say the distinction between the objective and the subjective is unwarranted. Our reality is the product of a dynamic interplay between what is given to us and what is contributed to by our cognitive system. The interplay is such that any simplistic division between the two determinants is futile. Likewise with Ayahuasca visions. Admittedly, interpretative processes are involved in these visions, yet this need not imply that what is seen is 'merely subjective'.

Thus, rather than direct the onus of our discussion to the question of how much of what is seen in Ayahuasca visions is due to drinkers' interpretation and how much to non-idiosyncratic determinants, I would focus on the distinct characteristics of what is seen in the visions and on the dynamics of their generation. Like ordinary perception, Ayahuasca visioning involves processes of interpretation, but these processes cannot be teased apart from the processes of visioning in any neat, modular fashion. Interpretation is not added to perception but rather, is ingrained in it. And so is meaning. Like music, which is not the product of the concatenation of acoustic notes one to another but is there from the very beginning of the act of playing, simple though it may be, meaning is there in the very foundations of whatever we experience. In both ordinary perception and in visioning, the material of perception is imbued with interpretation, laden with meaning. What distinguishes the two mental states are the particular semantics that they manifest and the dynamics governing their generation and their development. On this I shall say more in Ch. 21.

The foregoing discussion notwithstanding, one should not dismiss the puzzling nature of Ayahuasca visions, their content, and the knowledge they are experienced to impart. A strictly psychological account cannot exhaust the questions at hand; a full treatment of these has to involve epistemological and metaphysical considerations. Some of these are addressed in the last two chapters of this book.

Concluding Remarks

Taken together, the observations made throughout this and the previous chapter suggest a twofold pattern. In the previous chapter we noted that atemporality leads to the realm of meaning. Here we observed that the complement is also true: ultimate meanings are atemporal by nature. In line with these observations, as well as with my firsthand experiences of non-ordinary temporality, I would like to venture a hypothesis: the more powerful the vision one is experiencing (or perhaps, the more strength one has) the higher one's capacity to maintain one's experience of timelessness. With this, one will experience more and more in what are in fact shorter and shorter periods of time. In the limit, the actual duration (that is, the time measured by a clock) would be minimal whereas the subjective experience would be felt as everlasting. Reflection will reveal that this makes perfect sense. The atemporal realm is outside time. In its pristine manifestation, it does not depend on time at all.

In the present discussion, which focuses on Ayahuasca, the relationship between knowledge and atemporality is associated with highly non-ordinary experiences. Reaching the end of this discussion, let me note that in fact this relationship is also encountered in the context of everyday life. As a rule, the more knowledgeable and masterful a person is, the less is he or she dependent on time. For example, if a lecturer is totally proficient in a subject matter and very much involved with it, he or she need not rehearse the lecture verbatim before delivering it. It is enough for the lecturer to decide what he or she is going to talk about, reflect or perhaps meditate upon it, stand on the podium, open his or her mouth, and talk. The lecture would take, say, an hour to deliver but epistemically the knowledge of the lecture is with the lecturer all at once, in a manner that, in a fashion, defies time. What has been said about Mozart in the previous chapter is the most extreme case of this phenomenon known.

In closing, I cite two lines I have found in José Saramago's (1989/1996: 312–13) novel, *The History of the Siege of Lisbon*: 'Into one year there enter no more things than into a minute, just because this is a minute and this is a year. What is important is not the vase but what each one of us can put into it.'[3]

[3] This is my own translation from the original Portuguese. In Portuguese the word I have translated as 'vase' is *vaso*; in the English translation of the novel 'glass' is used, in the Hebrew translation 'container'.

16

Consciousness II

> The prophets Isaiah and Ezekiel dined with me, and I asked them how they dared so roundly to assert that God spoke to them; and whether they did not think at the time that they would be misunderstood, and so be the cause of imposition. Isaiah answer'd: 'I saw no God, nor heard any, in a finite organical perception; but my sense discover'd the infinite in everything, and as I was then persuaded and remained confirm'd, that the voice of honest indignation is the voice of God, I cared not for consequences, but wrote.'
>
> William Blake

> And I pray you, queen, who can do anything
> You choose to do: after this great vision
> Enable him to keep his affections sane.
>
> Dante

In Ch. 12 I surveyed modifications in consciousness encountered with Ayahuasca and in the three subsequent chapters I examined specific topics that relate to these modifications; in this chapter, I discuss several further issues pertaining to consciousness. On the one hand, I consider a series of independent issues that relate to this topic. These include paranormal experiences, spiritual and mystical experiences, and the question of sanity and madness. On the other hand, I examine phenomena surveyed in the previous chapters from a more abstract theoretical perspective. The theoretical suggestions made here will further serve to demonstrate the great potential the study of Ayahuasca (along with that of altered states of consciousness in general) has for the study of the mind.

Paranormal Experiences

Reports of paranormal experiences with Ayahuasca abound. Practically everyone who has had more than a rudimentary exposure to the brew reports having had telepathic experiences. Many such reports also appear in the anthropological literature (see, for instance, Rouhier, 1927, as well as others cited below). Indeed, the first term—coined by Zerda-Bayón in 1912–for the active ingredient in Ayahuasca was *telepatina* (see Morton, 1931). Similarly, many of my informants said that without overt verbal articulation they could pass messages to other people present in the Ayahuasca session, notably individuals who were personally close to them (e.g. a spouse). Likewise, many indicated that they received such messages from other persons or beings. Usually, in visions in which drinkers feel that they are receiving messages or instructions from beings and creatures, the communication in question is said to be achieved without words—directly from thought to thought. Interestingly, this is consistent with claims made in earlier centuries by Blake (1794/1961) and Swedenborg (1854) (see the epigraphs to this chapter and to Ch. 14 respectively). Both men explain that prophecy and communication with the Divine is conducted not by means of language but rather through the direct transfer of thoughts. Furthermore, many report having had the feeling that their minds or thoughts are being controlled by others who participate in the session. This experience is especially common with novice drinkers. Similarly, some feel that they could control other persons' thoughts and influence their behaviour.

Akin to telepathic communication is the feeling that one understands the personalities of other individuals and gains special access to their mental states and inner feelings.[1] Non-mediated communication is also experienced with animals. Indeed, experiences in which drinkers feel that they can converse with animals and understand them are very common with Ayahuasca. Also reported are similar experiences with plants. Such non-ordinary communicative abilities are also reported in the indigenous context of Ayahuasca use (see e.g., Luna and Amaringo, 1993).

Experiences of paranormal time travel are very common. These are often associated with experiences of a change in personal identity and, as noted earlier, are often interpreted as instances of reincarnation. Specifically, drinkers feel that they are taking on another person's identity, that this is taking place in a location and a time other than the here and now, and that the non-ordinary identity at hand is one's own in a previous life incarnation.

Related to these experiences is that of gaining information about other persons and other places that one encounters and visits in one's visions. For instance, in visions of ancient civilizations—which, as noted, are very common with Ayahuasca—drinkers feel that they actually inspect other places as they were at other times and that the perceptions they have depict these places as they had actually been in historical times. Similarly, visions of future events are taken to be instances of veridical precognition. Also reported are cases of so-called remote viewing. One of my informants, a European who lived in Brazil, said he had such experiences; specifically, he saw what was happening at the exact moment to his girlfriend back home. Alas, I have no objective verification that this case did, in fact, involve any paranormal achievement. For early, non-indigenous accounts of remote viewing and of precognition with Ayahuasca, the reader is referred to McGovern (1927), Morton (1931), Wilson (1949), as well as Bristol (1966), Prance (1970), Rivier and Lindgren (1972), and Kensinger (1973).

[1] I find it interesting to cite the following similar observations made by Huxley (quoted in Horowitz and Palmer, 1977: 82) in conjunction with mescaline: 'Among the by-products of this state of being the given fact of love was a kind of intuitive understanding of other people. . . . I found myself saying things about [another person present], which I didn't know but which, when I said them, turned out to be true. Which, I suppose, is what one would expect if one happens to be manifesting the primordial fact of unity through love and the knowledge of mutual immanence.'

As mentioned in Ch. 15, also reported by some is the phenomenon of glossolalia or speaking in tongues. As noted, none of my informants reported having had this experience, but half a dozen said that in sessions in which they participated they saw this happen with other drinkers.

Except for glossolalia, I have personally experienced all the above. Especially impressive for me was an exchange that took place at the very end of a session I had with a Peruvian healer. Conversing with the healer, I could anticipate each sentence he uttered just a moment before he actually did so. The experience was startling indeed.

Yet I resist interpreting these experiences as paranormal. This may be a sign of a lack of intellectual openness on my part. However, unless it is clearly demonstrated to me that the experiences are indeed paranormal, I opt to remain sceptical. In general, it seems to me that telepathic communication and special psychological insight and understanding of other persons can be explained by reference to heightened intuition and enhanced psychological sensitivity—both are definitely effects that Ayahuasca induces. In those cases where personal discretion allowed me to check in detail reports of telepathic experiences, I have done so. The bottom line is that I remain to be convinced. Surely, the feelings of the experiences are there. And apparently, there is a human tendency, and perhaps a need, a kind of imagined longing, to believe in the paranormal. As indicated below, I myself am no exception to this. In particular, the experience that I had with the Peruvian healer can be attributed to a heightened sense of intuition, a total immersion in the situation, and a strong coupling with that other person. Don't we all experience this in ordinary life when deeply engaged in activities such as couple dancing or teamwork? Mind control can be attributed to suggestibility (for empirical data indicating increased suggestibility in conjunction with altered states of consciousness see Dobkin de Rios and Grob, 1994). Encountering other persons that one identifies as oneself can be interpreted as encountering potential facets of one's persona and/or roles one would have liked to have played.[2] The experiences taken to be visions of other incarnations also involve the heightened imagining that is, of course, one of the key effects of the Ayahuasca inebriation. The same holds for actual non-ordinary performances that one may execute under the intoxication, as those surveyed in Ch. 13.

As for glossolalia, I once had the opportunity to listen to a recording made of linguistic utterances said to be Hebrew, my native tongue—they were nothing of the sort. This is not to deny the subjective feeling that the drinkers had of having spoken a language they do not know. Furthermore, a more systematic inspection might possibly reveal that in terms of structure the utterances manifest articulation and coherent morphological structure that lend the semblance of a language (for empirical data collected in other contexts and supporting this possibility, the

[2] Psychologically, such experiences are definitely valuable. They give a person an opportunity to be in touch with facets of his or her personality he or she is unaware of. Seeing oneself in hypothetical situations quite different from those of the context of one's actual life may be very instructive without these experiences being interpreted as other incarnations.

reader is referred to F. Goodman, 1972). They are not, however, expressions in any specific language that the intoxicated individual does not know. In other words, remarkable articulatory fluency in some sort of gibberish may be manifested, but no paranormal phenomenon is involved.

The one experience that really startled me and which, for a while, made me believe in some sort of paranormal information transfer happened in a private session in which I and three other persons participated. One of the participants opened his mouth and began the recitation of a long poem presented as an ode of an Egyptian pharaoh. The poem was long, about 250 verses, it was recited in Portuguese and the lyrics were of high poetic quality. The text contained many allusions to ancient Egyptian names. Subsequently, I examined some ancient Egyptian poetry—for instance, Akhenaton's famous *Ode to the Sun*. The poem I heard (and recorded on tape) seemed to be of the same genre. As I said, I was startled. My friend could not have read any ancient Egyptian literature, I surmised. However, I later found out that he was a Freemason. The way I see this episode now is as follows. The man had never read Akhenaton's *Ode* but he had heard all sorts of fragmentary information about ancient Egypt. Under the intoxication, he had the wonderful talent to assume a role and enact a pharaoh reciting a poem. The poem I heard was not a replica of a Pharaonic recitation, it was created *de novo* as I and my other fellow participants in that session were listening to it. The feat was indeed impressive, but it was not paranormal.

Lastly, let me mention a phenomenon which is not part of the psychological experience of the Ayahuasca inebriation but that is intimately related to it—coincidence. As noted in the previous chapter, reports of synchronicities taking place in conjunction with Ayahuasca sessions, some of which are indeed amazing, are extremely common. I have many such stories to tell myself. The truth is that I do not know what to make of them.

In closing, I must emphasize that my reluctance to accept paranormal interpretations (i.e. ones that counter the laws of physics) does not in the least invalidate the fact that with Ayahuasca there is a genuine feeling that one is having paranormal experiences. This feeling is reported by many who have partaken of this brew, and it is also common in first-time experiences. Thus, as psychological phenomena, these experiences are factual. Wonder and enchantment need not, however, counter the rules of physics. Events in the physical world proceed as they always do, and Ayahuasca does not change them. What happens under the intoxication takes place, I surmise, on the level of consciousness. Consciousness—even ordinary consciousness—is to be accounted for by psychology, not physics. And, indeed, much about consciousness is wondrous.

Spiritual and Mystical Experiences

For many, perhaps most, people the main reason to partake of Ayahuasca is spiritual. In all traditional and institutionalized contexts of Ayahuasca the consumption of the brew is a sacrament, a sacred ritual. It is not for nothing that psychoactive plants have been called 'plants of the gods' (Schultes and Hofmann, 1979) and that for many the term 'entheogen' (i.e. that generating the god within; see the references indicated in Ch. 1) is currently replacing the older and, to some, pejorative, terms 'hallucinogen' and 'psychedelic'.[3]

Personally, if I were to pick one single effect of Ayahuasca that had the most important impact on my life (there were many and the choice of one is not at all easy), I would say that before my encounter with the brew I was an atheist (I used to define myself as a nineteenth-century-middle-European-like intellectual who is a devout atheist with a strong affinity to Jewish history and its scholarly tradition) and when I returned back home after my long journey in South America, I no longer was one. Likewise, a significant number of informants I have interviewed indicated that the main lesson they received from Ayahuasca was religious or spiritual. 'Ayahuasca showed me that God exists,' 'I have come to appreciate the place of the sacred in human life,' 'I have encountered the Divine,' are all statements I have heard more than one person say. There are many individuals who, in direct consequence of their experience with Ayahuasca, underwent a radical religious or spiritual conversion. In the context of the UdV, towards the end of sessions, it is very common for members to tell the group of the great impact that the encounter with the 'tea' has had on their lives. Many times in such testimonies I have heard people proclaim that this encounter was the single most important event in their lives, that it had totally changed them, that with it they found healing and new meaning to their human existence. For some, the transformative impact of Ayahuasca is long-lasting and its effects remain throughout the course of the person's entire life. In this conjunction, it will be noted that in both the Santo Daime Church and the UdV, affiliation with the group is not confined to the Ayahuasca experience itself, but embraces a wide group commitment that includes financial contributions and various societal engagements.

The consideration of the spiritual and mystical experiences associated with Ayahuasca extends beyond the cognitive-psychological domain proper. A serious theoretical examination of these experiences should include analyses pertaining to personality theory, clinical psychology, metaphysics, and theology. This discussion is bound to involve observations and reflections of a personal and speculative nature, ones calling for a style of discourse very different from that I have attempted to adopt here. Hence, this topic is outside the framework of this book.

[3] In this conjunction, let me make reference to the hypothesis that many of the world's great religions actually originated in the use of such entheogens. The first to propose this was Wasson (1968) in conjunction with the use of soma in ancient Vedic culture. Subsequently, Flattery and Schwartz (1989) extended this proposal to the use of haoma in the Zoroastrian religion. Furthermore, with several other scholars, Wasson *et al.* (1978, 1986) presented the case that the nectar *kikeion*, served in the Greek mysteries of Eleusis, contained a psychoactive ingredient very similar to LSD. In Shanon (1999), I present the thesis that the ancient Israelite religion has entheogenic roots as well. Central to my thesis is the observation that in the Sinai peninsula there grow two plants each containing one of the two active molecules found in the two plants of which Ayahuasca is made. Independently, Merkur (2000) proposed that the manna the Children of Israel ate in the desert was an entheogen.

Here I would only like to make several brief phenomenological remarks and to relate the experiences encountered with Ayahuasca to those reported in the literature on mysticism; for pertinent general discussions the reader is referred to Bucke (1901/1991), James (1929), Underhill (1955), Stace (1961), Maslow (1970), and Scharfstein (1973), as well as Masters and Houston (1966), and Wainwright (1981) who specifically discuss substance-induced mystical and religious experiences.

Encountering the Divine

In Chs. 8 and 9 I have examined visions in which God, deities, and divine beings are encountered. The emphasis there was on the visual aspects of the visions; here I would like to focus on their spiritual aspects and on the experience people characterize as encountering the Divine.

Not surprisingly, it is not easy to describe experiences of this kind. Here, indeed, we reach the realm of the ineffable. There is no question in my mind that when people say that they have had 'an encounter with the Divine' they are referring to a genuine experience that they have had. I have had such experiences too. Explaining what this experience consists in, however, is less clear. Nor is it clear that all people employ the phrase with the same facility. I have asked several of my informants to explain what they meant by this term. Their answers included references to 'a presence which is full', 'the ground of all that exists', 'the source of all life', 'the fountain of all wisdom', 'the utmost perfection', and 'sublime happiness'.

An experience that for me was very powerful occurred at a *Daime* session:

> I was sitting in front of a white wall on which a golden star of David (the seal of Solomon) was inscribed [this was real—this symbol is central in the Santo Daime Church]. The star-shaped figure was shining and I felt a presence very strongly. The expression 'behind the Veil' (traditionally employed in Judaism to characterize indirect encounters with the Divine) came to my mind and I engaged in a silent communication with whatever power was there.

Subsequently, in a letter to a friend I wrote that I was washed by grace and that I understood that all Existence is infused with the Divine, with sacredness.

Mystical Experiences

As noted, various aspects of the Ayahuasca experience are reminiscent of patterns found in reports of mystical experiences encountered in many religious traditions. In particular, I would like to point out that all the classical characteristics of mystical experiences defined by Stace (1961) are encountered with Ayahuasca (see also Bucke, 1901/1991; James, 1929; Forman, 1990; Wade, 1996; and Merkur, 1999). Stace distinguishes between two types of mystical experience which he labels the 'extrovertive' and the 'introvertive'. Each of these is defined by a series of seven characteristics. However, out of the seven only the first characteristic significantly differs for the two types.

Stace further claims that only the introvertive type can be induced by external means. I disagree: all the experiences he associates with the extrovertive are also encountered with Ayahuasca. In the light of this disagreement, and since with respect to most features the two types are the same, I have opted to combine the two aspects in one paragraph and present just one list of characteristics, not two, as Stace does.

With Ayahuasca all the patterns associated with both types can be encountered. Following is a list of these patterns along with commentary on the comparable phenomena encountered with Ayahuasca.

1. *Unity*. In the extrovertive case, unity pertains to the mystic's perception of the world; in the introvertive case, unity pertains to the mystic's state of consciousness. Consequently, in the former case the mystic directly feels that behind the multiplicity in the world there is oneness that is apprehended as unitary consciousness devoid of sensual form and conceptual content. In the latter, the mystic feels that the boundaries of the self dissipate and that he or she becomes one with an existence larger than him- or herself. Both patterns are encountered with Ayahuasca.

2. *Transcendence of time and space*. The mystic feels that time and space are no longer relevant. In the context of Ayahuasca, such phenomena were discussed at length in Chs. 14 and 15.

3. *Noesis*. The mystic regards what he or she experiences as illumination or true knowledge. In particular, visions and ideations are taken to pertain to an objective, independent reality. The assessment that this is the case is grounded in a direct, intuitive feeling. In the context of Ayahuasca, the phenomenon was discussed both in the previous chapter and in Ch. 10; I shall further expand on this topic in the final chapter of this book when examining the epistemological issues that the Ayahuasca experience raises. I shall add that as mentioned earlier, one of the four pillars of what is referred to as the 'doctrine' of the Santo Daime is truth.

4. *Positive feelings of blessedness, joy, peace, happiness*. As pointed out in Ch. 4, all these feelings are very marked with Ayahuasca. Reference to the Santo Daime doctrine is in place here too: harmony is yet another one of the pillar of this doctrine. Furthermore, one of the three principles specified in the context of the UdV is peace.[4]

5. *A sense of sacredness*. This is manifested in whatever is being apprehended to be taken to be holy and divine. We have discussed this in several chapters, most notably Ch. 4, in conjunction with the general atmosphere that Ayahuasca induces.

6. *Paradoxicality*. Mystical experiences seem to defy the standard canons of logic. The medieval philosopher and mystic Nicholas de Cusa (1960: 161) said explicitly that in order to reach the higher realms of the Divine one has to leave rationality behind:

> The abode wherein You [God] dwell unveiledly—an abode surrounded by the coincidences of contradictories. And this coincidence is the wall of Paradise, wherein You dwell. The gate of this wall is guarded by a most lofty rational spirit; unless the spirit is vanquished the entrance will not be accessible. Therefore, on the *other* side of the coincidence of contradictories You can be seen—but not at all on this side.

[4] The other two principles of the UdV are light and love.

A summary conclusion that I have articulated following my experience with Ayahuasca with regards to the defiance of logic is presented in the section on philosophical issues in Ch. 10. An interesting discussion of the relationship between logic and the mystical experience is presented in Stace (1961: ch. 5).

7. *Ineffability*. As I have indicated on many occasions in this book, many facets of the Ayahuasca experience are described by drinkers as being beyond any verbal description. I reckon that the most fitting thing is not to say anything further here; for general philosophical discussion on this matter, see Stace (1961: ch. 6) and S. Katz (1992).

Lastly, in the literature it is also pointed out that mystical experiences often have concrete, pragmatic effects. Indeed, some have taken the *transformative impact* to be a defining property of these experiences (see Pahnke and Richards, 1966; Pahnke, 1972). This impact may manifest itself in religious conversion, changes of world-views and belief systems, and in new definitions of one's personal and ethical values. As indicated in the introductory comments to this section, with Ayahuasca such manifestations are common, and often they are quite radical. For general discussion of the long-term psychological and psychotherapeutical impact of psychotropic substances the reader is referred to Masters and Houston (1966) and Grof (1994).

In sum, all the paradigmatic characteristics of the mystical experience are encountered with Ayahuasca. Some aspects pertaining to these will be further discussed in later chapters.

By way of conclusion, let me point out that the foregoing comparative statements also bear on the more general question regarding the status, meaning, and value of religious and spiritual experiences induced by the ingestion of psychoactive agents. Are these comparable to the experiences of mystics attained without external agents? Are they as valuable? With respect to the first facet of the question, the phenomenological, my empirical study of Ayahuasca leads me to answer with a categorical 'yes'. The second facet, that of value judgement, is to be discussed elsewhere. For discussions of this topic in the literature see Zaehner (1957, 1972), Stace (1961), H. Smith (1964), Pahnke and Richards (1966), Masters and Houston (1966), Pahnke (1967, 1972), Clark (1969), Houston and Masters (1972), Wainwright (1981), and A. Smith and Tart (1998), as well as the anthology by Forte (1997) and the collection of articles by H. Smith (2000); several interesting firsthand accounts of substance-induced mystical and religious experiences are reported in Metzner (1968).

Sanity and Madness

The Ayahuasca experience poignantly raises the issue of sanity and madness. There is no question about it—inspected in isolation and out of context the patterns of consciousness surveyed above may be regarded as symptoms of sheer madness. Surely, mental health professionals lacking knowledge and experience of this special state of consciousness will tend to classify such patterns as psychotic (hence the term 'psychotomimetic', mimicking psychosis, for psychoactive substances; for a very interesting discussion of these terms, see Ott, 1996).[5] Yet, this is precisely the point— things cannot be taken out of context. The difference between sanity and madness need not be intrinsic, in the state of mind itself. Thus, one's attitude plays a critical role. Whether one is the master of a non-ordinary experience or is enslaved by it can make the entire difference. The madman is chaotically pushed around by the non-ordinary experiences he is subject to. Frightened, he is prone to enter into combat and struggle; these are likely to be futile and more than anything, more fear will be induced. In contrast, the solid, mentally stable person is, ideally, a master of his state of being. Undergoing the radical alterations of consciousness he is subject to, this person will appreciate the new experiences presented to him, and filled with awe and wonder, he will go on to explore and even enjoy (for further related discussion, see Murphy, 1976).

But the truth is that some differences between madness and the Ayahuasca experience may be intrinsic. Unlike madness, the Ayahuasca experience is usually felt to be structured and meaningful. When one overcomes its initial harsh stages, the experience can be most rewarding, uplifting, and also enjoyable. My experience is that with practically no exceptions, at the end of Ayahuasca sessions people feel significantly better—in a better physical shape, happier, greatly enriched, and more fulfilled—than they were just before they partook of the brew. None of this can be said in conjunction with psychosis (for further related discussion, see Walsh, 1990).[6]

Awareness and Reality Judgements

Drawing towards the end of my discussion of the modifications of consciousness induced by Ayahuasca, let me comment on the very state of being that one is in under the intoxication. The following remarks concern awareness and reality judgements. How real are the objects and states of affairs seen in Ayahuasca visions taken to be? To what extent are the visions differentiated from ordinary perceptions? Do drinkers confuse the two and do they fail to appreciate what is real and what is not? How is the distinction between visions and ordinary reality made? Directly related to these questions is the consideration of the notion of hallucination, a topic on which I shall further expand below.

[5] In fact, the *Webster's Third New International Dictionary* (1975) defines Ayahuasco (*sic*) as an: 'American vine... having roots that yield a drink which produces delirious psychosis that alternates with prolonged hallucinations and dreams.' Apart from the several factual errors that this single sentence contains, the definition reflects the general Western misconception of the non-ordinary states of mind induced by Ayahuasca and other psychotropic substances. This misconception, grounded both in ignorance and prejudice, reflects, of course, that of the culture at large.

[6] Interestingly, the Secoya—an Amazonian tribe of Ecuador and Peru—employ the same word, *quepe*, for both 'insane' and 'intoxicated with *yagé*'.

Often, Ayahuasca visions are strong and drinkers do indeed feel that what is depicted in them is real. The assessment of reality is due to several factors: the visual clarity and vividness of the visions, the powerfulness of the experience, the appraisal that what one sees surpasses anything that one could have possibly imagined on one's own, the unmediated feeling that what is seen has independent existence, and the noetic feelings associated with the experience.

Significantly, all the reasons indicated for the conferral of reality on visions are in essence not very different from those that underlie our assessment that the ordinary world of our perception is real (for pertinent experimental work, see Johnson and Raye, 1981). As Descartes pointed out (1641/1960), it is not simple at all to prove that what we normally see are true perceptions of an independent real world and not the fancies of our imagination. Yet, while philosophically the Cartesian doubt can always be invoked, psychologically this is not the case: Normal human beings do not doubt that what they perceive is real. In a fashion, the act (or rather, the existential state) of perception forces itself upon us and the moment we open our sensory faculties we are immersed in the world of our perception without any choice or room for doubt, like fish in water. Analytically we cannot prove that what we perceive is real, yet no normal person would doubt that this is the case (for a contrast between philosophical scepticism and psychological certainty see Shanon, 1983). Experientially, what is felt with powerful Ayahuasca visions is very similar.

With Ayahuasca people often feel that what they see is 'more real than real' (see Ch. 12). From a theoretical point of view this is especially significant. Often in the philosophical and psychological literature, non-ordinary notions of reality are associated with a diminishing sense of reality. Thus, many have argued that what distinguishes true perception, on the one hand, and memories, imagination, and dreams, on the other, is that the former is clearer, more distinct, and more coherent than the latter (for a classical analysis, see Hume, 1739/1978; for a modern treatment, see Casey, 1976). In contrast, the Ayahuasca experience presents cases of an enhanced sense of reality. This indicates that reality judgement is actually a parameter that can take different values. Dreams and imagination usually decrease the value assigned to this parameter; the Ayahuasca experience may increase it. This notion of an enhanced sense of the real as it manifests itself in sensation, perception, ideation, and noesis is, I find, potentially fruitful, and can serve as the basis for cognitive research not necessarily linked to Ayahuasca.

Real though the Ayahuasca visions may be deemed to be, they are not usually confused with the normal perceptions of the ordinary world. Rather, the feeling is that what is seen in the visions pertains to other, separate realms. Hence, in general, at least a degree of an awareness of the real world is maintained: Usually, drinkers appreciate that they are in a particular place (e.g. sitting in a chair), that they can handle objects (e.g. hold on to something in order to support themselves) and, most significantly, maintain acknowledgement of the presence of other people around them. Thus, then, Ayahuasca drinkers may experience themselves as perceiving states of affairs that they regard as real, but at the same time they can appreciate that these states of affairs are separate and distinct from those pertaining to ordinary reality. The patterns of behaviour drinkers manifest *vis-à-vis* these different realities reflect these differential attitudes. Especially important in the distinction between visions and ordinary reality is the presence of other people and the drinker's interaction with them; I shall elaborate on this in Ch. 19. I shall add that the ability to differentiate between what are taken to be different realities does not imply that the person under the intoxication is constantly in touch with this world.

The world of visions may be so captivating that the Ayahuasca drinker may wish to be totally absorbed in it. In particular, the drinker may keep his or her eyes closed and avoid contact with ordinary reality. He or she may even succumb to a kind of seduction and the idea that the eyes can be opened might not occur to him or her at all (even if what is happening in the vision is most unpleasant). However, were the drinker to open his or her eyes, awareness of the real, external world will be there in some form or another. I shall again make use of a cinematographic analogy. In the cinema, one may be totally absorbed in the film one is watching and not think about or pay any attention to what there is around one (e.g. the cinema hall and the people in it). Yet, the viewer can always, if she so wishes, turn her head, let her eyes accommodate to the darkness, and inspect things in the hall. Usually, this is also the case with Ayahuasca.

This is not to imply that the passage from the hallucinatory to the ordinary realm is necessarily instantaneous and perfectly smooth. It may take a moment, or several moments, for one to reorient oneself. The situation is not unlike that experienced when being woken up to the ringing of a telephone when one is soundly asleep or waking up in the middle of the night when not in one's normal bed or room (of course, these two experiences can be simultaneously joined).

The foregoing discussion focuses primarily on the point of view of the drinker and his or her reality judgement. We can also consider the issues at hand focusing on visions and their *epistemic status*. This term, which I have coined in my study of ordinary consciousness, denotes the reality value people confer on their mentations. As pointed out in Shanon (1988*b*), the cognitive material people entertain have different statuses. Thus, we all differentiate between perceptions and memories, as well as between these two and thoughts and reflections. In my phenomenological study of thought sequences (see Shanon, 1989*a*), *Gedankenexperimenten*—that is, hypothetical states of affairs entertained and checked in the theatre of one's mind are noted as still another case (see also Johnson-Laird, 1983, on mental models). Along with other non-ordinary states of consciousness, the Ayahuasca experience presents yet another kind of epistemic status—that of *hallucination*. My appraisal is that the notion of hallucination is much more complex than is usually appreciated. The following discussion comes by way of showing this and highlighting various respects whereby the study of the phenomenology of the Ayahuasca experience may further the cognitive understanding of hallucinations.

Hallucination

What is a hallucination? I cite two representative definitions of this term; the first is taken from *The Encyclopaedia Britannica* (1971), the second from *The Penguin Dictionary of Psychology* by Reber (1984):

> The perception of objects with no reality or the experiencing of sensations in the absence of the usual cause. In effect, a hallucination is a perception within or of a person's fantasy that is mistakenly ascribed, by the person who experiences it, to sources outside.

> A perceptual experience with all the compelling subjective properties of a real sensory impression but without the normal physical stimulus for that sensory modality. Hallucinations are taken as classic indicators of a psychotic disturbance and are a hallmark of various disorders like schizophrenia.

In the literature, a distinction is also made between hallucinations and pseudo-hallucinations. The latter are images manifesting visual vividness like actual perceptions, but which are not taken by the person having them to be real (see e.g. Horowitz, 1975). In general, pseudo-hallucinations are considered to be a weaker form of abnormal perception than full-fledged hallucinations are.

Having conducted a comprehensive inspection of the literature—psychological, philosophical, and clinical—I can attest that the definitions cited above are most representative (for general reviews of the phenomenon of hallucination see G. Reed, 1972, as well as various contributions in the anthologies by West, 1962; Keup, 1970; and Siegel and West, 1975). Several features of these definitions (as well as practically all definitions in the scientific literature) will be noted. First, the orientation adopted in these definitions is *physical*, not psychological: a mental state is considered to be an instance of hallucination if there is an actual physical mismatch between a person's perceptual experience and the putatively corresponding physical objects. Second, the evaluation of hallucination is *external*: it is taken from the perspective of an external observer who is inspecting the person having the hallucination. The external observer compares the contents of what his subject, informant, or patient is reporting to see with what he (the observer) detects in what is deemed to be the objective state of affairs in the real world. Finding a mismatch between the two, he concludes that the subject is hallucinating. Thirdly, in most definitions in the literature a *value judgement* is involved. Specifically, it is assumed that this state is a perturbation of normal cognitive functioning, that it is a symptom of a mind that is deranged, that it is a condition that, in terms of good mental health, one should hope not to be in.

In contrast, the phenomenological perspective taken here is *psychological*, *internal*, and *non-judgemental*. The present analysis is based on the features of the perceptual experience itself. What is of interest to us here is the psychological experience of the hallucinating (or, as I prefer to call it, visioning) person, not the value judgements of agents who base their evaluation on non-psychological considerations. Correspondence or lack thereof with the external world may, indeed, be of prime import for the philosopher; veracity or falsehood may be critical as far as the physician or neuropsychologist is concerned. But the interest of the phenomenological cognitive psychologist is different.

A phenomenological inspection of the kind pursued in this book reveals that these definitions and the features they specify do not address themselves to the interesting distinctions actually manifested by the psychological phenomenon at hand. They do not capture the richness and complexity of the domain under investigation, and they do not offer the proper framework for its cognitive analysis. As noted above, phenomenologically, Ayahuasca drinkers do not confuse their visions with states of affairs in the ordinary external reality. However, in contrast to the conception usually endorsed in the literature, this does not make the visionary experience less powerful: Ayahuasca drinkers do confer reality on their visions, but they conceive of them as presenting separate realities which cannot be apprehended without Ayahuasca.

As I see it, the main lesson of the Ayahuasca experience in this regard is that the issue of epistemic status cannot be dealt with without taking account of the semantics of visions. Ayahuasca drinkers determine that their visions are not perceptions because the content of these visions is special. Often the scenarios they present simply cannot be encountered in ordinary reality and visually they manifest magnificence that cannot be compared to anything seen in normal contexts. The ambience of the visions and the ideations associated with them are non-ordinary as well. As indicated in Ch. 4, Ayahuasca visions are usually infused with magic and enchantment and the ideations associated with them often pertain to topics that most people are not concerned with in their regular daily lives.

These characteristics of the phenomenology of visioning point to still another feature of the standard definitions of hallucination that is not warranted, namely, their being *content-free*. The standard definition of a mental state as hallucinatory does not depend on the types of contents experienced during that state. Empirically, we find that such a definition does not do justice to the actual phenomenology of visioning. Thus, a major moral of the study of Ayahuasca is that hallucinations have to be investigated from an internal perspective, one in which semantic considerations are crucial.

In conclusion, Ayahuasca visioning presents a special kind of epistemic status. This epistemic status is different from both ordinary perception and hallucinations as they are described in the context of psychopathology. This, I find, is a most interesting state of affairs and elsewhere I intend to examine it further both phenomenologically and conceptually.

Finally, let me comment on the experience of dreaming. As noted in Chs. 6 and 8, and as further discussed in Ch. 22 as well as in the Appendix, the Ayahuasca visionary experience is quite different from that of ordinary nocturnal dreaming. The differences are manifold: there is very little overlap in the contents associated with the two kinds of experiences, the affective states associated with them are totally different, and so are the ideations entertained in their course. Dreaming, it is very rare to see phantasmagoric scenes such as those encountered with Ayahuasca, let alone entertain metaphysical ideas, experience spiritual uplifting, and have mystical feelings. Yet, it is pertinent to note that, at times, similes to dreams are invoked by Ayahuasca drinkers to describe what they feel under the intoxication. A characterization that crossed my mind and that I have also heard other people use is that of dreaming while at the same time being awake. Coupled with the blurring of the boundaries between self and non-self, this feeling may further lead to an experience characterized as the sharing of a dream with another person or of being inside another person's dream.

A General Theoretical Perspective

Having surveyed the various effects Ayahuasca may have on people's states of consciousness, let me take a more general perspective and relate the patterns examined in the foregoing phenomenological survey to the theoretical study of consciousness at large. Specifically, I shall look at the non-ordinary states of consciousness induced by Ayahuasca in terms of a theory of consciousness I developed prior to and independently of my study of the brew (see Shanon, 1990*a*). This theory was constructed for the analysis of ordinary consciousness; the phenomenological data we have considered here will serve to extend this theory so as to cover non-ordinary consciousness as well.

On the basis of a comprehensive survey of the psychological literature, I have come to distinguish three main types of (ordinary) consciousness. The first, *Consciousness*$_1$ consists of the quality of sensed being. While we are all experientially acquainted with this quality, conceptually it is very difficult to define. This quality underlines all conscious experience, and it is shared by both human beings and animals. In essence, it defines the basic elemental quality of being sentient. Intrinsically related to this quality is the fact that cognitive agents are in touch with the external, real world.

Whereas Consciousness$_1$ is a basic undifferentiated quality, *Consciousness*$_2$ is a differentiated, well-formed state. It consists of the totality of our articulated mental life and it comprises all those well-defined internal events of which we are aware. Included in it are all thought sequences, mental images, dreams and daydreams, and, for some people, also musical ideations. Locke's classical definition of consciousness as being 'the perception of what passes in a man's own mind' (Locke, 1690/1964) is a good characterization of this state.

The third type, *Consciouness*$_3$, is what is often referred to as self-consciousness. It pertains to the second-order phenomenon of the mind's ability to take its own productions as an object and reflect upon them.

The three types of consciousness are not just three types. As detailed in Shanon (1990*a*), they are interrelated in an orderly fashion and together they comprise a coherent, unified system. Thus, these three types can be regarded as the three facets of one unified whole. Originally, I defined the three types in conjunction with what is the core of the phenomenon of consciousness, namely, human subjective experience. However, the three types may also be viewed from a more abstract and more general point of view. Thus, Consciousness_1 defines the basic, undifferentiated state of the system, Consciousness_2 defines the bulk of its differentiated, articulated state, and Consciousness_3 defines its reflective nature. Significantly, viewed in this fashion, the same distinctions apply to other facets of the system of consciousness. In particular, this is true of the Self. Specifically, $\textit{Self}_1$ defines the basic quality of living existence, $\textit{Self}_2$ defines the well-formed differentiation between Self and World and thus the distinct qualities of one's being a particular individual cognitive agent, and $\textit{Self}_3$ defines the special quality of one's being aware of one's selfhood and being able to reflect upon it.

A theoretical question that naturally comes to mind is whether there are any other types of consciousness. This question has intrigued me for a long time—even before I became interested in non-ordinary states of consciousness. (Indeed, this question is specifically raised in Shanon, 1990*b*.) With the study of the phenomenology of the Ayahuasca experience, I have indeed found such additional types.

The first such type to be noted consists of mentations which one directly experiences (as is the case with the mentations comprised under Consciousness_2) but which—unlike the ordinary mentations of Consciousness_2—are not experienced as being generated by one's own mind and seem to have an independent, external existence. I shall refer to this type as $\textit{Consciousness}_4$. The paradigmatic case of Consciousness_4 are the visions experienced with Ayahuasca.

Extending the system even further, another type will be noted that—like Consciousness_1—consists of a pristine, undifferentiated state without any specific, articulated mental contents yet—in contradistinction to Consciousness_1—is experienced as not pertaining to one's own individual self. I refer to this type as $\textit{Consciousness}_5$. The experiences of super-consciousness encountered with Ayahuasca are of this type, and so is the so-called cosmic consciousness described in the mystical literature (see, in particular, Bucke, 1901/1991), as well as some genres of conscious experience indicated under the broad label of the transpersonal (see e.g. Assaglioli, 1973). I should emphasize that here I am not making any metaphysical, ontological, or mystical claims. In a very empirical fashion, I am defining the different types of experience encountered in human psychological phenomenology and I am attempting to conceptualize them within one unified theoretical framework. It is an empirical, phenomenological observation that, in some situations, human consciousness encompasses mental contents that are experienced as transcending individual agency.[7] For a contemporary discussion of the metaphysical ramifications associated with supra-individual consciousness, see Goswami, Reed, and Goswami (1995).

Just as the original three types of consciousness are interrelated, so are the two next types in relationship both to each other and to the original types in the system.

[7] It is interesting to compare these observations with the classical discussion of consciousness in James (1890/1950: ch. 9). There, individuated agency is specified as the primary feature of consciousness.

As already indicated, Consciousness$_4$ corresponds to Consciousness$_2$ in that both consist of distinct, differentiated, and articulated mental material. Consciousness$_5$ corresponds to Consciousness$_1$ in that both consist of elemental undifferentiated experiential states. Both Consciousness$_4$ and Consciousness$_5$ share the feature of being experienced as independent of the individual cognitive agent at hand. As such, these two contrast with both Consciousness$_2$ and Consciousness$_1$ which are ingrained in individuated cognitive agency.

Thus, significantly, the new types of consciousness discovered with Ayahuasca are not just two new types. Rather, they integrate coherently into the system of consciousness that I have constructed independently on the basis of the phenomenological inspection of ordinary consciousness. The Ayahuasca experience also introduces one new distinction into the system, namely, mental contents of which the cognitive agent is directly aware but which are experienced as being independent of his or her own mental processes. However, the extension pertaining to non-ordinary consciousness does not alter the system of consciousness as such.

I have underlined the symmetry between Consciousness$_1$ and Consciousness$_5$, Consciousness$_2$ and Consciousness$_4$. What about Consciousness$_3$, the reflective faculty? This one stands apart; thus it does not have a counterpart. Furthermore, this second-order faculty may be likened to the splitting of the mental domain into two—a level of experience proper and a level of the observer inspecting this experience and reflecting upon it. As such, it may apply both to Consciousness$_2$ *and* to Consciousness$_4$, hence both its standing apart and its not having a counterpart in the system.

The foregoing discussion was taken from a structural perspective, one that defines the different types of consciousness and the structural relations between them. Also pertinent is the dynamic perspective, that focusing on the transition between the types. Fundamental to the system of consciousness is its being in a constant flux. This, I maintain, is true of both ordinary and non-ordinary consciousness. In our normal wakeful state we continuously float between the first three types of consciousness. In general, the movement between these is subtle and usually we are not cognizant of the processes by which it proceeds. I propose that the same is true with respect to the states of non-ordinary consciousness and the transition between them and the ordinary ones.

All this can be appreciated even in our ordinary state of being. In particular, the patterns at hand are salient with skilled performances. Dancing is a prime example. When dancing one often finds oneself in states of consciousness that are quite akin to ones normally labelled as 'altered'. Engaged in his or her movement, the dancer may be immersed in a separate existential environment (a reality, one might say) thereby, in effect, dissociate him/herself from whatever is outside the frame of the dance. In *pas de deux*, dancers may even feel their personal identity to be transcended and that it is the pair, not the individual, that defines their being. Yet, at the same time, skilled performance involves the ability to keep in touch with what is actually going on, both with oneself and in one's surroundings. While immersed in his or her performance, the accomplished dancer—like the player of musical instruments or the athlete—is able to keep a watching, monitoring eye and maintain a critical objective judgement of his performance. Thus, he or she will constantly and delicately flow into the dance and out of it. Indeed, one of the key features of masterful performance is precisely this: the ability to plunge oneself into the act and even lose oneself, so to speak, because one is confident that whenever the situation demands it one would readily be able to disengage oneself from it. To my mind, with experience, one realizes that identically the same obtains with the Ayahuasca inebriation.[8] More on both the dynamics of the Ayahuasca inebriation and the development of mastery will be said in Chs. 18 and 21.

[8] It is, of course, no accident that in many cultural contexts dancing is employed by way of inducing non-ordinary state of consciousness. Sufi Dervish dancing is a notable example.

17

Light

O soma-drinker, drink of the soma-wine; the intoxication of thy rapture gives indeed the Light.

The Rig Veda

And suddenly, it seemed that, to full daylight
Full daylight had been added, as if he
Who can had put another sun in the sky.

Dante

That the Lord actually appears in heaven as a sun, has not only been told to me by the angels, but has also been given me to see several times.

Swedenborg

In practically all the accounts found in the anthropological literature, visualizations of light are described as characterizing the first stage of the Ayahuasca inebriation. This is, I think, a very misleading portrayal. While it is indeed true that typically the first visual effects of Ayahuasca generate internal stimuli of light and colour, it is by no means the case that light is confined to the beginnings of sessions or to the stages in which the force of the intoxication is at its weakest. Actually, quite the contrary is the case. Not infrequently, the most powerful moments in an Ayahuasca session involve light.

In fact, light permeates the Ayahuasca experience throughout. Once, asking a UdV master why is it the case that in Ayahuasca visions one sees serpents and felines, I received the following answer: 'It is all people's imagination, the tea [the term by which the brew is referred to in the UdV context] itself gives one only light.' Furthermore, many experienced members of the UdV told me that with time, they have fewer and fewer visions. In contrast, they said, the sessions become replenished with light. Not all partakers of Ayahuasca will subscribe to or endorse this assessment, but again, it underscores the centrality of light in the Ayahuasca experience. Given this centrality, and risking some repetition, I devote this last chapter of the general phenomenological survey of the Ayahuasca experience to the systematic consideration of the various manifestations of light.

Global Luminosity Effects

1. Luminosity Effects Occurring with the Eyes Open

As noted in earlier chapters, under the effect of Ayahuasca colours gain unusual saturation and intensity. Often, people feel that things radiate energy and shine. For example, reading a printed text (such as the hymn books in *Daime* sessions), I have often seen the letters glitter with emerald-green, gold, and/or red. Looking at my fingers, I have seen them glow in a similar fashion. Many informants recounted similar experiences to me.

Again, I underscore similar observations made by Huxley (1971: 75) when describing his experiences with mescaline:

> First and most important is the experience of light. Everything seen by those who visit the mind's antipodes is brilliantly illuminated and seems to shine from within. All colours are intensified to a pitch far beyond anything seen in the normal state, and at the same time the mind's capacity for recognizing fine distinctions of tone and hue is noticeably heightened.

Identical effects are commonly encountered with Ayahuasca.

2. Luminosity Effects Occurring with the Eyes Closed

As is the case with open eyes, with respect to objects in the real world, so too in the case of closed eyes, in relation to entities seen in visions—objects shine, radiate, and glitter. When the intoxication is strong, the colours in the visions are remarkably luminous. Many people say that some of these colours are ones they had never seen when in the ordinary state of consciousness.

As noted in Ch. 8, commonly, objects in Ayahuasca visions are made out of gold, precious stones, and crystal. All these are materials that shine and glitter. On a couple of occasions I saw visions comprised exclusively of artefacts made of gold; the most salient of these were crowns. Truly, it was a golden experience.

3. Afterimages

Ayahuasca markedly enhances the likelihood of afterimages, that is, images seen when the external sensory stimulus is no longer present. This, I believe, is a corollary of the general extension of the time-span of short-term memory that the brew brings about. On many occasions under the intoxication, I would look around and then close my eyes. For a period of time much longer than the usual one I would still see parts of what I had just seen with my eyes open. At times, what would be left was one particular, especially luminous constituent of the original visual image. On other occasions, negative afterimages consisting of colours complementary to those of the original image would appear. In general, soon after closing the eyes the isolated visual elements—be they iconic or afterimages—would be multiplied and then serve as the basis for subsequent hallucinatory material.

Once I noticed that the afterimages remained projected in my visual field also when my eyes were opened. Experimenting with this effect, I observed that it was greatly facilitated by rapid winking.

4. Illumination

A common effect is that of part of the visual field being focally illuminated. Typically, it is as if a flashlight were illuminating it. Also very common is the seeing of the world—be it that around one or that of one's vision—as covered with a multitude of glittering starlike or gemlike elements. All these effects are encountered both when the eyes are open and when they are closed.

5. Coloured Visual Space

Coloured visual space is precisely that—the experience that one's entire internal visual field consists only of colour. This usually occurs when the effect of Ayahuasca has already subsided, and quite often even a full day after a drinking session. I would close my eyes and my internal visual field would consist of coloured space. The colouring would be bright, monochromatic, exhaustive, and homogeneous. On some occasions, only one colour would be seen—in most cases it was bright red. On others, a series of coloured visual spaces would appear in succession, one giving way to another. The colours included green, bright yellow, purplish-blue, and intense white. Invariably, red was the first in the series and white the last.

For me, the most special episode of this kind occurred during an intermission of a festive *Daime* session. I lay down on the ground and relaxed. Closing my eyes, I discovered that my entire inner visual field was orange. There were no forms or content elements, no other colours. In one sense it was all monochromatic and uniform, in another sense it was so dynamic. The orange field was vibrating with energy, it was alive. The thought that crossed my mind was that I was seeing my soul. In retrospect, this, I think, was the closest I have approached an experience that could be characterized as Nirvana.

Once, I had a vision whose theme was the colour white. My entire inner visual field was washed in white. But it was not flat. Inspecting the extended surface in front of me, I began to discern different shades and textures. The more I looked, the more I saw in it. And, again, it was all white.

6. Darkness

I have spoken of light, but Ayahuasca may also present darkness. This happened to me only once. It was during a festive *Daime* session taking place in daytime. The hall was fully illuminated, yet when I closed my eyes I saw only black. It was black in the absolute. I understood that here I was being presented with an opportunity to learn to see the dark. I observed the black and found it fascinating. There were so many shades and grades to it. All were black, but it was so far from being monotonous. The experience, I felt, was most instructive.

Patterns of Light and Colour Devoid of Content

1. Simple Light and Colour Elements

In the scientific literature—both that pertaining to the biomedical sciences and to anthropology—the first visual effects of Ayahuasca are characterized as phosphenes, that is, patterns like those that are produced in the inner visual field when one closes one's eyes and gently presses the eyeballs (see Oster, 1970). In the literature, these patterns have been attributed to a spontaneous activity of the visual cortex (see Ermentrout and Cowan, 1979; Bressloff *et al.* 2001). Here I prefer to avoid any causal, neurophysiological explanation, and I confine myself to the strict phenomenological level of description. Phenomenologically, it is indeed the case that the weakest, and least impressive, visual effects of Ayahuasca are similar to phosphenes. However, on the basis of extended personal experience coupled with repeated interrogations directed at this specific issue, I conclude that the patterns normally associated with phosphenes cover only a very minor facet of even this weakest stage of the Ayahuasca intoxication. In fact, the difference between phosphenes and the light patterns induced by Ayahuasca begins to manifest itself with the very onset of the intoxication. Unlike phosphenes, the most basic light and colour elements induced by Ayahuasca exhibit remarkable numerousness and extension. As soon as the force of the intoxication strikes, these elements multiply in number and concurrently the scope of the inner visual field extends dramatically. The common feeling is that the globe of the inner visual field explodes and opens up to achieve very large dimensions that not infrequently can be experienced as gigantic or even astronomical or cosmic.

I should emphasize that unlike what is often claimed in the scientific literature, it is not the case that all Ayahuasca sessions begin with simple light and colour elements. It may very well be that one does not experience any visual effect at all and then, all of a sudden, figurative elements or even entire scenes appear before one's eyes.

2. More Complex Arrays and Configurations of Light and Colour

Complexity may be introduced by the number and extent of the light and colour elements, by the configurations they compose, and by movement. Complexity of the first kind has already been noted in the previous subsection—the number of elements multiplies as if ad infinitum and the field of their extension expands dramatically. As indicated in Ch. 6, configurational complexity may be manifested in two major forms—repetitive clusters of elementary constituents and geometric patterns. Examples of the former type were given there; on the latter type I further comment in the next subsection. The configuration of light and colour may also enter into movement. This, in turn, may generate patterns that people describe as lightning, comets, and fireworks. As noted in Ch. 6, the configurations of light and colour may also pulsate and be invested with rhythm.

3. Geometric Compositions

Geometric patterns of light and colour may yield compositions that span the entire field of one's internal visual space. In their simplest forms, the geometric compositions become veritable symphonies of light and colour, like abstract works of art; an association that comes to my mind are the paintings of Kandinsky. The compositions are like infinitely extended arabesques, tapestries, or honeycomb structures. The basic elements of the latter are often pentagonal or hexagonal, but they can also be much more intricate. In their totality, the abstract structures may consist of endless repetitive basic elements or lattices of complex interlacing designs, unlike any seen in reality. The structured patterns may be two-dimensional, three-dimensional, or they may be experienced as having more than three dimensions. Not infrequently, the geometric complexities manifested in the visions defy one's ability to articulate description.

The geometric patterns are always coloured. Typically, the colours are very luminous and exceedingly beautiful. My personal experience is that the elements from which the lattices are composed are multicoloured. Colour combinations I have seen frequently include pink, green, and white; bright yellow, red, and black; and amber and gold. My experience, corroborated by reports of others, is that when the effect of Ayahuasca is especially strong, the colour elements composing the geometric structures are marked by accentuated, well-defined dividing lines, usually black, like those encountered in works of stained glass.

When powerful, the geometric compositions may flow and move in a kind of a cosmic dance. These compositions are experienced as constituting grand visions in their own right. The geometric patterns may be all-embracing and, engulfed by them, the Ayahuasca drinker may feel that he or she is being carried to transcendental realms. The experience is usually enchanting and defies both verbal description and conceptual analysis. An association that came to my mind several times was that this was sublime music whose embodiment was light and colour.

4. Transitions

In the literature, primarily on the basis of Reichel-Dolmatoff's work (for example, 1975), it is suggested that the progression from the non-figurative to the figurative passes from the bursts and splashes through the geometric, and then via the various proto-figures to the figurative. While I have experienced this very sequence in the course of my first experiences with Ayahuasca, soon this was no longer the case. The progression from the splashes to the figurative was without the mediation of the geometric. In contrast, the geometric patterns appeared usually when the effect of the brew intermittently intensified. The geometric patterns may, in fact, signal the onset of a new stage in the inebriation. Several times I noted that while in a high state of visioning a geometric pattern would appear. This could close the visioning experience or, if I dared, lead to new, and most often stronger visions. I have shared these ideas with several very experienced drinkers and their impression was identical.

Thus, contrary to what the literature seems to suggest, patterned geometric designs are by no means confined to the beginnings of sessions. On the basis of many experiences I would say that these designs and patterns are encountered when one stage of the inebriation ends and another may or may not begin. That is, at points in which there is a sharp change in the power of the intoxication. These are also moments that can potentially be openings to a new phase in the Ayahuasca experience. More on this will be said in the next chapter.

Viewed as such, the geometric patterns may be metaphorically regarded as fences or perhaps intercellular membranes. The fence or membrane marks a boundary that delineates between two domains. Unlike the fields or cells it demarcates, by itself the fence or membrane does not contain much in terms of content—there are no plants or animals in the fence itself nor much specialized biochemical activity in the membrane. By no means, however, does this mean that it is of secondary importance. Far from it! The very identity of the field or the cell is defined by it, and the only way to get to the field or the cell is to cross it. Likewise with regard to the patterned geometric designs seen in Ayahuasca visions. Seeing them, one may conclude that the visions have ended and withdraw; or else one may forge ahead, and—like Alice stepping through the looking-glass—get to new realms with even more powerful, more fantastic figurative visions. More on the transitions in Ayahuasca visions and their dynamics will be presented in Chs. 18 and 21.

Light as a Content Element

1. Light Elements Embedded within Figurative Visions

So far in this survey, light was depicted as abstract (in the sense of non-figurative). Usually, this is indeed so, but there are also many instances where this is not the case. Specifically, light can be a content element embedded within a meaningful figurative vision. Both in the course of my own personal experience with Ayahuasca and in the reports of my informants many scenes are encountered in which sources of light are salient content elements. These sources may be heavenly bodies such as the sun, the moon, or the stars, they may appear as fire or lightning, they may consist of artefacts such as candles, torches, or chandeliers, or they may be characterized as supernatural.

By way of illustration, here are three visions in which a source of light featured as a main content element; the first two are my own, the third was reported to me by an informant. The first vision seemed to take place either in the late Middle Ages or in the early Renaissance:

> The scene depicted the terrace of a palace and its main protagonist was an aristocratic lady. She came out to the forefront of the terrace holding, with both hands, a candle. The light that emanated from this candle was utterly white and extremely powerful. It was not like the flame of any normal candle. The scene had a special, holy allure. It seemed to me that by presenting the candle to whomever was in front of her, the lady was enacting a benevolent act of grace and blessing.

The second vision depicted Indians in pursuit of the constituent plants of Ayahuasca:

> The Indians were marching through the forest in the dark, and light was hovering in front of them. The light was white, and not unlike that emitted by fireflies. As the Indians were going forward, the light was in front of them indicating how to proceed. And then, the light descended on a plant. This indicated to the Indians that here was the specimen they were after.

The third vision—that reported to me—depicted a scene from an ancient civilization:

> Priests were performing a religious ritual. They were holding candles and I felt that the light of these blessed candles was the energy that sustains the world.

Other elements of light that feature in Ayahuasca visions are rainbows, comets, and various manifestations of fire.

2. Halos and Auras

A particular kind of light pattern is one that always appears when people are seen. Manifestations of this kind include halos around the head, auras and rays of light that radiate from the body, and cones or clouds of light that hover above people. Also common are trails and coloured emanations of light as well as matrices of light that cover people's entire bodies. All these may be seen both with open eyes in conjunction with real people or with closed eyes in relationship to people or entities appearing in the visions. Pertinent illustrations are depicted in Luna and Amaringo (1993: pls. 3, 13).

In addition to its being seen, light may be felt—emotionally, intellectually, and spiritually. Just as one can see other people shine and emanate light, one can feel oneself as enveloped by light and as radiating energy. Typically, such experiences are coupled with profound feelings of bliss.

3. Beings of Light

Beings of light are transparent, translucent ethereal beings which are made of light. Usually, they have human features. Many people report seeing them. I, too, have seen such beings on many occasions. Usually, I took them to be divine or semi-divine beings that rule the universe; some of these might be regarded as angels. In this conjunction let me note a comment I heard many times with regard to angels. On various occasions, in different places and contexts, informants explained to me that the wings of angels seen with Ayahuasca are not wings in the standard ornithological sense, but rather, vibrating emanations of light and energy that to us look like wings. For concrete depictions of beings of light, see pl. 36 in Luna and Amaringo (1993); figures shown there are practically identical to those I have seen in my own visions.

4. The Spider Web

As described in Ch. 5, one of the most special of all light patterns induced by Ayahuasca is the one I call 'the spider web'. As noted, this pattern is observed with open eyes only and is usually encountered towards the end of sessions. It consists of fine, almost transparent lines of light that connect things in the real world. In general, these are interpreted as manifestations of cosmic energy, the energy that is the source and fountain of all Existence, that sustains all creation, breathes life to all Existence, and provides nourishment to all beings. This energy may also be conceived as the embodiment of Divine wisdom and as the source of all knowledge and creative work. As recounted in Ch. 10, one informant told me that the seeing of this web was what made her appreciate that God exists. For good depictions of these, see the paintings of the Brazilian artist Alexandre Segrégio; photographs of some of these are to be found in Weiskopf (1995).

5. The Temple of Light

Entities that can be made of light are not confined to beings. The following is the narrative of a vision I once had, in which all that was seen was made of pure light:

> There I was, in one of God's chambers—the chamber of light. It was a vast, very spacious hall, all white with tinges of bright yellow interspersed in it. Harmoniously, the entire space exuded an air of dignified temperateness, unhurried and aristocratic. At first I thought that there were some pillars and arcades there, but soon I realized that the place was just light, nothing but light. There was no matter there, but still you could walk around. Indeed, placing one's foot on the ephemeral ground, by the very touch of it, matter would be created. Or rather, out of the light a well-defined form would crystallize—a multicoloured and pleasantly ornate flowerlike form which was exceedingly beautiful. And when the foot would again be lifted up, the magical pedestal would evaporate into the ether only for another magnificent form to be created out of the nothingness by the mere touch of the step. Thus, hopping from one crystallization to another, one could smoothly roam the entire space. In fact, only trust was needed—trust that one was indeed able to waltz through, to dance on an immaterial floor that was nothing but light.

Interactive Patterns

1. Interacting with Light

In addition to it being seen, light can be interacted with. Seeing light, drinkers may engage themselves in activities involving the patterns or elements of light presented to them. One may submerge oneself in the light and be washed by it. One may also take the light and bring it inside oneself. One informant reported that in doing this she carried the threads of light throughout the inner parts of her entire body. This resulted in a wonderful cleansing experience.

2. Light Transformation

There are also cases in which drinkers themselves transform into light. Two informants reported such experiences. Both noted that the light into which they transformed became a source that illuminated all that was around them.

Abstract Scenes and Visions of Light

1. Abstract Scenes

The content elements discussed above defined parts of visions or adjuncts thereof. However, abstract patterns and light may constitute the very content of visions. Just as in the plastic arts, where paintings and sculptures need not depict figurative contents but may also be abstract, so too in the case of Ayahuasca visions. None the less, in every relevant sense, such visions may be classified as scenes. Here is an example:

> In front of me was a composition of a precious silvery light. Like a sculpture, it constituted an object with a definite location and a well-bounded form and shape. This sculpture was embedded in a purple-violet ambience and it rotated in a movement that impressed me as most harmonious.

What is of note here is that the item seen was perceived neither as a decoration nor as a representation of something else, but rather as an entity in its own right. Just as in figurative visions, in which objects and/or creatures are seen which define a certain topic, so too here—an object presented itself, only it was an abstract, non-figurative one.

2. Supreme Light

Encountering what I refer to as the supreme light is a very powerful, practically ineffable spiritual experience. The following was recounted to me by a non-Brazilian who has become an adept of the Santo Daime Church as the most remarkable experience she had had with the brew:

> In front of me was a most brilliant source of light, like a lantern of many, many facets. I knew that were I to look at it straight ahead I would collapse and perhaps die. Thus, I turned my head. During this experience, I felt as if I was losing my sense of self and the faculty of memory.

This account readily brings to mind the biblical comments in conjunction with the human inability to see the face of God and live (Exodus 33: 12–23). Indeed, in the Bible it is noted that it was not even possible for people to see the face of Moses after he had encountered the Divine, for his face shone (Exodus 34: 30). Similar descriptions regarding the divine light that cannot be beheld are encountered in other cultural traditions (e.g. Zeus appearing before Semele, his beloved human wife and the mother of Dionysus, who, as a consequence, dies).

Some visions of light are interpreted as depicting the creation of the universe. As recounted in Ch. 8, in this cataclysmic event light is primary.

Seeing the supreme light often involves an experience of spatial uplifting. An experience that was, again, recounted to me as most remarkable was one in which a man saw a great source of light. Then he heard a voice telling him not to stop but to lift his gaze. Above the first he saw another, even more powerful source of light.

As noted above, the light characterized as supreme is typically associated with special meaning. Once I experienced a scene in which I was carried up to the high heavens. The skies opened, and there was the light that, I felt, was the source and fountain of all things. The association that came to my mind was that of *ha-or ha-ganuz*, the Hidden Light which in the Jewish mystical tradition is said to have been created on the first day of Creation and is kept for the righteous for the Days to Come (see the epigraph prefacing this book). The best way for me to describe these experiences is by quoting from letters I sent from South America to close friends back home:

> The essence of it all is the seeing of the light. The bountiful light of the sun as it reflects the grace of the *Shekhina*.[1] And I saw the different worshipers of this Sun—King Solomon on the one hand, and kings of the Inca on the other hand—and I myself became intoxicated with the grace of the Holy.
>
> And then I encountered the primordial point of light. The point from which all has been created. The point which is the guarding flame of all of creation, of life, of intelligence whatever form intelligence takes. In religious traditions this point of light has been called 'God'.

Indeed, visions involving the supreme light may be conceptualized as encounters with the Divine. Expectedly, these experiences are very private, and by their nature are beyond the realm of words. I have made some comments on these experiences in the previous chapter, here I shall cite only one report given to me by a young European man who, following his first Ayahuasca vision, in which I was present too, emotionally declared: 'I saw God!' To my query as to what that meant, he replied:

> I was in a region which was pure and perfect, full of light and love, and I was healed. I looked at the skies up there and the heavens parted. There came a ray of light and it struck my heart. With this I was healed.

A Final Comment

We have begun our phenomenological chart with light and we have ended it with light. As noted, both the most elementary Ayahuasca visualizations and some of the most powerful experiences that this brew generates pertain to light. Indeed, to a great extent, the Ayahuasca experience may be regarded as a grand composition of light and its manifold transformations. This is manifested not only in that most prominent facet of the Ayahuasca experience, the visions, but also in the general atmosphere of the inebriation, the moods and affective states associated with it, the ideas and reflections entertained during the course of the intoxication, and the religious and spiritual sentiments drinkers have. With Ayahuasca, people have visions, but they also feel that they are washed by light, that their minds are clear, that their mentations are enlightened, that they are privy to the gift of illumination.

[1] As noted in Ch. 8, in the Jewish tradition this term refers to the female aspect of the Divine.

The psychological experience of light is not particular to Ayahuasca and its visions. We see flickers of light when we close our eyes, when we physically press the eyeballs we experience what are known as phosphenes, engaged in mental imagery we see virtual scenarios, and, of course, we all dream. Thus, light is there not only in the sun, the moon, and the stars, not only in fire and the electrical bulb, but also in the privacy of human minds. All these phenomena are commonplace and seemingly trivial. Yet, when one comes to think about it, the fact that human beings experience light when their eyes are shut is far from being straightforward.

Significantly, language reflects (*sic*) the special status of light. It is no accident that in English—as in many other languages—words such as those ending the previous paragraph but one are derived from the term 'light' (cf. 'enlightened', 'illumination'). In Hebrew, a language not at all related to English, the noun *'or* is light, the noun *ora* is one of the terms for joy, the adjective *mu'ar* is illuminated, *na'or* is enlightened, *me'or panim* denotes happy welcomingness, and so on and so forth.

We further note that our very biological existence, as that of all living creatures, depends on light. The physical universe too is founded on light. In both the microscopic domain of quantum mechanics and in the immensely grand domain of the theory of relativity light is fundamental. Is it just a coincidence that the antipodes of the mind, they too, shine with light?

No wonder human beings, through the ages and across all lands, have revered, worshipped, and deified the Light. And it is not for nothing that the biblical story of Creation begins with the light. The sun, the moon, and the stars were created on the fourth day, but light, the primordial Light, was created on day one. The Jewish esoteric tradition is replete with reflections about the nature of this light, the major opus of the Kabbalah is the book of *Zohar*, 'splendour'. One citation from this tradition was given as the main epigraph of this present book.

PART III

Theoretical Issues

18

Stages and Order

> As you practise meditation you may see in vision forms resembling snow, crystal, wind, smoke, fire, lightning, fireflies, the sun, the moon. These are signs that you are on the way to the revelation of Brahman.
>
> *The Upanishads*

I now turn from the structural phenomenological questions of the first order to those of the second order. Whereas the former deal with the characterization of the different types that may be defined in the phenomenology of the Ayahuasca experience, the latter deal with the relationships that may hold between tokens of these types.

The consideration of relationships is of special cognitive import because relationships are more likely to present regularities than specific contents are. This is so because, as a rule, relationships are less amenable to contingent variations, be they intra- or interpersonal. Hence, while the particular contents of visualizations that different people, or the same person in different sessions, see are prone to vary, the relationships between the visualizations are more likely to reveal intrinsic, hence, more robust, cognitive regularities. This, I emphasize, is not at all specific to Ayahuasca visions; analogous differences between first-order and second-order patterns of regularity are encountered in other domains of mental phenomenology. For a detailed, general argumentation in this regard the reader is referred to Shanon (1984, 1989*a*).

One facet of Ayahuasca visions for which the topic of phenomenological regularity has received explicit treatment in the anthropological literature is that of stages. As further noted below, many have observed that as a session proceeds the visions change character, and that in the course of the Ayahuasca session drinkers pass through several stages of visioning. Furthermore, the progression of stages is orderly and similar across persons.

The classical anthropological treatment of the question of stages and order is that put forth in the writings of Reichel-Dolmatoff (see also ch. 6). In Reichel-Dolmatoff (1991) the claim is made that there are two sequential stages in Ayahuasca visualizations. In the first, geometric patterns are seen. In the second, there are irregular multicoloured forms in movement. These, it is noted, may be similar to clouds or simply consist of great splashes of colour. According to Reichel-Dolmatoff the visualizations of the first stage are akin to phosphenes and are neurologically determined. In contrast, argues Reichel-Dolmatoff, the visualizations of the second stage are culturally determined images that are projected onto a background of colour and form produced by the hallucinogenic agent (for extended further discussion see Ch. 22).

Elsewhere, Reichel-Dolmatoff (1975, 1978*b*) distinguishes between three stages in the progression of Ayahuasca visualizations. The first stage is characterized by small, brilliant, star- or flower-shaped elements that flicker in repetitive kaleidoscopic patterns. These luminous perceptions, he says, may appear as clusters of fruity or feathery leaves and bilateral symmetry is often exhibited. Geometric patterns that are seen in this stage consist of grids, zigzag lines, undulating alternating lines, multicoloured concentric circles, or endless clusters of brilliant dots. By Reichel-Dolmatoff's account, in the second stage the symmetry and the overall geometric aspect of these perceptions disappear and figurative pictorial images emerge. Large blobs of colour are seen and from them emerge different shapes looking like people, animals, or unknown creatures. With this, mythological scenes appear. In the third stage the images disappear. A soft yellow light prevails, and wandering clouds and soft music are experienced. All these lead to a state of blissful serenity. Reichel-Dolmatoff (1975) further explains that the sphere of hallucinations is one of subjective interpretations in which the person projects a set of pre-established, stored mental material upon a wavering screen of shapes and colours. Thus, he claims, the mythical scenes seen by the Tukano Indians during the second stage of the experience can be seen only by members of *their* society.

Another characterization of the progression of Ayahuasca sessions is presented in Deltgen (1978/9, 1993); it is based on a distinction made by an indigenous shaman between three stages. The first stage is that of bright colours and geometric designs. In the second stage one sees figures of persons, plants, and animals. These are not real, but they appear to be even more 'real than real'. The third stage is that of true visions. In it, one sees the heroes, gods, and demons of mythical history. Further, in this stage Ayahuasca lets one travell across the universe and it reveals the secrets of the world. For other pertinent data based on indigenous reports, see the two citations presented in Ch. 4.

The Stages of Ayahuasca Visioning and their Order

Types, Stages, and Order

One should not confound the question of types and those of stages and order. For there to be orderly patterns in the progression of visualizations, the visualizations indeed have to be of different types. However, visualizations may be of various different types without the types constituting stages that can be ordered along any well-defined dimension. Furthermore, a qualitative order may be defined between the different types, but this need not coincide with the temporal order by which different types of visualizations appear within a session. For instance, the types may be ordered according to their strength, and in that sense different stages in the intoxication may be defined. Yet the sequential, temporal order of visions within a session need not match the order of types according to their strength. Finally, various notions of ordering may be thought of, some more rigid and others more flexible; more on this will be said below.

In sum, three different notions are to be noted. Keeping the distinction between them is, I find, of paramount importance. First are *types*—these mark the structural variations of visualizations in the manner surveyed in Ch. 6. Second are *stages* of intoxication—these differentiate between the types along some ordered dimension with respect to which the types may be compared. It should be noted that there is more than one qualitative dimension by which the types can be ordered. Lastly, there is the sequential, *temporal order* of types along a session. This order may or may not match that of strength or that of any of the ordered dimensions.

Let me make one further methodological comment before I proceed. The classification of Ayahuasca visualizations in terms of stages—that is, the attribution to visualizations of a value or a degree along a certain ordered dimension—is not a mere theoretical consideration imposed by scientifically minded Western scholars. The differentiation is intrinsic to the Ayahuasca experience. A key, and to my mind remarkable, feature of this experience is the realization—reached by practically all people with whom I have conversed—that in partaking of Ayahuasca one is entering a school, and that this school has different classes. Such a characterization of one's long-term involvement with Ayahuasca is also explicitly made in the indigenous context (see e.g. Payaguaje, 1983 and Taussig, 1987). Thus, the notion of relatively ordered stages is ingrained both with respect to progression within sessions and across sessions, over the course of one's accumulative experience with the brew. The following remark made by an indigenous Yebamasa shaman to Deltgen (1978/9: 72) reflects this appraisal: 'If you see only *dibujos* [geometric designs] you are still very much behind.'

Indeed, the differentiation of stages in the course of Ayahuasca sessions is common throughout all indigenous Amazonian cultures that use the brew. An example is the following description of the Tatuyo Indians taken from *La Tête Dedans* (Baldran and Bareiro-Sanguier, 1980: 93–8, my own free translation from the French).

When a man drinks the *yahé*, when he is well intoxicated with *yahé*, one thing arrives here, another there, a sound here, another there, all over there are noises, sounds, colours. There, there, there. He first sees a torch, like a cross, black.[1]

He takes the *yahé* again, and once intoxicated, he vomits and he sees a necklace of pearls, but one does not distinguish the individual pearls, it is a white necklace.

He takes the *yahé* again, and once intoxicated he sees the beautiful feathers of the bird Cauralel. He sees the spot, the multicoloured lines and the wings which are meticulously painted on the trumpets, similar to the beautiful motifs of the feathers of the jaguar-eagle.

[1] This extract was also cited in Ch. 11.

It is the earth itself that forms these motifs of colour under the sun. From there, from far away, he hears the *yahé* speak.

Well intoxicated with the *yahé*, he sees all the things, the *yahé* shows him the things as they were at the beginning. It says: we have the things as they were at the beginning of all things.

He is now well under the effect. The visions arrive, they are enchanting, it is the good *yahé*.

All other stages follow, all is going well for him, he hears the noises, the words around him, the people who are talking around him, here and there, noises, forms, colours.

Furthermore, as noted by Gow (1988), some Amerindian tribes have developed linguistic terms for describing these differences between stages. In many cases, particularly in the south-western Amazon, there is a differentiation between the initial and later stages in terms of design versus images. Thus, Siskind (1973*b*) reports that for the Sharanahua, a Yaminahua-speaking people of the Purus river, the initial phase is associated with *Kunu*, 'design', and the second with *Rami*, 'images'. The stage-like nature of the Ayahuasca schooling is also emphasized throughout Payaguaje (1983) which is a candid autobiography of a *bebedor de yagé*, a drinker of *yagé*. A similar observation was made by Lagrou (1998) in the context of the Cashinahua Indians. They characterize the visual materials seen in the early stages of one's Ayahuasca schooling as 'visions of the vine', a term denoting visions comprised only of geometric designs; these are contrasted with more advanced visions that comprise of figurative images.

Some Relative Rankings

The foregoing survey suggests various dimensions along which Ayahuasca visualizations may be compared and ranked as higher or lower. Some of the features associated with these dimensions have already been encountered in the typological discussion in Chs. 6 and 7; there too these dimensions were associated with lines of progression that may be defined between different types of Ayahuasca visualizations.

First is the dimension of *perceptual strength*. To the reader without any firsthand experience of Ayahuasca, the epithet 'perceptual strength' may seem fuzzy and subjective, but to anyone who has partaken of Ayahuasca it is very clear. Some visualizations are indeed striking, literally so: one feels that one is hit by the visualization one sees. Perceptual strength is manifested in sensory acuity, richness and brilliance of colour, distinctiveness, well-definedness and well-formedness of forms, and extended visual scope, as well as increased stability and permanence of images. The progression from glimpses and shots through proto-scenes to scenes and eventually grand scenes reflects these various aspects of increase in perceptual strength.

The second dimension is that of *richness of content*. In general, the more semantically complex a vision is and the more intricate the story it presents, the higher it is to be ranked. In this sense, figurative images are ranked above non-figurative ones, scenes above single images, and scenes with extended narratives above scenes devoid of narrative.

The third and fourth dimensions to be noted pertain to the general impression that the content of visions have on their viewers. The first of these two is *extraordinariness*. Visions will be ranked higher when they reveal things that impress one as surprising, novel, magnificent and—most notably—pertaining to other worlds or realities. Images characterized as 'marvellous' or 'ineffable' are highest on this dimension; in contrast, ones pertaining solely to one's own life experiences will be ranked lowest. The other dimension is *objectivity*. Visions will be ranked higher when they impress the viewer as presenting an independent reality. The existence of this reality, it seems, is not dependent on the individual drinker, who, in his or her turn, feels privileged to witness it.

Fifthly, there is the dimension of *involvement*. The more absorbing and captivating visions are, the higher are they to be ranked. The same holds for the degree of a drinker's immersion in his or her visions. Specifically, the further one enters into the other realities that Ayahuasca presents one with and the more one distances oneself from the normal, real world, the higher the ranking of the vision. Similarly, the higher the degree of interaction the Ayahuasca drinker maintains with what he or she sees, the higher is the ranking of the vision. As noted earlier, the general progression is from visions which one merely views in the manner that one watches a film, through visions which one feels one is witnessing from the inside, to visions in which one becomes an active participant—or indeed, an actor. The most powerful case with respect to interaction is that in which the drinker assumes the role of director.[2]

The last remarks bring us to the next dimension, that of the *control* drinkers have over their visions and the manner they handle them. At first glance, it may appear that this dimension is opposite to that of involvement. Indeed, usually the visualizations that rank highest in terms of involvement are those in which the drinker has very little, if any, degree of control. However, when one reaches a high level of mastery with Ayahuasca, the two dimensions can operate *in tandem*—drinkers may find themselves extremely involved and at the same time manage, when necessary, to exert control on their visions.

In the previous paragraph, along with the term 'control' I have also used the term 'mastery'. As I use them, these two terms have distinct meanings. Control signifies a particular *ad hoc* relationship (specifically, one of influence) that one may have *vis-à-vis* the visions one has. Mastery signifies a long-term, accumulative acquisition of skill. Indeed, drinking Ayahuasca is a skill, or perhaps an art, that is to be mastered through lengthy training. Mastery has many manifestations; a heightened degree of control is one of them. The more salient manifestations of mastery are fixedness of images, high degree of interaction, and complex

[2] I appreciate that, strictly speaking, the dimension of involvement and the various features associated with it pertain to drinkers, not to visions as such. I include these here for they are directly related to the force of visions: as a rule, the more absorbing visions are and the more involved drinkers are with them, the more powerful are the visions experienced to be. This comment applies also to the dimensions of control, mastery, and coming to terms with Ayahuasca.

narratives. More on mastery will be said in the last part of this chapter, when long-term developments in the Ayahuasca experience are discussed; for control and mastery as they are exhibited by shamans in general, the reader is referred to Peters and Price-Williams (1980).

Related to both involvement and control is the drinker's *accommodation* or *coming to terms* with Ayahuasca. As a session proceeds, as well as with one's gaining further long-term experience with the brew, one feels more and more at ease with it. Consequently, visions are more stable, less fluctuating, and the atmosphere they present is more tranquil. Visions in which one flies and passes over landscapes often exhibit these qualities.

Next, visualizations may be evaluated in terms of their *psychological*, *intellectual*, and *spiritual impact*. As noted in earlier chapters, Ayahuasca is taken not only for experiencing cinematographic-like visions; rather, the visionary experience is coupled with ideation and reflection. Powerful visions are ones in which people feel that they do, indeed, learn something. The learning experience that Ayahuasca brings may be of a personal psychological nature, intellectual (that is, pertaining to ideas), or spiritual. Experientially, visions with a strong spiritual impact are deemed to be especially meaningful and powerful. These impart upon their viewers intense religiosity and an appreciation of the sacred.

Finally, visualizations may be ranked in terms of parameters pertaining to the drinker's *state of consciousness*. As noted in Ch. 12, Ayahuasca may bring its drinkers to situations in which their very notion of self is altered. The boundary between the self and the world may dissolve and some basic distinctions such as that between perception and memory or between memory and thinking may be blurred. In the end one may lose one's own individual identity and/or undergo a transformation to another person, an animal, or even an inanimate object. The more radical the shift from standard consciousness, the stronger the visualizations will be said to be.

Types and their Features

Bearing in mind the distinctions introduced above, let us turn to the various structural types introduced in Ch. 6 and attempt to order them along the foregoing dimensions as well as in terms of their position along the temporal sequence within a session. First I mark for each of these types its relative strength and the phase during the session in which it is likely to occur; afterwards, I further characterize each type in terms of the other dimensions I have introduced. It should be noted that the various types need not all appear during one session. Hence, it makes no sense to order them all in terms of one temporal sequence. Rather, what I do is indicate at which stage of a session the given type is likely to occur: beginning, early middle, middle, late middle, late. Also indicated are the associations of the different types with climactic degrees—some types are associated with moments in which the intoxication gains strength, others with ones in which the intoxication has reached a stable degree. Finally, comments regarding the overall frequency of types will be made.

Methodological comments are in order here. By and large, the following discussion is based exclusively on my own personal experiences. While I did question people on the structural types of visions, only on rare occasions could I obtain organized information from informants regarding the temporal progression of these types in the course of a session. Informative reports in this regard require not only extensive experience with Ayahuasca, but also high levels of sensitivity and reflection, as well as intellectual sophistication, critical power of analysis, and comprehension of psychological considerations and distinctions. I have reached an appreciation of the pertinent distinctions only after an overall inspection of the entire data I had collected and following extended reflection on them. Admittedly, generalizations based on data pertaining to only one person have their obvious limitations.

The following analysis is presented in terms of the list of types (∗) below. Structurally, this list consists of all the items indicated in Ch. 6. In addition, it includes two types—(11) and (12)—which are not distinct structural types as such but which are distinct from an experiential point of view and which subjectively mark definite stages both in the course of single sessions and in the context of one's long-term apprenticeship with Ayahuasca. Structurally, both types are scenes; often, they would constitute grand-scenes.

(∗)

1. Bursts, puffs, and splashes of colour.
2. Repetitive, multiplying non-figurative elements.
3. Geometric designs and patterns.
4. Designs with figures.
5. Rapid figural transformations.
6. Kaleidoscopic images.
7. Well-defined, stable, single figurative images.
8. Proto-scenes.
9. Full-fledged scenes.
10. Interactive scenes.
11. Scenes of flight.
12. Celestial and heavenly scenes.
13. Virtual reality.
14. Supreme light.

1. *Bursts and splashes of colour*. These constitute the weakest type of visualization. They appear in practically every session, usually only at the very beginning.

2. *Repetitive, multiplying non-figurative elements*. These are stronger than those indicated in (1), but they are still weak; they are encountered in practically every session in which the brew has an effect. Typically, they appear after (1) at the beginning of sessions. They may also appear when a new wave of strength strikes or immediately after an intermittent opening of the eyes or movement of the body. In comparison to (1), the items of this type are characterized by enlargement of the

scope of the inner visual field and by the sharpening of the contours of the elements that appear in the visualization.

3. *Geometric designs and patterns.* Albeit not figurative, perceptually these are likely to be very powerful. My experience is that these are much less frequent than the literature seems to suggest; surely, there are many sessions in which they do not appear at all. Furthermore, they do not necessarily appear at the beginning of sessions as, again, the literature suggests. Rather, it seems to me that they appear whenever the intoxication gains strength—in the early phases of a session but also in the middle. Often, geometric patterns are associated with a strong brightening of colours, as in neon lights. At times one may feel total immersion in the geometric matrix and with this may lose the sense of one's individual identity.

4. *Designs with figures.* These are, by definition, subsequent to (3). Usually they initiate the development of independent figures and thus they serve as the opening for new stages in the progression of a session. There are many sessions in which these do not occur at all.

5. *Rapid figural transformations.* This is a very common type. Usually it appears as a sequel to (1) and (2) and is associated with what may be labelled the early-middle stage of sessions.

6. *Kaleidoscopic images.* These usually develop out of the repetitive non-figurative elements of (2). In terms of strength, they may be regarded as constituting an intermediate stage between the rapid figural transformations and the well-defined, stable figural images. Usually, this type appears in the early-middle stage of sessions.

7. *Well-defined, stable figurative images.* These usually appear after (5) and (6); they may also appear after (4), in the middle of sessions.

8. *Proto-scenes.* These are usually a sequel to (5), (6), and/or (7), and they appear in the middle of sessions. Snapshots and glimpses may also appear after full scenes, in which case they are subjectively deemed as pertaining to sights of more distant realities or more remote periods in time.

9. *Full-fledged scenes.* These are a sequel to (7) and/or (8), and they appear at the late-middle part of sessions. However, sessions may end without the drinker reaching the stage of having visions of this kind.

10. *Interactive scenes.* Usually, these are encountered at the late-middle part of sessions. At times, however, interaction is induced by extremely powerful visions that occur in the first parts of sessions. This is especially true when drinkers experience especially powerful visions very early in the course of their training with Ayahuasca.

11. *Scenes of flight.* In these, the Ayahuasca drinker feels that he or she is being elevated and reaching heights and distances. Several subtypes of flight may be noted. The first is that of *flying*, where one either feels that one is being transformed into a bird or experiences flying like a bird. Furthermore, flying, one may experience oneself passing over natural landscapes and viewing them from above. The other subtypes are characterized with larger heights and greater distances and

are not associated with the experience of bodily metamorphosis. In *planetary* voyages drinkers feel that they are up above the skies and watching the planet. In *cosmic* voyages drinkers feel that they are even further away, travelling throughout the cosmos. Usually, these visions appear at the later stages of sessions.

12. *Celestial and heavenly scenes*. In these one feels that other, non-natural worlds are revealed. Such visions of this kind are often endowed with a profound religious atmosphere, and usually they have a strong psychological and spiritual impact. Some of these may be referred to as encounters with the Divine. Figuratively, visions of this kind are relatively mellow. In general, celestial and heavenly scenes appear at the late stages of sessions and are accompanied by feelings of peaceful serenity.

13. *Virtual reality*. Visions of this kind are powerful and rare. If they do occur, they tend to do so in the late-middle part of sessions.

14. *Supreme Light*. These visions may be independent or embedded within scenes; they may also be related to celestial and heavenly scenes. Usually, they appear towards the ends of sessions and in general they are encountered by individuals already having fair experience with the brew. Typically, they are associated with euphoria and are spiritually uplifting.

Taken in unison, the foregoing characterizations suggest that, first and foremost, the typology of Ayahuasca visualizations is too rich for it to be defined in its entirety in terms of a fixed progression of stages along the course of a session. Not all types appear in all sessions. Likewise, there is no single dimension that can serve as the unifying basis for the characterization of the progression of stages along a session. At the beginning of sessions the progression is best defined in terms of increasing perceptual strength. There are, however, dimensions other than the perceptual. Visualizations may be less powerful in terms of perceptual strength but more advanced in that they can be reached only at late stages in a session and only with increasing across-session accumulative experience. Furthermore, the increase in perceptual strength over a session need not be monotonic. Strength may reach a certain plateau, or even decrease, only to be followed by other bursts in which it is increased.

I appreciate that reference to items in (*) by numbers requires repeated flipping between the pages. Yet the alternative of specifying all the types in full is cumbersome, lengthy, and inconvenient. I have thus opted for the lesser of two evils.

As far as perceptual strength is concerned, the highest-ranking types are (9), (10), (11), (12), and (13), as well as certain instances of (3). In terms of complexity of content the highest ranking types are (9), (10), (11), (12), and (13). Both objectivity and extraordinariness are associated with types (9), (10), (11), (12), (13), and (14). In the case of (10) and (13) the objectivity is coupled with the judgement that what is seen is just as real as the ordinary world in which we normally live and act; with the other types, the objectivity marks the independent existential status of what is seen, but ordinary reality is not necessarily attributed to this material. In terms of involvement and absorption the highest-ranking types of visions are types (10), (11) and—more than anything—(13). In terms of control the highest-ranking is (10), whereas in terms of mastery the highest-ranking types of visions are (10), (11), and (12). Strong spiritual impact is primarily associated with types (11), (12), and (14). Alterations in consciousness are especially marked in types (11) and (13).

As further noted, not all types appear in all sessions. The most common types are (1), (2), (3), (5), (6), (7), and (8). Next in line is type (9), whereas types (10), (11), (12), (13), and (14) might not be reached at all. Types (4) and (6) do not always appear; I devote special comments to this issue below. Overall, types (1), (2), and (5) are characteristic of the beginnings of sessions. Types (6), (7), (8), and (9) are usually associated with the middle part of sessions. Interactive visions of type (10) usually occur at this stage as well. Types (11), (12), and (14) are characteristic of the later part of sessions. Type (3) seems to be associated with moments in which perceptual strength is increased—this may be either at the beginning or in the middle of sessions; type (4) is linked with type (3).

Taken together, all these patterns mark a shift in saliency of dimension in the course of Ayahuasca sessions. The beginnings of sessions are characterized by a progressive increase in perceptual strength. The strength may then taper off and lead to a development in content, both in terms of complexity and in terms of extraordinariness. As the session proceeds, control, and perhaps active interaction, may become more and more pertinent. In the last phase of a session visions with spiritual content may appear. Intermittently, there may be outbursts in the strength of the intoxication with patterned geometric designs exhibiting increased perceptual strength. Note that usually the scenes that are most complex in terms of content, interaction, and narration, as well as those that are most meaningful in terms of personal and spiritual impact, are also those that are most powerful from a strict perceptual point of view.

As noted, some types are closely linked with others. It seems to me that the most common progression is the following: (1), (2), (5), (6), (7), (8), (9), (12). Type (2) may develop into type (4), which in turn may develop into type (5) subsequently leading to (6), (7), (8), and (9). Visions of type (9) may develop into ones of type (10); visions of type (10) and (11) may also develop into type (13); visions of type (11) into ones of type (12), which in turn may evolve to ones of type (14).

Some Further Aspects

Lastly, I would like to mention three more specific aspects that seem to relate to progressions that Ayahuasca visualizations exhibit. The first has to do with the general *atmosphere* that the visualizations carry. My experience is that very often the early visions in a session have an air of bizarreness and even menace. Not infrequently the first figures seen in a session seem to be slightly malicious and/or somewhat seductive. As the inebriation gains strength, the bizarreness gives way to enchantment and majesty. With this, the visions may become dramatic. Later, with the decline of perceptual strength and the onset of the middle phase of the session, the overall atmosphere is likely to become more tranquil and more serene. Thus, visions can be viewed with a feeling of equanimity and be savoured with delight. The last phases of Ayahuasca sessions are usually marked by upliftedness, euphoria, and a spirited feeling of joy. One of my most experienced independent informants told me that as a rule—and like Dante's *Divine Comedy*—his Ayahuasca sessions begin in hell and terminate in paradise.

Second is *colour*. it seems to me that the different phases in the course of a session are also likely to be associated with different dominant colours. I do not have full, systematic records of this parameter for all my sessions, but overall, it seems that the non-figurative perceptually powerful images at the beginnings of sessions are mostly associated with amber and gold, bright green, and dark red. The figurative images noted at peaks of perceptual strength are often of very bright colours including reds, yellows, pinks, and emerald greens. The less powerful scenes of distant places and times appearing in the middle of sessions are characterized by colours of a more realistic flavour. The entire spectrum may be seen. The celestial and heavenly scenes at the end of sessions are marked by blue and white as well as the yellow of light. I should emphasize that these are broad generalizations, and there are very many exceptions to them.

The third aspect is the *stability and durability* of visualizations. Visualizations consisting of what in the terminology introduced above were defined as low—notably, glimpses, snapshots, and kaleidoscopic images—tend to change very rapidly. Not so the types characterized as more advanced—scenes and especially grand scenes. The stability and durability of these may pertain not only to the form and visual aspect of visions, but also to their contents: powerful visions are prone to be concerned with only one major theme. The cinematographic analogy may again be especially illuminating here. In the early days of cinema (as well as in the pioneering attempts that preceded cinema as we know it today) shows consisted of many short snapshots with minimal semantic content and with no thematic relationship between them. The main interest was to present spectators with fascinating modern (at the time) optical gimmicks. It is not for such superficial effects that the contemporary cinematophile goes to the movies. Rather, he or she goes to see a specific work of art. On one occasion, an individual may be interested in the rendering of a Shakespeare play into a film, on another in a Hitchcock thriller, and on still another in a documentary about a particular subject matter of his or her interest. Normally, people do not go to the movies in order to see a random mix of all these (as well as other) genres. The same holds true with Ayahuasca. Only here, what happens depends not only on external factors but also on the person him- or herself—he or she is in charge of the screening too. The more restless a drinker is and the more 'jumpy' his or her thoughts, the more scattered and haphazard are his or her visualizations prone to be. And conversely, the more relaxed the drinker is, the higher the likelihood is for the visualizations to be stable. In the teachings of the UdV it is emphasized that concentration is the key to advancement in the school of Ayahuasca. Likewise, on one of the very few occasions on which I was given specific instructions on how to handle myself within a session, the traditional Amazonian healer told me to sit still, breathe in a regular manner, maintain my mind quiet and relaxed, and concentrate. 'If your mind jumps around, so too will the visions,' he said. 'It will be as if you are switching from one TV channel to another; but, surely, you want to focus on only one channel.' When I listened to these instructions, I could not help but note that they were almost identical to ones I received, many years earlier, in the course of a meditation training I was following in a Buddhist monastery.

From Stages to Functional Progressions

I would not be surprised if the reader has found the foregoing summaries somewhat cumbersome. I too find them so. That this is the case is not without significance. One moral of the foregoing analysis is that while there are some correlations between types and phases along the course of a session, there is no single progression of stages that can characterize all sessions. Rather, what we have are certain qualitative progressions—a number of functions (in the mathematical sense) that change, each in its own characteristic manner and time course, along a session.

The stages, as experienced by Ayahuasca drinkers and as depicted by the phenomenological analysis, are the products of the joint concerted application of different functions. Thus, it is, I think, more meaningful to talk of different functions that govern the progression of the Ayahuasca inebriation than of stages that occur in the course of it. Indeed, it seems to me that once the shift from a terminology of stages to a terminology of functions is made, the analytical characterization of order effects becomes much simpler.

I should emphasize that the conceptual shift I advocate here is grounded in a general cognitive theoretical orientation that is not specific to the study of Ayahuasca. For reasons that have nothing to do with my study of Ayahuasca and altered states of consciousness, I have come to believe that psychological theory based on process terminology is more advantageous than one defined in terms of well-defined structures. In Shanon (1993*a*) I call for a theoretical paradigm change in cognition and advocate a shift from a structural-representational orientation towards a dynamical-procedural one. The proposals made here in the context of Ayahuasca visions are a special case of my general process-based cognitive orientation.

In sum, what is being contrasted here are two modes by which the different phenomenological patterns experienced in the course of a session may be looked at. One is as the successive presentations of different types of images and other experiences. The other is in terms of the progressive application of various functions, each having its own distinct time course. The various images and associated experiences are the end products of these functions. In the literature, analyses are invariably conducted in terms of the first mode; I propose to adopt the second. As for the specific functions that are at work—these are defined by the different dimensions with which our discussion began: perceptual strength, richness of content, extraordinariness and objectivity, involvement and interaction, control and accommodation, psychological, intellectual and spiritual impact, and measures pertaining to consciousness.

It should be emphasized that it is absolutely not the case that each Ayahuasca session manifests all these functions and runs throughout all the types that have been indicated here. In fact, many individuals never experience some of the types nor reach some of the stages we have surveyed. Many do not even pass beyond the rudimentary stages of non-figurative elements; some never have any visions at all.[3] Furthermore, even the temporal order of progression may deviate from the general scheme indicated. First, it need not necessarily be the case that all the steps indicated are followed. In mathematical terms, the order is partial—that is, it relatively ranks different types but it does not dictate the occurrence of all types.[4] Moreover, as already indicated in Ch. 4, the progression throughout a session need not be linear and monotonic; rather, it may exhibit ups and downs. Specifically, over the course of a session—as in the course of life in general—one may regress, retreat, even fall. Already having reached a stage that would be ranked higher in terms of the dimensions noted here, the Ayahuasca drinker may recede so as to go back to a stage that would be ranked lower. The ongoing, monotonically increasing progression is an abstract idealization, not a concrete, mandatory path that is universally followed.[5]

Global and Long-Term Developments

Global Characterizations of Sessions

Throughout the foregoing discussion I have drawn the relationship between the different visualizations in a session primarily in terms of progression. This,

[3] It is difficult to have statistical estimates of how many individuals actually experience visions. Very often, very little happens in the first and even second session with Ayahuasca. Given the great physical malaise that the brew is likely to induce, people may be reluctant to persevere and engage themselves in further trials. Obviously, those individuals whom I interviewed are those who did go on and eventually did find partaking of Ayahuasca rewarding. Further, the effect of Ayahuasca can also be related (although not in any simple linear fashion) to the quantity and quality of the potion consumed. One should also bear in mind that for good reasons, in all contexts of Ayahuasca use, leaders of sessions are prone to be cautious when serving novices.

[4] By way of clarifying the mathematical notion of partial ordering, let me give an example. Given the five ordered elements 1, 2, 3, 4, and 5, various sets may be constructed—some maintaining the order between the elements, some not; those that do are said to exhibit the partial ordering in question. Examples of the former are: (1, 2, 3, 4, 5), (3, 4, 5), (1, 3, 5), (1, 4), (4, 5), etc., and of the latter: (5, 4, 3, 2, 1), (5, 3, 1), (4, 2), etc.

[5] It is pertinent to add that in this regard the anthropological literature is especially misleading. Usually this literature presents summary descriptions characterized as portraying 'what the Indians see' or 'what indigenous persons report'. As indicated earlier, seldom, if ever, are the number of informants given or indications of range and variation furnished.

however, is not the only kind of relationship that may be defined between the different visualizations within a session. In particular, I would draw attention to features characterizing each Ayahuasca session as a whole and applying especially with experienced drinkers and when the visualizations are of the types characterized as more advanced.

Just as there are different types of visualization, so too are there different types of sessions. Some sessions are more powerful and have more impact on drinkers than others. Some are joyful, others are painful; some are easygoing, others are dramatic. In some sessions the primary experience is visual whereas in others the primary experience is ideational or affective. Sessions also differ in the topics they are primarily concerned with: some have a more personal flavour whereas others are more spiritual or metaphysical. Overall, I can say that those that had the most significant impact on me were, on the one hand, those that presented rich and powerful visions and, on the other hand, in which I felt I learnt something, gained personal and intellectual insights, underwent spiritual experiences, and, last but not least, managed to overcome difficulties and advanced in my Ayahuasca schooling. My experienced informants confirmed these generalizations.

Orderly patterns and functional progressions of the kinds surveyed in this chapter can also be related to drinkers' levels of preparation and readiness. The pertinent key factor may be referred to as cleanliness. As William Blake (1794/1961) said, 'If the doors of perception were cleansed, everything would appear to man as it is, infinite.' These verses were the basis for the title Huxley (1971) gave to his book on mescaline; recently, they were also employed by H. Smith (2000) in a book on psychedelics and the religious experience (the title of this new book is *Cleansing the Doors of Perception*). With Ayahuasca, the impact of cleanliness is concrete and evident. For a person to experience the wonders of Ayahuasca in all their magnificence, she or he has to be cleansed. If one's body is not clean (for instance, if one is not careful in observing the dietary precautions before the Ayahuasca session), one will purge and vomit. If one's psyche is not clean and one carries psychological matters that have not been confronted and properly dealt with, one's experiences under the intoxication will reflect personal conflicts and concerns. Likewise with regard to the contingencies of the particular session: if the drinker is frightened or unrelaxed, his experience will be idiosyncratically tainted. Only when both body and psyche are clean will the mind be ready genuinely to profit from what the brew can offer and experience what might be characterized as 'the world of Ayahuasca'.

The foregoing statement is corroborated by my data. It appears that personal, biographical materials are especially common in drinkers' first sessions with Ayahuasca. The more experienced drinkers are, and the more powerful their visions, the less idiosyncratic and personal the contents of the visions tend to be. The relationship between drinkers' personal characteristics and the nature of their experience with Ayahuasca will be discussed at length in the next chapter.

Long-Term Developments

The observations made at the end of the previous paragraph bring us to the consideration of long-term developments in the Ayahuasca experience. So far, we have primarily focused on patterns pertaining to the progression of visualizations within a session as well as the relationship between such visualizations. There are, however, also patterns of development that are encountered over a longer period of time—between sessions and over the course of a drinker's long-term experience with the brew. As repeatedly emphasized, taken seriously, Ayahuasca drinking is not a one-shot, single-occasion fun experience. Sooner or later, practically all drinkers discover that Ayahuasca drinking is a school, or a course of training. In this final section, I comment on structural patterns that pertain to time-scales larger than those considered so far in this chapter.

The relationship between the patterns discussed above and those pertaining to longer time-scales is direct. As a drinker gains experience with Ayahuasca, the likelihood increases of his or her experiencing visualizations of types that by the definitions introduced above are deemed higher or more powerful. Furthermore, drinkers may reach the more advanced stages of visioning more easily and more rapidly. Indeed, they may reach those stages directly and without necessarily passing through the lower, simpler types.

This last observation deserves emphasis, for I think the literature is deceiving in this respect. All discussions of stages and order in the anthropological literature give a picture of an orderly stage by stage progression whereby more powerful types of visualizations are reached only after passing through all the less powerful types. In particular, the universal picture is that phosphene-like patterns and geometric designs are the necessary first stages by which all Ayahuasca sessions begin. My experience is that this need not be the case and, indeed, it often is not. With extended experience (as well as with higher dosages of the brew), I discovered that I was having full-fledged visions directly, and without preceding visualizations of the geometric kind. It seems to me that the summary statements encountered in the literature are based on reports of individuals with elementary or moderate Ayahuasca experience. This is, I suspect, a systemic bias that should be borne in mind.

The Ayahuasca Schooling

On several occasions throughout the foregoing discussion, I have alluded to the partaking of Ayahuasca in terms of a skill that is developed over time. As indicated in Ch. 18, with accumulative, long-term experience drinkers' handling of the Ayahuasca inebriation changes. For, indeed, Ayahuasca is a course of training, a school (see also Gorman, 1992). Many Western individuals adopting a hedonistic self-centred attitude partake of Ayahuasca with the aim of having exciting experiences—either fun or great drama. Expressions I have heard include 'try Ayahuasca', 'do Ayahuasca', 'trip'. But Ayahuasca cannot be measured in terms of one-off, great experiences or in simplistic hedonistic terms. In this respect, it is no different from any other serious learning experience. If one wishes to know something about classical music, just stepping into a music shop and listening to one piece by Bach, one by Mozart, and one by Schubert won't do. Appreciating music is a long, accumulative process—at times it is dramatic, often it is just very pleasurable, and always it involves concentration and reflection. Similarly, rushing to a library, opening a couple of books, and leafing through them is by no means a way to acquire knowledge of what literature is or is about. For this, there is no way but to start reading. Moreover, the reading should be done in an orderly fashion—some books can be read only after others, some suit the interests and mentality of only certain types of people, some require further prerequisite expertise. All this is true with Ayahuasca.

Especially interesting are long-term changes in the contents and themes of visions and in the import they have for the Ayahuasca drinker. As one gains more experience with the brew, one discovers that what happens to one under the intoxication is not haphazard—it seems to have an internal logic and order. It is as if there is, within the brew itself, a wise teacher who decides what one should experience and learn in each session.[6] At the time, during each particular session, this may not be evident but in retrospect the picture becomes clear and then its coherence is often most impressive. My experience has been that after every so many sessions, I would appreciate that I had completed what may be likened to a *course* on a particular topic. Often, this appreciation dawned upon me only when I was already two or three sessions into the next course. Long clusters of courses comprised what I would characterize as *cycles*. At times, there were also special *classes*—like guest lectures in a university course—on topics that seemed to be appropriate at that particular juncture of my life but that did not belong to the general theme of the course in progress at the time. I have heard the very same overall characterization from other experienced drinkers.

Following is a schematic characterization of the topics of the courses comprising the first cycle of my Ayahuasca schooling. These span the 67 sessions constituting what I have been referring to here as my core corpus. (I am not getting into details here for these bear on the more personal facets of my story, which I leave to another context.) First was an *exposition*. The visions I had in that initial phase of my experience with Ayahuasca were some of the most powerful I ever had. It was as if the brew presented itself to me so as to give me a general idea of the wonders and marvels it could reveal. It was as if it wanted to show me how extraordinary these can be and to convince me that there was something there worthy of further pursuit. The topic of the second course was *discipline*. The strength of the visions subsided, and I realized that now it was my turn—to learn to handle the Ayahuasca intoxication and to be responsible for my own contribution to my experiences with the brew. Specifically, I learnt to manage my physical discomfort, to stay still in my

[6] Recall the indigenous conceptualization of Ayahuasca being a teaching plant (see Luna, 1984*a*).

place for long stretches of time, to follow the demands of the rituals in which I participated. The third course of my schooling was primarily concerned with *healing and disease*. On the one hand, I acquired knowledge about the curative powers of Ayahuasca and the medicinal practices of traditional healers. On the other hand, inflicted with malaria myself I came to learn about disease and healing firsthand. The grades that followed focused on the *sacred* and involved powerful spiritual experiences. Then I had a long period—coupled with my partaking of Ayahuasca with traditional Amazonian healers—that focused on *shamanism*. I learnt how shamans work, served as an apprentice to some, and developed my own insights regarding the personal qualities that make a good *ayahuasquero*. This phase of my Ayahuasca schooling terminated, it seemed to me, when patients were passed to me and I was involved in their healing. The subsequent course of my Ayahuasca schooling focused on a variety of more *specific issues*. These concerned knowledge and its relationship to faith, and the nature of the Ayahuasca experience. Then there came a point where it was clear to me that this cycle of training was complete. It was at this point that I began working on this book.

When the core corpus was closed, another cycle of training began and new topics came up. This second cycle extended over about 60 sessions spanning a period of four years. One characteristic of this cycle was a shift in perspective—from that of a passive participant to that of a director of a session. On the one hand, I participated in more than a dozen sessions that seemed to teach me the various failures that can be encountered in the way a session is conducted. On the other hand I myself began to conduct sessions, in private settings. In these, I realized that the focus of my attention shifted from the fascination of visions and other non-ordinary phenomena to the attentive observation of what is happening to the other individuals who participate in the session and a caring concern with them. Other topics included in my second cycle of training were music, mind and the nature of knowledge, consciousness, and mysticism. At present, it seems to me that I have embarked upon a third cycle.

The long-term developments in the Ayahuasca experience are manifested not only with respect to the structure, content, and themes of the visions. As indicated in the scheme sketched above, no less important are developments pertaining to the Ayahuasca drinker himself. With experience, drinkers learn to handle the Ayahuasca inebriation and their stance and demeanour under the intoxication change. One experienced person told me: 'It is like learning to handle a high speed race car.' More on this will be said in the next three chapters, which are devoted to the contextual and dynamical aspects of the Ayahuasca experience.

Concluding Remarks

Drawing to the end of this discussion, I return to the anthropological literature and once again comment on the analysis or stages expounded by Reichel-Dolmatoff. The foregoing discussion suggests that this analysis is faulty in several respects.

First, it seems to me that in his analysis Reichel-Dolmatoff identifies types as temporal stages; this results in a picture that is just too rigid. Second, as we have noted throughout Ch. 6, the number of distinct types of visualizations by far exceeds two or three (i.e. the number of stages suggested). Consequently, the variety of possible orderings is quite rich, definitely richer than that proposed in the anthropological literature. Third, it appears that the patterned geometric designs do not constitute a necessary step at the beginning of sessions, nor are they confined to the initial stage of visioning. Fourth, some of the features Reichel-Dolmatoff (along with other writers) gives by way of defining types and stages are too specific. These features are indeed associated with the types and/or stages in question, some may even be quite typical, yet by no means is it the case that they are defining. In other words, they need not occur with all people and on all occasions. In fact, no single feature ever does. Finally, I question Reichel-Dolmatoff's claims regarding cultural determination. Many facets of Ayahuasca visions are not culture-specific and striking commonalities are observed in the visions of individuals coming from very different personal and social backgrounds. On this last point I shall further elaborate in the next chapter.

Finally, let me point out that there are similarities between the stages noted with Ayahuasca and those indicated in the esoteric Tibetan traditions in conjunction with the *bardos*, that is, the stages that the soul undergoes after death and before being tied again to a new incarnation (see Evans-Wentz, 1960; Sogyal Rinpoche, 1993). Traditionally, three *bardos* are specified: the first is associated with all-encompassing luminosity, the second with undifferentiated radiance of colour and light, the third with particularized figurative visions, either hellish or heavenly. These stages should be read in a reverse order and the understanding is that the more evolved a soul is, the shorter is its passage. Specifically, highly evolved souls can go directly to the first *bardo*, lesser evolved ones go directly to the second (and subsequently to the first), lesser souls have to go through all three stages. All three stages have parallels in the Ayahuasca experience. Indeed, Leary, Metzner, and Alpert (1969) have presented an ingenious reading of the Tibetan *Book of the Dead* in which the *bardos* are viewed as the different stages in an altered state of consciousness, specifically that associated with LSD.

19

Contextual Considerations

Who shall ascend into the hill of the Lord?
And who shall stand in his holy place?
He that hath clean hands, and a pure heart
who hath not lifted up his soul into vanity
nor sworn deceitfully.

Psalms 24: 3–4

He appears to every one according to the quality of the reception of Him, and therefore in one way to those who receive him in the good of love, and in another to those who receive him in the good of faith.

Swedenborg

So far our discussion has been concerned primarily with the question of 'What?' We have examined what types of visualizations Ayahuasca produces, what the contents of Ayahuasca visualizations are, what themes they may be concerned with, what ideas are associated with them, what stages Ayahuasca visualizations may define and in what order they may appear within a session. Now we turn to the question of 'How?' In this chapter, I discuss factors and variables that pertain to the contingencies of the context in which Ayahuasca is consumed. In particular, I shall examine the various factors and variables that may influence what is happening to a drinker under the intoxication and those that contribute to the generation of visualizations. In subsequent chapters, I shall discuss dynamical aspects of the Ayahuasca experience and highlight several cognitive-psychological patterns pertaining to the development of visualizations and their progression in time.

Context—General Cognitive Considerations

Context is the totality of contingencies in which human activity takes place. As discussed at length in Shanon (1993*a*), one of the most fundamental characteristics of the human cognitive system is that it is context-dependent. However, context is not a given, externally defined setting imposed upon the cognitive agent. Rather, as argued in Shanon (1990*b*), the cognitive agent herself is a key determiner—in fact, a generator—of the context. Context comes into being and is constituted by the meeting of the cognitive agent and the environment—indeed, its very definition is a function of both. Thus, two different agents placed in what a physicalist may deem to be the same environment, are, from a cognitive point of view, actually situated in two distinct contexts (think, for instance, of the world as seen by an elephant and how it is seen by an ant). For further discussion of this theme, the reader is referred to the Gibsonian school of ecological psychology—for instance, Gibson (1966*a*, 1979) and the review by Michaels and Carello (1981).

Not only is the human cognitive system maximally context-sensitive, but its very functioning presupposes grounding, or situatedness, in context (see also Heidegger, 1962; Merleau-Ponty, 1962). In other words, cognitive activity is intrinsically tied to the external world, physical as well as social. Indeed, no clear-cut distinction between the inner and the outer domains can be made. Consequently, it may be said that the locus of cognition is not the internal, mental domain but rather the interface between the cognitive agent and the environment. Accordingly, rather than symbol manipulation and information processing, the basic capability of the human cognitive system is acting-in-the-world.

Manifestations of the crucial role of grounding in the external world in normal cognitive functioning are abundant. In general, it is found that when immersed in a real and meaningful situation, people perform better when asked to execute the same task in settings that are artificial. In Shanon (1993*a*) I have likened cognitive behaviour to swimming—just as one cannot swim without water, one cannot behave cognitively without the world. In the context of Ayahuasca, too, the external world, both physical and social, is of paramount importance, and it is crucial for keeping firmly in touch with reality and being able to maintain one's normal sense of self.

As mentioned earlier in this book, in the literature on altered states of consciousness the standard terms for contextual effects are 'set' and 'setting'. The first term denotes those effects that pertain to the individual person experiencing the altered state, the second term denotes those that pertain to the situation and the environment in which he or she is placed (see Leary, Metzner, and Alpert, 1969; Zinberg, 1984). Here I am introducing another division—that between *factors* and *variables*. The variables are well circumscribed and their impact is relatively local and well defined; the factors are more basic and of a broader nature. *Grosso modo*, the contribution of the variables may be localized in specific moments in a session whereas that of the factors is more general. Both factors and variables may pertain to either set or setting.

The following discussion begins with the consideration of the more general factors. These include the personality of the drinker and the socio-cultural framework in which he or she partakes of Ayahuasca. Afterwards, a series of local variables is examined. The chapter ends with the consideration of two more specific issues—the personal motivations involved in people's decision to partake of Ayahuasca and potential dangers associated with the use of the brew.

General Contextual Factors and their Effects

General Constitution and Personality

An individual's general constitution and personality have a great effect on what will happen to him or her during a session. The relationship between these and the particulars of the Ayahuasca experience are not simple to describe for, in principle, every and any personality trait and stance can be relevant. As a rule, people who are psychologically solid, whose souls are 'clean' and whose minds unperturbed are more likely to have good sessions. Characteristic of such sessions are powerful visions, revelatory insights and ideations, and spiritual uplift. In such cases nothing menacing happens and the session is felt to be meaningful and rewarding. By way of example, let me mention a good friend of mine, a very pious woman of about 60 who—in the good sense of the term—has kept her innocent inner child alive. She partook of Ayahuasca only twice; on both occasions I was also present. Invariably, she had fairy tale visions—enchanted gardens, beautiful singing birds, angels, and nature all in a dance. It seems clear to me that what this woman experienced in her visions is directly related to the type of person she is. The numerous conversations I had with people before and after Ayahuasca sessions repeatedly revealed to me how extremely important are parameters pertaining to drinkers' personality. As I have heard said repeatedly in all contexts of Ayahuasca use (including indigenous ones)—with this brew, each person receives exactly what he or she merits. Over the years, I have come to appreciate how perfect this match is. At times it manifests itself in ironic ways, and often its perfection might be characterized as cruel.

In the previous paragraph I have used the word 'clean'. This is very much in line with the traditional views associated with Ayahuasca. Both the brew and the notorious vomiting it induces are often referred to as *purga* (purge). Indeed, the Ayahuasca experience can be viewed as a multilevel process of self-cleansing. The most elementary level is the bodily one—this is achieved by the vomit. Vomiting, it should be noted, is neither necessary nor universal. Apparently, it occurs when the drinker's body has to be cleaned. Once drinkers adequately prepare themselves for sessions—in particular, when they keep a proper diet for several days prior to the session—the likelihood that vomiting will take place is greatly reduced. The second level of cleansing is the psychological. A drinker very much concerned with personal issues and problems will not have the freedom of mind to delve into the other worlds that Ayahuasca may reveal. Only when psychological cleansing has been achieved can drinkers disengage themselves from their daily concerns and be open for a full-fledged Ayahuasca experience. This usually results in the third level of cleansing, the spiritual one. I shall note that the following general picture is based on extensive observations, both of my own experiences with Ayahuasca and those of many other persons. When this book was near completion I heard the exact general characterization from an experienced indigenous *ayahuasquero*.

The interests and aims of the individual and his or her reasons for partaking of Ayahuasca are just as important. Drinkers cannot determine what will happen to them in the course of an Ayahuasca session. Indeed, in general, what does happen impresses drinkers as surprising and beyond imagination. Yet, on the basis of extended observation, I can attest that extraordinary as the Ayahuasca visions are, they do fit what might be characterized as the mindset of the individual drinker. A person who is religiously inclined is more likely to have visions pertaining to the Divine, one who is adventure-minded is more prone to have phantasmagoric voyages, one who has a light-hearted, playful character may have ludic-like experiences. Similarly, persons normally interested in philosophy are more likely to experience metaphysical visions and ideations. All these statements are based on actual, concrete cases I have interviewed and observed. Further, an inspection of the data furnished by the interviews I have conducted reveals that some of the richest visions reported were furnished by people who are either professional artists or artistically inclined. None of this implies that people who are not religiously or spiritually inclined may not, unexpectedly, experience visions pertaining to the Divine, or that individuals with no knowledge of philosophy cannot have visions pertaining to metaphysics or epistemology. Indeed, in my corpus of data there are many examples of precisely such cases. All I am saying here is that the non-personal character of Ayahuasca visions notwithstanding, when sensitively inspected they can none the less tell quite a bit about the individual who has experienced them.

Preparation

Partaking of Ayahuasca requires preparation on the part of the drinker. As explained in Ch. 1, in all contexts of Ayahuasca use special emphasis is placed on this preparation and its regulation. In particular, restrictions on one's diet are applied several days prior to the consumption of the brew. Sexual abstention is demanded as well. In addition, the preparation may take more personal aspects—solitude and rest, meditation and contemplation, concern with specific psychological and spiritual issues, prayer, and the like. All these have direct influence on the quality of the subsequent Ayahuasca session and on its success. At times drinkers learn this the hard way. The hurdles—both bodily and mental—they confront over the course of the intoxication carry a lesson that is both direct and evident: this is not the way to approach Ayahuasca, one should be prepared!

The Ayahuasca Ritual

Ayahuasca is not a solitary, one-person affair. The brew is generally taken in the company of other people, in the framework of a structured session, following a ritual which, in turn, is directed by one experienced individual who is in charge.

The first effects of the ritual I shall note are not at all specific to Ayahuasca or to any ritual as such. I refer to the comfort and security human beings generally feel when not alone. When in the company of other people concurrently involved in the same activity one usually feels more protected. This is due both to a general state of being and to the practical assurance of the presence of others whom, if the necessity arises, one can approach for help. Especially significant in the context of Ayahuasca is the comparative feedback thus afforded: each drinker can watch how the other drinkers are doing and gauge his or her own state of being and conduct accordingly. One may, for instance, see that in comparison to other participants one is actually faring pretty well. Alternatively, when in difficulty, one may see that other individuals are persevering despite the unpleasantness and managing to cope with it. Naturally, this gives one assurance and encouragement.

While the presence of other persons *qua* other human beings is important, that of a leader is paramount. In all contexts of Ayahuasca use, a good leader will radiate security and induce an ambience of trust. The one indigenous *ayahuasquero* whom I met that impressed me most of all exhibited the following personal characteristics—solidity, patience, a non-judgemental attitude, coupled with the sense that 'he has seen it all' and can accept people whatever their comportment might be, psychological sensitivity, the good sense of knowing how to say the right word at the right moment, and last but not least, a wonderful sense of humour. Clearly, these characteristics are not culture-specific and none is particular to Ayahuasca, yet practically they are all crucially important.

Indeed, as already mentioned in Ch. 16, in sessions of the UdV, as well as in those of the Santo Daime Church, it is usual for the presiding master or another person of responsibility to pass amongst the participants and ask how they are. The verbal interchange is brief and highly regulated. The expected response is that one is doing fine, and personal reports of one's state of mind are considered to be out of place. Yet, on the basis of my own repeated experience, I can vouch that the interchange is very meaningful, and most effective. The very fact that one establishes, even for a moment and at times with concentrated effort, a realistic contact with another person reaffirms one's sense of self and of reality. A single word, a light touch of the hand, a smile, or a mere glance of the eye all have an immediate calibrating effect that is always most reassuring and, in appreciating it and acknowledging the good presence of the other people around him or her, the drinker can return with renewed confidence to the non-ordinary, visionary realms. With this, drinkers trustfully know that ultimately they are grounded in the here and now. It is in place to note that the positive influence of the company of other people presupposes that the individuals concerned are, indeed, well suited. When this is not the case (for instance, when a person misbehaves in a session, or acts out his or her fantasies), the situation can be quite disturbing.

Seen from a general theoretical point of view, the assuring and comforting effect of the presence of other people is a manifestation of one of the most basic, and most general, features of human cognition, namely, its being grounded in the world. As argued at length in Shanon (1993*a*), the human cognitive system is designed so that for it to function it presupposes a concrete situated background. This background is constituted by a series of parameters, ones that orthodox cognitive scientists usually dismiss as not being purely cognitive. These parameters include the physical world, the presence of social others, and the ongoing flow of time. In the alternative view of mind developed in Shanon (1993*a*) these non-cognitive parameters are primary.

Practically, assurance, comfort, and assistance are all very important, but they are not primary. The primary contribution of a ritual is, I think, its offering of a framework of meaning and with this, both orientation and sense. In general, the non-ordinary effects that Ayahuasca induces are so powerful and so perplexing that it is only natural for drinkers to attempt interpretations of them. It is also the case that these experiences may be so rich that drinkers feel that unguided they may be led astray. In providing a set framework determining both what to do during a session and what to make out of what happens to one in its course, rituals offer such a guidance. This is especially marked in the works of the Santo Daime Church. These are highly structured religious services designed to lead people to the path of faith. Through the various facets of the rituals they employ—notably the lyrics of the hymns that are being sung—these works instil powerful religious and spiritual feelings in the hearts of participants. Correspondingly, in these services there is a strong likelihood that visions of figures such as Jesus Christ and the Virgin Mary will be seen.

Music

One major aspect of all Ayahuasca rituals is music (for a general review, see Dobkin de Rios, 1984; for specific anthropological reports see F. Katz and Dobkin de Rios, 1971; Siskind, 1973*a*; Dobkin de Rios and Katz, 1975; Bellier, 1986; Gebhart-Sayer, 1986; Luna, 1992; Buchillet, 1992; and Hill, 1992). Discussing the role of singing in Ayahuasca shamanism Luna (1986*b*: 104) concludes:

> Being a vegetalista is almost synonymous with being capable of mastering an often large repertoire of icaros. . . . Icaros may serve all the functions that we associate with shamanic endeavour: They may be used for evoking the spirits of a plant-teacher or a dead shaman, for travelling to other realms and dealing with the beings of those worlds, for modifying and tuning one's visions during ayahuasca sessions, for healing, for hunting certain animals or for fishing a certain species, etc.

Similarly, Rouget (1985: 126) states that 'shamanizing and musicating are two aspects of one and the same activity'. Indeed, in all contexts of Ayahuasca use, it is very common to hear drinkers of the brew declare that the music they heard sung during a session had a pivotal influence on their experience. Often, songs are characterized as being enchanting and the music is said to have the wondrous power to carry those who listen to it far away to other worlds.

Especially striking is the role music plays in sessions of the Church of Santo Daime. In these sessions the hymns constitute the very skeleton defining all that is experienced. As practically anyone who participates in the *Daime* rituals finds out for himself, the hymns and their progression perfectly suit the various stages of the

Ayahuasca inebriation. At the beginning, the hymns express anxiety and convey supplication, then they manifest the accumulation of energy and drive on into perseverance (*firmezza*, firmness, is a word that is frequently used at this stage), later the hymns reflect advancement and ongoing elevation, and towards the end of sessions the singing is always an extravaganza of bliss and euphoria. Invariably sessions terminate with a great sense of personal and group accomplishment and the songs clearly reflect it all.

Why is it that music affects Ayahuasca drinkers in such a powerful and enchanting fashion? It seems to me that in essence this question is no different from that concerning the powerful (and indeed, wondrous) effect music in general has on the human psyche. Why this is the case is not evident. Often, it is said that music has a direct, non-mediated psychological influence, but from a cognitive theoretical point of view this is just a statement of a fact, not an explanation (for some interesting discussion, see Langer, 1953, 1969). This is not the place to discuss this puzzling, and far from simple, cognitive-psychological question. Here let me just highlight a number of effects that I find especially salient in the context of Ayahuasca. Before spelling them out, I would once again say that I think all that will be said below is true of music in general. However, as it does with all other modalities, Ayahuasca highly accentuates both sensory perception and affective responses. Hence, all the effects to be noted here are more enhanced under the intoxication than they normally are.

It is often said that music is abstract. I gather that what is meant by this is that unlike other forms of art, music does not involve a clear, one-to-one mapping of form to content. The contrast is usually made to literature and the various visual arts that denote well-defined meanings and are associated with well-formed semantics. All this is true. Yet, to my mind, the characterization of music as abstract is very misleading. These comments notwithstanding, music is also extremely concrete. Specifically, in music the sensory medium counts to the utmost. Indeed, in musical expressions, unlike linguistic ones, any minute distinction pertaining to the medium counts (in fact, it is precisely these seemingly minor variations that make the most significant differences between the performances of different artists, all fully competent and all playing by the book).[1]

I would say that the reason music has such a special psychological impact is precisely because it is both abstract (in the sense of not being figurative or conventionally semantic) and, at the same time, very concrete. Being abstract, it is not tied to any special code and thus it can be effective in different contexts. Being concrete, it cannot fail to make an effect. This is definitely not true of the verbal arts, and is much less so for the plastic ones. I might go further and say that this coupling of the abstract and the concrete is a characteristic of human cognition in general (see Shanon, 1993*a*). In fact, it is perhaps because it embodies some

[1] This feature, whereby any distinction makes a difference, is referred to as 'density' by N. Goodman (1968), who singles it out as a basic characteristic of art in general; from a psychological point of view, density is the hallmark of concreteness, hence non-abstractness (see Kolers and Smythe, 1984).

basic features of human cognition at large that music has this special impact on people.

Especially important in this regard is time. More than any art form music is temporal.[2] In fact, not only is it the case that music is fully embedded in time, it actually orders and structures people's experience of time. Thus, when fast rhythmic music is heard, time seems to run faster and *mutatis mutandis* for music with an extended, slow pace. It is this temporal dimension that is especially pronounced with Ayahuasca. As described in Ch. 14, with Ayahuasca time perception may be greatly modified. Often, a moment seems to last an endless eternity. If what one is seeing at that moment is frightening, the experience can be most terrifying indeed. It is here that music has, I think, its most important specific effect. Subjectively, a drinker may feel that what he or she experiences will never end, yet at the same time, listening to a song that is familiar the drinker may gain some temporal perspective. For instance, the particular stanza that is being sung or heard may give the drinker a clue with regard to objective time estimation—both the time that has elapsed since the song has begun and that which remains until its termination.

In the context of Ayahuasca, music also has an interpersonal significance. Usually, in Ayahuasca sessions there is hardly any interaction between the participants. Conversation is not allowed. In most contexts of Ayahuasca use, the only interpersonal activity is the musical one. Each person is within his or her own world, each experiencing the separate realities that the brew introduces in his or her distinct fashion. Often, all this occurs in darkness and with the drinkers' eyes closed. In these conditions, the only percept that is distinctly real, that pertains to the ordinary physical world, that is shared with other persons, and is acknowledged to be so—is the music being sung and/or played.

Finally, let me propose a characteristic of music that I shall be defining here for the first time—its *homomorphism*. Homomorphism is a property of mathematical mapping. This characteristic obtains when the objects upon which a mapping applies and the products of the mapping are of the same kind; the function, we may say, maps upon itself. An example is translocation: moving a geometric figure in space does not affect the figure's shape or size. As a rule, except for position, the translocation function does not change the objects upon which it applies. Now, it seems to me that music is unique in that it exhibits homomorphism under perceptual distortion or hallucination. In a non-ordinary state of consciousness semantically well-defined figures may change their form and hence their identity (e.g. a flower is no longer a flower). In contrast, subject to any perceptual modification as it might be, music always remains music.

The foregoing cognitive observations also have more mystical ramifications. Because of the homomorphism noted, music is unique in that it pertains to both worlds—the ordinary physical one and whatever non-ordinary one(s) may be

[2] I am not saying that only music is temporal—after all, it takes time to read a novel or even inspect a painting.

experienced with Ayahuasca. Thus, in a most concrete fashion, music can function as the *axis mundi*—that is, the axis that joins the worlds (see Eliade, 1964)—*par excellence*.

In almost all discussions in the anthropological literature, music is considered from the point of view of those participants in the Ayahuasca session who are listening to it. But music—more specifically, singing—has an important effect on the person who produces it. The following comments are based on my own personal experience.

First, like those who listen to the singing, the one who produces it also needs grounding and guidance. Like them, he, too, gains these through music. But there are musical effects that are specific to the person who produces the music. Often, the Ayahuasca inebriation makes people wish to express themselves in action (as well as, in the psychological sense, act out). Some people simply cannot sit still—under the intoxication, they move about and are therefore more likely to be bothered by the adverse effects that the brew induces. Others may engage in overt behaviours that may bother the participants in the session—overexuberant dancing, the emittance of all sorts of sounds, and inappropriate interpersonal interactions. Singing offers a readily available channel of action and thus provides a solution. Having—as is usually the case in all contexts of Ayahuasca use—at one's disposal a set of songs that are already known affords a state of affairs that is especially comforting: the singer does not have to invest mental effort to decide what to do under the intoxication. This is especially true in *Daime* rituals in which the session is guided by a pre-established, fixed set of hymns which are listed in a booklet. It will be noted that not only does the singing not disturb the other participants, it actually helps them through the session.

Singing is also a wonderful outlet for emotions. In particular, it channels the emotions in a positive direction—notably that of praise and joy. Furthermore, singing is uplifting, quite literally so. When the voice is raised up, so too is one's soul and with it one's visions. Indeed, on many occasions I felt that as I was singing and my voice rose, I too was lifting up—to the skies, to the far reaches of the cosmos, to the heavens. Finally, let me recount that, personally, many times I have felt that music helped me to resist physical malaise during a session: instead of vomiting, I poured music out of my mouth.

Local Contextual Variables

I now turn to the more localized contextual variables. In particular, I examine those variations that have a direct effect on what drinkers may see under the Ayahuasca intoxication.

External Stimulation

The most evident and straightforward local variable is that of external stimulation. At times, what people see in their visions can be related directly to stimulation originating in the external world. The following are two typical cases; both took place when, under the intoxication, I momentarily opened my eyes and then closed them. In the first case, with my eyes open I saw a crack in the door. When I closed my eyes, a column of light mirroring the shape of the crack was seen. In the second case, I was sitting facing a woman; opening my eyes I momentarily gazed at this woman's face. Closing them, I saw a transfiguration of that face into one of a witch.

Often, the most notable external stimulation during Ayahuasca sessions is the musical one. As indicated above, the music heard during a session exerts great influence on what is seen in the visions. Many of the non-indigenous persons I interviewed who partook of Ayahuasca in a traditional Amerindian setting commented that the *ayahuasquero's* singing directly determined what they saw in their visions. Many felt that specific *ayahuasqueros* had induced, when they sang, special magical powers. Many further said that the visions they had lasted as long as the singing did, and when the singing stopped, so did the visions. Especially to be noted are the parameters of rhythm and movement. As already described in Ch. 11, it is very common for images to manifest movement consonant with the rhythm of the music one hears.

Similar effects are encountered in conjunction with motor perturbations. Here is an example that happened to me more than once. Abruptly, I moved my body (for instance, in order to check my tape recorder) and found myself out of my previous meditative balance. With this, my inner visual field was inundated with repetitive non-figurative elements whose overall contour traced the pattern of my movement.

Attitudinal Factors and Reactions

In the previous sections I have considered drinkers' general constitution and personality. Analogous statements can be made with respect to drinkers' frames of mind and the attitudes they take in any given session. In general, the more relaxed a drinker is, the higher the likelihood of him or her to have a good session. In particular, the more tranquil one is and the higher the level of concentration one maintains, the more stable are the visions and fuller the episodes that are revealed by them. Perhaps it is in this vein that the kaleidoscopic nature of Ayahuasca images may—at least in part—be explained. Specifically, it may be that the rapid transformation of the images, especially at the early phases of a session, actually reflect the drinker's own inquietude. Indeed, on the two rare occasions in which shamans have given me explicit instructions on how to behave during a session, the message was precisely this—to relax and concentrate. As mentioned earlier, the instructions were identical to those typically given in the context of Buddhist meditation—to sit quietly, to maintain a straight bodily posture, to keep the hands in a certain fixed, stable position, to breathe in a steady, regular fashion, to contemplate and not to be carried away by the flight of one's mentations. Then, I was told, the visions will not jump from one thing to another, but rather specific, continuous, and stable scenes will be seen.

In a similar fashion, the minute by minute reactions to what goes on during a session have an ongoing effect on what a person will experience under the intoxication. In particular, when one is afraid it is likely that what one sees will be menacing whereas when one is tranquil and relaxed what one sees will not be so. Personally, I have discovered that sitting straight and directing my mental attention upwards is likely to bring about more lofty images.

Epistemic Stance and Interaction

By the term 'epistemic stance' I refer to the cognitive attitude a drinker takes *vis-à-vis* the visions he or she has. The present discussion of epistemic stance is intimately related to that of epistemic status entertained in Ch. 16. By and large, epistemic status and epistemic stance may be regarded as the two sides of the same coin: the former is defined with respect to the visionary material while the latter applies to the drinker. Correspondingly, whereas the first pertains to the structural dimension, the second pertains to the contextual-dynamical one. Here I comment on aspects pertaining to the latter.

In general, the higher the degree that the Ayahuasca drinker takes the visions to be real, the higher the likelihood that the visions will indeed appear as such. In the limit, the most realistic visions are those in which the drinker fully immerses himself in the other reality that Ayahuasca seems to be presenting. Furthermore, a realistic stance often involves active interaction with what goes on in the vision. Comments made by the ex-shaman Amaringo (in Luna and Amaringo, 1993: 27) are most pertinent here:

> It is only when the person begins to hear and see as if he/she were inside the scene, not as something presented to him, that he is able to discover many things. There is nothing that he is not able to find out. I saw how the world was created, how everything is full with life, how great spirits intervene in every aspect of nature and make the universe expand. I was like a tourist, always asking the spirits what is this and that, asking them to take me from one place to the other, demanding explanations for everything. The world is multifaceted, so mysterious and unfathomable that it is beyond imagination. I also understood that human beings will never be happy until they realize their connection with the Creator and the spiritual dimensions.

Closely related to the foregoing considerations are those pertaining to interaction. As indicated in earlier chapters, interaction has different levels and degrees. In particular, the Ayahuasca drinker may step inside the world of the vision and move from the position of a witness to that of an actor, or perhaps a traveller. With this, he may find himself confronted with situations in which crucial decisions must be made. For instance, a trail may be reached and the drinker will have to decide whether to follow it or not, a door will be encountered and the viewer will have to decide whether to open it and enter or not, invitations may be extended and the decision made as to whether to accept them or not, instructions may be given and the drinker will have to decide whether to adhere to them or not. All these decisions will directly affect the further progression of what the Ayahuasca drinker will see.

The particular type of interaction in which a drinker may engage is especially revealing as far as his or her character and attitude are concerned. Seeing what may technically be (in a sense which is actually naïve and meaningless from a cognitive-psychological point of view) the same visualization, different drinkers may react in very different manners. Seeing a fantastic yet very foreign landscape the curious drinker may step in and begin to roam about and explore whereas a more reserved person may retreat. Confronting a ferocious animal the adventurous drinker may be intrigued and engage in some sort of interaction (a wonderful example is that reported in Ch. 7 of the informant who rode a jaguar). Other persons may feel overwhelmed or perhaps begin to fight and entangle themselves in sanguinary situations that they can hardly win. Inspecting the calamities of human history some drinkers may react with great despair whereas others may adopt a more metaphysical or even spiritual perspective and find some superior meaning in history as well as eventual reasons for faith and optimism.

In sum, attitudinal stances and the reactions and interactions that correspond to them may include the following: resistance and retreat, playfulness and adventurous engagement, quest for discovery and learning, fear and confrontation, psychological growth and healing, as well as those whose character is metaphysical, religious, and spiritual.

Interpretative Factors

How sensitive Ayahuasca visualizations are to a drinker's interpretation is a question of prime theoretical significance. As noted in the previous chapter, throughout his writings, Reichel-Dolmatoff makes a fundamental distinction between the first and second stages in the Ayahuasca inebriation. He notes that the images of the first stage are produced by brain activity, whereas those of the second stage are projected upon the images of the first by means of interpretative processes that reflect cultural preconceptions. On the basis of my own personal experience I can vouch that such processes are, indeed, operative. Seeing arrays of colours and light one starts to engage in interpretative speculations, not unlike those involved in the inspection of clouds or the dregs remaining in a cup of coffee. Abstract visual patterns pass in front of the mind's eye and one tends to confer identities upon them. With this, the images do indeed gain shape and form and they are seen as tokens of particular contents. Consonant with this is the finding, cited by Walsh (1990), that active, rich imaginative interpretation allowing one to see specific images in ill-defined patterns is a cognitive trait that is especially remarkable in shamans. These individuals, he notes, tend to see much in meaningless arrays of form and colour. For a more general discussion of the relationship between shamans' personalities and their visions, see Noll (1985). Furthermore, a correlation between disposition towards hallucination and vivid imagination was also found both in normal Western subjects (see Aleman, Bocker and De Haan, 1999) and in schizophrenics (see Mintz and Alpert, 1972). Finally, amongst my informants, a significant number of those who had powerful visions were individuals engaged in the arts.

As already commented on in Ch. 15, the key question is, how much of Ayahuasca visioning can be attributed to interpretative processes. I have been actively concerned with this question during many Ayahuasca sessions. I was trying to determine whether images I was seeing were in fact products of my own interpretation or not. When the intoxication is relatively weak and the images are somewhat hazy one can see how interpretative processes may be at work. However, when the level of intoxication is high, the images are experienced as imposing themselves upon one. There are visions that are so powerful and so remote from anything pertaining to one's own life experience that the feasibility of interpretation is highly doubtful. At least subjectively, it does not seem that an active interpretative process is responsible for what one sees.

Negative observations are pertinent as well. I have never managed to produce images at will. I never found myself wishing—during a session—to see an item and then having an image of it in my mind's eye. More on this issue of interpretation will be said in Ch. 21.

Other Factors and Variables

Closing the list of factors and variables, I would like to mark its *non*-exhaustiveness. Ayahuasca often induces powerful visions that do not seem to be related to any well-defined stimulus, interpretative process, or active involvement on the part of the drinker. These images may be astoundingly fabulous and they may catch one by utter surprise. Typically, they are described as being above and beyond imagination and as presenting states of affairs that have nothing to do with the drinker's background and previous life experience. Effectively, they seem to present another, non-ordinary, reality. How they are produced is, I find, a veritable puzzle. Acknowledging my inability to explain this phenomenon, I would say that either the brain or the mind presents one with images that under normal circumstances are not accessible to one. Whatever the mechanism at hand, the experience is that the images are created, or come into view, in ways that seem to transcend cognition as psychologists normally conceive it to be. I shall return to this issue in the last two chapters of this book.

Cultural Considerations

Throughout the foregoing discussion I have emphasized the paramount significance of individual factors pertaining to personality, interests and beliefs, background, and context of use. As I have repeatedly indicated, what happens to one during the course of the Ayahuasca inebriation is crucially affected by all of these. Naturally, a person's beliefs are also grounded in the cultural tradition to which he or she belongs. Yet, having acknowledged this, it is important to appreciate that many facets of the Ayahuasca experience are actually independent of personal and socio-cultural background. As explained at the beginning of this book, the first consideration that ignited my professional interest in Ayahuasca was the detection of common patterns between my visions and those I read about in the anthropological literature. How is it, I wondered, that I—a person knowing practically nothing about the indigenous cultures of the Amazon—saw things similar to those that the Indians see? Perplexed, I set myself to further my knowledge of Ayahuasca and, subsequently, to investigate it scientifically. These endeavours led me to appreciate that, indeed, while the visions each person has are significantly affected by all sorts of idiosyncratic factors and variables, the visions reveal marked cross-personal commonalities. As indicated in Ch. 8, my investigation further suggests that the more powerful Ayahuasca visions are, the more striking the commonalities in their contents tend to be.

In the light of all this, the direct linkage often made in the anthropological literature between the Ayahuasca experience and the psychology of indigenous people deserves special criticism. In particular, I would like to voice my strong objection to comments made throughout the literature with respect to special qualities that indigenous persons putatively have and which make them have the visions they do. Brüzzi Alvesi da Silva (1962: 262), for example, proposes that the effects of Ayahuasca reported by indigenous people are to be explained by the 'facile excitation of their imagination'.

Apart from the fact that claims of this kind seem to reflect at least a degree of anachronistic ethnic bias, they are also factually wrong. In the anthropological literature, analysing the effects of Ayahuasca in terms that are specific to the Indians is common (an example is the citation from Schultes quoted in the last chapter of this book). This is both understandable and legitimate given that the anthropological research in question is, as a rule, concerned with the indigenous Amazonian cultures. However, one should not draw the conclusion from this that what happens under the Ayahuasca intoxication is specific to Indians. Moreover, in the anthropological literature these statements are usually not accompanied by specifications regarding number of informants, frequency of phenomenon, and interpersonal differences and variance. In particular, much of Reichel-Dolmatoff's own analysis of the effects of Ayahuasca is based on the reports of just one informant, albeit a very experienced and knowledgeable one (see Reichel-Dolmatoff, 1971). Of course, historically the Amazonian Indians were the first to have used Ayahuasca and, obviously, contextual factors are paramount in the Ayahuasca experience. By no means, however, are the visions seen with the brew the special prerogative of the Indians or reflect special qualities that these people have. Indeed, the whole point of this book is that what happens under the intoxication reflects the nature of the human mind, by no means the particular indigenous, or 'savage', mind.

Just as it is not the case that visions, and by extension the most frequent content items and themes associated with them, are particular to Amerindian people, it is also not the case that the visionary experiences of all indigenous persons are the same. As is the case with Westerners, so it is with the Amerindians: they are not all alike and they, too, are subject to the various contextual considerations surveyed in this chapter. Indeed, in this respect the anthropological literature is extremely deceiving: not all Indians see the same things in their visions! The fact of the matter is that not all people see the same things in their visions, and not the very same things are seen by each person in all the sessions in which he or she participates—this is true of all people, whether indigenous or not.

In closing, let me tie these contextual and comparative considerations to an issue that has received much attention in the anthropological literature, namely, that of geometric designs (for general discussion, see Lagrou, 1991, 1996, 1998). We have noted that geometric designs are very commonly seen with Ayahuasca. In some Amazonian indigenous cultures such designs feature prominently. The Indians themselves link the use of geometric designs in their cultures to the seeing of designs in Ayahuasca visions. The drawing of such a causal link is an important theme in Reichel-Dolmatoff's writings (1975, 1978*b*). Specifically, he underlines the similarity between the designs encountered in the artefacts of Amerindian cultures and the entoptic visual patterns known as phosphenes (see the discussion in Ch. 6). Totally agreeing on this point with Reichel-Dolmatoff, I would like to respond to objections to him made in the anthropological literature. In particular, Gow (1988) proposed that the causal link goes the other way around and that geometric designs are seen *because* they are important in the indigenous cultures. Repeatedly and forcefully, my empirical data counter this statement. Geometric designs are extremely common in Ayahuasca visions, and they are also seen by individuals who have nothing to do with the traditional Amerindian cultures.[3]

An Aside on Anthropology

The reader will surely appreciate that many of the issues discussed in the preceding sections bear on what are normally regarded as anthropological topics. How does this square with the strong stance advocated at the outset of this book, by which the study of Ayahuasca belongs to psychology, not anthropology? Indeed, I totally stand by my original statement that the real mysteries of Ayahuasca pertain to the mind, not society or culture, and hence they primarily pertain to the province of psychology—more specifically cognitive psychology, not anthropology. Moreover, I would imagine that anyone who has pursued the reading of this text until now fully appreciates that it is impossible to study Ayahuasca without having substantive firsthand experience with the brew. Yet, this does not mean that there is no place for a socio-anthropological study of Ayahuasca. Such a study must, I think, be grounded in a solid psychological investigation, but once the

[3] In support of his claim Gow cites two personal experiences he had with Ayahuasca, once before he encountered the geometric designs in the traditional Amerindian context, one afterwards. Only in the latter did he see geometric designs in his visions. This may very well be the case, but psychological generalization cannot be made on the basis of a single case. Furthermore, as emphasized throughout this book, it is not the case that in each session each drinker passes through all the stages that may be associated with the Ayahuasca experience. Admittedly, seeing geometric designs is extremely common, but it is not an absolute, universal necessity.

psychological investigation has been carried out, the anthropological one inevitably follows.

Making this last statement, I am directed by my general view of human cognition, one which was developed in Shanon (1993*a*), long before I ever heard of Ayahuasca let alone began to study it professionally. By this, non-orthodox, view cognition is situated and not internal. Human psychological life takes place not in the inner province of the head, so to speak, but out there, in the environment where organism and world—both physical and social—meet. Against dominant views in contemporary cognitive science, my own is that the basic capability of the human cognitive system is not to process information but rather, to be and act in the world. Even our most private, most subjective experiences attest to this fundamental state of affairs (see Shanon, 1998*b*). This being the case, the internal and the external are intertwined and there cannot be a sharp divide between the two. Specifically, the mental is embodied in the corporeal and individual cognition is embedded in the matrix of social interrelationships. As the Soviet psychologist Vygotsky and his disciples argued, mind is *in* society (see Vygotsky, 1978). Hence, in a very fundamental fashion, even the most individualistic psychologist cannot ignore the societal.

As has been seen, in the context of Ayahuasca the involvement of the external (again, both physical and societal) is crucial. Having reviewed the various factors and variables pertaining to context, I would like to comment further on how, at this stage of my engagement with the brew (that is, almost six years since the first draft of the beginning of this book was written) I see the social-anthropological facets of Ayahuasca.

When one is first exposed to the effects of Ayahuasca the experience is so extraordinary, so singular that it is only natural for one to be captivated by it and direct all the focus of one's attention and interest to the experience proper. By its very nature, this experience is bound to be private. With extended history of Ayahuasca use, however, drinkers come to appreciate that what happens to them under the intoxication is a function not only of the brew itself, but also of the human context in which the brew is consumed. What one will experience climbing the Himalayas could not be experienced when climbing the Judaean hills. Yet, assuming that the climb of the Himalayas will not be a solitary affair, what will be experienced in the Nepalese high mountains will be, continuously and in every single respect, affected by who one is with on the expedition. Thus, when preparing for the quest, it is not enough to take care of the physical and navigational facets of the climb, one should also pay attention to the human and interpersonal ones. Just the same with Ayahuasca. Indeed, the more experience I have gained, the more appreciative I have become of the effect the context has on what was happening to me during the course of the inebriation. As time passed, I have also become more and more appreciative of the singular contribution of individual leaders and the way they directed the Ayahuasca session. As I familiarized myself with different contexts of Ayahuasca use I have begun to develop my own preferences. When joining a session, I have come not only to partake of the brew but also to be in the company of people I liked. I remember the first time I declined an invitation to a session—I intuited that I did not wish to undergo an Ayahuasca experience with the person who extended that invitation to me. At that moment, I knew, I advanced yet another grade in my Ayahuasca schooling.

Now, when I feel that I am familiar with the various aspects of the Ayahuasca experience and have a fair psychological understanding thereof, I find myself especially interested in the contextual and cross-personal facets of Ayahuasca use. This interest brings me to what may be regarded as societal, perhaps even anthropological, considerations, but essentially, the way I see it, these are all intrinsically related to the psychological. In a very concrete, very personal, manner, I have come to appreciate this on occasions where the effect of the intoxication was particularly strong, perhaps overwhelming. What can I do? How will I survive this? I asked myself. On many occasions, my only help would be the music, which I joined and sang myself. The musical score was presented to me, all prepared and ready, by the ritual. I did not have to struggle or wander—the course of conduct was already set and predefined. The predesigned ritual served as a structure of support and guidance. All told, the more familiar I have become with the various traditions of Ayahuasca use, the more have I found myself to be appreciative of—and interested in—the rituals in which they were embedded.

The state of affairs just indicated is similar to the one I (and surely, I am not at all special in this) have encountered in my long-term involvement with music. When first exposed to music, the listener's focus of attention is directed to what is being played: now one is listening to a Beethoven piano sonata, now to an opera. The more experienced the music listener is, however, the more attention he or she pays to the performance ('execution' in Hebrew, 'interpretation' in French). One learns to appreciate that even though different players play the same score accurately note by note, in essence the way the piece is rendered by the different musicians is diverse. With time, these renderings are as interesting—perhaps even more interesting—than the music itself. Eventually, as much as one likes Mozart one will go to the concert not merely to hear this particular sonata, but to hear how a particular pianist plays it. When the pianist is a master, one will enthusiastically go to a recital regardless of the identity of the pieces that are being played.

Using the linguistic terminology introduced earlier in this book, context and the socio-cultural factors associated with it mark a progression from the semantic to the pragmatic. The different rituals of Ayahuasca are different solutions that different people, and different cultures, have found to the problem of what to do under the intoxication. The Ayahuasca inebriation is too strong to be left unharnessed, undirected, perhaps (at least to some degree) undomesticated. Some structure has to be imposed upon it, and with this some sort of particularization, of decision, of choice, of interpretation. This holds true for each individual drinker in the course of a given session, and collectively for groups of people, societies, and cultures.

The Ayahuasca inebriation is also unboundedly rich, and therefore it both lends itself to and calls for different lines of conduct, use, and interpretation. If one wishes to find the religious and the spiritual in Ayahuasca, they are there for one to experience; if one is more inclined towards intellectual and metaphysical rumination, these will be there for one as well. If what interests an individual are psychological insight and personal development Ayahuasca will allow him or her to proceed accordingly. Likewise, for concerns with particular fields of knowledge and understanding. And surely, if one comes to the brew in search of bodily health, it will offer a cure. The Ayahuasca experience is all these and more (see comment on Divine knowledge in Ch. 10). The point I wish to make here is that given its unbounded richness, the Ayahuasca experience naturally calls for different modes of employment. When structured and socially shared, modes of employment crystallize in rituals, and when these are embraced interpersonally and kept over time they become the fabric of a culture. Anthropologists have directed the primary focus of their attention to the different forms and shapes of these fabrics, and there is much to be learnt from their investigations. As I see it, however, these fabrics are the answers to questions that are fundamentally psychological, decisions that we mortals inevitably have to make when confronted with a power that transcends the ordinary conditions of life and entrusts human beings with a direct recognition of the finitude of their very existence.

I shall not turn to metaphysical speculations here. Just as this book is not anthropological, it is not philosophical either. But I will drop some hints (just hints) of a more general perspective by which all this may be viewed. It should not come as a surprise if I say that the ideas to be hinted at here came to my mind on various occasions when I was under the intoxication. Whatever the context in which one partakes of Ayahuasca, one is likely to confront one's very self—discover who one is and learn something about the meaning of one's life. And then, as exemplified by the various examples in Chs. 10 and 11, the Ayahuasca teachings extend to human life in general. In many respects an Ayahuasca session is a frame in which the human condition is presented in a condensed fashion. Thus, the manner each individual drinker will be affected by the brew and the way he or she will navigate the course of the inebriation are, to a significant extent, a reflection, as in a mirror, of what one's life actually is. (More about this will be said in the next chapter.)

Obviously, different people direct their lives differently. This is true for all contexts of human life, but is especially salient in Ayahuasca sessions. As described in Ch. 5, inspecting the different ways participants behave in a session can be, for some, a most interesting enterprise. And what holds for individuals holds for societies and cultures. Just as it reveals the inner faces of the individual psyche and the drama of the human condition in general, Ayahuasca also reveals the grand odysseys which are the different cultures human societies have created as they struggled to survive and progress on this planet and invest the endeavour with meaning.

Whether on the individual or the societal level, the different courses of action involve choices—choices about what to do and what not to do. Choices are to be evaluated not only in absolute terms of their outcome, but—no less significantly—in terms of the options available in the contexts in which they are taken. These choices and the particular lines of action and interpretation associated with them are all responses, the best each person or society can furnish under the circumstances, when encountering a force that is so overwhelming, so awesome, and potentially so enriching. It is only with the examination of the particular choices made and the ensuing accomplishments of different individuals and groups in different contexts that one comes to appreciate the grandeur of the force that has demanded the choices and inspired the various lines of action and creation pursued. In their totality, all these present the anthropological story of Ayahuasca. One day perhaps I shall put in writing what my meditations on this story have taught me. These, of course, will reflect my own struggle with Ayahuasca, my own encounter with the limitedness of my being confronting the infinite power to which the Ayahuasca experience exposes its partakers. Articulating these meditations, however, I shall not pursue the role of the cognitive psychologist, nor that of the anthropologist or the philosopher. Rather, I shall merely utter the voice of one individual human being who has partaken of this brew. But for now, I return to my empirical cognitive study and consider a question that is actually very basic, very primary—'Why do it—why drink Ayahuasca?'

Why Do It?

Presenting this question I do not refer to first- or second-timers who, most understandably, often partake of Ayahuasca out of curiosity and without serious reflection. My question concerns, rather, those who have already drunk Ayahuasca a number of times and who make the conscious decision to go on doing so. This they do despite the bitter taste, the likelihood of great bodily discomfort, and the most fearful moments that the Ayahuasca experience may present them with.

Why then do people who have already partaken of Ayahuasca continue to do so? As explained in Ch. 1, in the indigenous context Ayahuasca was consumed for a host of reasons. Prominent amongst these was the gaining of knowledge, the foretelling of the future, the acquisition of the power to heal, and the transcendence of the ordinary human state of being (see e.g. Chaumeil, 1983). As indicated by citations in Chs. 1 and 18, traditionally Ayahuasca was regarded as a school or a course of study and many experienced drinkers underscore the lure to discover more and more as a prime motive for the further consumption of the brew; for further discussion see Langdon (1979*b*, 1992), Luna (1984*a*), Langdon and Baer (1992), and Luna and Amaringo (1993).

A pertinent personal testimony is the following unsolicited narrative recounted to me by a mestizo shaman:

The first time I drank I knew nothing about this at all, I drank because my brother invited me to join him for a session. The visions I had were so magnificent that I was extremely curious how it could be another time. The second time it was, again, marvellous, and very different from the first time. And so it went. The experience never repeated itself, and always it was wonderful and revelatory. That was some twenty years ago and I am still amazed, I am still learning.

My own feeling is the same.

I have presented the 'Why?' question to all my experienced informants. The answers cluster into almost a dozen main types. These, I should clarify, are not mutually exclusive: a person may be motivated by more than one of these reasons. Furthermore, at different stages in one's engagement with Ayahuasca the reasons and/or their relative weights may change. I, for one, identify with all the reasons that are indicated below.

The first reason I shall indicate is the less specific one, namely, the intrinsic *joy* or *aesthetical pleasure* of the experience. This reason is completely analogous to that which, I gather, motivates the great majority of people to go to concerts or to the movies—they simply enjoy music and love cinema. These are bona fide reasons in their own right, and nothing needs to be added to them. Indeed, the bottom line is that the Ayahuasca experience can be an extremely gratifying one. Personally, *post facto*, I have never regretted having partaken of the brew and invariably at the end of sessions I have seen practically all participants, in whatever context Ayahuasca has been consumed, manifestly content and happy. Admittedly, during the course of the intoxication there is a phase that may be excruciating and there are moments that can be very tough. Yet, these are always overcome. Indeed, most people who partake of Ayahuasca make the conscious and informed decision to do so even though on prior occasions they may have undergone unpleasant and difficult moments.

Akin to the aesthetical pleasure are several other reasons that entice people to go on and partake of Ayahuasca. My informants speak of the allure of magic and enchantment, the wonderment, the sense of perfect harmony, the all-embracing sentiment of love, as well as the thrill of adventure.

A second major reason is the *epistemic*. Many drinkers partake of Ayahuasca because they believe that the brew reveals true, and otherwise unattainable, knowledge and enables them to discover new, non-ordinary realities. Some individuals are especially enchanted and intrigued by these discoveries. Indeed, several of my informants specifically associated the quest for knowledge with the thrill of adventure. My enquiries suggest that these individuals are amongst those most likely to have powerful, full-fledged visions. Significantly, all the shamans I personally knew well specified discovery and the quest for knowledge as major reasons for their active involvement with Ayahuasca.

People may also partake of Ayahuasca because they want to gain *specific knowledge* about topics in which they are interested. This is most common with traditional healers. For them a major reason for consuming Ayahuasca is to gain knowledge about healing. In particular, it is said that Ayahuasca enables healers to diagnose their patients and instructs them how to go about treating them. Also common is the association of Ayahuasca with specific philosophical and metaphysical topics. Several of my informants—including ones with very little formal education—mentioned these as key reasons for their continuing involvement with the brew. The topics in question included consciousness, death and reincarnation, and the nature of the Divine. Other persons, in other contexts, may be interested in other types of knowledge. For example, several artists told me that Ayahuasca presented them with specific ideas that they eventually implemented in their work. I myself have gained some insights pertaining to cognition that I still find significant.

People may also partake of Ayahuasca in order to gain *psychological insight and understanding*. This may have to do with general self-understanding, or pertain to particular issues with which individuals are concerned at a particular moment of their lives. In fact, upon partaking of the brew, many experienced drinkers formulate specific questions for which they hope to receive answers. Often the insights thus gained serve not only for better personal understanding but also to direct people's actual course of action.

Another kind of motivation has to do with *well-being*. Again, this may be general, pertaining to one's general existential state, or specific—i.e. pertaining to a particular health consideration. Many, myself included, indicate that Ayahuasca allowed them to get to know their body better and to be in closer contact with it. This is especially advantageous, they noted, for self-healing.

According to many, the most important reason for partaking of Ayahuasca is *personal transformation*. Over the years, a great number of individuals told me that, in the long run, there is just one reason that justifies the continuous drinking of Ayahuasca—a serious wish to change the manner in which one is conducting one's life and to become a better person. All other reasons, especially curiosity, many said, cannot hold for long.

Still another major reason is the attainment of *transcendence*.[4] Several informants specifically mentioned that Ayahuasca allows them to be in touch with a special kind of consciousness which some further characterized as universal or absolute and attributed independent ontological status to. Significantly, I have heard such reasoning also from indigenous persons.

Related reasons are those having to do with *religiosity* and *spirituality*. In conjunction with these, several informants specifically underlined the fact that Ayahuasca brought them in touch with the *sacred* and made them appreciate its meaning and significance. Also related to this is a reason that for me, with time, has become increasingly cardinal, namely, to praise and to thank. As I have said, personally I would use the Hebrew term *Hallelujah*. Clearly, in the context of religious sects such as those many of my informants belong to, such concerns are

[4] The urge to transcend one's own regular existence is singled out by La Barre (1972) as a very basic, central human need and as a key factor in the promulgation of religion throughout human history.

primary. However, my interrogations reveal that they are definitely not confined to individuals who are members of these sects.

Finally, some may consume Ayahuasca because of the *social* dimension. This rationale was given to me by many individuals associated with both the Church of Santo Daime and the UdV. They explained that the reason for their participation in sessions is not just for the sake of drinking the brew. No less significant for them is the group membership, the solidarity with the sect as an institution, and the personal affection and ties with other members. Indeed, in both sects it is emphasized that the schooling that the brew offers is not confined to the experience of the inebriation. It extends to various social considerations such as solidarity with the group, participation in various community activities (e.g. charity activities, works with children, help of the sick and the poor), general moral conduct, and ethical stance. Considerations of social cohesion and cultural identity also feature prominently (see Siskind, 1973*b*; Reichel-Dolmatoff, 1975).

It is interesting to compare these reasons, provided by my informants, to those specified by Reichel-Dolmatoff throughout his writings (1971, 1975). Reichel-Dolmatoff specifies three reasons for the consumption of Ayahuasca in the indigenous context.[5] The first is to gain access to the mythological heritage of one's ethnic group. This epistemic reason also has an important social aspect. Gaining this knowledge, the young person being initiated into Ayahuasca gains direct access to the information he had been told about by the older members of the group. He appreciates that the stories related are, indeed, true and with this he becomes a full-fledged member of his culture. The second reason Reichel-Dolmatoff specifies is the gaining of psychological understanding. This may be regarded as the analogue to psychotherapy in Western culture. The third reason, one which he (1978*b*) discusses at length, has to do with artistic creation. As indicated earlier, Reichel-Dolmatoff noted that there is great similarity between Ayahuasca-induced visualizations and the geometric designs drawn by the Desana (the indigenous people he studied). Such designs are prominent in that culture and are encountered in face and body painting, house decorations, and designs made on various artefacts, especially ones considered sacred. On the basis of a detailed comparison, Reichel-Dolmatoff concludes that the source for the designs are the Ayahuasca visions. From this, he further concludes that one reason for the consumption of the brew is to facilitate the creation of these art forms.

Certainly, I do not have any qualms about the first two reasons, nor do I at all question the intimate connection between Ayahuasca visualizations and the indigenous designs. However, I do not know whether I would specify artistic creation as in itself being a reason for the consumption of Ayahuasca. I am not an anthropologist, and I do not pretend to be an expert on Amazonian culture. Yet, I wonder whether the singling out of this third reason is not, at least in part, a

[5] It will be noted that the reasons Reichel-Dolmatoff gives are his own theoretical formulation of the anthropological observations he made whereas the reasons that I present in this chapter were directly given by the Ayahuasca drinkers about their own involvement with the brew.

reflection of the investigator's own special interest in the arts. It would be further noted, that the analogies Reichel-Dolmatoff has found and discussed pertain only to the very first, and weakest, stages of the Ayahuasca inebriation. They are all of the type Reichel-Dolmatoff characterizes as phosphenes and which in terms of the typology presented in this book pertain to the most elementary types of Ayahuasca visualizations. Curiously, nothing is said about full-fledged visions and their putative relationship to indigenous artistic creation.

More noteworthy are the reasons indicated by my informants and which are not indicated by Reichel-Dolmatoff. In particular, gaining knowledge about healing and power to effectuate it is glaringly missing. This reason is to be found throughout the anthropological literature (see references above) and has been singled out as primary by all shamans I have interviewed myself. Second is the discovery of other worlds, not necessarily those specified in the mythologies of one's group. Personally, I have been struck by the great adventurous spirit of some *ayahuasqueros* and by their great zeal to venture into unknown territories. And last but definitely not least, I am struck by the absence of the spiritual aspect in Reichel-Dolmatoff's account. Again, I am not an anthropologist, but there is no question in my mind: perhaps the foremost reason for the consumption of Ayahuasca, for both Amerindians and others, is to transcend ordinary existence and come in contact with the Divine.

By way of conclusion, I offer the following summary statement. In essence, what Ayahuasca does is place drinkers in a special state of being. For a couple of hours the person partaking of the brew is granted the gifts of heightened sensitivity—both sensory and psychological, of increased intellectual and psychological insight, of enhanced vital stamina and well-being, of spiritual uplifting. It is up to each person to decide what to do during these hours of grace. He or she may employ it in order to question the inner self,[6] for direction with respect to specific problems that currently concern his or her life, to gain psychological or intellectual understanding and insight, to heal body and soul, to grow spiritually, to have a wondrous multi-modal aesthetical experience, to enjoy well-being and shared communion in the sacred, to be able, even if for a brief time, to transcend the normal confines of human existence.

Drawing towards the end of this discussion, I cite at length from an answer given to me by one informant, a Brazilian who is an experienced member of the Church of Santo Daime, in response of the last question in the structured questionnaire. The question was 'Why do you go on drinking Ayahuasca [*Daime*, in this particular case]?' The answer touches upon several different issues; respecting the honesty and sincerity of this answer, I decided to present it in an integral fashion and cite also those parts that do not pertain specifically to the topic under focal consideration here; some details of a personal nature are omitted:

[6] In the context of the Santo Daime doctrine the term 'the superior I' is used.

The *Daime* opened my eyes. With it, I have discovered an infinite world of which I am a part. A world which is at the very same time within me and outside. I have discovered that this life is a very serious business. That it is a very important opportunity to be here, with my own body, ready to work towards perfection. I cannot miss this opportunity, nobody should. But people are so often oblivious to the spiritual dimension of their lives. . . . The *Daime* shows all this to us because it is violent. The process is violent, tempestuous. So you cannot deny such a strength. You are forced to understand and to remember. And you are forced to understand that love is the only way. Of course, everyone can say this phrase and everyone can endorse it. It is obvious. But it is a very different thing to say this phrase and to really understand that if you don't work with love, if you don't respect other people, if you won't respect other people's limits, you shall pay a price for that. And this price can be very, very high. On the other hand, if you live with love, you shall have the heavens. And with the *Daime* you can see, or at least gain a perception, of what the heaven can be, what an *état de grace* can be. . . . Once I have understood the infinite that is in me, I want to go ahead, and go deeper and deeper in this infinite. I want to discover more and more, even knowing that the essential I have already got. . . . [Giving] this answer I really understand why I take the *Daime*. In the *Daime* works we are at the closest to the essence of our being as we can get, and at this state we are together with our brothers and sisters. This is beautiful.

Finally, let me cite from a hymn of the Barquinha:

> I am drinking [Ayahuasca] not out of curiosity
> but in order to change
> I am drinking not in order to aggrandize myself
> but in order to find the light.

Concluding Words of Caution

By and large, the characterization of the Ayahuasca experience presented in this book is a positive one. This reflects my genuine belief that this experience is a very valuable one. Unquestionably, I personally feel that I have greatly profited from this experience. This feeling was corroborated by my interviews and conversations with many other persons. Time and again, people have told me that Ayahuasca has greatly contributed to their lives. Many further indicated that as a consequence of their long-term experience with the brew, their lives have significantly changed for the better. Yet, is it all positive? Are there no negative aspects to the Ayahuasca experience? Does it not have any adverse effects? Yes, it may. There is no question about it—Ayahuasca is a very powerful substance, and one should not take it lightly. Having examined the reasons for partaking of Ayahuasca, I would now like to turn to some considerations against doing so.

First, let me emphasize that from a strict physiological point of view Ayahuasca is not addictive. The brew does not induce long-term bodily changes that result in either tolerance or dependence (see Schultes and Hofmann, 1979). Yet, the brew may be said to be psychologically, or spiritually addictive. Specifically, people who have had powerful Ayahuasca experiences are, in a fashion, stamped: they do not remain the same and often the brew becomes a very important constituent of their lives. I would liken this to the strong fascination people may develop in conjunction with music, puzzling intellectual matters, or perhaps a person with whom they may fall in love. In all these cases, whether the fascination is positive or negative, a blessing or a vice, is up to every person to judge in the context of his or her own life.

But the Ayahuasca experience may also present more concrete pitfalls. First, I would note what in the clinical psychological jargon is known as 'acting out'. This term denotes patterns of behaviour that overtly express a person's psychological wishes, fantasies, and/or conflicts that under normal circumstances are not thus expressed. Under the intoxication, people may engage in such behaviours and do so in exaggerated and inappropriate manners.

There is also the danger of performing, under the intoxication or soon after its termination, acts that the drinker will later regret having executed. These, I should note, are not necessarily bad or dangerous deeds in the standard sense of the term. Acts of this kind that I have committed include exaggerated generosity, inappropriate expressions of friendship and openness with other persons, and sharing thoughts and feelings that normally I would keep to myself. None of these deeds were disastrous in any objective sense, but prudence was definitely breached.

Serious psychological outbreaks are extremely rare. I have heard of one person who committed suicide under the effect of Ayahuasca (see Castillo, 1996) and of one case of violence of a criminal nature. The former was a young Brazilian who, covered with a blanket, stepped into a campfire (the context was the *Daime* midwinter festivities in which traditionally, a campfire is lit). Apparently, he had psychotic tendencies earlier, which were not related to Ayahuasca at all. Checking with Brazilian physicians and mental health workers experienced with the brew, my understanding is that psychotic outbreaks associated with Ayahuasca do not exceed the overall rate commonly found in the population at large. To my repeated enquiry as to whether persons suffering from serious psychopathology should partake of the brew, such professionals answered negatively. Apparently, the Ayahuasca experience may bring up too much psychological material and the person in question may have difficulty in dealing with the powerful effects associated with this.

Returning to the population at large, I shall note that one should be careful for one or two days after a session. In the Jewish esoteric tradition it is noted that a mystical experience is to be judged not only by the extent that one has been able to get into it, but also with respect to how well one has managed to come out of it and return to one's normal life. This is, I think, true of Ayahuasca too. In different contexts of Ayahuasca use it is advised to take it easy the day following the consumption of the brew. Exhaustion is common (although, often, marked vitality is experienced) and at times mild effects of depression are encountered.

More subtle, and more common, are long-term changes in people's belief systems. As noted above, many who have partaken of Ayahuasca feel that their lives have changed radically. In Brazil, many join the Ayahuasca religions and sects; this move, in turn, often involves geographical relocation and perhaps radical changes in people's marital and professional lives. But again, it is generally the case that significant things in life are gained with effort and involve a price. Whether the effort and the price are valuable is up to each person to judge in the context of their own life. And as with everything, some people are prudent and wise, and some are less so.

As I see it, the bottom line is that, definitely, drinking Ayahuasca is not a game. It should not be taken lightly, it should not be played with in an irresponsible manner. Like fire, the brew presents innumerable benefits that might not be achievable otherwise, but it also has its dangers. If not used with prudence, the brew can exact a price, even a heavy one. One should partake of it only when one is in a good state of mind, prepared both physically and psychologically, under the guidance of a competent person whom one trusts, and in the company of people one feels comfortable with. Furthermore, each person should clearly and honestly define his or her reasons and ascertain that they are good ones and justify the act. When deciding about dosages, one should be cognizant of one's limitations and remember the golden rule of moderation. These recommendations are in line with those made by Grinspoon and Bakalar (1979) in a general discussion of psychoactive substances. The authors recommend that these substances should not be taken alone, not for hedonistic reasons, and always in the context of a ritual. I fully agree with these recommendations. I shall add that those who administer Ayahuasca should thoroughly check who their clients are. Traditionally, in small social groups people knew each other. With the recent—and to my mind, potentially dangerous—expansion of Ayahuasca into Europe and North America unfortunately this is often no longer the case.[7]

[7] The foregoing discussion may be related to the question of whether Ayahuasca is a drug or not. Personally, I am neither judge, lawmaker, nor sociologist, and the legal and sociological aspects of the use of psychotropic substances are outside my professional expertise. Surely, they deserve books in their own right (see, for instance, Grinspoon and Bakalar, 1997; Doblin, 2001). Yet, given both my theoretical orientation in psychology and my firsthand experiences with Ayahuasca and the different contexts of its use, let me say that my feeling is that, by and large, the legal debate conducted in the West with regards to psychotropic substances is fundamentally misguided. Usually, this debate is grounded in a simplistic reductionist view, according to which the psychological effects of these substances—be they positive or negative—are the direct result of the biochemical action the substances have on brain and body physiology. Given my general theoretical orientation in psychology (developed prior to my work on Ayahuasca), I strongly oppose this view. By way of clarification, let me advance the example of sex. If we were to film the physical act of the most amorous marital conjugation we would see the very same anatomy and physiology we would detect in a the filming of squalid pornography. The difference between romantic love and pornography is neither in anatomy nor in physiology, but in the meaning people invest in the act that, so are the facts of life, involves particular bodily parts and bodily mechanisms. Precisely so with respect to psychotropic substances. Whether they are drugs (which, unfortunately, they certainly can be made to be) or sacraments (which, to my mind, they should be) depends on the people who use them. Every powerful tool involves potential risks and demands wise and responsible employment. But would we censor the use of fire for cooking and heating because with fire entire civilizations may be (and have been) destroyed?

20

Cognitive Parameters

> [Metaphor] is perhaps one of man's most fruitful potentialities. Its efficacy verges on magic, and it seems a tool for creation which God forgot inside one of His creatures when he made him.
>
> Ortega y Gasset

Having specified the various contextual factors and variables that may contribute to the generation of visions, let me attempt a definition of the major cognitive parameters that seem to be affected by Ayahuasca. In defining these parameters, in effect, I am attempting an answer to the major psychological question—what does Ayahuasca do to the human cognitive system? Theoretically, this is tantamount to the definition of modifications in the values of a series of specifications that together define what may be happening in the special state of mind that Ayahuasca induces.

The various parameters to be introduced here are interrelated and together they seem to define a cognitive state that exhibits functional coherence. None the less, the parameters differ in terms of scope and level of their application. In particular, four kinds of parameters may be distinguished. First are *global systemic* parameters. These are very basic and their psychological effects are across the board. They have to do with the general sensitivity of the cognitive system, the overall level of meaningfulness that envelops one's state of being, the speed of mentation, and a parameter I call energy. The parameters of the second kind are systemic as well but they are less basic and more *specific* than those of the first kind and functionally they are built upon them. They have to do with the relative salience of the medium of expression, iconic memory, metaphoricity, and synaesthesia. In some fashion all these parameters are closely related to one another. While all the parameters to be introduced here manifest themselves in conjunction with temporal changes and progressions, one—that of fluidity—pertains by its very definition to the cognitive system's overall functional state as it manifests itself in modifications in time course and temporal patterns; as such, it will be referred to as 'dynamic'. In the next chapter, the dynamic ramifications of the modifications surveyed here will become more apparent. Last are parameters that have to do with *consciousness*; functionally, these are also systemic.

The following discussion examines the various parameters in the order in which they have been introduced. Because of the integrative nature of the discussion, it includes several repetitions of topics treated in earlier chapters. Yet, the reader should appreciate the difference in perspective and level of discussion: in previous chapters the perspective of our discussion was phenomenological and descriptive; here it is theoretical and analytical. In other words, having surveyed the different phenomena, we shall now attempt to find the cognitive-psychological sense of what is going on.

Global Systemic Parameters

Overall Heightened Sensitivity

Ayahuasca induces an overall enhancement of sensitivity. As a result, drinkers' perception becomes extremely refined and they come to sense and feel things they would not be able to in their ordinary state of being. The effect is pervasive—it is manifested in the acuity of perception in all sensory modalities, in the sensation one has of one's own body and one's tuning in to it, in one's introspection and ability to appreciate the working of one's own mind, in one's general intuitive capacity, in the understanding of other people and in the appreciation of the aesthetic.

The other side of this very sensitivity is fragility. Under the Ayahuasca intoxication people are prone to be affected by all sorts of variations in the external world and as a consequence find that they are vulnerable in ways that normally they are not. Vulnerability is most pronounced in moments of change. For me, this was especially marked in sessions in which, for some contingent reason, I had to make a sudden move. Such moves occurred when I turned to talk to another person, when I abruptly changed my bodily position, or attended to a technical matter such as controlling an audio system (see also the discussion of motor activity in the previous chapter). Manifestly, the activities involved in these situations were all simple. But precisely because they were so banal, I took them for granted and executed them in the manner I normally would. In so doing, I overlooked the special state I was in. Thus, in essence, I was like a bull in a china shop. For my extremely fine state of being, normally mundane activities and movements were too rough. As a consequence, in all these cases I immediately experienced great malaise, a feeling that was totally out of line with what I had felt just before I undertook the action or movement in question.

Sensitivity and vulnerability are also manifest along the socio-psychological dimension. With Ayahuasca, people are usually extremely sensitive to the behaviour of others and may be highly affected by it. Another person's pain or sickness, his or her sadness, as well as his or her perceived negative personality traits may all evoke intense, even physiological, reactions in the observer. As pointed out by Dobkin de Rios and Grob (1994), Ayahuasca also increases people's suggestibility, that is, the degree to which they are influenced by others.

Interestingly, a characterization along these lines was presented to me by one Peruvian shaman. Insightfully, he explained to me that Ayahuasca opens all the channels of both one's body and one's soul. As a consequence, he said, people become very sensitive, but they also are prone to become more vulnerable, and therefore if they are not well prepared, they may fall.

Enhanced Meaningfulness

Closely related to the overall increase in sensitivity is the overall enhancement of meaningfulness. We have discussed this effect both at the very outset of our survey, in Ch. 4, and in Ch. 15. As explained there, Ayahuasca makes people feel that things are invested with great meaningfulness. Even the most mundane states of affairs gain special significance. It appears that things in the world are not there by mere contingency—rather, there is sense and reason for it all. Things and states of affairs that are not usually deemed to be related are experienced as being connected and interrelated, and together they cohere into one unified whole. In a fashion, the feeling is that all of existence is telling a story.

Enhanced meaningfulness has many manifestations. Salient amongst these are the phenomena of beautification and sanctification discussed in Ch. 4. Religious, spiritual, and esoteric modes of interpretation and understanding can also be viewed as such manifestations. So too can the animistic perspective already mentioned in earlier chapters. Essentially, by this perspective, inanimate objects are invested with personality: the sun and the moon or, in fact, the Ayahuasca beverage itself, are not merely astronomical bodies or a liquid with certain physiological and psychological effects. Rather, they are all animated agents having distinct identities. Under the effect of Ayahuasca, the animistic perspective is directly experienced even by persons who in their ordinary life are very distant from such a mode of thinking. The various facets of metaphoricity discussed below are further manifestations of the enhanced meaningfulness that Ayahuasca induces.

Speed of Mentation

As noted earlier, it is very common for Ayahuasca drinkers to report that the brew makes them think faster than they usually do.[1] Coupled with the heightened sensitivity of the cognitive system, quite extraordinary mental performances can ensue. Indeed, with Ayahuasca, many people feel that they are more intelligent and more creative. Many of the ideations surveyed in Ch. 10 attest to this. Especially impressive, I find, are the philosophical insights attained by drinkers without prior formal education.

Energy

Unlike the terms employed as headings for the previous subsections in this chapter, energy is not a technical cognitive-psychological term. It seems to me, however, that this term—which is universally and spontaneously employed by drinkers of Ayahuasca to describe the special condition they experience under the

[1] It would be most interesting to check these claims experimentally. Indeed, I think methodologically this is feasible. The only relevant experimental study I know of was conducted with psilocybin (Spitzer *et al.*, 1996) which showed that this substance increased the speed of semantic priming.

intoxication—is, in fact, most apt. Of course, the term is prevalent in ordinary daily discourse: expressions such as 'This person is so energetic,' and 'I just do not have energy today,' are commonplace. From a theoretical cognitive point of view, this can be conceived as a setting such as that of the idle speed of an engine or the gears in a car. Just as an engine or a car may be set to work at different speeds, so is the human agent and his or her mind. With Ayahuasca, the sensation of enhanced personal energy is felt quite literally. People do indeed feel that they are imbued with unusually high stamina and power to act. If they are solid, mature, and clean then the powerful energy that they gain will push them upwards and enable them to see and do so much that they normally cannot. I shall add that in a fashion, the parameter of energy may be regarded as complementary to that of heightened sensitivity with which we began this survey. Whereas sensitivity pertains to reception, energy pertains to production, action, and execution.

When a car is set in high gear it uses less power and runs at a higher velocity. The change of gear has, in other words, a very specific effect. Not so with the energy conferred by Ayahuasca. Practically anything can be done with it. In other words, the extra energy is given to the drinker, but it is up to the individual drinker to decide how to use it. In this respect, the energy in question is like money that is given to one with no strings attached so that one is totally free to make use of it and spend it as one fancies. More on the use of energy in the course of the Ayahuasca inebriation will be said in the next chapter when the dynamics of the Ayahuasca experience are discussed.

Specific Systemic Parameters

The Salience of the Medium

As explained in Ch. 15, the medium is the non-semantic aspect of cognitive expressions, the dimension that defines the specific forms in which such expressions are articulated or presented. Instantiations of the medium are the phonological forms of words, the graphological shapes of scripts, the contours and colours of mental images. In general, in the psychological literature the human cognitive system is characterized as being semantically oriented. In other words, as being attuned to the meaning of expressions, not the specific variables of the medium in which they are instantiated. Indeed, the common stance in cognitive psychology today is that, except for very brief periods of time following sensory encoding, the cognitive system is not sensitive to the medium (for a classical study, see Sachs, 1967). Admittedly, under normal circumstances human beings focally attend to the contents of what they see and hear, and the medium of articulation is neglected or even altogether ignored. Yet, on the basis of a comprehensive inspection of various facets of human cognitive performance (all independent of the domain of altered states of consciousness), I have come to conclude that the human cognitive system is actually much more sensitive to medium variations than most psychologists assume. In fact, in some contexts, the parameters of the medium can become most pertinent and have significant psychological effects. Theoretically, the import of the medium may be so crucial that, in essence, the distinction between medium and content loses much of its force and one might even argue that there is no reason to maintain it. (For further discussion and supporting empirical data the reader is referred to Shanon 1987, 1993*a*). What I propose here is that Ayahuasca greatly increases the salience of the medium.[2] Examples of this effect were given in Ch. 15; a specific pattern in which the salience of the medium is pivotal will be presented in the next chapter.

Iconic Memory

A specific manifestation of the salience of the medium as it pertains to the temporal dimension is the increase in the time span of *iconic memory*, which consists of the retaining in memory of information in a quasi-perceptual manner, as if a copy of the external perceptual stimulus is maintained. Normally, the span of iconic memory is very brief—it is estimated to be between 350 and 500 milliseconds (see Coltheart, 1983; Baddeley, 1990). With Ayahuasca, the time-span of iconic memory is significantly lengthened. One closes one's eyes and an image of what one has just actually seen is retained. The time of retention is much longer than normal. A related phenomenon is that of afterimages (see Ch. 17). These, too, are very pronounced when, during the inebriation, one closes one's eyes. Both phenomena result in a lengthening of the time that perceptual stimuli (or their derivatives, such as afterimages) are amenable to mental inspection. As a consequence, the scope of the mental transformations that these stimuli can generate is increased.

Coupled with the enhanced rapidity of one's mental functioning, the lengthened iconic memory and afterimages that Ayahuasca induces can bring about quite dramatic effects. In particular, very often the afterimages—which in themselves are in the shape of elementary geometric forms—immediately multiply and turn into fantastic kaleidoscopic arrays. Thus, what may actually be very simple visual stimuli can be extended in space and copied innumerable times so as to generate an infinite multiplicity of figures; furthermore, these may be echoed temporally so as to produce rhythmic patterns.

Metaphoricity

Strictly speaking, metaphor is a linguistic phenomenon. Yet, as will become apparent, metaphor may be viewed in a broader fashion and as such may be regarded as a general cognitive phenomenon, one that is not specifically confined to language. With this broader perspective in mind, I propose that an important cognitive aspect of the Ayahuasca inebriation is an increase in metaphoricity.

In the philosophical, linguistic, and psychological literature metaphor is usually defined as the characterization of one thing in terms of attributes or predicates normally pertaining to another (for pertinent discussions see the anthology by

[2] It is interesting to note that the salience of the medium is also marked in schizophrenia (see Arieti, 1974).

Ortony, 1979). The reader will note that, technically, this definition of metaphor presupposes nothing that is specific to language. Referring to a tree in terms normally associated with a human being and seeing a tree as a person are, in essence, the same cognitive feat.

Throughout the cognitive literature it is assumed that for metaphors to apply, the features defining the attributes or predicates employed in the metaphor have to be given and well-defined. By this view, metaphorical acts consist of non-standard selections of semantic features out of fixed sets of such features. Furthermore, metaphorical expressions are considered to be secondary, derived from the more basic literal ones. In Shanon (1992) and (1993*a*) I argue against this common view and suggest that for a metaphor to obtain it is not at all necessary that the semantic features or distinctions encountered in the metaphorical expression be given and fully defined prior to the articulation of that expression. Furthermore, on the basis of both empirical data and conceptual analysis, I claim that rather than being secondary, metaphorical processing is primary and non-derivative. This claim is supported by considerations of speed of processing in normal adults, ontogenetic patterns (it appears that metaphors are very common in the speech of young children), and the so-called primary (*sic*) processes encountered in dreams (these, note, are highly metaphorical; see Freud, 1900/1953). As I see it, the very essence of metaphoricity is the creation of new features. In other words, when producing or receiving a metaphor, cognitive agents draw new distinctions and induce new ways of looking at things. In this process, features are not selected out of prior, given semantic sets; rather, new semantic differentiations are made and new semantic features are generated. It is precisely this that makes metaphor cognitively so important—it is one of the most important mechanisms for novelty in cognition.

Viewing metaphor not as a process of selection but rather as an act of creative generation is suited very well to the phenomenology of the Ayahuasca experience. As I see it (and as argued at length in Ch. 22), metaphoricity pervades the Ayahuasca experience, and it does this precisely in that it increases the creative power of the mind. Following is a survey of several phenomena in which the creative metaphoricity induced by Ayahuasca is manifested. It should be appreciated that the great semantic openness exhibited in these phenomena presents further evidence against the selection view that attempts to account for metaphor in terms of given, well-defined features.

First is the phenomenon of *seeing-as* that we have noted in conjunction with visions experienced with the eyes open. Here I would like to expand on this phenomenon from a broader cognitive-psychological perspective. A universally accepted tenet in cognition is that our perception is not naïve (see, for instance, Gregory, 1966). In other words, perception is not the psychological analogue of photography or audio-recording. In perceiving the world we are not passive receivers translating physical stimulation into mental images. Rather, in perceiving we interact with the world in terms of our background knowledge, values, and needs. In general, perception is intertwined with interpretation and the assignment of meaning, with being in the world and acting in it. Consequently, there need not be one, universal mode of seeing that is to be referred to as the objective, absolute view of things. Surely, a roaming antelope will be perceived differently by a person, a lion, and a bull, and a string quartet is heard very differently by a person who is an aficionado of classical music and by one who is not. Thus, consider the following mapping function f relating objects o and percepts p:

$$p = f(o)$$

Even in day-to-day life this mapping is not necessarily a direct, simple one assigning single, universal perceptual values p to objects o. Framing and illumination, on the one hand, and mood, prior knowledge, and conceptualization, on the other hand all affect the function f. Ayahuasca, I suggest, extends the variability of the relationship. With it, for instance, a leaf may be seen as a human face. In the limit, anything could be seen as anything else. Indeed, under the intoxication I once had this very realization. I felt that I could see anything in everything. The function f, in other words, was entirely unconstrained.

A second phenomenon manifesting metaphoricity is exhibited by the *parables* mentioned in Ch. 7. Recall: in these one sees an image, interprets it as a metaphor, and draws a lesson.

Third, let me mention the phenomenon of *second reading*, namely, the finding of hidden readings in verbal utterances and linguistic expressions. As has been seen, this is very common with *Daime* hymns. Objectively, the texts of these hymns are very simple. Under the intoxication, however, people tend to find deep meanings associated with them. Simple descriptions of the sun or moon, for instance, will be read as revealing the eternal mysteries of the universe, whereas a song about a bird will be read as a parable about the human soul and its predicament. In various esoteric traditions—notably the Jewish Kabbalah and Moslem Sufism—such second readings are very common.

Last but not least is the phenomenon of *synaesthesia* that we have already considered in Ch. 11. Inspection of the definition of metaphor reveals that synaesthesia can be viewed as the sensory-perceptual counterpart of metaphor. Indeed, synaesthesia and linguistic metaphor are highly linked. Without having some synaesthetic sensitivity speakers of language would not be able to appreciate very common phrases such as 'Her face lit up,' or 'His heart was heavy.' Since in the context of Ayahuasca synaesthesia is so basic and in the light of the theoretical import of this phenomenon, I discuss it further in the section that follows.

Synaesthesia

Appreciating the cross-modal nature of some Ayahuasca images one can shrug and say 'Well, it is only synaesthesia.' By this one would mean that what the intoxicated person sees is a mere transcodification of what he or she has seen or heard in the real world, and hence nothing that merits special wonder. While, superficially, the facts of the matter may fit such a description, I would prefer to see things differently. Specifically, rather than say that Ayahuasca generates a process of cross-modal translation, I would say that what the brew does is to push drinkers' perception to a mode of operation that does not distinguish between the different sensory modalities. This mode of operation is, I believe, very basic to human cognition, but under normal conditions it is not very apparent.

In Shanon (1993*a*) I argue that synaesthesia reflects the fact that the basic human mode of viewing the world defies the differentiation between the different senses or perceptual modalities. This claim is based on the consideration of various empirical findings that have nothing to do with Ayahuasca or altered states of consciousness. In particular, let me mention that with children synaesthesia is much more common than with adults. Apparently, as they grow and become more 'mature', human beings lose their synaesthetic sensitivity (incidentally, men more so than women). Thus, it seems that Ayahuasca enhances a mode of cognitive functioning that is usually suppressed in adult cognition. For supporting psychological data and for further discussion see Hunt (1985) and Cytowic (1989, 1997), as well as Shanon (1992).

The foregoing observations highlight the intrinsic affinity between synaesthesia and metaphoricity. As indicated above, in cognitive-psychological discourse, the latter is generally linked primarily with language, whereas the former is regarded as sensory. I propose, rather, that they are to be regarded as the two manifestations of what is essentially the same basic cognitive phenomenon, namely, functioning in a mode that does not differentiate between domains that, from the perspective of normal mature adult cognition, are totally distinct. In metaphor these domains are semantic fields, while in synaesthesia they are sensory modalities, but otherwise these two cognitive phenomena are the same. Together, both may be regarded as manifestations of an enhanced degree of latitude with respect to priorly given, standardly established distinctions; this effect may be referred to as 'non-fixedness'. The moral of our discussion is that Ayahuasca greatly enhances both these manifestations of non-fixedness (for further discussion supporting this view in domains other than Ayahuasca, the reader is referred to Shanon, 1992, 1993*a*).

Metaphoricity, and by implication synaesthesia, may also be linked to the first of the specific systemic parameters, namely, the salience of the medium. Distinct though they are conceptually, it seems to me that metaphoricity and the salience of the medium may both be related to the same basic structural change that Ayahuasca induces, to wit—the overall increase in cognitive sensitivity and the relaxation of fixed codifications, distinctions, and divisions.

A Dynamic Parameter

Fluidity

I have underscored the non-fixedness that Ayahuasca induces. This non-fixedness manifests itself in an openness to novel ways of seeing—that is, in drinkers gaining relative freedom from pre-established, pre-determined modes of perception and schemes of ideation. This global cognitive effect also has a dynamical dimension, namely, *fluidity*. Essentially, fluidity is non-fixedness over time. Just as it makes its drinkers extremely sensitive to cognitive aspects and distinctions that in the ordinary state of consciousness people tend to ignore, Ayahuasca also makes its drinkers very prone to shift rapidly from one cognitive experience to another.

Overall, fluidity brings about a very rapid influx of constantly changing cognitive material. This is especially marked given the heightened level of energy that Ayahuasca induces. It depends on the individual drinker whether the rich and rapidly changing cogitations will be experienced as chaotic overflow or as a fantastic mental dance. The situation is like that of navigating a fast-moving vehicle: if the navigator is astute the drive will be wonderful, if not it is likely to be terrifying and even disastrous. I shall note that the same observation was made by an Amazonian *ayahuasqeuro* interviewed by Luna (1986*b*). Shamanism, the man explained, is like learning to navigate, only that the space being travelled is interior. This topic will be discussed at length in the next chapter.

Consciousness

How Ayahuasca affects the system of consciousness was the subject of our discussion in Ch. 12, and there is no reason to repeat that discussion here. I would, however, like to highlight some modifications of consciousness that directly relate to the overall functioning of the cognitive system which is the focal topic of our discussion in this chapter. In particular, I shall reconsider the parameters of differentiation, connectedness, and the conferral of reality discussed in Ch. 12, the topics of noesis and meaning discussed in Ch. 15, and that of transcendence discussed in Ch. 16.

The Relationship Between the Self and the World

With Ayahuasca, the divide between the self and non-self is blurred and the balance between the internal and the external changes. Thus, the differentiation between the cognitive agent and the world is weakened and a stronger connectedness between the two is experienced. Employing a biological analogy, one might say that the membrane constituting the barrier between the cognitive agent and the world becomes more porous, more permeable. Normally, we do not appreciate the existence of this membrane and the constraints imposed by it. But, surely, a barrier there always is—otherwise human beings would lose their individuation and would not be able to function in the world as independent, autonomous agents who maintain solid permanence through time. With Ayahuasca, however, this may change. Various manifestations of this have been described throughout this book. First and foremost, under the intoxication, drinkers feel that they are more connected to nature and the cosmos at large. The heightened sensitivity discussed above is directly related to this effect. Further, drinkers feel closer to other human beings and living organisms. The sentiments of love, empathy, and compassion common with Ayahuasca are corollaries of this effect. The closeness felt to the Divine and the Ground of all Being is another manifestation of the fundamental systemic change at hand. As we have noted, in the limit, the barrier between the self and the world may dissipate to such a degree that individuality is indeed lost and the mental and the real become one.

The visions Ayahuasca induces may be related to the changes in balance and connectedness. Specifically, the visionary experience may be conceived as resulting from a shift towards externalization. With it, what one sees is experienced not as the product of internal (that is, mental) creative activity but rather as the reflection of an external (hence, independent) reality. But perhaps a more apt conceptualization is one tying this experience to a mode of existence in which the distinction between the internal and the external loses much of its significance. The link between such a state of affairs and the basic systemic changes we are discussing here is evident; more on this topic will be said at the very end of this book.

Noesis, Meaning, and the Conferral of Reality

Related to the foregoing patterns of connectedness are the experiences of noesis and understanding. Respectively, these are linked to the feeling of connectedness with the world and with other persons. As noted in Ch. 12, knowledge may be conceived as involving some sort of contact between the knower and the known. Naturally the changes in differentiation and connectedness discussed above result in a feeling of a higher degree of contact, hence of deep knowledge and understanding. Also related to this is the enhanced meaningfulness discussed above. In order to appreciate this, let us consider the question 'What is meaning?' This is, of course, one of the central questions of modern philosophy. Throughout the twentieth century, different answers have been given to this question. The most important ones are, on the one hand, correspondence with states of affairs in the world (the notion of reference is central in this characterization), and, on the other hand, patterns of activity of the agent in the world (this is the core idea of the theory of meaning espoused in the later philosophy of Wittgenstein. These two theories of meaning are different from one another and standardly they are defined as antagonistic. Yet, interestingly, in both there is one common element, namely, some sort of connectivity between the cognitive agent and the world. Indeed, as explained in Ch. 12 (and as argued in detail in Shanon, 1993*a*), from a theoretical point of view, meaning can be viewed as the matrix by which the coupling between cognitive agents and the world is defined. In other words, meaning is the totality of patterns by which agent and world are connected. Thus, the noesis and the enhanced meaningfulness experienced with Ayahuasca may be regarded as direct ramifications of the diminished differentiation and increased connectedness with the world that Ayahuasca induces.

It is worthwhile to highlight an aspect discussed here that does not have a place in the semantic theories of meaning. I refer to the fact that meaningfulness may be a matter of degree. None of the prevalent theories of meaning, be they philosophical or psychological, allow for such a variation: meaning either applies or it does not. The phenomenology of the Ayahuasca experience shows that along the various parameters of consciousness, meaningfulness too may undergo modifications of weight.

Linking connectedness and meaningfulness is the conferral of reality. Experiencing oneself more connected to the world, one deems the world to be more real, and with this, one invests greater meaningfulness in whatever one encounters in the world. This holds with respect both to the natural world and to the social one. The propensity to find paranormality is, I think, a direct corollary of the enhanced meaningfulness attributed to the natural world. Suggestibility results from the effects noted here as they are associated with other people.

Transcendence

The present discussion is psychological and the parameters I have surveyed are cognitive; yet, it seems that my analysis would not be complete without mentioning one parameter that extends beyond the cognitive-psychological domain—transcendence. As noted in the previous chapter, one major reason for people to partake of Ayahuasca is to transcend their normal human condition. The experience of transcendence induced by Ayahuasca is intertwined with the overall energy already highlighted. The relationship is twofold: on the one hand, one symptom of the special energy at hand is an experience of transcendence; but it is also the case that when letting go and surfing along this energy, one finds that one is uplifted from one's ordinary human condition and thus reaches a degree of transcendence.

Concluding Remarks

I have underlined various respects by which the different parameters examined in this chapter are related to each other. In this concluding section, I will summarize the interrelationships between the parameters and consider them in a unified fashion.

All the global systemic parameters are interrelated. First and foremost, heightened sensitivity and enhanced meaningfulness are intertwined—when the cognitive system is more sensitive it detects more and thus is prone to find more meaning in things. The enhanced conferral of reality associated with both perception and ideation is directly related to this. The increased speed of mentation may be viewed as still another manifestation of this general effect. Energy, too, is intrinsically linked to the heightened sensitivity that Ayahuasca induces. In fact, it may be regarded as both the cause and the effect of this sensitivity. And the loosening of the divide between the self and the world and the change of balance between the internal and external are linked to all these effects.

The more specific systemic parameters are closely linked to one another too. Salience of the medium and the lengthening of iconic memory are, indeed, very similar. Synaesthesia involves an accentuation of medium effects as well. I have already underscored the affinity between metaphoricity and synaesthesia. Taken together, all these parametrical changes attest to an opening whereby what is normally fixed is made unfixed. This manifests itself in Ayahuasca drinkers being given the opportunity to see the world with fresh eyes, to be less bound to conventional categorizations and modes of perception, and to be more likely to make new cognitive distinctions and thus generate novelty in both perception and thinking. The increased speed of mentation and the dynamical fluidity observed with Ayahuasca amplify these structural effects as they manifest themselves in time.

Many of the patterns surveyed in this chapter are often associated with poetry and the arts. Indeed, it could be said that Ayahuasca induces what may be regarded as a poetic stance, that is, a mode of being and cognition in which people see the world as poets and other artists do. The overall enhanced sensitivity, the metaphoricity, synaesthesia, and fluidity all contribute to this. And of course, so does the last parameter I have introduced, namely, transcendence. Characteristic of the poetic stance are also two general factors we have discussed in Ch. 4, namely, heightened aesthetic appreciation and the increased conferral of meaningfulness.

Coupled with the dynamic parameters of speed of mentation and fluidity, the poetic stance is likely to result in enhanced cognitive novelty and creativity. This is manifested in the fantastic visions that the brew induces, in the ideations and novel insights entertained, and in the various performances—notably singing, dancing, and storytelling—in which drinkers may engage while under the intoxication. More on the topic of creativity will be said in the last two chapters of this book.

Let me add an evaluative comment here. In the psychiatric literature, states manifesting lack of differentiation of the type noted above are usually regarded as symptomatic of psychosis. The terms standardly employed in this context are 'disintegration of the self' and 'depersonalization' (see Nemiah, 1989). But this pathological evaluation is not necessary at all. The same phenomenological state can be viewed positively and be regarded as manifesting a unification of the self with facets of reality that are broader and/or higher than the individual—the world, the Ground of all Being, the Divine. Viewed (experienced) positively, this state may be taken not as a manifestation of psychological breakdown but rather as a welcome and rewarding condition in which the subject gains a fresh perspective on his or her existence and affords invigorating revitalization.

In sum, the moral of our discussion is that the different cognitive patterns that Ayahuasca induces are not a haphazard set of independent phenomena. They all seem to be the different manifestations of some very basic, and perhaps very simple, effects that the brew produces. In essence, Ayahuasca places people on a different, higher level of energy. To a great extent, the rest is up to the Ayahuasca drinker him- or herself. More on the contribution of the drinker and his or her interaction with the special state of mind that Ayahuasca generates will be said in the next chapter.

21

Dynamics

Whoso danceth not, knoweth not what cometh to pass.

The Acts of St John

I would believe only in a God that knows how to dance.

Nietzche

In this chapter I continue with the 'How?' question and consider a number of patterns which, to my mind, are important determinants of the way Ayahuasca visualizations develop. Following my general (that is, not specific to my study of Ayahuasca) radical phenomenological approach to cognition, unlike most investigations in contemporary cognitive science my analysis does not deal with putative underlying mental representations and computational operations modelled in conjunction with them. In line with my attempt to offer a systematic and comprehensive phenomenological account, I highlight patterns that are operative in the visionary experiences induced by Ayahuasca and that define configurations that are pivotal to their progression. This should be borne in mind even when I use terms such as 'process' and 'mechanisms'. In all cases these should be read only in a formal-phenomenological sense, not one that reifies underlying structures and processes.

The Double-Face Configuration

I begin with the consideration of a particular cognitive pattern that is specifically related to the parameter medium discussed in the previous chapter. I have discovered it in my investigation of thought sequences and I call it the *double-face configuration*. Thought sequences, recall, are wordlike trains of thought that spontaneously pass through people's minds. I am referring, of course, to a very common phenomenon that is familiar to every person in his or her daily life. A central topic of my research has been to characterize the structural patterns by which adjacent expressions in a sequence are related to one another and to define the dynamic patterns by which sequences progress. The double-face configuration is one of the main findings of this research. The following sequence is an example. Originally, it was entertained in Hebrew, and was triggered when the thinker saw a girl calling an agitated dog by the name of 'Doni':

1. He's[1] really frisky.
2. She should have called him 'Shedoni' [in Hebrew, a small devil].
3. Or for short, 'shed' [in Hebrew, devil].
4. That has a meaning in English too, 'shed'.

The move of interest here is that between the last two expressions in the sequence. This move is based on a commonality of phonological form, a commonality that from a semantic point of view is meaningless. Specifically, the Hebrew word for 'devil' happens to be homophonic to the (totally unrelated) English word 'shed'. This commonality is, of course, purely accidental; yet without it the sequence could not have progressed in the manner it did. The significance of this progression is the fact that it came as a surprise even to the person who entertained it. It was unplanned and unexpected, and with it the topic of mentation changed—it moved from the consideration of a dog to that of the word 'shed' in English.

The phenomenon of importance in the sequence at hand, and the key feature of the double-face configuration in general, is the special contribution of an aspect of expression which normally is secondary, and which as a consequence, is usually ignored. In our example, this aspect is the specific phonological form of the wordlike constituents in the thought expressions. Phonological form defines the medium of linguistic expressions and normally it is taken to be irrelevant from a semantic point of view. In this sequence, however, this form becomes relevant through what may be regarded as a local coupling and decoupling of medium and content. Normally, medium and content are tied together—each word defines a specific pairing of phonological form and meaning. Decoupling lends the medium local independence from content whereby the content is momentarily disregarded and the medium becomes autonomous. The decoupling is only momentary, since expressions consisting of medium alone are meaningless, hence devoid of any independent cognitive status. Yet, the decoupling is of great import. When the medium is again coupled with content, the cognizer may find him-or herself entertaining a new semantic domain. Thus, the decoupling has generated unplanned novelty in the thought process.

I reiterate that in this dynamic process the contribution of the medium is crucial. Were thought conducted without a medium and governed only by considerations of content, the cognitive agent could think only of what he or she intends to, and would be confined to his or her already established repertory of knowledge and belief. The articulation of thought in a specific medium and the introduction of aspects that are irrelevant from the perspective of content creates the potential for new directions for thought to progress. Specifically, the double-face configuration defines a basic pattern whereby the seemingly semantically irrelevant medium allows the cognitive system to pass from one semantic domain to another which it has neither planned for nor anticipated. I shall note that this pattern is especially common in puns and in jokes.

[1] In Hebrew animals are referred to by the pronouns 'he' and 'she', not 'it'.

Phonological form is not the only possible instantiation of the medium, nor is medium confined to the domain of language. Analogous progressions and interplays between medium and content are also encountered in mental imagery. In another investigation (Shanon, 1989*b*) I studied sequences of mental images. Subjects were instructed to close their eyes and picture in their minds a particular object or situation which the experimenter specified; the subjects were then asked to report the ensuing mental images as they progressed. The sequences thus reported revealed that the progression of images was directed not only by the subject matter being entertained but also by specific, contingent properties of the particular images pictured. For instance, the shape or colour of an image might bring to mind the image of an entity sharing that shape or colour, but pertaining to a totally unrelated semantic domain. Again, the basic dynamic pattern is that of the double-face phenomenon.

It seems to me that with Ayahuasca, dynamical patterns of this kind are especially pronounced. Given the overall heightened sensitivity that the brew induces, any dimension and aspect of the medium may become relevant. The rapidity of mentation encountered under the intoxication is likely to amplify this further. Thus, the shifts of relevant dimensions and aspects and the coupling and decoupling of mental material can be extremely fast and consequently lead to remarkable richness and complexity.

In particular, the double-face configuration is salient in conjunction with shifts between stages of the Ayahuasca inebriation. Here, the medium is instantiated in geometric forms that have no semantic reading. In the literature it is usually claimed that geometric forms appear in the initial stage of a session, hence when the intoxication is relatively weak. But as I have noted in Ch. 18, geometric forms also appear when the intoxication passes from one stage to another, stronger one. I have experienced this myself many times and have confirmed the observation with several experienced informants.

The double-face configuration may interact with the phenomenon of extended iconic memory and thereby define still another pattern by which momentum to the progression of visions may be introduced. As my experience with Ayahuasca accumulated I found myself developing the following practice. When, during a session (notably, sessions held in illuminated surroundings) I would wish to invest my visions with new momentum, I would open my eyes and look around. Closing my eyes again, the new perceptions from the real world would be incorporated into the images that appeared before me and further progressions in the visioning would be experienced. Engaged in this process, I often saw myself as a painter who dips his brush in the paints on the palette and then applies them to the canvas. Just as the painter uses the paint as material that he can then turn into a work of art, the person under the effect of Ayahuasca may use the perceptual stimulation of the world as material with which visions can be created. In formal terms, the sensory material may be regarded as a medium layer that is extracted from the semantics of objects and scenes in the real world and that in turn serves as the source for visionary figures and scenes that pertain to a totally different semantic domain. Synaesthetic association may further enrich the generative dynamics at hand.

An Anthropological Aside

Interestingly, corroborative remarks are found in the study of geometric designs as they appear in the facial and bodily paintings and on the decorations drawn on many artefacts of certain Amerindian groups. Anthropologists studying Amerindian cultures have repeatedly pointed out the close affinity between these designs and the designs that appear in Ayahuasca visions (Reichel-Dolmatoff, 1975, 1978*b*; Langdon, 1992; Lagrou, 1995, 1996, 1998). The following are comments made by Guss (1989: 122) when discussing the designs of the Cashinahua Indians:

> The abstract designs [are the] simultaneous portrayal of a dual reality. . . . What is really depicted is the dynamic relation between the two. Unlike the static minor images of the figurative designs, the kinetic structure of these forms create an endless movement between the different elements, drawing the spectator into them. Perception now becomes a challenge, with the viewer forced to decide which image is real and which an illusion. . . . It is . . . a constant interplay between the physical forms that are seen and the invisible that charge them.

On the basis of her work with the same ethnic group Lagrou (1999) makes the following observations:

> [The] designs function as paths to be followed by the eye-soul on its way away from and back to normal space and perception. When asked about the role of design in vision, one ayahuasca specialist told me, 'One should always stay inside the design in order to not get lost.' . . . In their quality of paths, these designs describe and map known space, while suggesting an infinite extension into unknown territories yet to be explored.

Similarly, marking the affinity between the geometric designs seen with Ayahuasca and indigenous bodily paintings, Gebhart-Sayer (1985, 1986) proposed that the latter serve as a barrier, but also a bridge, between the natural and the supernatural, the visible and the invisible. In the same vein, Reichel-Dolmatoff (1971) cites an indigenous informant who characterizes the Ayahuasca experience as the breaching of a wall so as to gain access to other worlds. Further theoretical suggestions along these lines are found in Siskind (1973*a*), Langdon (1979*a*), and Lagrou (1995, 1998).

Here, I would like to take the reverse line of argumentation and apply the cultural-anthropological ideas back to the phenomenology of the Ayahuasca experience. As already argued in Ch. 17, I propose that the psychological function of the geometric forms seen under the intoxication is analogous to that assigned to the bodily paintings and other geometric designs in indigenous cultures. Specifically, the visionary forms may be regarded as a kind of barrier that has to be crossed. It is as if a choice is being presented before the drinker: either to assume that there are 'only' geometric forms here, leave it at that and proceed no deeper, or else to immerse him- or herself in these forms, break through, and get to the other side, so to speak. Success in doing this will result in the drinker reaching a new stage in the inebriation, and advancing upwards on the Ayahuasca ladder.[2] Typically, figures will emerge from the geometric grid and eventually full-fledged scenes will appear. The reader will appreciate that this is exactly the configuration defined by the double-face phenomenon which originally I discovered in ordinary thought sequences.

The Mental pas de deux

Various observations made throughout the foregoing discussion indicate that the new worlds into which drinkers enter during the course of the Ayahuasca inebriation are worlds that are created and unfold as the drinkers progressively immerse themselves in them. The visions are not simple products of one's brain nor mere revelations of the hidden depths of one's psyche. Rather, the drinker's stance *vis-à-vis* the visions and his or her interaction with them is a crucial determinant of the generation of the visions and their development. Thus, (and as will be further argued at length in the last two chapters of this book) the Ayahuasca drinker is both an observer and a creator. Again, the analogy with the artist and his canvas may be useful. The artist is the one who creates what there is on the canvas, but the canvas is a partner in the act of creation as well. Rare is the artist who jots, in one swoop, a complete painting from his head onto the canvas. The canvas, with all the previous marks that have been assembled on it, has to be there in front of the artist for him to be able to continue his work. Admittedly, all that is there on the canvas is the product of the artist's own hands, but for the artist to make the decisions governing this process the canvas has to be there in front of his eyes. At the outset the canvas is empty, but progressively, the act of drawing consists more and more of reactions to the canvas and interactions with it. This, incidentally, is a major reason for the fact that the process of creation (like all other psychological processes) takes time. For a wonderful documentation of this process the reader is referred to the film '*Le Mystère Picasso*' (Clouzot, 1955; see also Quinn and Penrose, 1965) documenting the famous painter at work; for corroborative experimental data and further discussion, see Shanon (1993*b*). It seems to me that a similar state of affairs obtains with Ayahuasca visualizations. Initially, visualizations may be triggered by simple brain stimulation, as with phosphenes. However, more complex and powerful visual effects, it seems to me, cannot be accounted for in such a simplistic, reductionist manner. Just as one needs both paints and painter for a painting to be created, one needs both the psychoactive substance and the engaged, and often daring, cognitive agent for powerful Ayahuasca visions to develop.

In the light of all this, I would liken the Ayahuasca visioning state to music one plays with one's own mind or, alternatively, to a dance in which the observing self

[2] Here, let me note that on several occasions, in different locations, I heard people comment that the term aya-huasca (vine of the spirits) may be read metaphorically as meaning the ladder to the world of the spirits.

joins in a *pas de deux* with the visions that his or her brain (or mind) has created. The brain (or mind) may generate the visions, but the viewer's engagement plays a major role in their progression. Eventually it is meaningless to divide the contribution of the two. As in a good dance, the paired dancers constitute one entity that harmoniously joins in movement: it makes no sense to ask what is the contribution of one and what of the other.

By way of example, let me refer to my first major Ayahuasca vision, that of the dragons described in the Prologue. This was one of the most realistic visions I have ever experienced. Potentially, it was also one of the most frightening. In the middle of the night, in the midst of the forest, and without any prior preparation I was, with my eyes open, confronted by menacing dragons. The situation appeared to be totally real. At that moment, there was no other way for me but to accept it as such. And as with any reality that presents itself to one, I had to react. I decided to make peace with the dragons and to show them that I accepted them for what they were. With this, the dragons became less menacing, a peaceful coexistence between them and myself was established, and new avenues for the progression of the vision opened up. In retrospect, I can see that I could have reacted in a different way and dismissed what I saw as mere hallucination. Technically, I am not at all sure that this was, at that moment, a viable option. But even if it were, its price should be acknowledged—the vision would have come to an end. With this, I would have been deprived of one of the most astounding experiences of my life. The moral of the story is clear: if one likes to have powerful visions there is no alternative but to regard the visions as real. As I have said before, the higher the degree of reality that one attributes to the visions, the higher the likelihood that the visions will, indeed, appear to be real. This, note, amounts to a paradox: the one who determines the visions' level of reality is, to a great extent, the drinker himself, but for this to be successful the drinker must be unaware of his own great contribution. He should regard the visions not as reflections of his own mind, but as scenes conferred with a veritable reality of their own. Perhaps this is a reason for the fact that the most powerful Ayahuasca visions people have are often their very first.

Related to this is a similar pattern encountered in conjunction with consciousness and personal identity. Often, the Ayahuasca experience places drinkers in situations in which they feel that their individual identity is in jeopardy. They may try to dismiss the experience as unreal, as mere hallucination. The problem with this strategy is twofold. First, there is no guarantee that the drinker would be able to convince himself of the unreality of the experience. In this case, the experience is likely to be more, not less, frightening. Alternatively, the drinker may succeed in self-suggestion but then the experience will be lost. As several fellow travellers on this path have told me, the only viable option is to let go—suspend judgement and enter. Then one is most likely to have a meaningful experience.

Last but not least are the negative aspects of the drinker's contribution. The brew may offer the drinker the opportunity to undergo all sorts of non-ordinary experiences, but the drinker may not be ready for these and consequently may find that he is not able to profit from the chance presented to him and pass the experiences up. This state of affairs was reported to me by many, and poignantly I have also experienced it myself. The negative contribution of the drinker may be active or passive. Actively, the drinker may decide that he does not wish to soar with the special energy that Ayahuasca affords nor undergo the non-ordinary experiences it presents. For instance, feeling that he is about to fly, the drinker may break off and extract himself from the situation. Later, the drinker may very well regret having missed the special opportunity that Ayahuasca was presenting. The passive condition may be even more frustrating. The drinker may detect the effects of the brew yet feel that he just cannot savour them. For instance, the drinker may find himself bombarded with visions but discover that he is not in a state in which he can fully appreciate them. Be it because of ingrained personality shortcomings or circumstantial contingencies, the drinker discovers that he is not ready or not worthy of these visions. When one's body, mind and/or soul are not in proper shape, one is prone to find out that one cannot make use of the powerful energy that is bestowed upon one. One is aware that wonderful riches are available and presented to one, and it is only due to one's own limitations and deficiencies that one cannot benefit from them. Whether it is of the active or the passive sort, the realization that one is not up to it can be most frustrating. Again, the financial analogy may be instructive: imagine a person who had inherited a large sum of money only to discover that the will stipulates a condition that unfortunately he or she is not able to meet. I myself have learnt this lesson painfully in conjunction with the episode of the 'blue place' recounted in Ch. 9. Frustrating though they are, the negative moments are prone to be most instructive. They may define moments of truth in which drinkers are able to appreciate their real values, put their existential state of being into perspective, recognize the crucial importance of preparation, and appreciate their own contribution with respect to what is happening to them with Ayahuasca.

The Issue of Interpretation

The foregoing discussion is related to the issue of interpretation that I mentioned in Chs. 15 and 19. This issue is not specific to Ayahuasca; indeed, it is one of the most fundamental theoretical questions in cognitive science today. The key question is whether a clear demarcation line can be drawn between uninterpreted cognitive material and material that has been subject to interpretation. This question appears in various domains of psychology. In perception it is associated with the distinction between sensation and perception: adopting the distinction is tantamount to admitting the existence of 'naked', raw cognitive material. In linguistics and psycholinguistics, the question relates to the division between syntax and semantics: assuming that the division holds is tantamount to claiming that there is a clear separation between, on the one hand, purely formal structures and, on the other hand, those components of language that have to do with meaning. The distinction is also central in analyses having to do with what is currently referred to as the general architecture of the mind. Consonant with it is the division between automatic processes that are not subject to conscious intervention on the part of the cognitive agent and processes that do permit such intervention (these have been referred to as impenetrable and penetrable, respectively; see Pylyshyn, 1979, 1984, as well as the division drawn in Fodor, 1983, between peripheral processes and central ones). But perhaps no clear-cut demarcation lines can be drawn. Perhaps meaning and interpretation filter cognition throughout so that there is no sense in drawing a line between that which is interpreted and that which is interpretation-free. My own view on cognition, which has been greatly influenced by the pioneering ideas of James Gibson (see references in previous chapters), embraces this last stance.

While there are no proper cognitive-psychological analyses of the Ayahuasca experience as such, the assumed distinction associated with interpretation is actually reflected in the anthropological literature. Explicitly, it is noted in Reichel-Dolmatoff's (1975) analysis of stages which I have discussed in Ch. 18; implicitly, it seems to me that it is assumed by many others. Recall that Reichel-Dolmatoff attributes the first stage of Ayahuasca visioning (notably, the phosphenes) to physiological processes whereas the second (as well as third) he relates to interpretation. Essentially, the cognitive picture assumed is the following. The chemically stimulated brain generates senseless visual stimuli. Under the intoxication, it is proposed, drinkers' imagination is enhanced and hence they impose all sorts of fanciful interpretation upon these stimuli. Subsequently, these are interpreted as visions. If, however, the demarcation line defined above does not exist, then there is no sense in distinguishing between visualizations that are interpretation-free and ones that are not. Furthermore, there is no clear-cut differentiation between interpretation and creation. As will be seen further in the next chapter, the Ayahuasca experience involves both heightened internal stimulation *and* enhanced imagination. Coupled together, the two result in a state in which the mind is extremely creative. In the spirit of the *pas de deux* advocated above, it is senseless to factor the feat of creative imagination associated with Ayahuasca into that which is free of interpretation and that which is not. In essence, all is interpretative, all is creative.

The Dialectics of Immersion

In the previous chapter I highlighted the change of balance between self and world that Ayahuasca induces. This has important dynamic ramifications. First, while the effects of the brew are often very powerful, they can also be most subtle. In particular, under the intoxication, drinkers may move back and forth smoothly between radically different states of mind. Such movement need not be either erratic or chaotic, and for the experienced drinker it may even have the feel of playfulness. As I have suggested earlier, I would liken it to surfing the waves of the sea, or a bird's flight, or perhaps a musician's masterful playing (*sic*) of his or her instrument. Several persons with extended experience with both Ayahuasca and other psychoactive agents indicated to me that this feature is one that distinguishes the former from the latter.

Above, as well as in Ch. 16, I have referred to the Ayahuasca drinker in terms of skilled bodily performance. I also drew an analogy with artistic performance and suggested that the Ayahuasca experience may be likened to one's playing one's own mind in the manner one plays a musical instrument. As pointed out in Shanon (1993*a*), the master pianist is a person who can bring together two seemingly contradictory abilities. On the one hand, the master can immerse himself within the piano and be one with the instrument, thus expressing his entire personality in the playing. On the other hand, the master can keep a distance and maintain critical reflection of his own performance. Throughout the course of playing there is a constant flow between the two poles of total immersion and of critical judgement. The more astute a pianist is, the smoother is this flow. Eventually, the pianist may be said to be able to maintain a concurrent dual existence in which he can, at will, be both immersed in and reflective about his playing (for further interesting observations, see Sudnow, 1980 as well as Herrigel, 1985). I propose that this is exactly the talent the master *ayahuasquero* develops. Like the pianist, he exhibits two skills: in his mentation, he can soar infinitely high and be constrained by no boundaries and, at the same time, he is well grounded in reality. In general, the more grounded in this world one is, the more daring and free one's visionary soaring can be. The grounding is manifested both globally, in various personality traits, and locally, in one's conduct in the course of the given Ayahuasca session. Globally, grounding reflects one's basic mental stability, the solidity and integrity of one's personality, one's honesty with oneself, and one's overall moral and spiritual stature. Locally, grounding is achieved through straight bodily posture, stable breathing, relaxed yet concentrated psychological attitude, contemplation that resists distraction, and the finding of the golden path between fear and vanity (see below). I have met such masters. Manifestly they could immerse themselves in the otherly realms of Ayahuasca. Yet, when the situation demanded, they could immediately shift gear and take charge of affairs that presented themselves in the real world (e.g. attending a person undergoing tough moments or chasing away a dog that suddenly entered the terrain in which the Ayahuasca session was taking place). The shaman would take care of what had to be taken care of, and then resume his seated position, close his eyes and return to the other worlds of the Ayahuasca experience.

I will close this section with some comments on a factor that is critical in the Ayahuasca state of mind and which has already been alluded to. I refer to fear. Notoriously, the Ayahuasca experience can be a most frightening one. This is especially marked with novices, and in fact, one of the key features of long-term experience with the brew is the mastery of fear. Expectedly, the more experienced one is with Ayahuasca, the less the likelihood that one will be frightened by what one sees. By no means, however, does this mean that fear is always conquered and that experienced drinkers cannot undergo very frightening moments. Indeed, in the syncretic sects in Brazil on many occasions I have seen, just before the beginning of sessions, people cross themselves and pray in a very solemn and at times austere manner.

One has to master one's fear but at the same time one has to beware not to fall into the trap of vanity, of hubris. In the different contexts of Ayahuasca use, fear and vanity are very commonly singled out as the two 'big enemies', the two big pitfalls of which one should be aware.[3] Fear is, of course, natural; yet, it may generate positive feedback (hence negative, detrimental effects) whereby the more afraid one is the more horrible the Ayahuasca experience may be felt to be. As a consequence, one is also more likely to suffer physical discomfort and malaise. The way to handle frightening visions is precisely this—not to be afraid. I learnt this in my first powerful Ayahuasca vision: confronted with the ominous dragons, I realized that as long as I mastered my fear nothing bad would happen to me. When my attitude towards the dragons was one of benevolent acceptance, I found out that there was nothing frightening about these creatures. I have heard very similar reports from many other persons.

But once fear has been mastered, there is the danger of vanity and overconfidence. Indeed, appreciating that one can manage frightful situations, one is prone to plunge into haughtiness. This can render one less careful, less prudent. One may drive oneself beyond the limits of one's capabilities, and hence increase the likelihood of fall and even collapse. Thus, again, the moral is that a golden line of a balanced path has to be found. Indeed, have not the great sages of various cultural traditions singled out moderation as a general recommendation for a wise, hence happy, life?

Obviously, fear is neither pleasant nor desirable. Yet, some degree of fear may actually be beneficial, for then the sense of reality conferred on the visions is likely to increase. Indeed, it seems to me that the remarkable, dramatic quality many people experience in their first Ayahuasca sessions might be attributed precisely to the fear the inexperienced drinker may have of what is happening. The advised path that seems to suggest itself is a healthy balance between humble awe and valiant courage. This balance parallels the one suggested above between immersion and critical distance. One would be best advised to take the visions as real, interact with them, and when the situation presents itself step inside the visionary world. At the same time, however, one should always remember that one is an autonomous self who can have considerable influence on what goes on during the Ayahuasca session. With this, and in line with Eastern spiritual traditions, it is perhaps advisable to always keep a part of oneself functioning as an observing witness.

[3] It is interesting to note that Castaneda (1974) distinguishes between four enemies in the path of the 'man of knowledge': fear, clarity (which is akin to the vanity indicated here), power, and old age.

The Use of Energy

How can the special energy Ayahuasca affords be used? Again, this question may be likened to that confronted by a person who suddenly gains or inherits a large sum of money and who has to decide how to spend it. Following the analogy, and on the basis of my own firsthand experience with Ayahuasca, let me place myself in the position of a hypothetical financial adviser. Or rather, a financial adviser who is at the same time also a good psychological counsellor.

Perhaps the best advice that can be given is identical to that which is likely to be given by the financial-psychological adviser: 'Spend the resources allotted to you on what is really meaningful to you!' The heightened level of energy that Ayahuasca affords one is so special that, in principle, anything can be done with it—hence, indeed, one had better be prudent. The fact that the time-span during which this energy is available is relatively short, coupled with the fact that, for most people, the opportunity of partaking of the brew is limited, makes this basic recommendation even more poignant.

The financial adviser may also recommend saving or investment. On the basis of my personal experience with the brew, I can attest to an analogous pattern with Ayahuasca. As explained in Ch. 19, visions are extremely affected by drinkers' state of being and frame of mind. If one wants to have good visions one should not expend one's resources when one is not yet ready for them. Rather, one had better build up energy and make use of it only when one is up to it. Personally, I have discovered that in the *Daime* context I have often found myself confronted with the choice either to close my eyes and have visions or keep them open and go on with the dancing and singing (I needed to keep my eyes open in order to follow the text of the hymns). Inasmuch as I wished to have wonderful visions, I opted for the latter; at least, for the time being, until the moment came and I was ready. When I felt I had, indeed, collected enough energy I would draw aside, close my eyes, and let myself be seduced by the enchantment of the visions. The delay, and the collection of energy, were major contributors to the meaningful, rewarding visionary experiences that followed.

This very choice also has an interpersonal dimension. Withdrawing to a bench at the back of the hall so as to enjoy my visions would imply leaving the lines of people dancing and singing. Staying in this line and going on with the dance and the song is not easy (the festive *Daime* sessions last an entire night), and the presence and co-participation of other persons serve as a crucial support in this respect. Leaving the line is, in a fashion, not being loyal to one's companions in an arduous enterprise which is, in many important respects, communal. Thus, even though my erstwhile reason for the participation in these sessions was primarily focused on my own personal enjoyment of the visions, during the sessions my values changed and I felt obliged (in the good sense of the term) to the people I was with. By no means was this a sacrifice. Definitely, the one who profited most from this was me, but in a different way from what I had envisioned beforehand.

Under the Ayahuasca intoxication people often discover what is really important to them; which, incidentally, may be different from what they would have deemed as such before the experience. I, for one, am extremely fond of visions and most interested in them. Yet, in the course of *Daime* sessions many times I appreciated that singing *Hallelujah* was no less meaningful and important for me. I am not a member of the Santo Daime Church (and never have I had the slightest inclination to become one) nor am I a Christian. Yet, with the *Daime* I have discovered the extraordinary benefit of the singing of the *Hallelujah*. I am not expressing any religious dogma here. Rather, I am referring to straightforward observations I have had with regard to my personal well-being. In the light of these, and even though I cherish the visions so much, on many occasions I have opted to keep my eyes open and go on with the singing.

Let me add that singing is an optimal way to deal with the unpleasant bodily discomfort that Ayahuasca often induces. As told in Ch. 19, I have discovered that instead of vomiting I could sing. Quite literally so: instead of throwing up, I pour out melodies.[4] An episode related to this is the following. Once during a *Daime* session I felt I had to urinate, but I was apprehensive about leaving the hall. What I did instead was concentrate on the liquid that accumulated in my bladder and direct it upwards. In my vision I channelled it towards my head, and then it all came out from the top of my skull emerging as a wonderful multicoloured fountain. In doing this, I ceased to be bothered by my bodily needs.

Having visions, on the one hand, and singing and dancing, on the other hand, are not the only options for the investment of the Ayahuasca energy. There are all sorts of other things one can do with the special powerful energy that the brew presents one with. Amongst these are the activities of healing, storytelling, discussing intellectual matters, and observing nature and other people. Together, all this marks the pragmatic, performative facet of the Ayahuasca experience; I shall return to the theoretical consideration of this later in the chapter.

Considerations pertaining to the channelling of energy are also encountered with respect to memory. Usually, I remember my visions and these memories remain with me for a long time. Due to my professional interests, I have a special penchant for remembering what I see and experience. Yet there were occasions that it was clear to me that I had to make a choice—if I really wished to undergo the experience presenting itself to me, I would have to forego my future recollection of it and give up any thought of ever talking about it. It was around the time that my core corpus was sealed that I changed attitude and more and more often opted not to remember (which does not necessarily mean that I did not in fact remember).

A similar trade-off is encountered between delving into the Ayahuasca world and maintaining a critical attitude towards what is happening to one under the intoxication. As mentioned in Ch. 10, on several occasions in the course of the

[4] This pertains not so much to the *Daime* context, in which the participants are all expected and encouraged to sing, but to other contexts in which this is not so.

Ayahuasca inebriation I experienced this same dialectics as is manifested in the historical and the cultural context. Specifically, in a very direct fashion I came to appreciate the conflict between two modes of being in the world, between two modes of knowledge—the intuitive one in which the knower is directly related to the things known, and the analytic one founded on a distance between the knower and the known. The latter is that of my rationalistic and scientific-minded Western culture; the former I associated with Ayahuasca and its indigenous users. On the one hand, I appreciated the advantages our culture presents. In particular, I felt, directly intuited knowledge has to be experienced by each person, and each generation, *de novo* on their own—it cannot be passed and transmitted out of context in an 'objective', packaged manner. Yet, very poignantly, I realized how limited the scientific approach is. It was evident to me that pursuing this stance, there are realms of knowledge that can never be attained. I further comprehended that there are levels of knowledge that demand one to let go and relinquish all critical, distanced analysis. Otherwise, the knowledge in question could simply not be reached. Different persons differ in the extent to which they can maintain the analytical stance, but there always comes a stage where this can no longer be done. In this respect, despite all its limitations in terms of sociological power and cultural permanence, the indigenous stance has the upper hand.[5]

The negative side of the energy afforded by Ayahuasca should also be noted. By and large I have spoken of this energy primarily in positive terms, but one should always remember that if one attempts to resist the energy, one's experiences can turn out to be quite awful. The situation is similar—in fact, identical—to that encountered in surfing on the sea. High waves are ideal for great surfing; however, if in the midst of her being in the water the surfer suddenly decides that she does not wish to surf anymore, then the encounter with the waves can be disastrous. So too with Ayahuasca.

Transcendence

As in the previous chapter, here too I comment on a factor that extends beyond the cognitive-psychological discussion proper, namely, transcendence. In earlier chapters, transcendence and spiritual uplift were indicated as specific effects pertaining to the phenomenology of the Ayahuasca inebriation. Here I would add that transcendence—or rather, the quest for transcendence—can have a crucial pragmatic contribution in that it sustains the momentum of the experience. Earlier, I pointed out the crucial importance of the process of letting go. This cannot be achieved in a vacuum. It should be grounded in some fundamental sense of trust. Since in the non-ordinary state of the Ayahuasca inebriation reliance on one's own powers and resources is likely to be of no avail, one had rather place one's trust in something that extends and transcends one's own existence.

[5] Significantly, I have reached analogous conclusions in my study of the different modes of cognition—the *representational* and the *presentational*; see Shanon (1993*a*).

It is not the case that all drinkers, at all times, experience transcendence. But it does seem to me to be the case that powerful and meaningful Ayahuasca sessions are bound to involve such a sense. This subjective experience is also the best measure possible to combat fear, to protect one from vanity, and to secure one from the hazard of losing one's mind. Cognitively, I might venture the hypothesis that in reaching levels of consciousness higher than the ordinary ones (in particular, $Consciousness_4$ in the terminology of Ch. 16) *ipso facto* the special energy discussed in this and the previous chapter is manifested. With this, one may feel that the Muses descend upon one. I shall return to these ideas towards the end of the book.

A Parable

By way of summing up the foregoing discussion of the dynamic aspects of the Ayahuasca experience, let me recount a parable once presented to me in a vision. The intoxication was very strong and I was in the midst of a powerful matrix of interlacing, seemingly hyper-dimensional abstract patterns. The predominant colours were blazing red and shining yellow. It all felt like a transcendental fire. At one point I opened my eyes and then quickly closed them. I noticed that whatever was actually seen by my eyes when they were open was incorporated into the vision and was immediately transformed into a pattern within the glowing geometric matrix. For a moment, the colours of that external percept were reflected in the vision but it was all very short, and very soon the red and yellow prevailed as before. All this, I realized, was just as it is in real life with fire. A flame is nourished by whatever stimulus is presented to it. If a branch of wood is thrown into the fire, for a short period the contours of the branch will be seen glowing in the flame. Very quickly, however, the wood will be part and parcel of the fire. In the context of a disaster, as when a blaze threatens a forest, this branch will no longer be one of its brothers and sisters the trees that are about to be devoured; rather, it will be an indistinguishable part of the flames that devour them.

The vision of the fire could be viewed as a parable explaining the general dynamics of Ayahuasca visualizations. These can be influenced by whatever stimulus—perceptual or ideational, affective or social—that is presented to the person during the inebriation. If the drinker opens his or her eyes and sees a flower, flowers may be seen in his or her visions; if music is heard, the visions are likely to move by the rhythm of the melodies; if one is heavily preoccupied with thoughts about some particular topic, the contents of the vision may be affected accordingly. All these, however, are secondary. As in the case of the (real) fire, the primary thing is the fire itself, not the elements that contribute to it. These elements do have their effect, but in perspective, they are only local contingent perturbations. For the fire to remain alive, combustibles are needed, but as long as they are in supply, the particular identity and form of these resources is immaterial.

This ties in very well with the ideas regarding the mental *pas de deux*. Consider a fast dance. Suppose the dancers encounter an obstacle to their movement (for instance, a child suddenly runs through the dance floor, or it becomes too hot and humid in the hall). If they are experienced, the dancers will rapidly respond to the obstacle (they will hold the child's hand and lead him away, or take off their sweaters or shirts) but the flow of the dance will be only minimally affected.

All this should be borne in mind when the contents and themes of Ayahuasca visions are discussed. Very early in my exposure to Ayahuasca I was told by many that one has to learn to distinguish between what is intrinsically of the brew and what is due to one's own contribution (similar comments made by indigenous users of Ayahuasca are reported in Taussig, 1987). The former is essence, the latter is projection, interference, or even noise. On many occasions, I asked for clarification and instruction— 'How does one distinguish between the two?' 'In time, you will learn,' I was repeatedly told. Indeed.

The Pragmatic Perspective

Up to now, actional, performative aspects of the Ayahuasca experience have been highlighted. Here, I will consider these from a more general theoretical point of view. In particular, following a terminology employed in contemporary cognitive science, I propose that the Ayahuasca experience may be approached from a *pragmatic* perspective. The epithet 'pragmatic' is to be contrasted with the epithet 'semantic'. In linguistics, semantics is the dimension (or discipline) concerned with the meaning of verbal expressions; pragmatics is that concerned with the use of these expressions (see Morris, 1938; Lyons, 1977). A major debate in the modern philosophy of language is whether meaning can be characterized solely in semantic terms or perhaps the phenomenon of meaning cannot be accounted for without taking into consideration the specific context in which linguistic activity takes place. Wittgenstein (1953, 1958) argued that language gains its meaningfulness not because sentences reflect some conceptual or ideational structure of which they are the overt articulations, but rather through a manifold of action in the world. Following these insights, modern students of language have argued that language should be investigated from an action-based, performative orientation (see Austin, 1962; Searle, 1969).

While the terms 'semantics' and 'pragmatics' originated in the study of language, they are extendible to cognition at large. In fact, the major theoretical debate in contemporary cognitive psychology regarding the representational view of mind is, to a great extent, a conflict between the semantic and the pragmatic perspectives as applied to the study of mind. Representationalism sets itself to account for behaviour (not just verbal behaviour) by reference to internal representations of knowledge that are stored in the mind; as noted in Ch. 2, these are referred to as 'semantic representations' (see Fodor, 1975; Pylyshyn, 1979). By this view, the basic capability of the human cognitive system is the manipulation of symbols by means of computational operations applied upon the mental representations. Against this establishment view, several independent dissenting theoretical lines argue that the basic capability of the human cognitive system is action in the world and that mental performance is achieved by means of operations akin to those encountered when people function in the external world. Especially to be noted are the work of Vygotsky and his followers in the Soviet school of activity theory (see Vygotsky, 1978 as well as Wertsch, 1981, 1985) and the ecological school of Gibson and his followers. My own work in cognition also subscribes to the pragmatic orientation (see Shanon, 1993*a*).

It seems to me that many facets of the phenomenology of the Ayahuasca experience favour an action-based account. As already pointed out, what is experienced in the course of the Ayahuasca inebriation is not just something that 'happens' to one as the result of ingesting a psychoactive agent. Rather, in important respects what takes place during the course of the inebriation is a function of the drinkers' interaction with their own visions. The inebriation makes people not only see or hear things they normally do not, it also enables them to act and perform in distinct, non-ordinary manners (in Ch. 13 I used the term 'role-play'). Under the intoxication people both act in ways they usually do not and reach levels of performance that significantly exceed those they normally attain. The philosopher of language Austin (1962) claimed that we do not just *say* things with words— rather we *do* things with them (saying being one of these things). My work on ordinary consciousness has led me to posit that with the silent mentations in our minds (i.e. thought sequences) we do not entertain thoughts but rather do things and act in the theatre of our minds (see Shanon, 1998*b*). I have further argued that what consciousness affords is a kind of virtual reality whereby human beings can act even when actual action in the external world is not possible. My claim has been made on the basis of ordinary consciousness. In the case of non-ordinary consciousness the case is even more extreme. I would like to propose that with Ayahuasca the human propensity of world creation is increased manifoldly. Pursuing the pragmatic perspective, I would say that under the intoxication, drinkers act in an extremely creative, non-standard fashion. Imagination is one such activity, but, as we have already seen, there are other activities—singing, dancing, storytelling—where the performative aspects are more manifest and even publicly observable.

A Last Comment on Control and Mastery

Closing this chapter, I would underline a main moral of the foregoing discussion, namely, how significant may be the contribution drinkers may have with regard to what their experience and how substantial can be the influence they may have on it. Novices usually feel that they do not have any control of what happens under the intoxication. Apparently, the Ayahuasca experience is so foreign and so overwhelming that the natural conclusion to be made is that one has no control over what is happening. With time, however, many Ayahuasca drinkers come to appreciate that, in fact, they can have a significant influence. Naturally, one cannot, by mere wish, stop the intoxication. One has imbibed a powerful psychoactive agent which has its own time course—this one cannot change. Granted this, however, the range of control a drinker can have is quite substantial. Again, I would liken the situation to that encountered in the driving of a vehicle, especially a fast one. If one is flying an airplane one cannot simply, at a whim, declare a pause and jump out. This does not mean, however, that one cannot have any control on the course of the flight. And surely, the more experienced one is as a pilot, the greater one's navigational control would be. Eventually, one could even do aerobatics. The situation with Ayahuasca is identical. As explained in the previous chapter, the brew places one in a very powerful energetic state. It is upon each individual to find the right manner of handling this state and navigate within it. In general, if one is to fight the power one risks the danger of a breakdown. But if one accepts the situation one can comfortably swim in it, surf with it. Then, one is very likely to find the experience both enjoyable and extremely gratifying (for a general discussion of the control shamans have over their trance experiences, see Peters and Price-Williams, 1980). All this may be summed up in the wise words of a good friend of mine, a person with extended experience with Ayahuasca—'partaking of this brew is an art to be mastered'.

22

A General Theoretical Perspective

> [I have appeared] to you in several forms. At the sea I appeared to you as a warrior engaged in war. . . . At Sinai I appeared as an elder teaching Torah. . . . At the Tabernacle I appeared as a bridegroom entering his nuptial chamber.
>
> The Jewish *Midrash*

Preliminaries

Our charting of the phenomenology of the Ayahuasca experience is now complete. First- and second-order, dynamical, and contextual questions have been examined. I turn now to the theoretical cognitive-psychological level. Taking a broad, unified perspective, I shall attempt to make sense of the special state of consciousness that Ayahuasca induces. What happens to the mind under the intoxication? How should the visions and ideations experienced with Ayahuasca be conceived?

Before presenting my own view on the matter, I shall review the various theoretical accounts that may be invoked by way of explaining the Ayahuasca experience—accounts proposed in the scientific literature (or rather, accounts phrased in a terminology like that proposed in the literature on altered states of consciousness) as well as ones suggested by lay persons, both traditional and contemporary, partaking of the brew. In considering the various theoretical accounts and in dismissing them one by one, I shall prepare the ground for the particular account that I favour.

Non-Naturalistic and Spiritualistic Accounts

The first approach I shall consider is that which attributes what is happening over the course of the Ayahuasca inebriation, most notably the visions that the brew induces, to an independent, non-ordinary reality. This reality is said to be different from the ordinary one but just as real. I present this approach first, for this is the one held by the original users of Ayahuasca, the indigenous peoples of the Amazon (see e.g. Harner, 1972; Langdon, 1979*a*; Chaumeil, 1983). As we have seen in earlier chapters, these people believed that the brew enabled them to displace themselves in an immaterial fashion and allowed them to travel across time and into realms that transcend the natural, thereby bringing them to the world of the spirits. Thus, they maintained that what is seen in the visions exists—in other places, other times, other realms. In particular, they interpreted visions of persons and places in terms of reincarnation and paranormal travel. Similarly, they regarded ideations entertained under the intoxication as veridical apprehensions of objective truths and/or as messages originating from external sources. All the new syncretic sects using Ayahuasca endorse these credences as well. Indeed, on the basis of the interviews I have conducted, my assessment is that, at least to some extent, practically all who have had more than rudimentary experiences with the brew hold similar beliefs. As indicated earlier, in the various non-indigenous contexts the realm of non-ordinary reality is referred to as the astral; henceforth, I shall use this epithet to refer to all non-naturalistic accounts of the Ayahuasca experience, whatever the contexts in which they are invoked.

If the astral is to be conceived as a realm endowed with some sort of concrete, quasi-physical reality (as contrasted with abstract, ideational status) which is not dependent on the minds of human beings, then I categorically object to it. I do not believe that there are physical places other than those in the ordinary natural world, I do not believe that there are beings and creatures just like us who reside elsewhere in other realms, I do not believe in reincarnation and paranormal travel.

Many fellow travellers upon the Ayahuasca path, some of them very close friends, often reproached me for not sharing their beliefs in the astral. In my sceptic stance they saw the expression of dogmatic scientificism and rigid unwillingness to be open and accept what is beyond the reach of the contemporary established Western world-view. Still, as explained in Ch. 16, despite numerous attestations I have heard from many persons, I do not feel that I have encountered solid evidence that convinces me of the paranormal actuality of the astral. As I have already indicated several times in previous chapters, I recognize that experientially, under the intoxication people do, indeed, feel that what is seen in the visions are wondrous realms that have actual, independent existence. I have experienced this firsthand. However, I am a Western university professor, a psychologist, and a philosopher, not an Amerindian shaman nor an adherent of any Afro-Brazilian cult. Thus, acknowledging the enchanted nature of the Ayahuasca experience, I am trying to account for it while at the same time respecting frames of thought and canons of judgement that define my own cultural and professional heritage. Admittedly, not always is this easy.

However, there might be other conceptualizations of mind-independent realities that are less contaminated by spiritualistic and/or paranormal assumptions. In the common philosophical terminology, such conceptualizations are labelled idealistic or realistic. In the Western philosophical context, that which first comes to mind is Platonism. This is the metaphysical view postulating a mind-independent realm which, essentially, is that determining the meanings of all things. Paradigmatically, this realm is populated by Ideas as well as the objects of mathematics. Since Platonistic realism involves the postulation of realms of reality that are not physical, most modern-minded scientists feel rather ill at ease with it (although, it seems that many mathematicians do endorse it, see, for instance, Penrose, 1990,

1994). Furthermore, since this option takes us outside the scope of psychology and leads us to that of metaphysics, there is not much I can say about it in the context of the present cognitive discussion. (I shall return to it, though, in the next chapter, that devoted to philosophical reflections and speculations). There is, however, one implication of realism that should be noted here: if realism is adopted, then in principle, it is possible that the information entertained under the intoxication is independent of the Ayahuasca drinkers and is not created or invented by them.

So much for the paranormal and the spiritualistic. I shall turn to science and consider the various explanatory options that it may offer. As shall be readily seen, I have my qualms with many of these as well.

Biological Reductionism

Obviously, the Ayahuasca experience is the result of the ingestion of specific chemical substances that have been demonstrated to have specific effects on human brain physiology. It is therefore most natural to attempt to explain this experience in terms of brain biochemistry and neurophysiology. By this approach, whatever happens in the course of the Ayahuasca inebriation is fully explainable in terms of the biochemical and neurophysiological effects that result from the pharmacological action of the substances at hand.

The causal relationship between psychoactive substances and the psychological effects encountered in the course of the Ayahuasca inebriation is incontestable. In itself, however, this relationship does not imply that biochemistry and neurophysiology offer the proper level for explaining the psychology of the Ayahuasca experience. As explained in Ch. 2, I do not subscribe to biological reductionism. As further explained, my position reflects my general approach to psychology and the study of mind and it is totally independent of my study of Ayahuasca. In general, I do not believe that understanding of the human mind is to be achieved by means of the theoretical framework of biology (or chemistry, or physics).

My view of psychology is grounded in the belief that psychological phenomena should be explained by theoretical means that are genuinely psychological. I do not deny that for human beings and other organisms to behave and cogitate the brain and the nervous system are necessary. Of course they are. However, this does not mean that neuroanatomic and neurophysiological accounts can in themselves serve as explanations in psychology. I think they cannot. Scientific explanation concerns lawful regularities. In principle, I think it is hopeless to look for lawful regularities between semantic patterns and intentional states on the one hand, and specific brain structures and patterns of activity on the other hand. A simple and clear argument against reductionist explanations is that offered by Polanyi (1968); the argument is based on an analogy with the meaning of written texts. Polanyi points out that it is ludicrous to analyse the meaning of a text in terms of the chemical constitution of the paper and the ink in which it is printed. Indeed, it is also futile to do so in terms of rigid geometric analysis of the script. Strictly speaking, the shapes and forms of the letters can vary greatly without there being any alterations in the meanings that they convey. Those changes that do affect meaning are precisely those that are pertinent on the cognitive level. There is absolutely no direct correlation between these and variations on the lower levels of physics and chemistry.

Fundamental to my anti-reductionist stance is the appreciation that the goal of all scientific pursuit is the definition of lawful regularities. In doing so, each scientific discipline has to demarcate that level on which regularities pertinent to its perspective of interest are to be found. With this, distinctions are made and categorizations drawn so that groupings of things which, from the perspective of the discipline in question, are regarded as being the same are established; technically, these groupings are known as 'equivalence classes'. The equivalence classes vary with the disciplines: what the physicist or the biologist regards as different, may not be deemed so by the psychologist who attempts an understanding of the human mind. In fact, it is generally the case that patterns of behaviour that from a cognitive point of view are the same are actually generated by means of different brain patterns of activity. As I see it, the subject matter of scientific psychology is the definition of lawful regularities in the forms and contents of patterns invested with meaning (for a similar view, see Searle, 1980). The present study of Ayahuasca follows this general orientation. For further arguments against reductionist explanations in psychology, see Putnam (1973) and Fodor (1975).

The foregoing objections to biological reductionism were based on arguments that are not at all specific to Ayahuasca. In the context of Ayahuasca, further, specific objections can be invoked. Since these relate to the issue of cross-personal commonalities, which is the main topic of discussion in the next chapter, I defer their presentation to there.

Psychological Accounts Based on Unconscious Processing

General Considerations

Having dismissed biological reductionism, I turn to psychological accounts. There are no theoretical cognitive-psychological accounts of Ayahuasca yet—that attempted here is a pioneer. However, some such accounts have been offered in conjunction with other psychoactive substances. There are not many such accounts in the literature, and most were proposed in conjunction with LSD. This was done in the period when LSD research was still legal, that is, before the mid-1970s, hence historically before the advent of modern cognitive science (especially to be noted in this conjunction is the extensive work of Grof (in particular, see Grof, 1994) as well as that of Masters and Houston, 1966).

Almost all psychological accounts of altered states of consciousness in the literature are based on the postulation of some sort of unconscious processing that the non-ordinary state is purported to uncover. The recourse to unconscious structures and processes is standard in modern psychology. Practically all theoreticians in the field maintain that, by and large, human information processing is achieved without any involvement of consciousness. Hence, in effect, the view of mind assumed by contemporary cognitive psychology is that most of our cognitive life is unconscious. I do not subscribe to this view (for a general discussion as well as a presentation of my own position, see Shanon, 1998*b*). Again, my reasons pertain to my general view of mind, not to the study of Ayahuasca. My critique of the unconscious involves theoretical considerations whose exposition lies well outside the scope of the present investigation. Here, let me just point out that it is a direct result of my critique of the representational-computational view of mind (which defines the establishment view in cognitive psychology today).[1]

My critique led me not to accept the postulation of covert underlying mental structures and processes. But then, if explanation in psychology consists not in the modelling of mind by means of underlying computational operations, what else can it be? The answer I have come up with is that what is left for the psychologist to do is the systematic study of the surface, so to speak, and the establishment of lawful regularities in it. This is tantamount to saying that for me, the domain of the psychological coincides with that of conscious experience.[2] In this domain, the unconscious does not exist.[3] Like William James (1890/1950), I maintain that mental activities and processes are conscious, and they cannot be achieved outside of consciousness. It is in the light of this fundamental theoretical conclusion that I try to understand the Ayahuasca experience.

Parenthetically, let it be noted that in objecting to the postulation of the unconscious I am not claiming that we human beings know everything about ourselves. Undoubtedly there is much that we do not know. However, not everything that is unknown should be labelled unconscious. I do not know how my liver functions, but the functioning of my liver is not an unconscious process. In invoking unconscious knowledge, the psychologist postulates the existence of psychological material that is essentially of the same type as material that *is* conscious, only that it is not. For instance, he would postulate representational structures that are in all respects like the words and expressions of articulated natural language, with the sole difference being that they are not articulated. In the spirit of Wittgenstein (1953, 1958), I deny the existence of such entities.

Lastly, my critique of the representational-computational view of mind also leads to a non-orthodox stance as to what the basic capabilities of the human cognitive

[1] Technically, the critique presented in Shanon (1993*a*) focuses on orthodox representationalism, which defines mental representations in terms of well-formed symbolic structures. More recently, the paradigm of connectionism has emerged which does not presuppose such well-formed structures. Connectionist models, however, still assume that cognitive activity is achieved by means of computations operating in a covert, underlying substrate. Thus, disregarding specific details, in principle, my critique of the representational-computational view of mind applies to these models as well.

[2] This also includes what can potentially become conscious.

[3] I say 'in this domain', because I do accept that there are activities in other domains of which we are not aware, and that the psychological is enabled by the biological. The underlying biological activities, however, are outside the psychological domain, hence the attribution of consciousness (that is, characterizing them as either conscious or unconscious) does not apply to them.

system are. Almost all contemporary frameworks in cognitive science today assume that the basic capability of this system is the processing of information (exceptions are the Gibsonian school of ecological psychology and the psychologist Vygotsky and his followers in the Soviet school of activity theory). In contrast, my view is that the basic capabilities of the human cognitive system are acting-in-the-world and the generation of novel articulations. Rather than being an information processor, the mind is the creator of novelty. Obviously, novelty and creation are not solely the privilege of especially talented persons—they are there with all of us every single moment of our existence. When we utter sentences we do not—as the orthodox representational-computational view of mind would have it—produce expressions that are the reflections of underlying internal semantic representations and are generated by means of covert computational processes. Rather, talking is a *de novo* creation, there in the world, of something that does not correspond to anything on an underlying, hidden level. As will become apparent both below and in the next chapter, in my view of the Ayahuasca experience creativity is central.

Ayahuasca and the Unconscious

From the general comments regarding the unconscious I now turn to Ayahuasca. The recourse to the unconscious in accounting for the Ayahuasca experience can be of three types, each associated with a different conceptualization of the unconscious—the cognitive informational, the Freudian, and the Jungian. (Again, let me note that accounts of all three types are encountered in the literature about altered states of consciousness but not specifically for Ayahuasca).

The *informational unconscious* is that assumed by cognitive representationalism as discussed above. Explanations in terms of the informational (or cognitive) unconscious posit that, under the effect of the psychoactive agent, information that is usually not accessible to consciousness is revealed. Unlike the unconscious postulated in psychoanalytical theories, the informational unconscious does not involve hidden wishes, desires, or threatening conflicts, nor does it involve any defence mechanisms or censorship. Rather, the reason for its being unconscious is a functional matter: In the ordinary state of mental function, this information is not accessible to consciousness, in the non-ordinary state induced by the psychoactive agent it is. Technically, this may be attributed to a release from inhibition, a shifting of the balance between externally oriented and internally oriented processing or to the failure of screening and control mechanisms; in addition, some concurrent excitation (e.g. on the sensory level) may be said to be operative. Proposals along these lines were offered by West (1962, 1975); Jarvik (1970); Hartmann (1975); Horowitz (1975); Asaad and Shapiro (1986); Asaad (1990); for a recent review, see Bentall (2000).

The explanation in terms of the *Freudian* unconscious is similar to that in terms of the informational unconscious, only that in addition, it is assumed that the unconscious psychological material is held in that status because it exhibits some special qualities—notably, its content being menacing and subject to censorship by the ego. That the psychological material is unconscious is not a functionalistic matter (as in the case of the informational unconscious) but rather, is the result of active censorship. In altered states of consciousness, the force of censoring is assumed to be reduced. Consequently, the unconscious material is uncovered (for instance, see Savage, 1975; Hartmann, 1975).

Unlike both the informational and the Freudian unconscious, that postulated by Jung (1969) is not personal but *collective*—it pertains not to the realm of the individual person, but rather to that of the human species at large. Associated with this unconscious is psychological material that is very specific. This material—the archetypes of the collective unconscious—reflects the general predicament of human life and the history of the species. More specifically, the Jungian archetypes may define stages in people's lives (e.g. the Youth, the Old Man), major events they may undergo (notably, the different stations in the Hero's journey), and roles they may play (e.g. the Trickster, the Sage). In non-ordinary states of consciousness the archetypes come to the fore and are expressed in specific forms; artistic creations are also such expressions.

Counter-Arguments

Since in general I do not endorse psychological explanations based on the postulation of unconscious structures and processes, in the context of Ayahuasca I do not endorse such explanations either. In the case of Ayahuasca, however, I have additional, more specific reasons for my stance.

The first counter-argument I would like to raise applies to all explanations invoking the unconscious, regardless of the particular theoretical framework in which they are couched. If the primary effect of the psychoactive agent were to be the release from inhibition and the bringing to consciousness of what normally is unconscious and therefore hidden, then there should be no specificity in contents associated with the various non-ordinary states of consciousness people may be subject to: since the unconscious is one and these various states all uncover the information represented in it, the particular contents associated with all should be the same. Factually, it appears that this is not the case. The contents manifested in different drug-induced states of consciousness vary, and, in turn, they differ from the contents presented in dreams. More on this last point will be said below.

I now turn to considerations that specifically address the different conceptualizations of the unconscious indicated above. Essentially, the main cognitive process associated with accounts based on the informational unconscious is that of release from inhibition; enhanced activation and sensory acuity may be proposed as well. This account is, I believe, extremely limited: the rich phenomenology of the Ayahuasca experience we have surveyed just cannot be straightjacketed to these factors. Most of this phenomenology is simply left unaccounted for, indeed—unaccountable, in terms of them. In particular, I highlight four facets of the Ayahuasca experience for which this is the case—the overall *atmosphere* and general feelings, enhanced *meaningfulness* and the new modes of seeing and interpretation associated with it, modifications in *consciousness*, the identity transformations and temporal modifications that they entail, and non-ordinary *performance* abilities. More important of all, the informational account ignores the *creative* aspect of the Ayahuasca experience. Its main theoretical conceptualization is information processing. Under the intoxication, much more than information processing seems to take place. Rather than reveal what has hitherto been hidden, Ayahuasca seems to induce creativity and the generation of novelty. As indicated above, my personal stance is that even ordinary cognition cannot be accounted for in terms of strict information processing—there too generation is fundamental. With Ayahuasca, this is all the more so.

Second, consider the Freudian approach. By this approach, Ayahuasca would be thought of as bringing to the fore what is normally hidden in the depths of one's psyche. Therefore, we should expect that what is to be experienced with the brew will be the reflection of drinkers' conflicts and concerns, wishes and hopes. Empirically, this is usually not the case. Usually, the contents of Ayahuasca visions do not depict the particular life histories of the individuals in question nor do they seem to be related to them—either in a direct or symbolic fashion. Furthermore, there is so much in the Ayahuasca experience that does not seem to have anything to do with the personalities and life histories of the individuals reporting them.

Further supporting the last claim is the manifest difference between Ayahuasca visions and dreams. An independent analysis I have conducted comparing the corpus of my Ayahuasca visions with a corpus of my dreams, with those reported by Freud in *The Interpretation of Dreams* (Freud, 1900/1953), as well as with summary content analyses of dreams in the literature (Hall and Van de Castle, 1966) shows that the distribution of contents in Ayahuasca visions and dreams are quite different (for details, see the Appendix). In fact, there is almost no overlap between the two. In the main, the contents of dreams are personal, reflecting dreamers' concerns, conflicts, and wishes, as well as their recent past (details of the past day or two are especially common). In contrast, the contents of Ayahuasca visions usually have very little to do with drinkers' individual lives and the states of affairs that they depict are usually characterized as otherworldly and phantasmagoric. Were the effect of Ayahuasca merely a release of inhibition, such differences should not be.

Another argument against the Freudian account has to do with the characterizations of dreams and the dream work. By Freudian psychoanalytical theory dreams are the products of processes of condensation and displacement and they manifest chaos, breaches of logic, and disregard of temporal distinctions and constraints. First, Ayahuasca visions do not exhibit chaos nor are they illogical or paradoxical in the way dreams may be. Certainly what is seen in the visions often defies anything people normally experience in their lives, but this does not make the visions chaotic or illogical. In particular, consider cases of double and transformed identity. In dreams it is quite common to see a person that in all respects looks like one individual and at the same time to know that this person is actually another individual. Such a phenomenon does, indeed, defy the canons of logic and is paradoxical. Superficially, the patterns of seeing-as and transformation of identity encountered with Ayahuasca may seem to be similar, but actually they are quite different. In these cases, the entity or being presented to one is actually seen in a non-ordinary fashion and/or is taken to have undergone metamorphosis. Admittedly, these effects are extraordinary and often experienced as being magical, but they are not paradoxical in the sense that patterns in dreams are. Lastly, consider the temporal dimension. Indeed, as surveyed in Ch. 14, Ayahuasca presents a whole spectrum of patterns that defy the ordinary constraints of temporality. Yet, these patterns are different from those encountered in dreams. In dreams temporal relations may be mixed up so as to produce disorderly patterns analogous to those defying normal reason and logic. In contrast, with Ayahuasca past, present, and future are transcended. Rather than feeling that they collapse into meaningless temporal chaos, drinkers of Ayahuasca feel that they are freed from the dominion of time and reach the realm of the eternal which—as we have seen—is characterized by enhanced meaningfulness.

Prima facie, the situation with the Jungian collective unconscious seems to be quite different. Of all theoretical frameworks in psychology the Jungian one seems to be especially fitting for the analysis of the Ayahuasca experience (see Furst, 1990). There is, indeed, much in common between Ayahuasca visions and the visions investigated by Jung (see Jung, 1976).[4] Yet my assessment is that the Jungian archetypes cannot be directly applied by way of accounting for the Ayahuasca experience. Since in Jung's theory the topic of psychological universals is central, I defer my discussion of this theory to the next chapter, which focally deals with cross-personal commonalities exhibited in Ayahuasca visions and the psychological universals to which they possibly attest.

All that has been said above was critical and negative. The positive question remains: How are students of mind to account for the host of activities that have classically led to the postulation of unconscious processes? I am referring to slips of the tongue, *faux pas*, and other actions that reveal 'hidden' feelings or desires, of which the actor him- or herself is not aware (see Freud, 1901/1960). Again, the issues at stake are complex fundamental ones that cannot be discussed in full here. Let me just say that as I see it, any alternative explanation of the phenomena at hand should involve a radical change in the conceptualization of some of the key notions of psychology, notably knowledge, memory, and time. The common conceptualization of both knowledge and memory in the psychological literature is in terms of engrams stored in the mind (or the brain). Since I do not ascribe to the postulation of covert, underlying internal mental representations, I have to conceive memory in a different fashion (for a rare and wonderful critique of the common conceptualization of memory, see Bursen, 1978). The conceptualization of time has also to be rethought (see Shanon, 1998*a*). Unlike physical time,

[4] A quantitative analysis now in progress corroborates this statement.

psychological time is not a chain of successive temporal atoms. Rather, following Bergson (1944), I think it should be regarded as an unbreakable flow of living. Further, following Gibson (1966*b*), I would say that psychological time has to be conceived in terms of events in the context of meaningful action. Thus, psychologically speaking, the ultimate temporal unit is a person's entire life. At this very moment, I—a person who is now 50 years of age—*am* the sum total of all I have lived throughout all the years of my living. Rather than document-like mental structures stored in my mind, all my memories are *de facto* part and parcel of what I am now, in the present. Had I been different as a child I would have been a different middle-aged man.

This last line of thinking will serve as the basis for the alternative proposal I shall put forth by way of accounting for what is happening with Ayahuasca. However, before presenting this proposal, let me consider other psychological proposals, ones which do not revert to the postulation of unconscious information.

Other Psychological Accounts

As noted, almost all the psychological accounts of altered states of consciousness in the literature involve, in one way or another, reference to unconscious information. The only exceptions to this rule are, first, models that characterize what happens under the effect of psychoactive substances in terms of cognitive impairment. Second is an account published by Merkur (1998), a student of religion, just as this book was nearing completion. As I see it, it is the only one in the literature that actually presents a positive cognitive characterization of so-called altered states of consciousness and which does not involve unconscious information. Third is a model that was not proposed in the psychological literature as such—it is my reconstruction of the view assumed by Reichel-Dolmatoff (1975) in his anthropological study of Ayahuasca.

Cognitive Impairment

The first family of theoretical accounts to be considered here is directly related to those formulated in terms of information processing. These accounts characterize what happens with psychoactive substances as the result of cognitive *impairment*. In particular, the kinds of impairments that have been suggested involve thinking and reasoning, memory and attention, cognitive and motor control (see e.g. Jarvik, 1970; Horowitz, 1975).

Of course, if one wishes to find these effects one will surely find them. But this, I think, misses the whole point about the non-ordinary states of consciousness in question. In investigating the Ayahuasca experience (along with other non-ordinary states of consciousness), one should appreciate what is of essence and what is peripheral. Of essence are the special effects that people seek and for which they partake of a substance that induces them. At times, the intoxication may also induce what an external observer may deem to be cognitive impairments; these, however, are both incidental and peripheral. Especially to the point in this regard, I find, is a comment by an anonymous discussant I encountered on the Internet in a debate on marijuana. In response to scientific studies showing that this substance causes impairment in arithmetic performance, this person commented: 'But this is not what one wishes to do under the effect, this is not the dimension by which one would judge whether the marijuana experience was good or not.' I would add that if, in the midst of a concert performance, investigators would subject the pianist to a test probing his or her arithmetic or memory skills, they would also find severe impairment in performance. Does this mean that piano playing should be characterized as a cognitive impairment? Thus, it seems to me that the modelling in terms of impairment misses much of what is of essence about psychoactive substances, and by implication, about Ayahuasca. It is not that factually what is stated by these models is wrong—what is significant is what is lacking in them. The cognitive impairments may very well be there, but by no means do they capture the essence of the experiences at hand. The rich phenomenology sketched in this book is, in itself, a testimony of this.

I shall add that often the characterization in terms of impairment is grounded in clinical work and is associated with the viewing of psychoactive substances as psychotomimetics, that is mimickers of psychosis (see the terminological discussion in Ch. 1). Psychosis is a disease and a very sad human condition. Psychotic patients are not happy with their condition, and they find neither meaning nor order in it. Further, they feel tormented by the non-ordinary states into which they are plunged, and even enslaved to them. In contrast, Ayahuasca drinkers find meaning and order in the special state of being that the brew induces in them, they feel that they become happier and better persons with it, and when experienced they are masters of the inebriation, not its slaves. Clinical-based investigations are also biased in that they do not deal with experts. Would we judge piano playing by the discordant tones produced by novices? I would further venture that as a rule, accounts by reference to impairment are not based on firsthand experiences with the substances in question. In all likelihood, they also reflect judgemental bias against the states of mind at hand.

Imagination—Merkur's Proposal

Essentially, apart from the Jungian, all the psychological models considered thus far were negative: They characterized the effect of psychoactive substances either in terms of release from inhibition or in terms of cognitive impairment—they did not specify any positive cognitive mechanism as such. The only proposal in the literature that does make a positive specification of a cognitive activity is that recently offered by Merkur. On the basis of the consideration of the literature on LSD, mescaline, and psilocybin Merkur (1998: 3) proposes:

> My . . . contention is that psychedelic drugs induce an alternate state—not restricted to consciousness—that consists of intense fantasying. Depending on the dosage, the fantasies

may or may not reach pseudohallucinatory intensity. Beyond the induction of a state of intense fantasying or, to introduce a synonymous term, a state of reverie, the drugs themselves do nothing. They do not alter perception, or release forgotten memories, or induce psychoses, or anything else. All of the mental phenomena that are produced by psychedelics are fantasies.

I applaud Merkur's proposal for its being—by the terminology coined in Shanon (1993*a*)—genuinely psychological. In other words, it accounts for psychological phenomena by means of terms drawn from psychology proper, not by means of terms adopted from other disciplines such as computer science or biology. Yet, applying Merkur's proposal to what happens under the Ayahuasca intoxication, my assessment is that it is not fully adequate. The characterization of what happens exclusively in terms of imagination is, I think, on the one hand too broad and on the other too narrow.

The characterization is too broad because, as defined in Merkur's text, imagination is a cognitive faculty that allows practically anything—just imagine something, and it will be imagined. Is this all that can be said on the matter? Can we not be more specific? The moral of the entire discussion in this book is that we definitely can. First, we have seen that there are interesting relationships between, on the one hand, drinkers' individual personalities and, on the other hand, what happens to them under the intoxication. Second, the finding of content items, themes, and ideations that tend to appear more frequently indicates that there are intrinsic constraints on the space of all possible experiences with Ayahuasca. Admittedly, this space is infinite, yet it does exhibit regularities and differential patterns—in other words, it is subject to constraints. Third, the state of mind induced by Ayahuasca (along with those induced by other psychoactive agents) is only one of many cases in which the imaginative capabilities of the human mind express themselves. Other such cases are dreams, confabulations, and artistic creativity—all encountered in ordinary life, without any externally induced alteration of one's state of consciousness. Yet, as noted above, the contents of the visions seen with Ayahuasca and of dreams are quite different. If we were simply to say that both are products of the human mind let loose to imagine, we—as students of cognition—would clearly fail in our task, for we would miss those characterizations, or constraints, that specifically define the cognitive faculty at hand. All this is tantamount to saying that, from a cognitive-psychological point of view, the Ayahuasca experience is subject to constraints. The definition of these is precisely what the cognitive-psychological investigation of Ayahuasca is supposed to do.

Merkur's characterization is also too narrow. Imagination is only one of the various alterations of cognition that Ayahuasca induces. Indeed, the foregoing survey of the Ayahuasca experience was precisely this—a specification of these various and multifaceted alterations. In this survey we encountered increased perceptual acuity and sensitivity, acceleration and enhancement of thought, increased intuition and psychological insight, reduced social inhibition and enhanced suggestibility, facilitation of various types of performance, religious, spiritual, and metaphysical sentiments and reflections, enhanced meaningfulness and aesthetic sensibilities, identity transformations, modification in the sense of time, as well as a host of affective changes. Thus, the list is long, and it is not confined to imagination or fantasy.

But perhaps the most crucial criticism of Merkur's proposal is that his conceptualization of imagination (in his words, fantasy) is essentially derogatory. I have given Merkur the credit that his proposal is the only positive cognitive one in the literature. Nevertheless, this proposal suffers from some shortcomings too. Reading the citation above we note that his focus on fantasy is, in the main, by way of contrast. Merkur's primary point is to show that psychoactive substances do not produce all sorts of drastic effects that may have been attributed to them. What they produce is 'only fantasy'. For the cognitive psychologist, however, fantasy (or, in the terminology employed here, the power of imagination) is the very subject matter of interest. Yes, indeed, psychoactive substances may not do all the things that have been associated with them and to which Merkur objects. But saying that it is only imaginative fantasy is, from a cognitive-psychological point of view, just an opening. As far as the cognitive psychologist is concerned, the real work still remains to be done.

In the light of these last remarks, let me take a broader perspective and consider the very notion of imagination. Usually, imagination is conceived of as an active process in which cognitive agents set themselves to engage. With Ayahuasca, this is not the case. Specifically, drinkers are engaged in creative imagining without being consciously determined to imagine, let alone create the specific images in question. Prima facie, imagining without deciding or wishing to do so may sound like a contradiction in terms. However, there are some considerations that may suggest that it need not be.

First, consider dreams. Ordinary nocturnal dreams are precisely this—feats of imagination that take place in the mind without there being an active decision to induce them and without any deliberate involvement on behalf of the dreamer. The dreamer is responsible neither for the fact that he or she is dreaming nor for the specific content of his or her dreams. The very possibility of dreaming shows that the above characterization in conjunction with Ayahuasca is perfectly plausible.

Second, consider accomplished performances such as dancing or the playing of a musical instrument. When immersed in a dance, people can find themselves accomplishing steps and movements that they themselves are surprised they could execute. When it flows, a dance can seem to be above and beyond the dancers who are enacting it. Rather than the dancers executing the dance, it appears that the dance carries the dancers (for an interesting phenomenological account of dancing, see Fraleigh, 1993). *Mutatis mutandis*, the same can be said with regards to musical improvisation. Traditionally, such states of being have been referred to by the expression 'the Muses descend upon one'. But surely, the Muses do not descend upon just anyone, at any moment. They do so only on those who are prepared. Only an accomplished musician can be carried by the Muses so as to play in a masterful fashion without any rehearsal or even prior composition or planning. Not everyone, not always—but yes, for some, sometimes unplanned creative activity which is neither consciously solicited nor consciously monitored *is* possible.

In the previous paragraph, the innovative power of the human mind is underlined. With Ayahuasca, the accomplishments of this power are often stupefying. I poignantly appreciated this at the very moment I had the vision, recounted in Chs. 8 and 9, of an exhibition presenting what appeared to be an entire, unknown culture. I was thinking to myself: 'If this is not real, if my mind is creating all this, then the human mind must be much more amazing, much more mysterious than standardly assumed by psychologists. Indeed, if my mind is creating all this,' my thinking went on, 'then cognitive psychologists just know nothing about the mind.' Thus, to the suggestion that the effect of psychoactive substances is, as Merkur (1998) claims, just 'intense fantasying' I retort: Perhaps, indeed, this is all that is happening, but this should not be taken in a dismissive, half-derogatory fashion. It may very well be that it is the creative ability of the mind but, if so, the mind's ability to create surpasses anything we cognitive scientists ever think of.

The Hybrid, Two-Stage Model

The last account to be considered here is based on a hybrid model consisting of two stages, one biological and the other psychological; these two stages are ordered—the biological precedes the psychological. I have already considered this account in previous chapters in conjunction with the issue of interpretation. In the following discussion I approach the subject from a different angle, but eventually I come to conclusions that converge upon those proposed in that earlier chapter.

The picture proposed by the hybrid, two-stage model is that the chemically stimulated brain generates senseless stimuli which are seen as phosphenes and lack semantic interpretation. Under the intoxication, however, people's imagination is enhanced, and thus they tend to impose all sorts of fanciful interpretations upon the visual stimuli presented before their mental eyes. These interpretative processes define the second (psychological) stage of the model at hand and are responsible for all the visualizations that have well-defined semantic content. The interpretative processes reflect both the psychology of drinkers and their socio-cultural background. Applying this distinction to the structural typology drawn in this book, bursts and splashes of colour and light as well as non-figurative geometric designs and patterns are conceived by the hybrid model to be the direct products of brain activity; all the other visualizations that Ayahuasca produces are taken to be the products of psychological processes of imagination and interpretation. Curiously, the one place in the literature that the hybrid model is spelled out in some fashion is in an anthropological, not a psychological, context; I refer to Reichel-Dolmatoff's (1975) theory already mentioned in Chs. 15 and 21. In this theory, the second stage reflects both the psychology of drinkers and their socio-cultural background.

Examining the two-stage hybrid model from a cognitive-psychological point of view, I shall make three points. First, the two-stage model strictly equates the neurophysiological component with the earlier, and weaker, stages of the Ayahuasca inebriation and the psychological (and socio-cultural) component with the later, and more powerful, stages. Furthermore, the model assumes a necessary tie, on the one hand, between the early stages and phosphene-like geometric patterns, and, on the other hand, between the later stages and figurative, fanciful visualizations. One major conclusion of the present investigation is that empirically such an identification cannot be made. As underlined in Ch. 18, it is not always the case that Ayahuasca sessions begin with phosphenes and other simple non-figurative elements, nor is it the case that patterned geometric designs are always associated with early and weak stages of the Ayahuasca inebriation. Endogenous brain activity may very well constitute an important input to imaginative psychological activity, but the relationship between the two need not be a universally ordered and rigidly structured linear one. Second, it should be appreciated that, psychologically speaking, not much is being explained by the hybrid model. The model establishes a division between the biological and the psychological, but this having been said, it does not offer much by way of a cognitive account. In particular, how are the commonalities in Ayahuasca visions to be accounted for? How are the various non-ordinary effects pertaining to the Ayahuasca experience to be conceived and explained? Third, some more general, and more abstract, theoretical considerations may be invoked against the two-stage model. This model, like many others that have been proposed in various domains of psychology, assumes a clear-cut dividing line between processes deemed simple and ones deemed complex. Furthermore, this division presupposes a clear demarcation line between that which is interpretation-free and that which is subject to interpretation. In contemporary cognitive science, such a division is the hallmark of several theories concerning the architecture of mind which assume cognitive penetrability and modularity (see Fodor, 1983; Pylyshyn, 1984). However, some cognitive theoreticians (notably, James Gibson and his followers) have argued strongly against the feasibility of such a division and have claimed that cognitive activity is laden with meaning throughout. Personally, I am sympathetic to this latter approach. Again, this stance of mine reflects my general views regarding psychology and the study of mind, has nothing to do with my study of Ayahuasca, and cannot be further explained and defended here.

The foregoing criticism notwithstanding, I would not go as far as to say that there is nothing of value to the two-stage model. Admittedly, Ayahuasca induces patterns of brain activity that are experienced as internal visual perceptions which are subject to subsequent semantic interpretation. But this is just one scenario, not a sole, universal chain of processing. As we have noted many times throughout this book, during the course of the Ayahuasca inebriation the mind is in a heightened energetical state. It is exceedingly sensitive and anything that it happens upon can be incorporated into the imaginal matrix it is generating. Simple, brain-produced, non-semantic elements are one such input, but so are external sensory stimuli, as well as full-fledged semantic imaginary visualizations, and ideas and reflections entertained by the drinker during the course of the inebriation. When the richness and the multifariousness of the Ayahuasca phenomenology and dynamics are appreciated, one realizes that the state of affairs defined by the hybrid model is neither necessary nor general. With this, the model loses much of its point. Surely, from a cognitive point of view the model does not capture the essence of what is going on under the intoxication.

Lest I be misunderstood let me reiterate once again what by now should be obvious: without the neurophysiological input nothing non-ordinary will happen, but in itself, brain activity does not tell the story of the Ayahuasca experience. To my mind, this story should be recounted in terms of the *pas de deux* suggested in the previous chapter. In this *pas de deux*, the mind dances, so to speak, with the products of both brain and mind. Viewed from this perspective, the picture drawn is, despite some important similarities, quite different from that presented by the hybrid model. Central to the hybrid model is the separation and division between the brain and the mind, as well as between, on the one hand, direct effects of Ayahuasca proper, so to speak, and, on the other hand, the products of drinkers' fanciful imaginations and cultural beliefs. In Reichel-Dolmatoff's perspective, only the former seem to be considered to be genuine, proper effects of Ayahuasca; the latter seem to be the reflection of either cultural biases of the aboriginal mind or the fanciful imagination of some romantic souls (cf. Reichel-Dolmatoff, 1975: 33). While, as repeatedly noted, I do accept that the state of affairs specified by the hybrid model can actually happen, the spirit of my picture of the Ayahuasca experience is categorically different. In it, there is no clear-cut separation between brain and mind, and there is no sense in characterizing some effects of Ayahuasca as 'real' and others as pertaining to 'mere imagination'. All that happens under the intoxication is the product of brain activity, just as all that happens is the product of the dance of the mind. Likewise, all is real *and* all is the product of imagination. For psychologically speaking, imagination (I am talking here of the activity, not the contents of its products) is as real as any other activity of the mind—just as speaking, thinking, and remembering are. As argued at length in Shanon (1993*a*), the task of the cognitive psychologist is precisely that: to outline the story of the mind's narrative, to draw the choreography of the mind's dance. In my view, making sense of the Ayahuasca experience is no different.

The Mirror of the Soul

But, of course, the dance in which the mind engages during the course of the Ayahuasca inebriation is a very special one indeed. What I would like to propose is that the Ayahuasca inebriation is a time of grace during which ordinary human beings can, with some constraints that I shall spell out in a moment, be like dancers or musicians when the Muses descend upon them. What happens over the course of the inebriation cannot be conceived solely in terms of an activation of the mind or the brain by means of an externally administered chemical substance. Rather, as proposed earlier in this book, what happens is the product of a joint intimate interaction between the drinker and the brew, between brain and mind. A fundamental feature of this enterprise is that whatever happens with it is a perfect reflection of who one is. As is said in the context both of the Santo Daime Church and the UdV—each person receives according to his or her merit, each according to the level of his or her understanding.

What is one's merit? It is the sum total of what one is, of *everything* about one—one's personality, dispositions, and attitudes, and, in fact, the entire biographical history that one carries. But look at this definition of merit. It is precisely the same definition that I have proposed above for what memory, ordinary human memory, is. Thus, just as the way one's body now moves, the way one's face smiles, the way one's hands gesture—are all perfect embodiments of who one is, that is—expressions of the entire path one has traversed in the course of that unit of time which is the step (as in a dance) or flow of one's entire life, so too is it the case with regard to that which is experienced with Ayahuasca. Whatever one sees in the visions, whatever one feels, whatever ideas one entertains over the course of the inebriation is determined by all that there is to one's self and being.

What I have just said is, of course, true of any moment of a person's life. What is special about Ayahuasca is, as detailed in Ch. 20, that with it, all sensitivities are enhanced and intensified so that people think faster and better, imagine (as Merkur says) in a richer and fancier way, and engage in all sorts of overt performances (notably singing) at a level they would not normally be able to attain. But, again, as with dancing and the playing of musical instruments, here too, the Muses descend only upon those who are trained and ready. It is naïve (and actually stupid) to expect a person who has never practiced the scales to suddenly improvise on the piano in a manner that will both surpass all his or her previous performances and touch the hearts of the audience. Likewise here. The merit of one's personality encompasses one's entire past including the cumulative training that one has had with Ayahuasca itself. The cognitive enhancement that the brew can grace people with is without bound, yet, to be privy to it one has to be both worthy as a person and trained as a drinker.

Interestingly, assessments along these very lines, ones that are identical to those entertained in the context of the Church of Santo Daime and the UdV, are found in other cultural traditions. In particular, let me refer to my own Jewish one, using three excerpts from the kabbalistic literature, all cited in Wolfson (1994). The first is taken from *Sefer Levush Malkhut* (The Book of Kingly Garments) written in the sixteenth century by Obadiah Hamon concerning those mystics who managed to ascend in spirit and see the higher realms of the Divine. The second and third are from *Sha'arei Kedusha* (Portals of Holiness) written by Rabbi Hayyim Vital around 1570; the third excerpt deals specifically with the nature of visions, in this particular case that of birds:

They imagined in their minds divine matters until it seemed to them as if all these things were revealed before their eyes so that they saw them with their eyes. Each one in accord with the greatness of his intellect and the breadth of his heart comprehended what he comprehended concerning what is good and what is evil. (p. 111)

The Holy Spirit rests on a person when he is awake, when the soul is in his body and does not leave it [as in sleep]. But [the prophetic state involves] the matter of separation [of the soul from the body], for he removes [from his mind] all [mundane] thoughts entirely. And the imaginative faculty in him, which is a faculty that derives from the elementary animal soul, prevents him from imagining or thinking about any matter pertaining to this world, as if his soul had left him. Then his imaginative faculty transforms his thoughts until he imagines that he ascends to the upper worlds to the roots of his soul . . . And the forms of all the lights will be inscribed in this thought as if he imagined and saw them, as in the way of the imaginative faculty to imagine in his mind things of this world even though he does not [actually] see them. . . . The [divine] light and influx reaches the rational soul that is in his body, and from there it reaches the vital soul and the corporeal form in the imaginative faculty so that [the prophet] can comprehend them as if he actually saw them with the [physical] eye. (p. 320)

It is not an actual creature or bird, but only appears and is seen in this way, and he reveals secrets of Torah. Indeed, according to the level and stature of the person who sees them, they appear before him, and on occasion they appear before two people at once, and before one person it appears in the form of a bird or another creature, and before the second person it appears in another manner; it all depends on the level of the person who sees them. (p. 320)

Note how similar these comments are to those made by the eighteenth-century Swedish scientist-turned-mystic Swedenborg and cited as one of the epigraphs to Ch. 19.

Experientially, I came to this realization following a powerful vision I had, just as the writing of this book was coming to an end, exactly eight years (to the day) since I first partook of Ayahuasca. It was a vision I had often longed to have, and later, appreciating that visions do not come by command, I no longer thought about or hoped for—that of the Temple in Jerusalem, before the city and the Holy Mount were destroyed by the Romans, in AD 70. How come I had this vision? Obviously, had I not known about the Temple, I would not have had that vision. Had there not been some sort of fascination in my heart towards that special period in Jewish history (specifically, the last century before and the first century after the beginning of the Common Era), the likelihood of my having this vision would have been reduced as well. But I have been a Jew since my birth, I have known about the Temple since my childhood, I was interested in the said historical period since my adolescence and I have been drinking Ayahuasca for eight years, and I never had that vision, which I *did* wish to have. Apparently, I was not ready for that, I did not yet merit this vision. The vision appeared when I was ready.

Put in formal, mathematical terminology, the process of vision generation can be conceived as a function whose argument (that is, the variable on which the function applies, like the term x in the expression $f(x)$) is the sum total of *everything* about the drinker at hand. This includes one's entire past, the present state of both one's body and psyche, all the facets of one's personality, one's dispositions and one's talents (including that of imagination), one's knowledge and beliefs (this list is, of course, not exhaustive). In sum, what a drinker will experience is the reflection of that person's full being at the moment at hand.

A word of clarification is in place here before we go on. In saying that I have seen the Jerusalem Temple, I am not saying that what I have seen was an image, in a photographic sense, of what the Temple looked liked in Jerusalem two thousand years ago. To say this would be tantamount to the admittance of reincarnation and/or paranormal time and space travel. As I said at the very outset of this chapter, in these I personally do not believe. What I have seen was surely affected by *my* conception of the Jerusalem Temple. Yet, it was not 'just' an imagination. It is not possible for me, now, to sit down, close my eyes, wish to see the Temple, relax, concentrate, and then—see. The feats of imagination we are talking of in the context of Ayahuasca are neither solicited nor planned. They come to be when one is ready for them, when one is in the state of being that perfectly matches that very vision.

Also to be underlined is the fact that I saw a temple. While the seeing of the Jewish Temple has to do with my own life history, the seeing of a temple has not. As indicated earlier in this book (and as corroborated by the statistical analyses in the Appendix), palaces and temples are one of the most common content items seen in Ayahuasca visions. Thus, what is seen in the visions reflects both the contingencies pertaining to the particular drinker, and at the same time, patterns whose universality transcends individual idiosyncracies. More on this seemingly paradoxical state of affairs will be said in the next chapter.

Thus, the fact that it is *my* vision of the Temple, that the picture I see in the vision is *in the image* of my personality, does not at all detract from the value and import of this vision. After all, the architect who actually designed the Temple was also a human being. He too construed the Temple in the image of *his* personality. And who knows, perhaps with sufficient knowledge of the historical period, with sufficient intuition with regards to people's psychology, a person partaking of Ayahuasca in 1999 could direct his or her creative powers so that they would resemble those of the architect who lived and worked more than two thousand years earlier. The more forceful such tuning-in, the more similar would the image of the Temple in the vision be to its actual form. No paranormal time travel or reincarnation need be involved.

By way of further explanation, I will use again the metaphor of music. Think of a flute: whatever sound the flute emits is a perfect function (again, in the formal, mathematical sense) of the material the flute is made of (compare wood to metal or to cane), its length, width, and the like. And of course, in order for sound to be emitted, a breath of air has to be blown into the flute. In our case, the flute (the musical instrument, the vehicle) is the drinker's being (psyche, mind, faculty of imagination) and the breath of air is the special energy that Ayahuasca induces. What happens in the course of the inebriation is the product of this energy, on the one hand, and the sum total of the drinker's state of being, on the other. In sum, I would say that what happens with Ayahuasca may be likened to music played on the instrument (in the musical sense of the term) which is the human soul. Potentially, the spectrum of tunes and melodies that can be played is infinite. Actually, at any given moment, the tune played is exactly that which expresses the drinker's being. Thus, for example, in my own case, my seeing of the Temple was the tune, the only possible one, that could emanate once the Ayahuasca energy vibrated in the instrument that was me (remember, this encompasses everything about me, including my entire past) at the moment. I might add that one special characteristic of the 'music' of Ayahuasca is that it is primarily articulated in images. This might be related to the primacy of the visual faculty in human cognition in general. In principle, however, this music brings all senses together (recall the phenomenon of synaesthesia) and the instrument played both reflects and is reflected in a person's entire being. It is in this sense that the Ayahuasca experience may be said to be the mirroring of the soul.[5]

But who is playing the flute? Experientially, especially when the force of the inebriation is strong, drinkers of Ayahuasca feel that the source of what happens is not themselves. Often drinkers attribute what is happening to forces above them, to a personification of the brew (this is the traditional Amerindian conception, see Luna, 1984*a*), to cosmic consciousness. These terms are outside the psychological framework adopted here. What answer can the cognitive psychologist give?

Well, who is playing the flute when ordinary cognitive behaviour is concerned? A lay person would say 'me' or the 'self', but the truth of the matter is that contemporary academic psychology has no answer to this question. Scientific cognitive psychology defines how cognitive performances are executed, but it does not say how they are generated. Indeed, it cannot. Cognitive theories model the purported machinery of the mind, but in principle they can say nothing about whoever puts this machinery into action. The bottom line is that as things stand today, cognitive psychologists cannot account for the generation of behaviour. They, in other words, cannot explain how the music of the mind is played. That this is the state of the art in cognition at large puts the scientific account of the Ayahuasca experience in a more balanced perspective. Thus, I say that the Ayahuasca experience is like music played on an instrument which is the soul and that this music is a perfect mirroring of one's entire being. Beyond that, as a psychologist, I cannot say. Yet, if allowed to delve into more speculative philosophical reflections, there are some further things I could say on the matter. This I do in the next, final chapter of this book.

A word of clarification is in place here. The foregoing analysis may give the impression that the Ayahuasca experience is one of total passivity. True, one cannot push this experience and visions cannot come by demand. Furthermore, the optimal conditions for powerful experiences are achieved by serene contem-

[5] After having written this chapter, I received Lagrou (1998) in which an indigenous Cashinahua describes the state of a powerful Ayahuasca intoxication as one in which his body is singing.

plation and an attitude of letting go. Yet, as we have seen, these very conditions can also produce performances in which the Ayahuasca drinker actually engages in manifest activity. Whether the drinker is overtly passive or active, however, does not affect the mathematical function proposed here. The seemingly passive reception, as in the case of visions, *and* the manifestly active performances such as singing and healing, are alike—both are the tunes the soul produces when the Ayahuasca energy breezes through it.

Let me close with another metaphor—that of light passing through a prism. The entering light is white, that which comes out is deflected and it exhibits a certain spectrum of colours. Where do the colours come from? On the one hand, they are all incorporated within the entering white light. On the other hand, the particular spectrum is totally dependent on the nature of the prism. Just as they are the product of the entering light, the colours are the product of the prism. Likewise with the mind. When it lets go and is fully receptive to what it experiences as being external to it, at this very moment it is prone to manifest the very best of its potentialities and its creative capabilities soar. This is what is happening with Ayahuasca.

Inspecting the prism further, we note that the particular pattern of light emitted by the prism is, in the perfect mathematical sense, a function of the structure of the prism, of all the contingent values that define what the prism is. The cleaner and the more polished the prism, the richer and the more beautiful the spread of colours that it emits. A tainted glass would dim the light, a marred one might altogether block it. A perfect crystal would display the entire spectrum of light in all its magnificence. Again, the same holds true of the mind and Ayahuasca. As indicated earlier in this book, for drinkers to experience the wonders of Ayahuasca in all their magnificence they have to be cleansed. When this is not the case, what is experienced under the intoxication is prone to be tainted by personal matters and idiosyncratic contingencies. Only when both body and psyche are clean will the mind be ready genuinely to profit from what the brew can offer and experience what might be characterized as 'the world of Ayahuasca'. With this, the mind will also manifest the best of its own creative potential; indeed, it will reveal the quintessence of its own very nature.

But what would happen with a perfect prism, a prism devoid of any contamination, a prism that is absolutely pure? In such a case, there would be neither deflection nor refraction: the prism would be like a transparent glass (or a perfect mirror) through which the light would pass (or by which it would be reflected) without any alteration or transformation, just as it is. In the limit, the light and the prism would unite and be one. This is exactly what is encountered with Ayahuasca. Under the intoxication, the cleansed mind will see richness whose content is unbounded and whose scope is truly infinite. With this, lucidity will be in its utmost and the power of mentation and reflection will be immensely increased. Yet, in the peaks of the intoxication, a stage will be reached in which the visualizations will no longer be bound by any content or definite form, when neither language, reason, nor memory apply, when no conceptualization or rational reflection is possible. When the light reigns supreme understanding ceases.

I have introduced light and the prism as a parable, but I ended up talking of the Ayahuasca experience itself. In my parable, light was, of course, a metaphor. Yet, with Ayahuasca, light is part and parcel of the phenomenon itself. Indeed, it may very well be the case that the metaphor fits precisely because in the story of Ayahuasca light is of the very essence. With this, the metaphorical and the literal coincide. By way of further reflection, the reader is invited to look back at the comments closing the phenomenological part of this book (the end of Ch. 17) as well as at the teaching of a master of the UdV described in Ch. 5.

Summary and Further Meta-Theoretical Considerations

By way of summary I will review how the Ayahuasca experience is conceived by each of the perspectives considered here. By the non-naturalistic approach, the effects of Ayahuasca come about by means of *paranormal perception* of non-ordinary realities. Spiritualistic connotations may be associated with this approach as well. While I do not accept this line of thinking in the context of this psychological analysis, in the final chapter of this book, which is devoted to philosophical reflection, I will say something about it. Biological reduction dictates that what is happening with Ayahuasca is the direct result of underlying *brain activity*. Whether this is satisfactory as an explanation is up to each reader to judge. As noted, the judgement at hand is not specific to Ayahuasca and will depend on one's general view of cognition and the study of mind. By the various approaches based on the unconscious, what is happening under the intoxication is the *uncovering of information* that is stored in the mind but which in ordinary states of consciousness is covert. Associated with this may be a release from inhibition coupled with enhanced sensory activation. Because of my general stance in cognition, I cannot accept this line of explanation. I have also presented specific arguments (that is, ones directly associated with the study of Ayahuasca) against it. Other psychological explanations characterize non-ordinary states of consciousness in terms of *cognitive impairment*. These do not do justice to the rich phenomenology at hand and are, apparently, tainted with biased value-judgements. By both Merkur's proposal and mine what is happening is a non-ordinary mode of cognitive activity. According to Merkur this activity consists in *imagination* or enhanced fantasy. While I fully acknowledge the importance of imagination, I feel that it does not capture the full richness of the Ayahuasca phenomenology. In my own proposal I characterize the Ayahuasca experience as a *total expression of the state of one's being* and I liken it to the playing of a musical instrument or a dance.

The different models we have surveyed also assign different statuses to the Ayahuasca drinker. By the paranormal and the spiritualistic views, he is the *discoverer* of other realms; by the psychological views involving the unconscious the drinker is the subject of a passive process of *dis-covery*, that is, the uncovering or the bringing to the fore of psychological material that normally is covert and hidden. The biological, reductionist account can be consistent with either of these two conceptualizations. On the one hand, this account may be conceived in terms totally analogous to the psychological ones involving the unconscious; in this case, the activity of the brain induced by Ayahuasca is to be viewed as a passive process of dis-covery. Alternatively, being non-psychological, this activity may be deemed as manifesting a realm that is, in a fashion, outside the sovereignty of the individual self. In this sense, what happens under the intoxication may be regarded as a process of discovery—of that which is hidden in our brains. Arguments and reservations have been presented against all these conceptualizations. By and large, the view put forth here is that the Ayahuasca experience is one of *generation* and *creation*. Rather than being a discoverer or subject to a passive process of discovery, the Ayahuasca drinker should be viewed as an imaginative creator. More on this will be said in the next chapter.

In fact, the characterization of the Ayahuasca experience as a mirror of the soul points out that in addition to dis-covery (or revelation) and creation still another possibility may be envisioned. Before introducing it, I will return to the meta-theoretical level and consider the various options, or lines of explanation, that may be pursued by scientists when accounting for regularities observed in the world, be it psychological or natural. The first option is to view observable regularities as the *reflection* of some more basic covert structures. In psychology, this logic underlies the postulation of the various forms of the unconscious as well as the language acquisition device postulated by Chomsky (1972, 1975). Outside psychology this line of explanation is encountered in the Platonian postulation of the suprahuman realm of Ideas. The second option is to view observable regularities as the outcome of a process of *generation* subject to a set of *constraints*. In the psychological domain, such constraints have been defined, for instance, in theories of child development; a paradigmatic case is Piaget's (1983) theory of cognitive development. Outside psychology, generative models are encountered in theories of biological evolution and auto-organization (see Atlan, 1979). The third option I would single out is one I shall call *deterministic*. In this case, the regularities are regarded as the product, in the mathematical sense, of the function of all variables pertinent to the system that is exhibiting the regularities being accounted for. Explanations in physics are of this kind. The falling of the stone is neither the overt reflection of the underlying rules of physics nor is it the product of a generative process. Rather, it is determined by the sum total of all the qualities of the stone and the environment in which it is situated. The present metaphor of the mirror of the soul is of this kind.

Yet, as emphasized, the Ayahuasca drinker is different from the stone in that he or she does have his or her own contribution. The drinker is not fully passive, he freely interacts with both the context and with his visions, he creates. Thus, the model I am proposing here is a hybrid, generative-deterministic one. What happens to a person with Ayahuasca mirrors all about this person, hence the state of affairs fits a deterministic account. Yet, given that what distinguishes human beings is their ability to act freely in the world and to create, the present perspective of the Ayahuasca experience is, at the same time, also generative.

I have emphasized the fit—one I characterized as perfect—between what a person experiences with Ayahuasca and the sum total of his or her personality, constitution, history, and disposition. Throughout this entire book, however, I have also underlined the remarkable similarities between the experiences that different persons have with Ayahuasca. Is there not a contradiction here?

Although at first glance it would appear that unique singularity and the encounter of cross-personal similarities contradict each other, in fact they do not. Consider, for instance, the example given above of my vision of the Temple in Jerusalem. Above I have underscored the relationship of this vision to my life history, interests, and contingent disposition. At the same time, temples are one of the most common content items encountered in Ayahuasca visions. Thus, there is a non-personal factor that makes the seeing of temples very likely. Given this, the choice of a particular temple may reflect various idiosyncratic variations of the specific drinker at hand. There is thus no contradiction between the two patterns, and together they span a vast space for variation. More on the cross-personal commonalities in Ayahuasca visions will be said in the next chapter.

Taking a more general perspective, the sum total of a person's individuality also includes all the knowledge the person has acquired through his or her interaction with other people, from what he or she has been taught, from his or her social milieu and culture. In this formulation, then, the dividing line between the personal and the collective, the psychological and the socio-cultural, blurs. Prime manifestations of this are great works of art. These are, at the same time, both the highest expressions of individual genius *and* the manifestations of the most universal, the most eternal aspects of humanity.

Lastly, I should comment on the limitations of the discussion pursued in this chapter. What I have attempted is to gain some overall psychological understanding of the Ayahuasca experience. The question of how visions come into being was focal in this undertaking. All along, it was crucial for me that my analysis both do justice to the phenomenology of the Ayahuasca experience and be consistent with my general theoretical stance in psychology. The account proposed here more or less reaches the limits of what I could do while respecting these two criteria *in tandem*. Yet, maybe it is not possible to gain true understanding of the Ayahuasca experience and to account for it while still holding to a perspective of thought that can be endorsed within the framework of scientific psychology. Albeit non-orthodox, my view both of cognition and of psychology keeps to the fundamental assumption that the subject matter of the science of cognition is the human mind and that it is in the province of this mind that all matters cognitive take place. But perhaps this assumption is not warranted. Perhaps consciousness, especially non-ordinary consciousness, cannot be conceived while maintaining this assumption. For this, it might be necessary to resort to some sort of a supra-personal mind, to some intelligence that transcends the human, to a cosmic consciousness of which our individual consciousness is only a pale, derivative reflection. Recently, such non-orthodox conceptualizations have been proposed by several scientists. Remarkably, not one of them is a psychologist—practically all are physicists (see Goswami, 1990; Wolf, 1992; Close, 1997). Significantly, what people feel when under the Ayahuasca intoxication is very much in line with these non-orthodox ideas. As noted in the review of ideas, insights, and reflections in Ch. 10, many drinkers come to believe that Ayahuasca brings them in touch with the *anima mundi* or the Divine consciousness which is the Ground of all Being, the source of all knowledge, the fountain of all wisdom. In a direct, non-mediated fashion drinkers also feel that it is this consciousness that is the source of the visions and the insights associated with them. When the force of the inebriation is especially strong, drinkers feel that the boundaries between this consciousness and their own individual one are less and less defined. In the limit, I and God become one. All that can be known is part and parcel of the Divine mind, hence also of my mind.

This *cosmic* account of the Ayahuasca experience is, of course, reminiscent of ideas that have been proposed in the mystical literature throughout the ages. It is in line with what has been called the perennial philosophy (see Huxley, 1944). Many may find the tenets of this philosophy very familiar, some may even consider them obvious. Indeed, it could very well be that the supra-psychological approach is the only one that can actually make sense of the Ayahuasca experience in its entirety. In particular, this account may give definite answers to those issues that the proposal outlined here leaves open—notably, what the source of the visions is and how they are actually produced. However, to defend this approach in a scientifically acceptable manner is no simple matter at all. Surely, such an enterprise involves ontological and metaphysical analyses that go well beyond the scope of the cognitive investigation attempted here. I shall leave the matter at that.

23

Concluding Philosophical Reflections

> The world of imagination is the world of eternity; it is the divine bosom into which we shall all go after the death of the vegetated body. This world of imagination is infinite and eternal, whereas the world of generation, or vegetation, is finite and temporal. There exist in that eternal world the permanent realities of every thing which we see reflected in this vegetable glass of nature. All things are comprehended in their eternal forms in the divine body of the saviour, the true vine of eternity, the human imagination, who appeared to me as coming to judgment among his saints and throwing off the temporal that the eternal might be established.
>
> William Blake

Two Philosophical Questions

As I have made abundantly clear, this is a psychological book, not a philosophical one. The philosophical questions—or rather, puzzles—that Ayahuasca raises are manifold and they deserve to be treated separately. Yet some philosophical questions are so directly linked to the psychological data that has been reviewed that I cannot conclude this book without at least making some mention of them. As indicated in Ch. 10, philosophical reflection is part and parcel of the Ayahuasca experience. Furthermore, hearing about the psychological data I have collected, many who have not had any firsthand experience with the brew have presented questions of a philosophical nature to me. Thus I turn to some philosophical reflections. By no means does the following discussion intend to do full justice to the questions at hand. It is only meant to give a general flavour of my position on these questions and to lay the grounds for further consideration of them.

The spectrum of philosophical questions that Ayahuasca raises is very broad. Here I shall focus only on ones that are directly related to one major empirical finding revealed by the present investigation, namely, the common patterns exhibited in the experiences different people have with Ayahuasca. As described at the outset of this book, what impressed me greatly when I first partook of the brew were the similarities between the experiences I had with it and ones described in the anthropological literature. Some of the images that appeared in my visions were similar, at times identical, to ones reported by indigenous persons and by the first European explorers who encountered Ayahuasca. The extensive empirical research I have subsequently conducted corroborated these first impressions and led me to conclude that there are striking cross-personal commonalities in the contents of Ayahuasca visions, in their themes, and in the ideations that are associated with them (for supporting data, the reader is referred to the Appendix).

Typically, Ayahuasca visions—notably those that are most powerful—present scenarios that are totally unrelated to drinkers' lives, daily concerns, or past experiences. Often, drinkers (myself included) feel as if a veil is lifted and they are presented with another world. This world, it seems, exhibits internal consistency and coherence and subjectively it impresses those who view it as having an independent existence of its own. In fact, it is very common for Ayahuasca drinkers to feel that this world which has been unknown to them beforehand is, in some sense, even more real than the world of their everyday lives. This world exhibits an air of eternity, and it appears to maintain its existence independently and regardless of all that is taking place here in the ordinary realm of human beings and mortals in general. Moreover, subjectively, drinkers feel that what is being revealed to them in the visions is invested with great meaning. When the visions are powerful they are not mere images, or even scenes as in a video show. Rather, the visions reveal and instruct, and in witnessing them drinkers feel that they establish contact with a realm of existence that is both real and sublime (the second section of the previous chapter, 'Non-Naturalistic and Spiritualistic Accounts', reflects this state of affairs).

All this points to two main issues of interest—*cognitive universals* and *status and veracity*. The first concerns the prevalence of cross-personal commonalities in the Ayahuasca experience and the various possible ways by which to account for them; the second concerns the very nature of the mental materials experienced with Ayahuasca—the extraordinariness of their content and form, the meanings that they might convey, the truths that might be associated with them, their reality. While conceptually these two issues are distinct, their theoretical analyses are not unrelated. With respect to the first, we ask how the commonalities in the experiences of different people are to be explained. The most immediate and natural way to account for such commonalities would be in psychological terms, invoking universals of the human mind. However, as suggested above and as shall be further elaborated upon below, at least experientially, some aspects of the Ayahuasca experience seem to defy standard psychological explanation. It is here that the second issue arises. If an account cannot be achieved in strict psychological terms, then the subject matter of the discussion turns from the psychological to the metaphysical. With this, questions of ontological status and of epistemic veracity have to be considered. These address the putative nature of the other realities that might be revealed by the visions, the truth value of the information they seem to depict, and the import of the messages they seem to convey.

As attested to by the foregoing exposition, the question of universals and the questions of status and veracity are not on the same par. Universals may be encountered without there being any special ontological or epistemological import to the visions. In contrast, if the visions do attest to the independent existence of some realms or dimensions, this very fact can define a theoretical framework by which the commonalities of experience may be accounted for. Thus, as noted, the two questions are distinct but not unrelated. In particular, if the second, ontological, question is to be addressed positively, then the two questions, as well as the epistemological one of veracity, all tie together so as to compose one whole.

Bringing the two questions together, I would define the key issue as follows: Is the Ayahuasca drinker a discoverer or an inventor? In line with what has been said in the previous chapter, there are three ways of looking at the situation. According to the first view, Ayahuasca uncovers the deeper layers of the unconscious, and the drinker is the *dis-coverer* (or un-coverer) of things that normally lie hidden—either in the mind or in the brain—and that the brew brings to the fore. According to the second view, Ayahuasca induces the perception of other realities, ones that cannot be perceived in the ordinary state of consciousness; the drinker, therefore, is a *discoverer*. According to the third view, the Ayahuasca experience is generative and creative throughout; this is to say that the drinker is an *inventor*. Under the intoxication, people's imagination and creative powers are greatly enhanced. Thus, their minds are prone to create the fantastic images they see with the brew. As already indicated in the previous chapter, it is this last view that I shall favour here.

Before turning to the discussion itself, let me note that many of the ideas to be presented here are reminiscent of ones already presented in the previous chapter. This is no accident—both that chapter and this one are devoted to the integrative theoretical analysis in which I attempt to make sense of the Ayahuasca experience as a whole. The previous chapter did this from a psychological perspective; this one from a philosophical one. The perspectives are two, but the substance is one. Thus, it is only natural that the significant facets of the theory of Ayahuasca will appear, seen from two different angles, in both. It turns out that this is also very much in line with what, towards the end of this chapter, will surface as the most important philosophical lesson of this discussion, namely, that with Ayahuasca various well-established distinctions and divisions have to be relinquished and surpassed.

Cognitive Universals

I begin with the question of universals, and shall consider some of the theoretical options already examined in the previous chapter, beginning with the biological, reductionistic one. There (and in Ch. 2) I argued against biological reductionism in terms of my general view of psychology. Here I will consider this option further, focusing on the specific topic of cognitive universals. A reductionist would argue that the commonalities in the Ayahuasca experience are directly due to neuroanatomical structures and brain processes and hence should be explained in biological terms. While at first glance such an explanation may seem most natural, further reflection reveals that it is far from being obvious. Would we say, for instance, that the reoccurrence of visual elements such as serpents and jaguars, palaces and works of art are due to there being specific brain centres in which such information is stored? Would we say that there are brain centres that specialize in metaphysical ideas or in spiritual and religious concerns? It seems to me that, at least in the present state of our understanding of the brain and nervous system, the postulation of such specific centres simply does not make sense. As commonly conceived, brain physiology and anatomy are not defined on the level of the semantic specificity exhibited by the commonalities encountered in Ayahuasca visions and the ideations associated with them.[1]

In this conjunction, I quote one of the few allusions to this issue in the scientific literature—comments made by the ethnobotanist Schultes in his introduction to Reichel Dolmatoff's anthropological book *The Shaman and the Jaguar* (1975: p. xiii):

> It has been said that the reason for the importance of huge snakes and large felines in native cultures is simply that the stealth and strength of these animals strike awe and fear into the Indian. But it is clear that such an explanation cannot withstand the scrutiny of any examination in depth. There have even been attempts to offer 'biochemical' explanations: that the great frequency of jaguars and other felines and snakes in descriptions of caapi-intoxication [i.e. Ayahuasca intoxication, B.S.] might be due to an ability of the active chemical constituents of the hallucinogenic plant to induce visions of definite kinds of subjects. But there is not a shred of evidence that such specificity of action exists. One is left, then, with the near certainty that underlying the primacy of the jaguar in cultures are ancient, deeply ingrained and highly positive experiences that are now a part of the mind of aboriginal man.

Alternatively, the cross-personal commonalities may be accounted for in universal, psychological terms. As indicated in the previous chapter, two principal

[1] After this book was completed, such a thesis was actually put forward. In their book *The Mystical Mind*, Newberg and D'Aquili (1999) suggest that religious and mystical sentiments are due to the activation of brain centres specifically responsible for such sentiments. The book had an immediate great impact and was even the topic of a *Newsweek* cover story (see Begley, 2001). The empirical data on which the suggestion is based involve the new sophisticated technology of brain imaging. When a meditating person is inspected with imaging devices specific patterns of brain activation are observed. Much has been made of this finding, and even the bombastic term 'neurotheology' has been coined. Brain imaging technology is indeed most impressive, yet conceptually, its impact is, I think, less radical than many believe. After all, everything a human being does involves brain activity. The finding of particular brain activation in conjunction with religious experience should thus be expected. By no means, however, does this finding imply that the brain causes the religious experience or that religious experiences (or religion, or God . . .) are to be accounted for in a biological, reductionistic manner. When one speaks the brain is acting as well; this does not mean that language and linguistic creations (think of Shakespeare's sonnets, for example) are reducible to brain activity and are to be explained in neurological terms. What the new findings may be indicating is that, like languaging, spirituality is ingrained in our very being, that the propensity for religiosity is a fundamental feature of what it is to be a human being. But do we need the supermodern technology of brain imaging to know that? Incidentally, I should say that I myself would never put Ayahuasca drinkers into an fMRI machine. The experimental manipulation, its artificiality, and the discomfort associated with it are very likely to affect the very experience it is set to examine. I doubt that any really powerful Ayahuasca session can be entertained under such circumstances, let alone the supreme spiritual experiences that are the subject of discussion here.

psychological orientations may be noted—that pertaining to the realm of the individual person and that pertaining to the realm of the human collective. Arguments against accounting for Ayahuasca visions in terms of the individual unconscious were presented in that chapter; here I focus on the second orientation. This orientation is paradigmatically linked with the Jungian notions of the collective unconscious and the archetypes associated with it (see Jung, 1969). Jung postulated the existence of the collective unconscious in the light of considerations very similar to those we are discussing here in the context of Ayahuasca. Specifically, discovering that various manifestations of psychic life—dreams, visions, fantasies, patterns of insanity, and works of culture—exhibit commonalities that cannot be accounted for in terms of the particular life histories of the individuals at hand, he proposed that there is a level that is psychological and at the same time collective.

Indeed, when attempting to account for cross-personal commonalities encountered with Ayahuasca, Jung's is the first theoretical approach that comes to mind. In fact, the scientific psychological literature seems to offer no other potential resource for explanation. Appreciating this, let me point out three failures of the Jungian account. The first two are general, and have nothing specifically to do with the phenomena investigated here; the third is directly related to the commonalities exhibited in Ayahuasca visions. First, it will be noted that there is much less in the notion of 'collective unconscious' than may seem at first glance. The postulation of the collective unconscious is tantamount to saying that there are psychological phenomena that are (like those studied by Jung's mentor, Freud) unconscious but which are not individualistic, but collective. But, note, this is nothing more than a restatement of the very phenomena that puzzle us—except for stating that these phenomena do in fact exist, the term 'collective unconscious' does not explain anything. Second, it should be appreciated that the nature of the Jungian archetypes is far from being clear. Are they mere theoretical abstractions on the psychological level? Are they reflections of biological structures that have been shaped in the course of the evolution of the species? Or perhaps they are Platonic forms that have special ontological status? The indeterminacy might not be commonly appreciated by those who read Jung and/or follow him, but there is no question about it—Jung himself was fully aware of it. A careful reading of Jung's writings reveals that he himself was very concerned with these questions (see, in particular, Jung, 1954/1969). Moreover, Jung was aware of how insufficient his account of the archetypes was. If these are not merely theoretical abstractions then, by Jung's own explicit acknowledgement, their very author could say very little substantive about them. If the proper level of analysis is the biological, one has to specify the biological structures and mechanisms that are at play. Jung was fully cognizant of the fact that the state of the art of the life sciences in his time did not allow for that. Furthermore, he was apprehensive that a biological account of the archetypes would lead to Lamarckism, a view that practically all biologists consider to be unscientific. On the other hand, if the proper level is ontological, one is led to a host of metaphysical problems that have puzzled philosophers for more than two millennia and which have still not found satisfactory answers. I shall return to the consideration of these in a moment, but first let me point out the third failure of the Jungian account, that which is specifically related to the phenomenology of the Ayahuasca experience.

The cross-personal commonalities encountered in Ayahuasca visions are very different from those that are associated with the Jungian archetypes. The Jungian archetypes can indeed be associated with the common heritage that is shared by all human beings and which may very well have evolved throughout the history of the species. The Great Mother, the Old Sage, the Hero, the Eternal Youth, the Trickster—these are all roles that are either played by people in the course of their lives or are portrayed by famous figures in all human societies and as such are personally known to all their members. Not so some of the commonalities encountered in Ayahuasca visions. Serpents, palaces, and objects of gold cannot be reduced to commonalities in human experience and to the predicament of life. They are specific contents that cannot be reduced any further.

Significantly, as revealed by an independent investigation I am now carrying out, the visions reported by Jung (1976) do exhibit the same types of common specific content items that are found in the content analysis on the micro-level reported in this book. Similar content items are also encountered in myths and fairy tales that Jung and his associates studied (see Jung, 1964) as well as in various non-ordinary psychological experiences. The latter include mystical experiences, out-of-body, and near-death experiences (see Green, 1973; Moody, 1975; Siegel, 1980). Thus, an ironical state of affairs obtains. Empirically, the Jungian data parallel those revealed by Ayahuasca visions; yet theoretically, the Jungian archetypes fail to explain the concrete commonalities of contents in both Ayahuasca visions and the visions and other materials Jung himself investigated.

The only scholar I know of who does address the topic of particulars of content in visionary experiences is Aldous Huxley. Huxley tackles the question from a perspective which is, in a way, complementary to that entertained here. He does not begin with the visionary experience, but rather ends with it. The question Huxley (1972: 35–7) raises is 'Why should precious stones have always been regarded as extremely precious?' This is his treatment of this wonderful question:

> Why should human beings have spent an immense amount of time, energy and money in collecting coloured pebbles? There is no conceivable economic value in this. . . . Plotinus, the great neoplatonic philosopher . . . say[s], 'In the intelligible world, which is the world of platonic ideas, everything shines; consequently, the most beautiful thing in our world is fire.' This remark is significant in several ways. First of all, it [shows] that a great metaphysical structure, the platonic and neoplatonic structure, was essentially built up on a quasi-sensory experience. The world of Ideas shines, it is a world which can be seen; and this curious fact that the ideal world can actually be seen, can be discovered also in Plato himself. In the *Phaedo*, Socrates speaks about [a world in which] . . . everything shines . . . the very stones of the road and on the mountains have the quality of precious stones; and he

> ends up by saying that the precious stones of our earth, our highly valued emeralds, rubies, and so on, are but infinitesimal fragments of the stones which are to be seen in this other earth; and this other earth, where everything is brighter and clearer and more real than in our world, is, he says, a vision of blessed beholders. Here again is another indication that a great metaphysical idea, the platonic Idea, the platonic system of an ideal world, is also based upon a world of vision. It is a vision of blessed beholders, and I think we now begin to see why precious stones are precious: they are precious because in some way they remind us of something which is already there in our minds. They remind us of this paradisal, more-than-real world which sometimes is glimpsed consciously by some people and which I think most people have had slight glimpses of, and which we are all, in some obscure way, aware of on an unconscious level. And as Plotinus says, it is because of the existence of this other world, this luminous other world, that the most beautiful thing on earth is fire. . . . [Thus,] I think quite definitely, that the reason why precious stones are precious is precisely this, that they remind us of this strange other world that is at the back of our heads to which some people can obtain access, and to which some people are given access spontaneously.

I would say that *mutatis mutandis* this is also why people have the propensity to create works of art, why they build palaces and temples, are moved by music, why they have an affinity towards the transcendent and the religious.

If the commonalities in the Ayahuasca experience cannot be accounted for in ordinary psychological terms, then perhaps we have no choice but to shift from the internal domain to the external one and consider the possibility that these commonalities reflect patterns exhibited on another, extra-human realm. Perhaps the commonalities are encountered because they are reflective of something else, something that exists independently of the human—some other realm of reality. With these speculations, we are led to the second question defined at the outset of this discussion, that of status and veracity.

Status and Veracity

As noted in the previous chapter, most users of Ayahuasca conceive of the Ayahuasca experience in extra-psychological terms. This is true both of indigenous persons and non-indigenous ones, and surely it is true of all the new syncretic Brazilian sects that employ the brew. The specific attributions offered vary, but in essence, all refer to non-naturalistic sources of knowledge and/or non-human entities and beings that impart knowledge.

The explanations usually invoked by users of Ayahuasca highlight the supernatural, paranormal nature of the other realities seen in the visions. As noted earlier, these realities are generally conceived of in a very concrete manner and often referred to as constituting 'the astral'. At times, an eternal storehouse of information is invoked in which all information—past, present, and future—is kept. According to this hypothesis, Ayahuasca brings those who partake of it to realms that are not dependent on time. In these, it is believed, all that has been known and all that will be known is co-present and potentially available for people in the visionary state to access. Significantly, the first time I ever thought of this option was under the effect of Ayahuasca. Afterwards, in the course of my interviews, several informants reported having the very same realization under the intoxication.[2]

As I have said, the recourse to the paranormal is foreign to my thinking. However, non-psychological explanations can be more abstract and formulated without paranormal and spiritualistic assumptions. In the European civilization, the paradigmatic theory in which recourse to other realities is made, is Plato's theory of the Ideas.[3] Plato postulated the independent existence of Ideas in order to account for both meaning and lawful regularities in this world. Could the cross-personal commonalities in the Ayahuasca experience be accounted for in a similar fashion? Furthermore, could it be that what Ayahuasca does is to allow human beings to gain access to the realm of the Platonic Ideas or to one similar to it? In his *Doors of Perception*, Huxley (1971) specifically suggests this in conjunction with mescaline. Several of my informants, ones with academic education, have explicitly made the same suggestion. This they did both on the basis of a direct, firsthand experience and subsequent intellectual reflection.

By way of dealing with the Platonistic account of universals in the Ayahuasca experience, let me consider the domain that lends itself best to a Platonistic analysis, namely, mathematics. What is the ontological and epistemological status of mathematical statements? Is mathematics created or is it discovered? If mathematics is created, how are we to explain the seemingly objective veracity of its claims and its not being dependent on the particular individuals who make them? If mathematics is discovered, then we have to postulate all sorts of entities existing in other—indeed, Platonic—realms. The first option seems to be at a loss to account for the objective veracity of the state of affairs at hand whereas the second option involves non-trivial metaphysical commitments, hence there are many who are not comfortable with it (for recent works that do embrace this option and are also concerned with cognitive issues, see Penrose, 1990, 1994). Mathematicians and philosophers of mathematics have not resolved the debate between the two views and I would imagine that the positions each scholar adopts reflect personal tastes and general ideological and philosophical inclinations. It is only natural to assume that the debate is to remain unsettled for the foreseeable future.

The ontological status of mathematics is, indeed, one of the central problems of philosophy. Einstein commented that the perplexing mystery is not that the world is incomprehensible, but precisely the opposite—that it *is* comprehensible, and more specifically—that it can be accountable by means of mathematics.[4] By and

[2] Just as this book was finalized, I encountered this hypothesis put forward also in Polari (1998).

[3] The reader will recall that, as noted in Ch. 8, when referring to ideas in the Platonic sense, I use the upper-case initial. When ideas in the regular cognitive sense are referred to the lower-case initial is employed.

[4] Following are two pertinent citations; unfortunately, I do not have the details for their original sources: 'How can it be that mathematics, being after all a product of human thought independent of experience, is so admirably adapted to the objects of reality?' 'The wonder is not that mankind comprehends the world, but that the world is comprehensible.'

large, three main positions with regard to the status of mathematics may be distinguished. The first position is *realism* or Platonism. By this view, mathematical truths are real and their ontological status is independent of the works of human minds. The second position is *inventionalism*. By it, the contents of mathematics are the inventions of the human mind and the reflection of the human mode of thinking. The third is *formalism*. It assigns no meaning to mathematics and avoids the attribution of ontological status to it. It regards mathematics as the set of all possible legitimate deductions from all possible sets of consistent axioms using all possible rules of inference. Unlike the first two positions, the third has no counterpart in the present context of Ayahuasca (or, for that matter, in that of the other domain to be discussed in this chapter, music). For further discussion the reader is referred to Shapiro (1997) and Schirn (1998).

The point I would like to make here is that regardless of the answer to the philosophical questions we are now considering in conjunction with Ayahuasca, the problematics they present is analogous to, I would even dare say identical with, that encountered in mathematics. If in mathematics the situation remains unresolved, so perhaps it is not so bad to accept it as such in the case of Ayahuasca too. At any rate, the appreciation is definitely a non-trivial advancement in the path of gaining some understanding of the puzzles that concern us here.

Mathematics is not the only domain in which Platonistic realism may be endorsed. Another such domain is music. Why do two tones differing by an octave sound 'the same'? Why do certain melodic progressions—such as those involving fifths and the thirds—sound right to our ears? Some 'neat' musical patterns may be attributed to our perceptual system being sensitive to relations that can be expressed in terms of integers. However, lawfulness in music reaches complexities that are not readily accountable for in such psychological or physiological terms. With this, questions completely analogous to those encountered in mathematics may be raised. The modern ear may find it dissonant to suggest that music is discovered, rather than created. Not so for the Pythagoreans who revered both mathematics and music and regarded the 'music of the spheres' as sacred (see references in Ch. 15). A modern exposition of this topic that I find most ingenious is that presented by Kivy (1987). In particular, Kivy shows that the distinction between discovery and creation is more delicate than is commonly assumed and he—very cogently, I find—presents the case for how Beethoven could be said to have discovered, not created, his symphonies.

One other domain in which Platonic Ideas have been invoked is linguistics. The fact that children acquire the grammars of natural language with apparently no direct instruction has mystified human beings for ages. In our day, this has triggered much of contemporary linguistic research. In the Chomskian school of linguistics the cognitive and philosophical issues associated with the knowledge of language and its acquisition are central. Chomsky adopts a radical nativist position (1965, 1972, 1975; see also Fodor, 1975) according to which knowledge of grammar is part and parcel of the biological endowment of human beings. While to my mind

Chomskian nativism shares much in common with Platonistic realism, Chomsky himself does not see it this way (see Chomsky, 1986). However, at least one contemporary philosopher of language, J. Katz (1981) has explicitly adopted a Platonistic position for language. By it, the rules and regulations of Universal Grammar reflect not human psychology but rather a realm endowed with an independent existence of its own.

Especially related to our discussion is still another domain in which similar considerations may apply—that of myth. Do universals in myths reflect only commonalities in the conditions of the life of human beings, their psychological make-ups, and the structure and mode of functioning of their minds? Can myths be fully accounted for in empiricist terms? Can they be explained in terms that are strictly psychological and individualistic? Or perhaps, as in the case of language and music, some aspects of myth cannot be accounted for in such terms? Lévi-Strauss (1964) asserts: 'Nous ne pretendons donc pas montrer comment les hommes pensent dans les mythes mais comment les mythes se pensent dans les hommes, et à leur insu.' As pointed out by Leach (1976), this sentence is ambiguous. Translated into English it could be rendered as 'We are not, therefore, claiming to show how men think the myths, but rather how the myths think themselves out in men and without men's knowledge.' This would suggest a Platonistic autonomy of myth, hence a view that is not a reductive psychological one. However, 'comment les mythes se pensent dans les hommes' can also be translated as 'how myths are thought in men'. In this case, the degree of autonomy implied is reduced. While Lévi-Strauss himself denied that he was an idealist, the exegesis of this text is controversial. For us here, what is of significance is the very consideration of a non-psychological account of myth as a possibility.

I have stressed the similarities between the problematics raised by Ayahuasca and that encountered in several other domains. But the special character of Ayahuasca should be emphasized as well. Mathematics, music, and grammar all involve commonalities of form; not so the visions seen with Ayahuasca. Here we are confronted with commonalities in specific contents. Can it be that there is a level of reality that either pertains to other realms or is an innate psychological one that pertains to specific contents, not just forms? In their classical sense, Platonic Ideas are, of course, of this kind—the Idea of the Horse is semantically specific yet, according to Plato, its existence is not dependent on that of any specific horse (nor, for that matter, of any individual human being).

As for the fourth domain mentioned above, myth, it may very well be the case that the problems presented by it are identical to those presented by Ayahuasca. In both cases the commonalities involve specific contents. It may also be the case that the two are causally, not only conceptually, linked. If myths were discovered by their originators through the use of substance-induced altered states of consciousness, then the world of myth *is* actually the world of Ayahuasca. I tend to believe that this is indeed the case.

In closing this discussion of Platonism, I should note that the link between the Platonic Ideas and the non-ordinary experiences associated with Ayahuasca may be more direct than it seems at first sight. Philosophers normally regard the Platonic Ideas as philosophical theoretical concepts, but perhaps this is not the right way to look at them. Perhaps the Ideas Plato was referring to were not the intellectual creations of the academician but rather entities of the very same kind reported to be seen and experienced by partakers of Ayahuasca. The mediating link for this hypothesis are the mysteries of Eleusis, rituals that lasted for more than two millennia and in which Plato too participated. It is known that in Eleusis, a sacred potion was drunk and that 'the Gods were seen'. Orthodox scholars of classical Greece have invariably interpreted this statement metaphorically. However, some non-orthodox reflection suggests another possibility. Wasson and his associates (Wasson, Hofmann, and Ruck, 1978; Wasson *et al.*, 1986) argue that what was consumed in Eleusis was a psychotropic potion whose chemical composition was very much like LSD. If so, it stands to reason that Plato's Ideas are not mere theoretical constructions and that the various visions described throughout his writings are not metaphors. Rather, they are all visions seen and experienced in the same manner that Ayahuasca visions and ideations are experienced. No wonder, then, that a person who underwent a psychedelic experience, and who was professionally a philosopher, adopted a radical realistic position and postulated the existence of the Ideas. As Wasson (1961: 155) says: 'It is clear to me where Plato found his Ideas, it was clear to his contemporaries too. Plato had drunk of the potion in the Temple of Eleusis and had spent the night seeing the great Vision.'

Mind and Creation

To return to the main line of our discussion: we have considered two options—the psychological and the Platonic. By the first, Ayahuasca visions reflect the hidden depths of people's unconscious; by the second, they capture other, non-human realities. Both options are problematic. The first simply does not seem able to square with the facts. The second is philosophically too difficult for most contemporary educated persons to swallow. Can there be any other option?

Yes—a *creational* one. According to that, Ayahuasca visions reflect neither what is hidden in the mind of the drinker nor mind-independent, Platonic-states of affairs, but rather, they are the works of creation. While the recourse to creation salvages us from Platonic realism, psychologically it is much more radical than may seem at first glance. Specifically, if we are to account for the particular contents that commonly appear in Ayahuasca visions and ideations in terms of enhanced cognitive abilities, including that of imagination highlighted by Merkur (1998), then the notions of 'human creativity' or 'power of imagination' turn out to be much more fantastic than they are usually thought to be.

Earlier, I have recounted the occasion when, under the intoxication, I first came to this realization. It was in conjunction with my vision of the exhibition of another culture. Here let me discuss another example which also bears on the topic of cognitive universals. The example is of subject matter that is very common in Ayahuasca visions—ancient Egyptian civilization. As noted in earlier chapters, seeing places, persons, deities, buildings, and objects pertaining to that old civilization is very common in Ayahuasca visions. How are we to explain this? Practically all people I have discussed this matter with would retort with some sort of explanation that, by the present typology, would be classified as paranormal. The most common explanation that drinkers offer is in terms of reincarnation. They maintain that under the intoxication they gain access to previous incarnations of their self. In the past, one such incarnation was in Egypt, hence visions of ancient Egypt are seen. As already indicated several times in this book, I do not accept this explanation.

I too had my taste of ancient Egypt. As recounted in Ch. 7, one of the most impressive visions I have had—both in terms of sheer visual magnificence and otherwise—was the encounter with two rulers that I took to be Egyptian pharaohs. As I have said, in real life I have been to Egypt, I have visited the Egyptian galleries of the British Museum and of the Louvre, and I have seen pictures and read books about that culture—yet there was just nothing like that vision. The palace in which I found myself under the Ayahuasca intoxication was so real, it was as if it was recently constructed and its appearance was of a lived-in place with regular entertaining. And subjectively, I felt that indeed I was in Egypt. If I were not carried by a kind of time machine to the past, how else could I explain what I saw? It could not be a simple retrieval of information stored in my memory. Even if some reliance on past knowledge was operative here, overall the visioning was extraordinary. Even if no discovery of other realities—be they past, present, or eternal—was involved, even if it was all the creative accomplishment of my own mind, then still it was stunning. I apply this epithet here not to the visions themselves (which, indeed, were extraordinary) but to the power of the mind that created them.

Dealing with the Egyptian case I have focused on the question of status, but again, our first question, that of universals, remains. Why is it that visions of ancient Egypt are so common? Again, to say that Egypt is seen because people often think of Egypt is not a satisfactory explanation. First, it is not true that I or my informants think of Egypt frequently. Second, even if this were the case—why doesn't Egypt appear in regular dreams? Let me venture a speculative hypothesis. According to it, things Egyptian are seen in Ayahuasca visions because the Egyptian civilization was especially good in producing works of art that touch upon the transcendent. Under the intoxication, drinkers neither visit Egypt nor recreate it, but rather, they are in a state of being directed towards the transcendent and their minds are creating sacred works of art. In other words, they are engaged in the very same enterprise in which the artists of ancient Egypt were engaged. The ancient Egyptians invested much effort in the production of sacred works of art, and they were especially good at it. Furthermore, their civilization spanned longer than any other that has existed in history. Consequently, in the hypothetical space of all possible artistic creations attempting to express or invoke transcendence, Egypt has an exceptionally large share: either the Egyptians actually created a large proportion of all the sacred works that potentially could have ever been created by human beings, or the likelihood that their works succeed in grasping the ultimate ideal of the sacred is markedly high. In the course of the Ayahuasca inebriation, drinkers' minds, in effect, attempt to reach the same ideal. Therefore, the likelihood that their visionary creations will bear resemblance to the Egyptian ones is strong. Again, it is not that the visionary creations are Egyptian; rather, being directed at an ideal goal to which Egyptian works are especially close, Ayahuasca visions depict works of art similar to those that the ancient Egyptians created. Given the all too human proclivity to attribute whatever is wondrous to the paranormal, perhaps for lack of any better point of reference, drinkers of Ayahuasca—myself included—tend to label what they see in their visions as Egyptian.

Let me emphasize once again: accounting for the apparently Egyptian visions in terms of creative imagination does not in any way detract from the extraordinariness of these visions. On the contrary. Here am I, an individual human being, whose mind is creating in a couple of minutes what for the Egyptian artists would have taken ages. That we are reaching similar results makes sense—the Egyptian artist, too, was a human being and he, too, was attempting to reach the transcendent. What Ayahuasca does is twofold. First, it orients one to the transcendent. Second, it makes one's mind operate extremely fast and on an extraordinary level of performance. Under the intoxication, the individual mind can achieve wonderful creations. Some have already been created, by other persons at other places, in the past, some will be created in the future, some will never be created as such in this world. Some of these creations resemble, or are even very much like, the Egyptian ones. As noted, the long history of the ancient Egyptian civilization and its great concern with the transcendent and the eternal make this more likely to be the case.[5]

Let us take stock. Avoiding the strictly individualistic psychological option on the one hand and the Platonistic one on the other hand, I have suggested we favour one whose key notion is creation. Creation is, of course, a standard cognitive-psychological notion. Yet, the creative accomplishments discussed here are far from being standard. First, the creative power at hand is extraordinary. Second, while unbounded it is subject to constraints: there are some specific contents that are more likely to be created than others. These—such as serpents and jaguars, palaces and artistic objects, objects made of gold and of precious stones, angels and beings of light—do not reflect the life history, knowledge, or concerns of the individual drinker and they even seem not to be specific to members of any particular cultural group. Thus, it appears that human beings, while not storing

[5] There is also the possibility that the Egyptian artist himself received his creative ideas through the consumption of some magic potion. For pertinent discussion the reader is referred to Shanon (1999).

in their memories ideas of these specific semantic contents, are built in such a fashion as to be likely to imagine them. On the level of the individual person this conclusion may seem to be quite, if not too, radical, but on that of the species this may not be the case. Along with language and the use of tools, art and creation are key defining features of what it is to be a member of the species *Homo sapiens*. In other words, human beings have—as a species—been made to develop art and music, seek the transcendent, and worship the holy. None of these accomplishments were there the day *Homo sapiens* first appeared on the stage of this planet. But were anyone to look at the species from a larger time perspective, perhaps from the perspective of the eternal, it would appear that human beings were made precisely in this fashion—to create the very specific things that they have actually created in the course of history—palaces and temples, paintings and sculptures, scientific theories, and metaphysical ideas. What Ayahuasca seems to be doing is to let time pass exceedingly fast and to let the mind be creative to the utmost. When the cognitive capacities of the individual are thus heightened and when the temporal constraints are greatly reduced, individuals may create, spontaneously and without apparent effort what, in the actual course of history, it has taken entire cultures countless generations to accomplish.

What is suggested here is to be confused with neither the traditional notion of innate ideas nor with the Jungian collective unconscious. What is being suggested rather are specific characteristics of human creative imagination. Unlike innate ideas, these characteristics are not stored in the mind of the individual as such, unlike the Jungian archetypes they cannot be reduced to existential aspects of human life, and unlike universals of music, grammar, and logic they pertain to content, not to form. Thus, in some respects these characteristics resemble the Platonic Ideas—they involve specific contents and they do not reflect the life history of any individual. Where do these characteristics belong, then? My suggestion is that they are part and parcel of the make-up of human beings. Just as bees are made to produce honey and beavers to construct dams, human beings are made to build pyramids and compose piano sonatas. Surely, not all members of the human species will ever do this, but the species as a whole is designed so that some of its members will eventually achieve these accomplishments. With Ayahuasca, the likelihood of individuals to be wonderfully creative is dramatically increased.

Concluding Remarks

Throughout the entire foregoing discussion I have accentuated the contrast between created, hence time-dependent, contents and the Platonic entities that are eternally fixed. But perhaps the difference between the two is not as it may appear at first sight. Creation unfolds in time, but whatever draws human beings to create and makes it possible for them to do so could perhaps be predetermined and thus might be time-independent. This manner of viewing things is in line with the ideas presented in Ch. 14 in conjunction with the notion of eternity and the way in which it is experienced with Ayahuasca. With it, a new perspective of looking at the Platonic Ideas suggests itself. Perhaps the Platonic Ideas are not as Plato himself conceived them to be. Perhaps they are not inert semantic contents posed in another realm, kinds of semantic representations that instead of lying underneath hover above, but rather they are to be regarded as blueprints imprinted in all creation, including the human mind, all human minds. Conceived in this manner, the Ideas seem to exhibit qualities that are more in line with the view of that other giant of classical philosophy, Aristotle. If so, still another dichotomy might be abolished—that between Plato and Aristotle.[6] But since so much of philosophy consists of mere footnotes to these two great men (see Whitehead, 1978), perhaps so many debates that have followed in the course of human history may lose their significance as a consequence.

And so also the distinction between the inner realm and the external one. In the citation above Huxley explicitly talks of other worlds and he endorses metaphysical realism yet at the same time the other world is placed at the very back of the human mind (the reader is invited to reread the second opening epigraph of this book). Perhaps there indeed comes a point where the distinction between the internal and the external has to be given up. Strikingly, experientially this is what Ayahuasca leads people to feel in a direct and very clear manner—with this brew, the boundaries between me and the world, between me and other human beings, between the human and the Divine are blurred and even transgressed. Furthermore, as detailed in Chs. 14 and 15, when ordinary temporality is breached, one experiences that one has embarked upon the realms of the eternal. There meaning reigns. That the philosophical view we have reached here step-by-step by means of a rational, analytical discourse matches what drinkers of Ayahuasca intuit, feel, and see in a direct fashion is, I think, both phenomenologically striking and intellectually significant. Perhaps this in itself is an indication that the 'perennial philosophy' both advocated by mystics throughout the ages and cultures (see Huxley, 1944) and suggested by various individuals who have partaken of Ayahuasca is actually the true metaphysics. But this is precisely the kind of speculation that I have promised not to discuss in this cognitive-psychological book. (For similar conclusions, see Polari's introduction to Cunha, 1986.)

Let me return to cognitive psychology. In Shanon (1993*a*), my critique of the representational-computational approach to the study of mind, I argue against the separation between cognition and the world, between the mental and the bodily, between the realm of the individual and the public domain, between pure and impure cognition. In the same vein, I argue against all sorts of theoretical boundaries and demarcation lines proposed in psychological theories—that between semantics and pragmatics, between cognition and action, between sensation and perception, between perception and so-called higher mental functions, between the pre-interpretative and the interpretative. The Ayahuasca experience seems to

[6] The reader is advised to reread the citation from Huxley (1971) at the end of Ch. 15.

suggest that there are modes of being or realms of existence where still other, more basic distinctions seem to break down—the distinction between creation and discovery, between psychology and ontology, between mind and world, between internal consciousness and the *anima mundi*, between the human and the Divine.

This is perhaps the moral of the whole story. The cross-personal commonalities exhibited in Ayahuasca visions, the wondrous scenarios revealed by them, and the insights gained through them are perhaps neither just psychological, nor just reflective of other realms, nor are they 'merely' a creation of the human mind. Rather, they might be psychological *and* creative *and* real. But when we appreciate this, so much of the fundamental notions by which we view both mind and world have to be considerably altered.

This brings us back to the player of the flute mentioned in the previous chapter. Recall, I characterized the flute as the sum total of the drinker's persona, present and past. The identity of the flutist, however, was left undetermined. In the light of the foregoing remarks, I would characterize the identity of the flutist as *both* psychological and supra-psychological. As a rule, Ayahuasca drinkers' creativity is at its best in two situations: when they surf the special energy that the brew induces and channel it in active performance *and* when they let go and totally immerse themselves in this energy and the effects it generates so that the visionary is experienced as real. Thus, here too, we must relinquish distinctions that are so much a part of us—that between creation and reception, between self and non-self, between hallucination and reality. With Ayahuasca the human mind is under a spell, in a state of grace, and washed by non-ordinary energy so that it, the mind, can perceive what normally cannot be seen or heard, and create in manners and levels that greatly surpass what it can accomplish when in its ordinary state. But, again, isn't this true also of dancers and professional musicians?

Significantly, drinkers' immediate experiences corroborate these observations.[7] On the one hand, drinkers strongly feel that the visions seen with Ayahuasca are not the products of their own individual minds and that many of the ideas and insights entertained under the intoxication are received from external sources. On the other hand, with experience drinkers learn to appreciate how much freedom they can have with the brew and how much that happens during the inebriation is in their own hands. With this realization, they will learn to navigate with the special energies that the brew affords them and channel them to concrete performances in this world—singing, healing, and teaching being the most notable

[7] In principle, I think that drinkers' experiences should be taken very seriously. It is not that drinkers' reports of their experiences should be readily taken without scrutiny and adopted as a psychological theory. Yet, at the same time, the student of mind should beware that his or her theory does full justice to the richness of the phenomenological domain at hand. I have made this appraisal long before studying Ayahuasca, in a general call for a phenomenological approach to psychology (see Shanon, 1984). With the non-ordinary phenomenology of Ayahuasca, this is even more marked. In this domain, psychologists do not have the prior direct acquaintance they normally have in matters cognitive. Therefore, in this case, special attention and respect should be accorded to drinkers' reports and the distinctions drawn in them.

manifestations of this. Further still, under the intoxication people feel that the boundaries between themselves and the world blur and dissipate. And as mystics throughout history have discovered, so do the differences between the individual self and the totality of Being, between individual consciousness and cosmic consciousness, between me and God.

And still there are so many questions the answers to which continue to elude me. Were there ever *ayahuasqueros* who knew about the natural world as much as, or perhaps more than, any modern physicist does? (Under the intoxication, I have felt that this was indeed the case, but I have not even the slightest evidence in support of this.) Can drinkers of the brew gain access to specific factual information? (Again, most Ayahuasca drinkers feel that this is so, but I have still not encountered demonstrative proofs for this.) Perhaps here, too, the only way to make progress is to give up some of the most rooted notions we have of both mind and world and to discard the standard notion of knowledge. In doing so, we are prone to find ourselves outside the province of normal scientific discourse. Here, however, what I have wished to do is study the Ayahuasca experience academically, as a psychologist and as a philosopher. Thus, I can only say that Ayahuasca and the philosophical puzzles it raises mystify me now no less than when this brew first presented me with its wonders and struck me with its mysteries. Now, after almost a decade of questing, I am more accepting of this as an option. Perhaps this appreciation of what I cannot know, what I shall never know, is one of the main lessons that I have learnt from Ayahuasca. Yet, from a cognitive-psychological point the moral of the story is clear: The antipodes of the mind reveal a geography that is much more amazing, much more wondrous than most, if not all, contemporary cognitive scientists seem to surmise.

Epilogue

> Although at times drowned by the roar of jet planes crossing the night sky, the magic chants of vegetalistas still travel to places Western technology will never reach—to the realm of the spirits, a world accepting modern wonders but encompassing the past and the future as well.
>
> Luis Eduardo Luna

I have completed what I set myself to do in this book, but by no means have I said all that there is to say about the Ayahuasca experience. Engaged in the enterprise I have defined for myself, I have deliberately avoided many aspects and topics that extend beyond the realm of the psychological and the scientific. Some involve metaphysical questions, some have to do with religious and spiritual issues, some are very private. The discussion of these must be at times rather speculative, and may call for styles that are distinctly different from the analytical one adopted here. Some of the matters at hand pertain to the ineffable, on others I have actually written quite a bit both in my diaries and in various essays and tales. Included in these are reflections on the nature of reality and the meaning of existence, on science and knowledge, creativity and the arts, about the human condition and the conduct of life, about health and insanity, mysticism and the transcendent, about the mystery that Ayahuasca is. I could also recount about serpents and birds, about histories that may or may not have happened, about music and words of praise, about healing and faith, about my own schooling with Ayahuasca. Confronting these topics and trying to deal with them I am, however, no longer the professional cognitive psychologist I have defined myself to be when engaged in the present investigation. Rather, I am one person, one human being who, along with so many before and like so many afterwards, has sipped of this brew, been affected, and is attempting some understanding.

In a deep sense, the topics not discussed in the chapters of this book may be by far more important and significant than those that are. Correspondingly, the aspects of the Ayahuasca experience on which the former bear may, ultimately, be more significant than the latter, strictly cognitive, ones. More important and significant to drinkers of Ayahuasca in general, and also more important and significant to me. After all, all throughout my engagement with Ayahuasca I have been deeply interested in the psychological issues I have discussed here, and while—as I have indicated at the beginning—my work in cognition is actually of prime importance in my personal life, clearly, it is not just for professional and intellectual interests that I have partaken of Ayahuasca, and continue to do so, while investing substantial energy and resources of all sorts. As further indicated in the Prologue, my engagement with this *bebida* has been a most personal odyssey, one which has greatly affected my world-view and beliefs, my values, my life.

But, again, here I wanted to write a professional, not a personal, book (although, certainly, there are personal aspects to this book too—how could it have been otherwise?). As a student of cognition I set myself to chart the phenomenology of the Ayahuasca experience and to lay the grounds for its cognitive analysis. This project, I feel, is now complete. My story of Ayahuasca is not. The other chapters of this story should be the topic of another book, perhaps books. Very possibly, some chapters may never be written, but quite a few have, in fact, already found their proper articulation in private conversations as well as in song, in prayer, and in silence.

In closing, I would like to express what logically should have been said at the very beginning—namely, my thanks. The sole reasons for my doing this here, at the very end, are technical, editorial ones: the text of the following acknowledgements includes many terms—in particular, names of places and the various contexts of Ayahuasca use—with which many readers may not have been familiar beforehand and which have been defined in earlier chapters of this book.

Substantively, the bottom line is clear: my journey with Ayahuasca was a personal odyssey which could not have been realized without the amiable help of many good persons. Indeed, no less than a voyage to other worlds, this odyssey is the story of a series of meaningful personal encounters. Through Ayahuasca I have had the great fortune to meet many people, coming from varied personal socio-cultural backgrounds, the encounter with whom has greatly enriched my life. Many of these were remarkable human beings, many became good friends. Without them all, my Ayahuasca odyssey would not have been possible. Listing the names of all the persons who helped me on my path would run to many pages. Thus, the following written acknowledgements are presented in a group fashion. In my heart, I thank all the men and women involved personally and individually.

I shall begin with all the community leaders of the Church of Santo Daime and the Masters of the União do Vegetal throughout Brazil as well as Dona Chica, the president of the Barquinha community in Rio Branco. Amiably, they all opened the doors of their houses of worship and extended a warm welcome to me. This they did even though it was clear to them that I did not share their religious beliefs and that I did not intend to become a member of the groups over which they presided. Time and again, special help was given to me as a visitor—transportation was arranged, lodging and food were offered, all sorts of practicalities were taken care of. Many of the leaders shared with me insightful reflections and wise words, from which I have greatly learnt and profited.

But I could also begin with the traditional *ayahuasqueros* who shared with me the treasures of their cultural traditions. Some of them have blessed me with what I discovered to be one of the most wondrous, most loving, most intimate of human contacts—the healing chant and the cleansing breath. Some invested extra effort to tell me something about their life experiences and to pass some of their knowledge to me. In this group I shall also include Dr Jacques Mabit who kindly invited me to stay in the Takiwasi Rehabilitation Centre and participate in some of its activities.

In all places and contexts I have met many individuals who shared experiences, thoughts, and feelings with me. Many helped me—with a hand, a warm glance, an assuring nod, a smile, or a good word—in the tough moments when the 'force' strikes strong. Many helped with information, the establishment of further contacts, as well as in all sorts of practical matters. *Muito obrigado* and *Gracias* to all.

My quest also led me to various interesting professionals—botanists, pharmacologists, physicians, anthropologists, students of Kabbalah, as well as artists. Special mention is due to the venerable Prof. Richard Evans Schultes who opened for me the doors of the incomparable library of the Harvard Herbarium, where I spent a couple of months before leaving for my extended journey in South America. I should also note the helpful hospitality of the Department of Social Anthropology of the Federal University of Rio de Janeiro who served as a convenient basis for my Amazonian excursions in Brazil.

Especially dear to my heart are all the independent drinkers I have met—fellow seekers of the mysteries of Ayahuasca who trod the uneasy road without the support of an established institutionalized framework. With many of these individuals I have had many long conversations from which I have greatly profited. With some I still maintain permanent friendly contact. By name, I shall mention only one, Eulisio, for he is no longer here—this robust black sailor with a golden heart has been untimely vanquished by AIDS. In the Hebrew tradition of which he was very fond, I wish to use this opportunity to bless his memory.

In Brazil, Peru, and Ecuador, on the one hand, and Israel and Europe, on the other, many people were kind enough to serve as informants for me. To all who were willing to share with me experiences that are most personal and at times delicately intimate—my appreciation and thanks.

There were also persons who had never had any experience with Ayahuasca—in fact, all but one had never been to South America—whose assuring presence I carried with me as I went along this odyssey. These were members of my family and my friends back in Israel—notably Lydia Aran, Amnon Levav, Ilana Nadir, Cariel Pardo, Batya Segal, and Yoni Shimoni.

Last but not least, I would like to mention the very beginning of my involvement with Ayahuasca. It was a person who had never partaken of the brew, my Israeli friend Amnon Levav who introduced me to Brazil and gave me the very first lead in my quest. It was through him that I made contact with Antonio Aranha who, in turn, gave me the name of Padrinho Raimundo Nonato of the Colônia Cinco Mil. There I drank for the first time, and my life changed.

In the different stages of my academic study of Ayahuasca I was helped by several diligent student assistants. Amit Hagar participated in the first phase of this research—a period which was especially difficult. David Carmel, Yaron Bromberg, and Ronen Golan assisted with the compilation of the data. Yoel Strimling, with his splendid sensitivity to language and extraordinary encyclopedic knowledge, was in charge of the editing and text processing. (Would you imagine an Orthodox rabbi who could find a passage in Christian literature that a Jesuit scholar said did not exist, and an Islamic one that a professor of Arab culture did not know where to locate? Reb Yoel did all this and more.) Yoel read many drafts of the manuscript and offered innumerable useful comments and suggestions. Generously, he continued giving his good council even when he was no longer a student at the University. He was also meticulously engaged with the many facets of the data analysis, as well as the statistical and computer work associated with them. Yossi Schellas joined in at the very last stages of the project. He carefully helped with the management of the bibliographical data and with the final typesetting and proofreading of the manuscript. But the one person who was involved in all stages of this project and whose dedicated engagement was indispensable for its completion was Nurit Shacham. On the one hand, Nurit toiled with the compilation of the data, their codification, and their analysis; doing this, she prudently introduced work procedures that were essential for these parts of the project to proceed. On the other hand, Nurit read all the many versions of this text and constantly contributed with thoughtful criticism and wise advice. Exhibiting exemplary care and responsibility, she kept doing this until the very last moment, when the final proofs were sent to the publisher. Furthermore, during all those years she was a wonderful interlocutor with whom I could share many thoughts and reflections, ones that found their place in this book and ones that did not. This book would not have reached its present form had Nurit Shacham not been there. At various stages of my writing, I also profited from the good comments made by Anne Simpson, the house editor of NIAS, my students David Carmel and Zvi Carmeli, and my friends Lydia Aran and Nathan Horowitz. And of course, at the very end of the process, the professional staff of Oxford University Press—Anne Ashby, Sarah Holmes, and Sylvia Jaffrey—provided their good services in turning a manuscript into a published book. To all—I am deeply grateful.

It has taken me almost six years to write this book. A significant part of this writing was carried out during my stay as a Golestan fellow at the Netherlands Institute of Advanced Study (NIAS). For the facilities that NIAS offered and for the wonderful assistance of its entire staff I am very grateful.

This project was supported in part by the Grace and Milton Rabson Endowment Fund and the Research Authority of the Hebrew University. The Multidisciplinary Association for Psychedelic Studies (MAPS) generously helped with the acquisition of computer equipment and with the bibliographical materials, both indispensable for the research on which this book is based.

Often when surfing the seas of Ayahuasca, I wished my parents could have a taste of this brew and of the marvels it presents. Let this book allow them to at least appreciate what it is all about.

APPENDIX

Quantitative Data

By and large, the discussion in this book has been qualitative in nature. In this Appendix, I will supplement this discussion with some quantitative information. The information presented here is far from being exhaustive—the data I have collected are extensive and they contain many details. The statistical analyses of these data are complex and multifaceted and their full consideration lies beyond the scope of this book. Thus, what is presented here is not a complete quantitative analysis but rather a preliminary reference summary that supplements the qualitative survey drawn throughout this book. A full quantitative and comparative analysis will be presented elsewhere (for some further discussion, see Shanon, 1998*c*).

The Data

In this Appendix I examine the various content items most commonly seen in Ayahuasca visions and investigate the relative frequency of their occurrence in several subject populations. These populations differ in place of residence, socio-cultural background, and past experience with Ayahuasca, as well as in the locales and contexts in which they have consumed the brew. The analyses reported here were triggered by the observation I made at the very outset of my encounter with Ayahuasca, namely, the detection of commonalities between the types of contents that appear in the visions reported by different persons. Here, these commonalities are examined by comparing the relative frequencies of different content items that appear in the visions. In particular, the present discussion focuses on what I have defined as the micro-level of the content analysis of visions. As underlined throughout this book, this is only one aspect of the Ayahuasca experience. In this analysis I focus on it because it is the aspect that is most amenable both for empirical investigation and for quantitative analysis.

The following discussion is based on the examination of my own core corpus, the interviews I have conducted, and reports in the literature. (Further information regarding the data collection and the distribution of the informants is presented in Ch. 3.) All told, twelve sets of data pertaining to Ayahuasca visions were analysed. These sets are described in detail below; by way of easy reference, Table 1 (see next page) lists the different sets of data examined here. In addition, by way of comparison, reports of dreams are also examined.

1. My Own Core Corpus

Set 1 consists of what I refer to as my own 'core corpus'. As explained in Ch. 3, this corpus consists of diary reports of the first 67 sessions I had with Ayahuasca. The reader is reminded that in the qualitative (i.e. not quantitative) survey in this book, references are made also to sessions that took place after that date. The core corpus includes Ayahuasca sessions that took place in various locales and in different contexts. These include 20 sessions with the Church of

TABLE 1. *List of data sets*

1.	The Core Corpus
2.	The Polari Corpus
3.	The Amaringo Corpus
4.	Structured Questionnaires—Independents I
5.	Structured Questionnaires—Independents II
6.	Structured Questionnaires—Affiliates
7.	First Sessions—Non-indigenous
8.	First Sessions—Indigenous
9.	Best Sessions
10.	Specific Sessions
11.	Reports from the Anthropological Literature—Whites
12.	Reports from the Anthropological Literature—Indians

Santo Daime, 11 with the UdV, 1 with the Barquinha, 18 in indigenous or mestizo Amazonian ceremonies, 12 with independent drinkers in Rio de Janeiro, and 5 in which I consumed Ayahuasca alone.

2. Data Furnished by Means of Structured Interviews

Three sets of data are based on structured interviews, each furnished by a different group of informants. The first, set 4, is based on interviews conducted with the group of 18 independent drinkers with extensive experience (at least 40 sessions) described in Ch. 3; the second, set 5, is based on interviews with 10 independent drinkers with moderate experience (between 10 and 40 sessions); the third, set 6, is based on interviews conducted with 21 members of the Church of Santo Daime and the UdV, all with extensive experience with Ayahuasca. Henceforth, these groups and the data sets corresponding to them will be referred to as Independents I, Independents II, and Affiliates, respectively.[1]

In the structured questionnaires informants were asked about their lifelong experience with Ayahuasca. The informants were presented with different content categories and were asked whether they had ever seen items pertaining to these categories. An affirmative answer indicated that the item in question was seen at least once by the informant, throughout his or her entire experience with the brew. Of course, this measure is sensitive neither to the actual distribution nor the relative frequencies of the contents in question. It is also bound to result in relatively high numerical values. Yet in other respects the data at hand have a special merit: given that they tap a large base of experience, the data collected by means of the structured questionnaires are likely to reveal a good general picture of the types of content encountered in Ayahuasca visions. Indeed, the total cumulative corpus of visionary experiences these questionnaires reflect is very large. My estimate is that, all told, the data discussed here are based on about 2,500 Ayahuasca sessions.

The questionnaire was constructed when the core corpus was closed and following cursory exchanges with various persons regarding their experiences with Ayahuasca. The queries presented in the questionnaire reflected the spectrum of content items indicated in these experiences and in the literature.

[1] I combine the members of the Church of Santo Daime and those of the UdV together, for otherwise the groups are too small.

3. Data Pertaining to Single Sessions

Several sets of data were based on interviews in which informants were asked about the first time they experienced visions with Ayahuasca, those sessions that they regarded as the most powerful or meaningful they have had and details regarding a specific session in which the person had participated. Henceforth, I refer to these three sets as 'first', 'best', and 'specific', respectively. The informants queried with respect to these three types of sessions included indigenous Amerindians, non-indigenous residents of South America (henceforth, South Americans), and foreigners (that is, persons from outside South America). Some of the informants queried about their first and best sessions—but not the specific ones—also answered the structured questionnaires.

For the first sessions, the data were separated into two sets—that of the non-indigenous informants and that of the indigenous ones, the former being comprised of the non-indigenous residents of South America and the foreigners. For the best and the specific sessions the number of informants did not warrant further separation according to the identity of the informants. All told, then, four sets of data pertain to single sessions: sets 7 and 8 for the first sessions (non-indigenous and indigenous informants, respectively), set 9 for the best sessions, and set 10 for the specific sessions.

Of the 152 individuals queried about their first sessions, 15 were indigenous and 136 not (of the latter 81 were urban South Americans and 55 foreigners). Reports of first sessions are especially interesting because they pertain to a stage where informants' experiences are least likely to be affected by expectations, prior knowledge, shared information, and/or group indoctrination.

A total of 51 individuals were asked about their best sessions The rationale for this query was twofold. Quantitatively, in one powerful session there may be more content details than in several non-powerful ones. Thus, focusing on best sessions is prone to entail a maximization of information. Qualitatively, the contents that appear in powerful sessions are likely to be richer, more extraordinary, and have greater personal and/or spiritual impact. The individuals included 16 members from the first group of the independent drinkers, 7 from the second group of independent drinkers, 14 members of the Affiliates, 6 members of the UdV not included in the Affiliate group, 2 indigenous shamans, 2 non-indigenous South Americans, and 4 foreigners.

Specific sessions were investigated whenever I had the opportunity to ask people about their experiences; usually this was immediately following sessions in which I myself had also participated. A total of 22 individuals were interrogated in this fashion. They included 8 South Americans, 8 indigenous Amerindian persons, and 6 foreigners. Taken together, the data from the specific sessions may be regarded as a survey based on random sampling, similar to that conducted in the social sciences when information regarding populations is tapped.

I shall further explain that the data pertaining to the single sessions differ from those collected by means of the structured questionnaire. Unlike the latter, these data are not retrospective assessments of cumulative experiences but rather the actual things that informants have experienced in a given, specific Ayahuasca session. Thus, the structured questionnaires highlight the kinds of contents that may be seen in Ayahuasca visions whereas, in contrast, the data pertaining to single sessions reflect the actual distribution of these contents. Given this basic difference, from a strict technical point of view the data collected by means of the structured questionnaire are not directly comparable with those based on either full accounts of sessions (as in my data and in the Polari and Amaringo corpora; see below) or those collected by means of the non-structured interviews.

4. Corpora from the Literature

In the literature there are only two corpora in which a sizeable number of Ayahuasca visions experienced by one individual are described. The first corpus (set 2) is that of all the Ayahuasca sessions recounted by Polari—one of the leaders of the Church of Santo Daime—in his autobiographical *O Livro das Mirações* (Polari, 1984). This corpus consists of 41 sessions; henceforth it is referred to as the 'Polari corpus'. The second corpus (set 3) is of the paintings made by the Peruvian shaman-turned-artist Pablo Amaringo and presented in Luna and Amaringo (1993). My analysis covers 45 of the 48 paintings reproduced in the book; three paintings were discarded, two because they do not portray visions but rather the context in which Ayahuasca sessions take place, and one because it represents a mythological legend and is of a quite different style from all the other paintings in the corpus. Henceforth this set of data is referred to as the 'Amaringo corpus'. The Amaringo paintings were coded while consulting the accompanying text in Luna and Amaringo (1993).

The Amaringo paintings are very rich and each contains a great number of items; it is also important to underscore their uniqueness. Producing such a corpus requires a conjunction of talents, each of which is rare in its own right. One has to have had extensive experience with Ayahuasca and the privilege of many rich visions. Not many individuals have visions at the level exhibited by Amaringo. Then one has to have the talent of recording the visions in an artistic medium. To be able to do so in the manner Amaringo has is, indeed, unique.

5. Data Drawn from the Anthropological Literature

In the course of my research I have examined all the firsthand accounts of Ayahuasca visions that I could find in the literature. Here I include two sets of such reports. The first, which defines set 11 of the present analysis, consists of 22 reports of non-indigenous persons who partook of Ayahuasca up to 1980. This set includes more or less all the firsthand reports by early scholars and travellers. My impression is that the early reports are very different from many that are appearing in print today. Personally, I find them similar in style to the data I have recorded myself. In contrast, many of the contemporary reports about Ayahuasca visions seem to me to be very embellished; they also seem to be contaminated by prior acquaintance with other reports in the literature.

The reports in this set are: the account of the Ecuadorian civil servant Villavicencio (1858, see Ch. 1), that of the anthropologist Koch-Grünberg (1921), four accounts by subjects tested by the physician Fischer Cardenas (1923) in his doctoral dissertation (apparently, the first scientific study of Ayahuasca ever conducted), the account of Colonel Morales cited in Rouhier (1924), the accounts of the writers Burroughs and Ginsberg (1963) as they appear in *The Yagé Letters* (the accounts for the two authors are treated separately), the account of the colonist merchant Kusel (1965, see Ch. 1), four accounts of white persons dating from between 1950s and mid-1970s cited in Taussig (1987), as well as the firsthand reports of the investigators Reinburg (1965), Waisbard (1969), Rivier and Lindgren (1972), Reichel-Dolmatoff (1975), Deltgen (1978/9), Harner (1980), Taussig (1987), and Luna (Luna's first experience with Ayahuasca which took place in 1980 is reported in Luna and Amaringo, 1993).

The second set from the anthropological literature consists of verbatim reports furnished by individual *indigenous* persons; in the context of the present analysis, it is labelled as set 12. Remarkably, there are very few such reports in the anthropological literature. Most are summary descriptions presenting no actual reports nor specific details on the number and identity of the informants being investigated. The 15 reports analysed here are all that I could find. They include that by a Jivaro shaman (Harner, 1973*a*), by Chaves (1958, quoted in Harner, 1973*c*), those cited in Der Marderosian *et al.* (1970), Lamb (1971), and Kensinger (1973), three by Langdon (two from 1979, and one from 1992), the cumulative descriptions presented in the memoir of the Secoya shaman Fernando Payaguaje (1983), a report cited in Luna (1986*b*), the description of an apprentice's journey with his master recounted by Arévalo Valera (1986; this description is cited in Ch. 1), and four accounts reported by Taussig (1987).

Codification and Semantic Considerations

I begin with some logical-semantic considerations. A fundamental tenet of psychological cognition is that categorization is a human projection, not a feature of the things in the world as such.[2] As a rule, things in the world can be categorized—and correspondingly, labelled or named—on different levels of specificity; for example, *Dumbo* is—all at the same time—an elephant, a mammal, and an animal. Furthermore, some categories are contained, as subsets, within others (e.g., *elephants* is a subset of the set *mammals* which, in turn, is a subset of the set *animals*). This defines a logical-semantic hierarchy in which the more general categories define higher levels whereas the more specific categories define lower levels. Thus, a priori, there is a built-in bias for the categories that pertain to higher semantic levels to rank higher when ordered in terms of frequency.

In the present context, the 'things' are the items seen in the visions. These were coded in terms of the most specific level other than that of the individual (that is, the most specific level along which a set could be defined). Any more specific codification would have been neither practical nor constructive for the analysis (whose aim, after all, is the abstraction of stable patterns out of the plethora of details). In the example just given, this level is that of elephant, not Dumbo. All told, the data were coded in terms of 213 elementary codification labels. This level of specificity proved adequate in that all the data reported by informants or cited in the literature was covered by one label or another.

For the purpose of the analysis, the elementary content items were grouped into the following 27 semantic clusters: personal and autobiographical material,[3] royalty and religious figures, mammals, birds, reptiles, sea creatures, insects, non-naturalistic animals, creatures, and beings (that is, ones that are neither human beings nor animals), landscapes, forests and gardens, cities, palaces and temples, objects of art and magic, household objects, vehicles of transportation (including spaceships), symbols and scripts, scenes and images pertaining to ancient civilizations, historical figures and scenes (history, for short), mythological figures and scenes (mythology, for short), celestial and planetary scenes, heavenly scenes, scenes of creation and evolution, divine and semi-divine beings, encounters with the Divine, scenes of supreme light, and death. The list presented here is somewhat different

[2] The following discussion is grounded in modern cognitive study of categorization, notably that of Rosch and her associates (1976, 1978).

[3] From a semantic-logical point of view, this is a super-category (see below); it is included here because of the paucity of items pertaining to it in the data (more on this later in this text).

from that presented in Ch. 8. The reason for the differences are editorial considerations having to do with the length and order of the discussion which were applicable in the chapter presentation, but not here. For further characterization of the categories and for examples, see Ch. 8.

Except for the first cluster (personal and autobiographical materials; see below), all the codification clusters just introduced constitute the *basic semantic categories* of the present analysis;[4] at times, these are referred to simply as categories. From a logical-semantic point of view, the basic categories define a middle level: above them are the larger and more general *super-categories* and below them are what are being referred to here as *details*.[5] (In other words, the basic categories are subsumed under the super-categories, whereas the instances of the basic categories are the details.) In a general sense, the super-categories, the basic categories, and the details are all *categories*.

The basic categories of the present codification are related and together they may be grouped into the following 6 *super-categories*: human beings, naturalistic animals, plants and botanical scenes, mythological and phantasmagoric creatures and beings, cities and architectural complexes, and objects. From a logical-semantic point of view, the category of personal and autobiographical material is a super-category as well. It is included in the list of categories because of the paucity of items pertaining to it in the data. Some elementary categories in the present codification system are not associated with any super-category.

Whenever pertinent, each item reported was coded in all the elementary labels that could be applicable. Thus, for example, a king's golden crown was coded as an object, as an object pertaining to royalty, and as an object made out of gold. All the data were coded by three people—the author and two research assistants. In cases of disagreement, which in general were minor, a discussion was held until a consensus between the coders was reached.

It should be emphasized that the present codification—and the analyses based on it—is not sensitive to whether a content item appeared once or more within a given session (or a painting, in the case of Amaringo). In other words, if an item appeared just once in a session or many times in several visions within a session, the count is the same. Thus, the numerical values presented here denote percentage of sessions out of the total number of sessions surveyed in which items pertaining to given content categories were seen, not the relative occurrence of such items within any single Ayahuasca session (and by implication, in the corpus as a whole).

Analysis and Comparison

Logically, meaningful comparisons can be made only for content items pertaining to categories of the same semantic level. In the main, the following discussion focuses on the middle level, that of basic categories. In some cases, where frequencies associated with some lower-level items are especially high, numerical values pertaining to the details are furnished as well. As for the super-categories, it should be noted that, for most, the frequency values

[4] The term 'basic category' is taken from Rosch (1978).

[5] The details defined here are distinct from the 'small details' surveyed at the end of Ch. 8. The definition made here is a conceptual, semantic one whereas that made in the chapter is impressionistic, denoting contents that may be semantically peripheral but which stand out in the visions in terms of salience and/or frequency. Some of these 'small details' happen also to be details from the semantic point of view defined here; many were actually one semantic level below the details considered in the analyses presented here.

are high; this reflects the logical structure of the semantic hierarchy, not specific patterns in the data as such. Obviously, if a super-category subsumes several (basic) categories, the numerical values of frequency associated with it are bound to be higher than those associated with any of the categories that are subsumed under it. However, if a low-level category is ranked as high as a high-level category, a significant pattern in the empirical data is indicated. Thus, most of the significant numerical values in the data pertain to the level of the basic categories.

Since the methods of data collection for the different sets differed, what are defined as data points for the analyses that correspond to them vary for the different sets; by implication, the meaning of the numerical values (percentages and rankings) computed for these sets varies as well. For the corpora of single individuals (my core corpus, the Polari reports, the Amaringo paintings) the numerical values specify the number of sessions (paintings, in the case of Amaringo) out of the total number of sessions (paintings) in the entire corpus in which items pertaining to a given content category were reported. For the structured interviews, the numerical values specify the percentage of individuals out of the total number of informants interviewed who reported having seen items of a given content category. For the single sessions, since each informant reported on only one session, the numerical values are defined over both informants and sessions. Thus, in the analysis of the sets pertaining to single sessions, *ipso facto* these frequency values indicate the percentage of individuals (out of the entire group of informants) who reported a given content category. As for the data collected from the anthropological literature, the numerical values in the tables indicate the percentage of reports that included items pertaining to the various content categories at hand.

I have emphasized that direct comparisons are feasible only for sets of data collected by the same method. By way of further clarification, I will summarize the considerations on which this general methodological statement is based. First, as noted above, the definition of the frequency values varies with the method of data collection. Second, it should be borne in mind that by design, the sets of data collected by means of structured interviews are expected to present higher numerical values than the other sets. This is so because in the structured questionnaires informants were asked about their lifelong experience with Ayahuasca. It was sufficient for an informant to have seen an item once throughout his or her cumulative experience with Ayahuasca for this item to be coded. In contrast, in the sets of data pertaining to both corpora and single sessions, the numerical values of frequency indicate the percentage of sessions out of the total number of sessions in the corpus in which a given item was seen. Obviously, the numerical values obtained here are expected to be smaller than those in the first. Third, while, technically, the sets of data pertaining to single sessions are directly comparable to those pertaining to the corpora (that is, mine and those of Polari and Amaringo) in that in both cases, the data comprises a set of reports each specifying items seen within one Ayahuasca session, the two differ in that the former involve a between-subject, the latter a within-subject comparison.

In the light of the methodological explanation made above, in presenting the results I focus on the basic categories. The presentation of the results in the text is coupled with a presentation in a series of tables. For reasons of brevity, for some categories the terms employed in the tables are shorter variants of the terms employed in the text. The relative frequencies for the different sets of data are presented in groups, according to the method by which the data were collected. Table 2 presents the data for my core corpus.

TABLE 2. *The Core Corpus*

Set 1. The Core Corpus (n=67)	
Super-categories	%
Human beings	52.24
Natural animals	50.75
Phantasmagoria/supernatural	47.76
Architecture	40.30
Objects	37.31
Plants	26.87
Personal biography	11.94
Categories	
Mammals	37.31
Objects of art and magic	35.82
Birds	28.36
Royal and religious figures	28.36
Landscapes	26.87
Palaces and temples	26.87
Non-natural animals	25.37
Heavenly scenes	25.37
Reptiles	23.88
Divine beings	23.88
Cities	23.88
Vehicles	22.39
Details	
Felines	22.39
Waterscapes	19.40
Flowers	19.40
Objects of gold	17.91
Serpents	17.91
Processions	16.42
Dancing women	14.93
Forests	13.43
Temples	13.43
Semi-divine beings	13.43
Royal figures	11.94
Enchanted cities	11.94
Open landscapes	11.94
Palaces	11.94
Angels and transparent beings	10.45
Gardens	10.45
Royal objects	10.45

TABLE 3. *The Polari and Amaringo Corpora*

Set 2. The Polari Corpus (n=41)		Set 3. The Amaringo Corpus (n=45)	
Super-categories	%	***Super-categories***	%
Phantasmagoria/supernatural	46.34	Natural animals	97.78
Human beings	31.71	Phantasmagoria/supernatural	95.56
Objects	31.71	Human beings	82.22
Architecture	26.83	Plants	73.33
Plants	14.63	Objects	66.67
Natural animals	12.20	Architecture	42.22
Personal biography	12.20	Personal biography	0.00
Categories		***Categories***	
Divine beings	26.83	Reptiles	91.11
Palaces and temples	26.83	Creatures and beings	84.44
Creatures and beings	24.39	Royal and religious figures	73.33
Landscapes	21.95	Mammals	71.11
Objects of art and magic	21.95	Vehicles	68.89
		Birds	64.44
		Objects of art and magic	62.22
		Non-natural animals	55.56
Details		***Details***	
Palaces	21.95	Serpents	88.89
Open landscapes	17.07	Nymphs	62.22
Angels and transparent beings	14.63	ETs and spaceships	57.78
Royal objects	14.63	Royal figures	53.33
Forests	12.20	Flowers	42.22
Pre-Columbian civilizations	12.20	Royal objects	40.00
		Chimera and winged beings	37.78
		Enchanted cities	35.56
		Religious figures	35.56
		Angels and transparent beings	26.67
		Waterscapes	26.67
		Objects of gold	24.44
		Forests	24.44
		Armoury	24.44
		Guides and guardians	22.22
		Felines	22.22

In the next table, Table 3, because they are directly comparable to mine, are the data for the Polari and Amaringo corpora. Table 4 presents the data for the three sets that are based on structured questionnaires. Table 5 presents the data pertaining to single sessions. Table 6 presents the data based on reports in the anthropological literature.

For my corpus and that of Polari the tables specify those categories that appear in more than 20% of the sessions in the corpus, and those details that appear in more than 10%.

TABLE 4. *Structured questionnaires*

Set 4. Independents I (n=18)		Set 5. Independents II (n=10)		Set 6. Affiliates (n=21)	
Super-categories	%	***Super-categories***	%	***Super-categories***	%
Phantasmagoria/supernatural	100.00	Phantasmagoria/supernatural	100.00	Phantasmagoria/supernatural	90.48
Architecture	100.00	Human beings	90.00	Architecture	76.19
Natural animals	94.44	Objects	90.00	Objects	76.19
Human beings	88.89	Natural animals	80.00	Human beings	66.67
Plants	88.89	Plants	80.00	Natural animals	66.67
Objects	88.89	Personal biography	70.00	Plants	47.62
Personal biography	61.11	Architecture	50.00	Personal biography	33.33
Categories		*Categories*		*Categories*	
Palaces and temples	100.00	Divine beings	90.00	Palaces and temples	66.67
Death	94.44	Death	80.00	Divine beings	57.14
Birds	88.89	Birds	70.00	Birds	52.38
Divine beings	88.89	Ancient civilizations	70.00	Landscapes	52.38
Creatures and beings	83.33	Celestial scenes	70.00	Ancient civilizations	52.38
Reptiles	83.33	Objects of art and magic	70.00	Objects of art and magic	52.38
Forests and gardens	77.78	Mammals	60.00		
Royal and religious figures	72.22	Reptiles	60.00		
Landscapes	72.22	Encounters with the Divine	60.00		
Vehicles	72.22	Royal and religious figures	60.00		
Celestial scenes	72.22	Landscapes	60.00		
Symbols	72.22	Creatures and beings	50.00		
Objects of art and magic	72.22	Palaces and temples	50.00		
Mammals	66.67	Symbols	50.00		
Ancient civilizations	66.67				
Heavenly scenes	61.11				

TABLE 4. (*cont.*):

Set 4. Independents I (n=18)		Set 5. Independents II (n=10)		Set 6. Affiliates (n=21)	
Cities	55.56				
Non-natural animals	50.00				
Mythology	50.00				
History	50.00				
Details	%	***Details***	%	***Details***	%
Serpents	83.33	Felines	60.00	Waterscapes	52.38
Angels and transparent beings	83.33	Serpents	60.00	Palaces	52.38
Palaces	83.33	Semi-divine beings	60.00		
Forests	72.22	Personal acquaintances	60.00		
Felines	61.11	Flowers	60.00		
Flowers	61.11	Angels and transparent beings	50.00		
Temples	61.11	Open landscapes	50.00		
Waterscapes	55.56				
ETs and spaceships	55.56				
Guides and guardians	55.56				
Egyptian scenes	50.00				

TABLE 5. *Single sessions*

Set 7. First Sessions: Non-indigenous (n=136)		Set 8. First Sessions: Indigenous (n=15)		Set 9. Best Sessions (n=51)		Set 10. Specific Sessions (n=22)	
Super-categories	%	***Super-categories***	%	***Super-categories***	%	***Super-categories***	%
Human beings	33.09	Natural animals	73.33	Human beings	43.14	Human beings	50.00
Natural animals	31.62	Architecture	60.00	Objects	39.22	Phantasmagoria/ supernatural	50.00
Phantasmagoria/ supernatural	29.41	Plants	40.00	Phantasmagoria/ supernatural	33.33	Natural animals	40.91
Objects	22.06	Phantasmagoria/ supernatural	33.33	Natural animals	27.45	Objects	22.73
Plants	20.59	Personal biography	33.33	Architecture	17.65	Architecture	18.18
Personal biography	19.85	Objects	20.00	Personal biography	9.80	Plants	18.18
Architecture	16.18	Human beings	20.00	Plants	5.88	Personal biography	9.09
Categories		*Categories*		*Categories*		*Categories*	
Landscapes	17.65	Reptiles	66.67	Objects of art and magic	23.53	Reptiles	27.27
Reptiles	16.18	Landscapes	33.33	Divine beings	17.65	Divine beings	22.73
Divine beings	15.44	Cities	33.33	Royal and religious figures	17.65	Creatures and beings	22.73
Ancient civilizations	15.44	Palaces and temples	33.33	Creatures and beings	15.69	Objects of art and magic	22.73
Mammals	14.71	Mammals	26.67	Mammals	15.69	Royal and religious figures	22.73
Death	13.97	Birds	26.67	Reptiles	13.73	Landscapes	18.18
Creatures and beings	13.97	Divine beings	26.67	Death	13.73	Mammals	18.18
Objects of art and magic	13.24	Royal and religious figures	20.00	Encounters with the Divine	11.77	Scenes of light	13.64
Birds	11.77	Heavenly scenes	20.00	Ancient civilizations	11.77		

Table. 5 (*cont.*):

Set 7. First Sessions: Non-indigenous (n=136)		Set 8. First Sessions: Indigenous (n=15)		Set 9. Best Sessions (n=51)		Set 10. Specific Sessions (n=22)	
Palaces and temples	11.03	Death	20.00	Heavenly scenes	11.77		
		Creatures and beings	13.33	History	11.77		
		Vehicles	13.33				
		Objects of art and magic	13.33				
Details	%	***Details***	%	***Details***	%	***Details***	%
Serpents	14.71	Serpents	60.00	Serpents	13.73	Serpents	27.27
Personal acquaintances	11.03	Felines	26.67	Semi-divine beings	13.73	Semi-divine beings	22.73
Semi-divine beings	11.03	Open landscapes	26.67			Processions	13.64
		Flowers	26.67			Religious figures	13.64
		Temples	26.67			Flowers	13.64
		Angels and transparent beings	20.00			Objects of gold	13.64
		Religious figures	20.00				
		Royal figures	13.33				
		Chimera and winged beings	13.33				
		Waterscapes	13.33				
		Enchanted cities	13.33				
		Boats	13.33				
		Processions	13.33				
		Royal objects	13.33				
		Dancing women	13.33				

TABLE 6. *The anthropological literature*

Set 11. Anthropological reports: Whites (n=22)		Set 12. Anthropological reports: Indians (n=15)	
Super-categories	%	***Super-categories***	%
Natural animals	59.09	Natural animals	93.33
Phantasmagoria/supernatural	45.46	Objects	80.00
Human beings	40.91	Phantasmagoria/supernatural	80.00
Objects	45.46	Human beings	60.00
Architecture	36.36	Plants	33.33
Plants	36.36	Architecture	26.67
Personal biography	18.18	Personal biography	6.67
Categories		***Categories***	
Mammals	40.91	Creatures and beings	73.33
Reptiles	36.36	Reptiles	66.67
Landscapes	36.36	Mammals	46.67
Objects of art and magic	31.82	Birds	46.67
Celestial scenes	27.27	Death	46.67
Creatures and beings	22.73	Sea creatures	40.00
History	22.73	Objects of art and magic	40.00
Death	22.73	Insects	33.33
		Encounters with the Divine	33.33
		Celestial voyages	33.33
		Non-natural animals	26.67
		Landscapes	26.67
		Cities	26.67
		Mythology	26.67
		Symbols	26.67
		Divine beings	20.00
		Royal and religious figures	20.00
		Vehicles	20.00
		Heavenly scenes	20.00
Details		***Details***	
Serpents	31.82	Serpents	66.67
Open landscapes	31.82	Objects of gold	26.87
Felines	22.73	Felines	20.00
		Angels and transparent beings	20.00
		Waterscapes	20.00
		Boats	20.00
		Processions	20.00
		Royal objects	20.00
		Armory	20.00

For the Amaringo corpus higher cut-off points were chosen—50% for the categories and 20% for the details. This was done because the paintings are so rich in detail: the application of the cut-off points employed with my corpus and that of Polari would have resulted in exceedingly long (and statistically not very meaningful) tables. For the structured questionnaires, the tables present those categories (basic categories and details) that were responded to affirmatively by at least 50% of the informants in the groups in question. The cut-off points for the corpora (mine, Polari's, and Amaringo's) are lower than those for the structured interviews; this is due to the fact that the latter taps informants' long-term experiences, whereas the former are based on actual counts of a given set of sessions. As for the single sessions, the cut-off points for both categories and details were 10%.

Substantively, the cut-off points employed here are high. This reflects the fact that the most common items seen in Ayahuasca visions are represented in relatively high frequencies. Serpents, for instance, may be seen in a third of the sessions. In absolute terms, this is most remarkable. To appreciate this, the reader is asked to consider the frequency of those content items that he or she considers to be the most common in his or her dreams. As revealed by the survey conducted by Hall and Van de Castle (1966), the frequency of even the most common content items in dreams reaches just the 15% level. In fact, such a level is reached by only very few items (e.g. parents, houses, and automobiles); in general, the frequency of the most common items that appear is between 3% and 6%.

In addition to the percentages, also presented in the tables are the rank orders of the different categories within each set of data. In contradistinction to the percentages, the rankings do allow for cross-set comparisons. I return to these comparisons below.

The tables are divided into three parts according to the semantic level of the items: super-categories, categories, and details. As indicated above, however, in the main, the focus is on the middle level of categories.

As explained, in most cases there is no sense in examining the values pertaining to the super-categories: the values associated with these are high by construction—they reflect the semantic level of these categories, not the empirical findings *per se*. Furthermore, by and large, the pertinent information pertaining to these values is indicated by the percentages associated with the basic categories subsumed under them. On the other hand, cases in which the frequency values associated with the super-categories are small are especially informative.

Regarding details that are especially frequent, obviously, in general the associated values are expected to be lower than those associated with the basic categories. In contrast, when this is not the case and the frequency associated with a category on the level of detail is higher than the established cut-off point for the basic category, the findings are telling.

Main Findings

1. My Core Corpus (Set 1)

Except for items pertaining to personal biography, all super-categories are represented in more than 25% of the sessions; the highest ranking super-categories are human beings and natural animals, which appear in more than half the sessions.

Eight categories occupy the top five ranks in this set: mammals, objects of art and magic, birds, royal and religious figures, landscapes, palaces and temples, non-natural animals, and heavenly scenes. Items pertaining to these categories appeared in at least 25% of the sessions in this set.

The most frequent details in the core corpus are felines, waterscapes, flowers, objects of gold, serpents, processions, and dancing women; these appeared in at least 15% of the sessions in this corpus (to be precise, dancing women appeared in 14.93% of the sessions).

Interested in how the contents of Ayahuasca visions change over time, I further divided my data into three subsets according to the chronological order of the sessions—my first 22 Ayahuasca sessions, my next 23 sessions with the brew, and the 22 sessions that followed. The most remarkable difference between these three subsets had to do with serpents: While in the earlier sessions these were very prominent (appearing in more than a third of the sessions), their frequency diminished drastically in the later sessions. This pattern parallels observations made by Polari to the effect that novices with the brew see serpents, but with increasing experience, often this is no longer the case (see Richman, 1990/1). I might add that, in terms of locale, it was the second subset that contained the highest number of sessions taking place in an indigenous context; the first subset included just two such sessions. Thus, it is not the Amazonian context *per se* that induces the appearance of serpents in Ayahuasca visions.

2. *Polari (Set 2)*

Of the super-categories, the highest ranking is that of phantasmagoric and supernatural beings—these appear in almost half of the sessions in this corpus. The lowest ranking of the super-categories are those of personal biography and natural animals.

Five categories occupy the top three ranks in this set: divine beings, palaces and temples, creatures and beings, landscapes, and objects of art and magic.

The most frequent details in this set were palaces, open landscapes, angels and transparent beings, royal objects, forests, and pre-Colombian civilizations.

3. *Amaringo (Set 3)*

Of the super-categories, the highest ranking are those of natural animals and that of phantasmagoric and supernatural beings; items pertaining to these two appear in more than 90% of the paintings in the corpus. In contrast, items pertaining to autobiographical material do not appear in this corpus at all.

Eight categories appeared in more than half the paintings in the corpus: reptiles, creatures and beings, royal and religious figures, mammals, vehicles, birds, objects of art and magic, and non-natural animals. These occupy the top eight ranks in the data.

The most frequent details in the Amaringo corpus are serpents, nymphs, extraterrestrials and spaceships, royal figures, flowers, and royal objects. These appeared in at least 40% of the paintings.

4. *Structured Questionnaires—Independents I (Set 4)*

The highest ranking super-categories are those of phantasmagoric and supernatural beings and of architecture. The lowest super-category is that of autobiographical material.

Seven categories were reported positively by more than 75% of the informants in this group: palaces and temples, death, birds, divine beings, creatures and beings, reptiles, and forests and gardens. These occupy the five highest ranks in the data.

The following details were reported positively by at least half the informants: serpents, angels and transparent beings, palaces, forests, felines, flowers, temples, waterscapes, extraterrestrials and spaceships, guides and guardians, and scenes pertaining to ancient Egypt. Significantly, the first three were reported by more than 80% of the informants.

5. Structured Questionnaires—Independents II (Set 5)

The highest ranking super-category is that of phantasmagoric and supernatural beings.

The six most frequent categories in this set are divine beings, death, birds, ancient civilizations, celestial scenes, and objects of art and magic; these occupy the three topmost ranks in the data and were reported by at least 70% of the informants.

Seven details were reported by half or more of the informants: felines, serpents, semi-divine beings, personal acquaintances, flowers, angels and transparent beings, and open landscapes.

6. Structured Questionnaires—Affiliates (Set 6)

The highest ranking super-categories are phantasmagoric and supernatural beings, architecture, and objects. The lowest is that of autobiographical material.

Six categories were reported by at least half the informants in this group: palaces and temples, divine beings, birds, landscapes, ancient civilizations, and objects of art and magic.

Two details were reported by at least half the informants in this group: waterscapes and palaces.

7. First Sessions (Sets 7 and 8)

For the indigenous persons these super-categories are present in at least a third of the reports: natural animals, architecture, plants, phantasmagoric and supernatural beings, and items pertaining to personal biography. For the non-indigenous persons no super-category reached this threshold; those who almost did were human beings, natural animals, and phantasmagoric and supernatural beings.

The categories reported by at least 10% of the non-indigenous informants were landscapes, reptiles, divine beings, ancient civilizations, mammals, death, creatures and beings, objects of art and magic, birds, and palaces and temples. Since the number of informants in the indigenous group was very small, for them I list those categories that were reported by at least a third of the members of this group; these include reptiles, landscapes, cities, and palaces and temples.

For both the indigenous and the non-indigenous informants, the most frequently reported detail was serpents. For the indigenous informants the second most common detail was felines. Incidentally, the reports of first sessions furnished by foreign informants reveal the same pattern.

8. Best Sessions (Set 9)

The super-categories reported by a third or more of the informants describing their best sessions were human beings, objects, and phantasmagoric and supernatural beings.

The categories reported by more than 15% of the informants were objects of art and magic, divine beings, royal and religious figures, creatures and beings, and mammals.

Details that appeared in more than 10% of the reports were serpents and semi-divine beings. In this data set experiences characterized as encounters with the Divine feature rather highly.

9. *Specific Sessions (Set 10)*

Of the super-categories, two were indicated by half the informants—human beings and phantasmagoric and supernatural beings.

As for the categories, since the number of specific sessions is relatively small, only those that were reported by at least 25% of the informants are presented. These include divine beings, mammals, reptiles, and landscapes.

The details reported by at least 25% of the informants were semi-divine beings and serpents.

10. *The Anthropological Literature: Reports of Whites (Set 11)*

The most common super-category was that of natural animals—it featured in almost 60% of the reports. The least common super-category in this data set is that of personal biography.

The five most frequent categories were mammals, reptiles, landscapes, objects of art and magic, and celestial scenes; these occupied the four highest ranks and were mentioned in at least a quarter of the reports.

The most frequently reported details in this set were serpents, open landscapes, and felines.

11. *The Anthropological Literature: Reports of Indians (Set 12)*

Of the super-categories, that of natural animals is represented in all but one of the reports. The other super-categories featuring in at least half the reports were objects, phantasmagoric and supernatural beings, and human beings. Materials pertaining to informants' personal biography were the rarest; indeed, such items were reported by only one person in this group.

The following categories appeared in at least 40% of the reports: creatures and beings, reptiles, mammals, birds, death, sea creatures, and objects of art and magic; these occupy the four highest ranks in the data.

The most frequently reported detail by far was that of serpents—it appeared in two-thirds of the reports. Next were objects made of gold, which appeared in a quarter of them.

Comparative Analysis

As has been explained, for methodological reasons, in their raw form, not all these data are directly comparable. A mode that does allow for comparison across the sets is that of ranking. Thus, for each set, all categories were ranked in descending order of frequency and received rank values accordingly (in other words, rank 1 marks the most frequent category for the given set).

For the rank-based comparisons to be meaningful the populations involved should not be too small. Obviously, if the difference between ranks amounts to a difference of just one or two informants the difference is not very meaningful. And when based on a large population, the same rank value carries a quite different meaning from when it is based on a small one. In the light of these considerations, I have decided to exclude the data on specific sessions (set 10) as well as the first sessions reported by indigenous persons (set 8), and to group together the three sets based on questionnaires (sets 4, 5, and 6).[6] For similar considerations of sample

[6] The ranks discussed here are the averages of the ranks for the three data sets (4, 5, and 6), not the ranks of averages computed over the combined values associated with the different content categories for these sets.

size, I have also excluded the sets pertaining to reports from the anthropological literature (sets 11 and 12) from this analysis as well. This produces results in six sets (my corpus, that of Polari, that of Amaringo, the questionnaire data, first sessions, and best sessions) each of which is based on at least 40 sessions and/or informants. The ranks pertaining to these sets are presented in Table 7.

Before turning to the data let me point out that since it is possible for more than one category to receive the same numerical value of frequency (in other words, some categories may be tied), within a set, several categories may share the same rank.

The ranks were subject to two comparative measures. The first measure indicated the number of sets in which a given category appeared in the top five ranks—the highest possible value for this measure is, of course, the total number of sets under comparison, namely, 6 (out of 6). The second measure indicated the average rank of each category across the six sets; for this measure the highest possible value is 1. The values for the two measures are shown in the final two columns of Table 7. In terms of the first measure, the highest value obtained was 4 (i.e. the category appeared in one of the top five positions in 4 out of the 6 sets under comparison). The categories for which this value was obtained were the following: royalty and religious figures, mammals, reptiles, landscapes, objects of art and magic, ancient civilizations, and divine beings. As for the second measure, the highest value obtained was 4, and the following categories received values between 4 and 5 (inclusive): objects of art and magic, mammals, reptiles, divine beings, and royal and religious figures. The intersection of these two lists comprises of the categories royalty and religious figures, mammals, reptiles, objects of art and magic, and divine beings. This last statement constitutes a concise summary of what are the most common contents seen in Ayahuasca visions.

Similar analyses were conducted also for the details. The most salient finding is that serpents were the most common detail throughout. Except for the Polari corpus, this item ranked at the top of the list for all the data sets examined here: It was first for all sets except set 10, where it ranked second.

I would also like to focus on some specific two-way comparisons. Such comparisons are possible only when the methodology of data collection is the same. First are the three sets consisting of detailed session reports (paintings in the case of Amaringo) furnished by one individual—by me, Polari, and Amaringo. First, I consider my data and those of Amaringo. Since the Amaringo paintings are extremely rich in detail (by far richer than an average session in my core corpus) the cut-off point for percentages and ranks discussed above do not do justice to this corpus: in Amaringo's corpus more categories receive high frequencies than in my corpus; thus, in the Amaringo corpus, some categories that do not reach the cut-off points considered so far are still very frequently represented. Given these considerations, in making the direct comparison between my data and Amaringo's, I lower the cut-off point for the latter from 50% to 25% (the cut-off point for my data remains as above in Table 2). With this change in criterion, in the Amaringo data there are 14 categories above the cut-off point (instead of 8 that appear in Table 3). Of these 14 categories, only 3 do not appear above the cut-off point in my core corpus—creatures and beings, mythology, and forests and gardens. Conversely, out of 12 categories that appear above the cut-off point in my corpus only 1, heavenly scenes, does not appear above the cut-off point in Amaringo's corpus. I find this similarity between the two sets remarkable indeed.

TABLE 7. *Frequencies and averaged ranks*

	Set 1 (n=67)	Set 2 (n=41)	Set 3 (n=45)	Sets 4, 5, 6 (n=49)	Set 7 (n=136)	Set 9 (n=51)	Number of times in the Top 5	Total rank average
	Rank	Rank	Rank	Rank	Rank	Rank		Rank
Personal biography	12	7	19	5.7	1	6	1	8.5
Royal and religious figures	3	5	3	5.3	11	2	4	4.9
Mammals	1	9	4	5.7	5	3	4	4.6
Birds	3	12	6	3.0	8	8	2	6.7
Reptiles	6	10	1	4.0	3	4	4	4.7
Sea creatures	15	12	15	10.0	17	9		13.0
Insects	15	11	14	10.3	13	9		12.1
Non-natural animals	5	12	8	10.3	12	9	1	9.4
Creatures and beings	11	2	2	5.7	6	3	3	5.0
Landscapes	4	3	12	4.3	2	7	4	5.4
Forests and gardens	8	7	13	6.0	10	8		8.7
Cities	6	12	10	6.7	15	8		9.6
Palaces and temples	4	1	11	2.3	9	6	3	5.6
Objects of art and magic	2	3	7	4.0	7	1	4	4.0
Household objects	15	11	12	10.7	16	10		12.5
Vehicles	7	10	5	7.0	16	8	1	8.8
Symbols	15	6	18	5.3	10	6		10.1
Ancient civilizations	8	4	14	4.3	4	5	4	6.6
History	12	10	19	8.3	12	5	1	11.1
Mythology	9	12	9	9.7	16	9		10.8
Celestial scenes	10	6	16	5.3	12	7		9.4
Heavenly scenes	5	7	17	8.3	14	5	2	9.4

TABLE 7. (*cont.*):

	Set 1 (n=67)	Set 2 (n=41)	Set 3 (n=45)	Sets 4, 5, 6 (n=49)	Set 7 (n=136)	Set 9 (n=51)	Number of times in the Top 5	Total rank average
Creation and evolution	13	12	19	10.0	16	6		12.7
Divine beings	6	1	13	**2.0**	**4**	**2**	**4**	**4.7**
Encounters with the divine	14	8	19	8.7	12	**5**	1	11.1
Scenes of light	16	9	19	10.7	14	8		12.8
Death	11	6	14	4.3	6	**4**	2	7.6

Note: Figures in bold indicate categories above the criterial threshold.

Similarities between my corpus and that of Amaringo's are also encountered on the level of details. In the two sets of data, there are 17 and 23 details that reached at least the 10% level (again, the cut-off point for the Amaringo data has been lowered so as to be the same as for my data) respectively. Of these, 13 are present in the both sets.

Turning to the comparison between my corpus and that of Polari, the cut-off points employed above (i.e. in Tables 2 and 3, respectively) are kept. For Polari's data there are 5 categories above the cut-off point; for mine there are, as before, 12. Of the categories that appear above the cut-off point in Polari's data, only 1, creatures and beings, does not appear above the cut-off point in mine; of the categories that appear above the cut-off point in my data, 8 (that is, two-thirds) do not appear above this point in Polari's data. Manifestly, this picture is very different from that revealed by the above comparison between my data and those of Amaringo's. I tend to attribute this difference to Polari's data being, indeed, somewhat different from all other sets of data examined here. First, of all the sets examined, in Polari's the number of categories appearing above the cut-off point is the smallest. Second, Polari's is the only set in which neither serpents nor felines feature prominently; in fact, the two are actually rare in this corpus. Indeed, this is the only set in which the super-category of animals is lowest in frequency. In contrast, in Polari's corpus the category of divine beings ranks highest. It would make sense to attribute this to the religious context in which Polari has partaken of Ayahuasca (as well as to Polari's remarkable spiritual qualities). While not denying this, I draw attention to the fact that divine beings are common in all data sets examined here, and in the great majority of them this category is one of the highest ranking. Thus, the religious context may very well have its effect, but the appearance of religious and spiritual materials in Ayahuasca visions are by no means dependent on such a context (or, for that matter, on prior experience with the brew).

As for the details, again the Polari data is less spread out than mine: there are 6 details above the 10% cut-off point in his data versus 17 in mine. Of these former 6, 5 are also present in the subset of details in my data.

When the data collected by means of questionnaires (sets 4, 5, and 6) are examined, we note 20, 14, and 6 categories above the cut-off point for the Independents I, Independents II, and Affiliates, respectively. The intersection between all three sets is maximal in the sense that *all* 6 categories in the smallest set of the three, namely that of the Affiliates, appear in this intersection (in other words, these 6 categories also appear in the other two sets). Another 7 categories (that is, a total of 13 out of the possible maximum of 14) are shared by the two groups of Independents.

Also allowing for a direct comparison are the sets based on single sessions (sets 7, 8, 9, and 10, respectively). Respectively, there were 10, 13, 11, and 8 categories above the cut-off point for these four sets. The intersection of these four subsets includes five categories (that is, about half or more of the categories in each of these sets). These include mammals, reptiles, creatures and beings, objects of art and magic, and divine beings. It shall be further noted that the categories appearing above the cut-off point only in the best sessions were encounters with the Divine, heavenly scenes, and historical figures and scenes. In other words, sessions characterized by drinkers as their best tend to include items with religious and/or spiritual content.

When the details of the four sets are compared we note that all sets include serpents. In the set of the first Ayahuasca experiences, the detail of personal acquaintances is also common.

Significantly, the data pertaining to the first sessions, for both indigenous and non-indigenous informants, reveal content categories that are typical of Ayahuasca visions in general. In particular, these categories include items of a phantasmagoric nature which do not relate specifically to informants' life experiences and cultural background. Amongst these are reptiles, palaces and temples, objects of art and magic, and divine beings.

Still another possible direct comparison is that between the two sets of data from the anthropological literature—the reports of the Indians and those of the whites. Above the cut-off point there are 19 categories for the former and 8 for the latter. Of the latter, only 1, historical figures and scenes, does not appear amongst the former; of the former, 12 do not appear amongst the latter. Evidently, there is a difference here, but the similarity should not be missed either. The difference between the distribution of the categories may be related, at least in part, to the difference in the number of categories above the cut-off point. Bearing this in mind, we can conclude that what the white people have seen resembles what the Indians saw, but that the visions the Indians had were richer and more varied. This pattern may also be related to the fact that some (but not all) of the Indians were more experienced with the brew than the whites. A different state of affairs would have occurred had the overlap between the two groups been smaller. Had this been the case, we would have said that the Indians see many things the whites do not, and the whites—because of their different socio-cultural background—see many things the Indians do not. Clearly, this is not the picture depicted by the data.

I shall add that the comparison of the details of the two sets of data further marks the similarity between them. For both, the highest-ranking detail was serpents.

General Discussion

By way of conclusion, I will summarize some of the salient patterns revealed by the present comparative analyses. First, some content categories appear high in all, or most, data sets examined here. Significantly, these include items that have nothing to do with drinkers' individual life histories. Notable amongst these are all sorts of animals, phantasmagoric creatures and beings, royalty and religious figures, objects of art and magic, and divine beings. It should be underlined that these contents appear also in the reports of first-timers with no prior acquaintance with or knowledge about Ayahuasca. At times, even the specific descriptions of the items that different informants give are very similar. Furthermore, in their totality, the most common content items reported seem to define one coherent picture. Much of this relates to the world of the fantastic, the marvellous, and the enchanted (see also the concluding remarks in Ch. 8). By and large, this picture appears in all the sets of data analysed here: it is manifested in reports furnished by persons who are unrelated, come from different places, and have varied personal and socio-cultural backgrounds.

Some specific patterns can be highlighted. First, with the exception of the Polari corpus, items relating to the natural world rank very highly in all the sets of data I have examined. These include some of the most frequent items in the corpus—serpents, felines, and birds. Notably, items pertaining to nature are very frequent even with individuals (including myself and most of the informants discussed here) whose personal and cultural background is distinctly urban (hence, also non-Amazonian).

Of all the details, that which is the most frequent is that of serpents. These rank highly in all data sets except those of Polari and the Affiliates.[7] Visions of serpents are very common

[7] Like Polari, the Affiliates are members of Christian-oriented sects, the Church of Santo Daime and the UdV.

even for people with no prior experience with the brew. Across sets, the second most common detail is that of felines. They are the most common item for me and the Independent II group, and they rank very highly also with the Independent I group, the indigenous first-timers, the specific sessions, as well as in the reports of both Indians and whites in the anthropological literature. The third most common are angels and transparent beings; these feature prominently in seven of the twelve sets of data discussed here.

Yet at the same time, some of the most common items that appear in the data include items that pertain not to nature but rather to culture. Notably, these include objects of art and magic (often precious ones), and various architectural complexes. These items are very prominent also in the Amaringo paintings and in the reports of indigenous persons. I cannot vouch that these people have never seen buildings or pictures thereof, but surely, it is very unlikely that they have ever, in real life, seen the kind of wonderfully elaborate edifices with richly ornate works of art that Ayahuasca has presented to them in the visions.

Also relevant is a comment on those items that are relatively rare in the visions. While some aspects of the Ayahuasca experience may be related to the individual's biography and personal concerns, most often this is not the case. Significantly, items of personal and autobiographical nature were common only in the reports of first-timers, both indigenous and non-indigenous. With experience, it seems, such content becomes increasingly less frequent. Further, the category of household objects appeared very low (often, lowest) in all data sets. As we shall see, in dreams a very different picture emerges: there, items pertaining to the individual's biography and life concerns appear frequently, and household objects are extremely common. Also relatively rare were items pertaining to the categories scenes of light, creation and evolution, history, and encounters with the Divine. I would attribute this to the fact that usually visions that contain these content items are considered to be 'grand scenes', and as such they represent only a small fraction of the visualizations that drinkers have with Ayahuasca.

Dreams

In addition to the various sets of Ayahuasca visions, I have also examined dreams. These are interesting for the sake of comparison and have been commented upon several times in this book. In particular, two corpora of dreams were examined. The first is a corpus I have collected of *my own dreams*; it consists of 33 reports. Second is the corpus of all 60 dreams recounted in Freud's *The Interpretation of Dreams* (Freud, 1900/1953)—of these 25 are Freud's own and 35 furnished to him by other people. The first corpus is interesting as it allows a direct comparison with my core corpus of visions: Here are the Ayahuasca visions and ordinary dreams of the same individual—how do they relate to each other? The second corpus is, of course, a classic; the corpus of Freud's own dreams serves by way of comparison to the corpora of Ayahuasca visions of a single individuals such as those recounted by Polari. The dreams recounted by Freud that are not his serve as further independent data of the dream experience.

The analysis of my dreams and those of Freud's was carried out in a manner identical to that of the reports of Ayahuasca visions. However, the data demanded that several additional categories be added for the dreams. These categories are found to be especially common in dreams, but not in Ayahuasca visions. They include family members (as distinct from people one knows), events of the past day or two (henceforth, recent events), seemingly non-significant details such as phone numbers or particular words uttered a day or two prior to the dream. Also added is a category labelled 'paradoxical patterns'; these are patterns that defy logic—the paradigmatic case is that of a person that looks like one individual but who is acknowledged to be another individual.

Table 8 presents the most common content categories in the three sets of dreams in the same manner employed for the Ayahuasca visions above.

Focusing on the super-categories, we note that by and large, these percentages fall into two groups: on the one hand, percentages that are very high and on the other hand, percentages that are very low. The super-categories of the first cluster are the same in all three sets—indeed, the ordering of the super-categories in all the data sets (mine, Freud's own, and Freud's informants) is identical; these super-categories are human beings, personal biography, and objects. The super-categories of the second cluster are animals and phantasmagoric and supernatural beings. Significantly, this pattern is opposite to that encountered with the visions: the super-categories that are most infrequent in the dreams are some of the most common in the visions. Conversely, items pertaining to the subject's personal biography are very common in the dreams but relatively infrequent in the visions.

Turning to the categories, we note that those appearing in more than 20% of my dreams are non-significant items, cities, and household objects. Those categories that appear in more than 20% of Freud's dreams are daily events, family, paradoxical patterns, non-significant items, and household objects. Those that those that appear in more than 20% of the other dreams reported are family members, daily events, non-significant items, household objects, and landscapes. Amongst these categories, all but cities and landscapes are not encountered in Ayahuasca visions. Indeed, salient amongst these categories are those that were especially added to the present codification system for the analysis of the dreams: daily events, family, childhood, non-significant times, and paradoxical patterns. Also encountered in the dreams but rarely in the visions are household objects.

As for the details, in all three sets personal acquaintances ranked highest, appearing in more than 50% of the reports in each of the sets. Again, this picture is quite different from that encountered in the visions.

Because they allow for direct comparison, I will focus further on my dreams in comparison to my visions. There is hardly any overlap in the distribution of the content items in these two. Except for the super-category of human beings, the most common content category in the dreams are items pertaining to personal biography, a category that is especially low in the visions. The content category architecture is common in both my dreams and my visions. However, whereas in the Ayahuasca visions most items pertaining to this category are special buildings—indeed, almost all are either palaces, temples, or magical buildings—in the dreams none of the buildings are of these types. All other categories that are prevalent in my visions are rare or non-existent in my dreams.

Thus, the main lesson of the foregoing comparison is clear: Ayahuasca visions are very different from dreams. In general, dreams have to do with the subject's current life events and concerns whereas the visions—especially when powerful—have to do with worlds that seem to be quite foreign to one's regular life experiences. I reckon that some people have dreams that are of a more phantasmagoric nature and that such dreams are especially common amongst lucid dreams. I, however, almost never had such a dream; nor are such dreams prevalent for Freud and his informants. More on the comparison between Ayahuasca and dreams, and as well as between these and visions not induced by Ayahuasca is left for future research.

TABLE 8. *Dreams*

Author's dreams (n=33)		Freud's dreams (n=25)		Reported by Freud (n=35)	
Super-categories	%	***Super-categories***	%	***Super-categories***	%
Human beings	81.82	Human beings	84.00	Human beings	82.86
Personal biography	60.61	Personal biography	72.00	Personal biography	60.00
Objects	54.55	Objects	68.00	Objects	57.14
Architecture	48.48	Architecture	56.00	Architecture	31.43
Plants	12.12	Plants	12.00	Plants	20.00
Natural animals	3.03	Natural animals	4.00	Natural animals	11.43
Phantasmagoria/supernatural	3.03	Phantasmagoria/supernatural	4.00	Phantasmagoria/supernatural	2.86
Categories		*Categories*		*Categories*	
Non-significant items	45.46	Daily events	44.00	Family	40.00
Cities	21.21	Family	32.00	Daily events	31.43
Household objects	21.21	Paradoxical patterns	28.00	Non-significant items	25.71
Symbols	18.18	Non-significant items	20.00	Household objects	20.00
Paradoxical patterns	18.18	Household objects	20.00	Landscapes	20.00
Objects of art and magic	12.12	Landscapes	16.00	Vehicles	17.14
Family	12.12	Vehicles	16.00	Paradoxical patterns	17.14
		Cities	16.00	Sexual scenes	14.29
		Symbols	16.00		
		Death	16.00		
		Royal and religious figures	12.00		
		History	12.00		
		Objects of art and magic	12.00		
Details		***Details***		***Details***	
Personal acquaintances	54.55	Personal acquaintances	72.00	Personal acquaintances	60.00
				Open landscapes	14.29

References

ALEMAN, A., BOCKER K. B. E, and DE HAAN E. H. F. (1999). Disposition towards Hallucination and Subjective versus Objective Vividness of Imagery in Normal Subjects. *Personality and Individual Differences*, 27: 707–14.

America Indígena (1986). Special Volume, 46.

ARÉVALO VALERA, G. (1986). El Ayahuasca y el curandero Shipibo-Conibo del Ucayali (Perú). *America Indígena*, 46: 147–61.

ARIETI, S. (1974). *Interpretation of Schizophrenia*. New York: Basic Books.

ASAAD, G. (1990). *Hallucinations in Clinical Psychiatry*. New York: Brunner/Mazel.

ASAAD, G., and SHAPIRO, B. (1986). Hallucinations: Theoretical and Clinical Overview. *American Journal of Psychiatry*, 143: 1088–97.

ASHLAG, Y. (1956). *Talmud Eser HaSefirot (The Study of the Ten Spheres)*. Jerusalem: M. Klar.

ASSAGLIOLI, R. (1965). *Psychosynthesis: A Manual of Principles and Techniques*. New York: Viking Penguin.

—— (1973). *The Act of Will*. New York: Viking Press.

ATLAN, H. (1979). *Entre le Cristal et la Fumée*. Paris: Seuil.

AUSTIN, J. (1962). *How to Do Things with Words*. Oxford: Clarendon Press.

AYALA FLORES, F., and LEWIS, W. (1978). Drinking the South American Hallucinogen Ayahuasca. *Economic Botany*, 32: 154–6.

BADDELEY, A. (1990). *Human Memory: Theory and Practice*. Hove and London: Lawrence Erlbaum Associates.

BAER, G. (1992). The One Intoxicated by Tobacco: Matsigenka Shamanism. In E. J. Langdon and G. Baer (eds.), *Portals of Power: Shamanism in South America*. Albuquerque, NM: University of New Mexico Press.

BAKER, J. (1994). The Old Woman and Her Gifts: Pharmacological Bases of the Chumash Use of *Datura*. *Curare*, 17: 253–76.

BALDRAN, J., and BAREIRO-SANGUIER, R. (1980). *La Têté Dedans*. Paris: François Maspero.

BALZER, C. (1999). Wege Zum Heil: Die Barquinha. Unpublished Master's thesis. Freien Universitat Berlin.

BARON, R., and ZEBROVITZ-MCARTHUR, L. (1983). Towards an Ecological Theory of Social Perception. *Psychological Review*, 90: 215–38.

BARON-COHEN, S., and HARRISON, J. (eds.) (1997). *Synaesthesia: Classic and Contemporary Readings*. Oxford: Blackwell.

BARWISE, J., and PERRY, J. (1983). *Situations and Attitudes*. Cambridge, MA: MIT Press.

BEGLEY, S. (2001). Religion and the Brain. *Newsweek*, 137: 44–9.

BELLIER, I. (1986). Los cantos mai Huna del Yajé (Amazonia Peruana). *America Indígena*, 46: 129–45.

BENTALL, R. P. (2000). Hallucinatory Experiences. In E. Cardeña, S. J. Lynn, and S. Kripner (eds.), *Varieties of Anomalous Experience: Examining the Scientific Evidence*. Washington, DC: American Psychological Association.

Bergson, H. L. (1944). *Creative Evolution*. New York: The Modern Library.

——(1950). *Time and Free Will*. New York: Macmillan.

Blake, W. (1794/1961). The Marriage of Heaven and Hell. In G. Keyens (ed.), *The Poetry and Prose of William Blake*. London: The Nonesuch Press.

Brentano, F. (1874/1973). *Psychology from an Empirical Standpoint*. London: Routledge & Kegan Paul.

Bressloff, P., Cowan, J., Golubitsky, M., Thomas, P., and Wiener, M. (2001). Geometric Visual Hallucinations, Euclidean Symmetry and the Functional Architecture of the Striate Cortex. *Philosophical Transactions of the Royal Society (London): Biological Sciences*, 356: 299–330.

Brissac, S. (1999). A Estrella do Norte Iluminado até o Sul: Uma Etnografia da União do Vegetal em um Contexto Urbano, Master's thesis, Universidad Federal do Rio de Janeiro.

Bristol, M. (1966). The Psychotropic Banisteriopsis among the Sibundoy of Colombia. In *Botanical Museum Leaflets (Harvard University)*, 21: 113–40.

Brüzzi Alves da Silva, A. (1962). *A Civilisação Indígena do Vaupes*. São Paulo: Linografica Editora.

Buchillet, D. (1992). Nobody is There to Hear: Desana Therapeutic Incantations. In E. J. Langdon and G. Baer (eds.), *Portals of Power*. Albuquerque, NM: University of New Mexico Press.

Bucke, R. (1901/1991). *Cosmic Consciousness*. New York: Penguin Books.

Burroughs, W., and Ginsberg, A. (1963). *The Yagé Letters*. San Francisco, CA: City Light Books.

Bursen, H. (1978). *Dismantling the Memory Machine*. Dordrecht: D. Reidel.

Callaway, J., McKenna, D., Grob, C., Brito, G., Raymon, L., Poland, R., Andrade, E. N., Andrade, E. O., and Mash, D. (1999). Pharmacokinetics of Hoasca Alkaloids in Healthy Humans. *Journal of Ethnopharmacology*, 65: 243–56.

Casey, E. (1976). Comparative Phenomenology of Mental Activity: Memory, Hallucination, and Fantasy Contrasted with Imagination. *Research in Phenomenology*, 6: 1–25.

Castaneda, C. (1971). *A Separate Reality*. London: Arkana.

——(1972). *Journey to Ixtlan: The Lessons of Don Juan*. New York: Simon & Schuster.

——(1974). *The Teachings of Don Juan: A Yaqui Way of Knowledge*. New York: Pocket Books.

Castillo, A. (1996). *Santo Daime: Levagem Cerebral*. Rio de Janeiro: Imago.

Centro de Memoria e Documentação (1989). *União do Vegetal: Hoasca—Fundamentos e Objectivos*. Brasilia: Centro Espiritual Beneficiente União do Vegetal.

Chango, A. (1984). *Yachaj Sami Yachachinu*. Quito: Abya-Yala.

Chaumeil, J. (1983). *Voir, Savoir, Pouvoir: Le Chamanisme chez les Yagua du Nord-Est Péruvien*. Paris: Éditions de L'École des Hautes Études en Sciences Sociales.

Chen, A., and Chen, K. (1939). Harmine: The Alkaloid of Caapi. *Quarterly Journal of Pharmacy and Pharmacology*, 12: 30–8.

Chiappe, M., Lemlij, M., and Millones, L. (1985). *Allucinógenos y Shamanismo en el Peru Contemporaneo*. Lima: Ediciones El Virrey.

Chomsky, N. (1965). *Aspects of the Theory of Syntax*. Cambridge, MA: MIT Press.

——(1972). *Language and Mind*. Rev. edn. New York: Harcourt Brace Jovanovich.

——(1975). *Reflections on Language*. New York: Pantheon Books.

Chomsky, N. (1980). *Rules and Representations*. New York: Columbia University Press.

——(1986). *Knowledge of Language: Its Nature, Origin, and Use*. New York: Praeger.

CLARK, W. (1969). *Chemical Ecstasy, Psychedelic Drugs and Religion*. New York: Sheed & Ward.

CLOSE, E. (1997). *Transcendental Physics*. Jackson, MO: Paradigm Press.

CLOUZOT, H. G. (1955). *Le Mystére Picasso* [Film].

COHEN, D. (1983). Birdcalls and the Rules of Palestrina Counterpoint: Towards the Discovery of Universal Qualities in Vocal Expression. *Israeli Studies in Musicology*, 3: 96–123.

COLTHEART, N. (1983). Iconic Memory. *Philosophocal Transactions of the Royal Society, London B.*, 302: 283–94.

COSTALL, A. (1995). Socializing Affordances. *Theory & Psychology*, 5: 467–81.

CUNHA, B. (1986). *O Império do Beija-Flor*. Rio Branco: Fundação Narciso Mendes.

CYTOWIC, R. (1989). *Synesthesia—A Union of the Senses*. New York: Springer Verlag.

——(1993). *The Man Who Tasted Shapes*. New York: Putnam.

——(1997). Synaesthesia: Phenomenology and Neuropsychology. In S. Baron-Cohen and J. Harrison (eds.), *Synaesthesia: Classic and Contemporary Readings*. Oxford: Blackwell.

D'AQUILI, E., and NEWBERG, A. (1999). *The Mystical Mind: Probing the Biology of the Religious Experience*. Minneapolis: Fortress.

DAVIS, W. (1991). Towards a New Synthesis in Ethnobotany. In M. Dobkin de Rios and H. Borgtoft Pedersen (eds.), *Las Plantas y el Hombre*. Quito: Abya-Yala.

——(1996). *One River: Explorations and Discoveries in the Amazon Rain Forest*. New York: Simon & Schuster.

DE CUSA, N. (1960). *The Vision of God*. New York: Frederick Ungar.

DELTGEN, F. (1978/9). Culture, Drug and Personality—a Preliminary Report about the Results of a Field Research among the Yebamasa Indians of Rio Piraparana in the Colombian Comisario del Vaupes. *Ethnomedizin*, 5: 57–82.

——(1993). *Gelenkte Ekstase*. Stuttgart: Franz Steiner.

DENNETT, D. (1987). *The Intentional Stance*. Cambridge, MA: MIT Press.

DER MARDEROSIAN, A., KENSINGER, K., CHAO, J., and GOLDSTEIN, F. (1970). The Use and Hallucinatory Principles of a Psychoactive Beverage of the Cashinahua Tribe. *Drug Dependence*, 5: 7–14.

DER MARDEROSIAN, A., PINKLEY, H., and DOBBINS, M. (1968). Native Use and Occurrence of N,N-dimethyltryptamine in the Leaves of Banisteriopsis rusbyana. *American Journal of Pharmacy*, 140: 137.

DESCARTES, R. (1641/1960). *Meditations on First Philosophy*. Indianapolis: Bobbs-Merrill.

DE SOUZA, F. (1996/7). The Ritual Use of Hoasca: Comments and Advice. *Bulletin of the Multidisciplinary Association of Psychedelic Studies*, 7.

DOBKIN DE RIOS, M. (1972). *Visionary Vine: Hallucinogenic Healing in the Peruvian Amazon*. San Francisco: Chandler.

——(1973). Curing with Ayahuasca in an Urban Slum. In *Hallucinogens and Shamanism*. Oxford: Oxford University Press.

——(1984). *Hallucinogens: Cross-Cultural Perspectives*. Albuquerque, NM: University of New Mexico Press.

——(1992). *Amazonian Healer: The Life and Times of an Urban Shaman*. Bridport: Prism Press.

DOBKIN DE RIOS, M., and GROB, C. (1994). Hallucinogens, Suggestibility, and Adolescence in Cross-Cultural Perspective. *Yearbook of Ethnomedicine and the Study of Consciousness*, 3: 113–32.

DOBKIN DE RIOS, M., and KATZ, F. (1975). Some Relationships between Music and Hallucinogenic Ritual: The Jungle Gym in Consciousness. *Ethos*, 3: 64–76.

DOBLIN, R. (2001). *Regulation of the Medical Use of Psychedelics and Marijuana*. Public Policy Ph.D. dissertation, Harvard University.

ELIADE, M. (1964). *Shamanism, Archaic Techniques of Ecstasy*. Princeton, NJ: Princeton University Press.

Encyclopaedia Britannica. (1971). Chicago: William Benton.

ERMENTROUT, G., and COWAN, J. (1979). A Mathematical Theory of Visual Hallucination Patterns. *Biological Cybernetics*, 34: 137–50.

EVANS-WENTZ, W. Y. (ed.) (1960). *The Tibetan Book of the Dead*. 3rd edn. Oxford: Oxford University Press.

FERICGLA, J. M. (1998). *Al Trasluz de la Ayahuasca*. Quito: Abya-Yala.

FESTI, F., and SAMORINI, G. (1996). 'Ayahuasca-like' Effects Obtained with Italian Plants. In *Proceedings of the 2nd International Congress for the Study of the Modification of States of Consciousness*. Barcelona: Institut de Prospectiva Antropologia.

FISCHER CARDENAS, G. (1923). *Estudio Sobre el Principio Activo del Yage*. Bogota: Universidad Naciónal.

FLATTERY, D. S., and SCHWARTZ, M. (1989). *Haoma and Harmaline: The Botanical Identity of the Indo-Iranian Sacred Hallucinogen 'Soma' and its Legacy in Religion, Language, and Middle Eastern Folklore*, Near Eastern Studies, 21. Berkeley: University of California Press.

FODOR, J. (1975). *The Language of Thought*. New York: Crowell.

—— (1983). *The Modularity of Mind*. Cambridge, MA: MIT Press.

FORMAN, R. (ed.) (1990). *The Problem of Pure Consciousness: Mysticism and Philosophy*. Oxford: Oxford University Press.

—— (1998). What Does Mysticism Have to Teach Us About Consciousness? *Journal of Consciousness Studies*, 5: 185–201.

FORTE, R. (ed.) (1997). *Entheogenes and the Future of Religion*. San Francisco: Council on Spiritual Practices.

FRAISSE, P. (1964). *The Psychology of Time*. London: Eyre & Spottiswoode.

FRALEIGH, S. (1993). Good Intentions and Dancing Moments: Agency, Freedom, and Self-knowledge in Dance. In U. Neisser (ed.), *The Perceived Self*. Cambridge: Cambridge University Press.

FRANKL, V. (1975). *The Unconscious God: Psychotherapy and Theology*. New York: Simon & Schuster.

FREUD, S. (1900/1953). *The Interpretation of Dreams*. Standard edn. London: Hogarth Press, iv, v.

—— (1901/1960). *The Psychopathology of Everyday Life*. Standard edn. London: Hogarth Press, vi.

FROES, V. (1986). *Histôria do Povo Juramidam: A Cultura do Santo Daime*. Manaus: Suframa.

FURST, P. T. (1976). *Hallucinogens and Culture*. San Francisco: Chandler & Sharp.

—— (ed.) (1990). *Flesh of the Gods: The Ritual Use of Hallucinogens*. Prospect Heights, IL: Waveland Press.

GACKENBACH, J., and LABERGE, S. (eds.) (1988). *Conscious Mind, Sleeping Brain*. New York: Plenum Press.

GEBHART-SAYER, A. (1985). The Geometric Designs of the Shipibo-Conibo in Ritual Context. *Journal of Latin American Lore*, 11: 145–75.

——(1986). Una terapia estetica: Los diseños visionarios del Ayahuasca entre los Shipibo-Conibo. *America Indígena*, 46: 189–218.

GHISELIN, B. (1952). *The Creative Process*. Berkeley: University of California Press.

GIBSON, J. J. (1966*a*). *The Senses Considered as a Perceptual System*. Boston: Houghton-Mifflin.

——(1966*b*). The Problem of Temporal Order in Stimulation and Perception. *Journal of Psychology*, 62: 141–9.

——(1975). Events are Perceivable but Time is Not. In J. T. Fraser and N. Lawrence (eds.), *The Study of Time II*. New York: Springer Verlag.

——(1979). *The Ecological Approach to Visual Perception*. Boston: Houghton-Mifflin.

——(1982). The Concept of Affordances in Development: The Renascence of Functionalism. In W. Collins (ed.), *The Concept of Development*. Hillsdale, NJ: Lawrence Erlbaum.

GINSBERG, A. (1977). *Journals: Early Fifties Early Sixties*, ed. G. Ball. New York: Grove Press.

GINZBERG, L. (1909). *The Legends of the Jews*. Philadelphia: Jewish Publication Society of America, i.

GLICKSOHN, J. (1986/7). Photic Driving and Altered States of Consciousness: An Exploratory Study. *Imagination, Cognition and Personality*, 6: 167–82.

GOLDMAN, I. (1979). *The Cubeo: Indians of the Northwest Amazon*. Urbana, IL: University of Illinois Press.

GOODMAN, F. (1972). *Speaking in Tongues: A Cross-Cultural Study of Glossolalia*. Chicago: Chicago University Press.

GOODMAN, N. (1968). *Languages of Art: An Approach to a Theory of Symbols*. Indianapolis: Bobbs-Merrill.

GORMAN, P. (1992). Journeys with Ayahuasca, the Vine of the Little Death. *Shaman's Drum*, 29: 49–58.

GOSWAMI, A. (1990). Consciousness in Quantum Physics and the Mind–Body Problem. *The Journal of Mind and Behaviour*, 11: 75–96.

GOSWAMI, A., REED, R., and GOSWAMI, M. (1995). *The Self-Aware Universe: How Consciousness Creates the Material World*. New York: Jeremy P. Tarcher.

GOW, P. (1988). Visual Compulsion: Design and Image in Western Amazonian Cultures. *Revindi*, 2: 19–32.

GRAHAM, G., and STEPHENS, L. (1994). Mind and Mine. In G. Graham and L. Stephens (eds.), *Philosophical Psychopathology*. Cambridge, MA: MIT Press.

GREEN, C. (1968). *Lucid Dreams*. London: Hamish Hamilton.

——(1973). *Out-of-Body Experiences*. New York: Ballantine.

GREEN, C., and MCCREERY, C. (1994). *Lucid Dreaming: The Paradox of Consciousness During Sleep*. London: Routledge.

GREENEWALT, C. (1968). *Bird Song: Acoustics and Physiology*. Washington, DC: Smithsonian Institute Press.

GREGORY, R. (1966). *Eye and Brain: The Psychology of Seeing*. London: Weidenfeld & Nicolson.

GRINSPOON, L., and BAKALAR, J. (1979). *Psychedelic Drugs Reconsidered.* New York: Bantam Books.

GROB, C., MCKENNA, D., CALLAWAY, J., BRITO, G., NEVES, E., OBERLANDER, G., SAIDE, O., LABIGALINI, E., TACLA, C., MIRANDA, C., STRASSMAN, R. J., and BOONE, K. (1996). Human Psychopharmacology of Hoasca, a Plant Hallucinogen Used in Ritual Context in Brazil. *Journal of Nervous and Mental Diseases*, 184: 86–94.

GROF, S. (1972). LSD and the Cosmic Game: Outline of Psychedelic Cosmology and Ontology. *Journal for the Study of Consciousness*, 5: 165–95.

——(1994). *LSD Psychotherapy.* Alameda, Calif.: Hunter House.

——(1998). *The Cosmic Game: Exploration of The Frontiers of Human Consciousness.* Albany, NY: State University of New York Press.

——(2000). *Psychology of the Future: Lessons from Modern Consciousness Research.* Albany, NY: State University of New York Press.

GROISMAN, A. (1999). *Eu Venho da Floresta: Um estudo sobre o contexto simbólico do uso do Santo Daime.* Florianópolis: Editora UFSC.

GROISMAN, A., and SELL, A. (1995). Healing Power: Neurophenomenology, Culture, and Therapy of Santo Daime. *Yearbook of Cross-Cultural Medicine and Psychotherapy*, 6: 241–56.

GROSSINGER, R. (1990). *Planet Medicine: From Stone Age Shamanism to Post-Industrial Healing.* Berkeley: North Atlantic Books.

GUSS, D. (1989). *To Weave and Sing: Art, Symbol and Narrative in the South American Rain Forest.* Berkeley: University of California Press.

HALL, C., and VAN DE CASTLE, R. (1966). *The Content Analysis of Dreams.* New York: Appleton-Century-Crofts.

HAMPEJS, H. V. (1994). *El Extasis Shamanico de la Consciencia.* Quito: Abya-Yala.

HARNER, M. J. (1962). Jívaro Souls. *American Anthropologist*, 64: 258–72.

——(1972). *The Jivaro.* London: R. Hale.

HARNER, M. J. (1973*a*). The Sound of Rushing Water. In M. J. Harner (ed.), *Hallucinogens and Shamanism.* Oxford: Oxford University Press.

——(Ed.). (1973*b*). *Hallucinogens and Shamanism.* Oxford: Oxford University Press.

——(1973*c*). Common Themes in South American Indian Yagé Experiences. In M. J. Harner (ed.), *Hallucinogens and Shamanism.* Oxford: Oxford University Press.

——(1980). *The Way of the Shaman.* New York: Harper Collins.

HARTMANN, E. (1975). Dreams and other Hallucinations: An Approach to the Underlying Mechanism. In M. J. Harner (ed.), *Hallucinations: Behaviour, Experience, and Theory.* New York: J. Wiley & Sons.

HEFT, H. (1989). Affordances and the Body: An Intentional Analysis of Gibson's Ecological Approach to Visual Perception. *Journal for the Theory of Social Behaviour*, 19: 1–30.

HEIDEGGER, M. (1962). *Being and Time.* New York: Harper & Row.

HENMAN, A. (1986). Uso del ayahuasca en un enctro autoritario: El caso de la UdV en Brasil. *America Indígena*, 46: 219–34.

HERRIGEL, E. (1985). *Zen in the Art of Archery.* London: Arkana.

HESSE, H. (1969). *Magister Ludi: The Glass Bead Game.* New York: Bantam Books.

HILL, J. (1992). A Musical Aesthetic of Ritual Curing in the Northwest Amazon. In E. J. Langdon and G. Baer (eds.), *Portals of Power: Shamanism in South America.* Albuquerque, NM: University of New Mexico Press.

HILLMAN, J. (1985). *Anima: An Anatomy of a Personified Notion*. Dallas: Spring.

HOLMSTEDT, B., and LINDGREN, J. (1979). Chemical Constituents and Pharmacology of South American Snuffs. In D. Efron, B. Holmstedt, and N. Kline (eds.), *Ethnopharmacologic Search for Psychoactive Drugs*. New York: Raven Press.

HOROWITZ, M. (1975). Hallucinations: An Information Processing Approach. In R. Siegel and L. West (eds.), *Hallucinations: Behaviour, Experience, and Theory*. New York: J. Wiley & Sons.

——(1978). *Image Formation and Cognition*. 2nd edn. New York: Appelton-Century-Crofts.

——and PALMER, C. (eds.) (1977). *Aldous Huxley—Moksha: Writings on Psychedelics and the Visionary Experience*. New York: Stonehill Publishing.

HOUSTON, J., and MASTERS, R. (1972), The Experimental Induction of Religious-Type Experiences. In J. White (ed.), *The Highest State of Consciousness*. Garden City, NY: Anchor Books.

HUGHES, G. E., and CRESWELL, M. (1996). *A New Introduction to Modal Logic*. London: Routledge.

HUME, D. (1739/1978). *A Treatise of Human Nature*. Oxford: Oxford University Press.

HUNT, H. (1985). Relations between the Phenomena of Religious Mysticism (Altered States of Consciousness) and the Psychology of Thought: A Cognitive Psychology of States of Consciousness and the Necessity of Subjective States for Cognitive Theory. *Perceptual and Motor Skills*, 61: 911–61.

HUXLEY, A. (1944). *The Perennial Philosophy*. New York: Harper & Row.

——(1969). *This Timeless Moment: A Personal View of Aldous Huxley*. London: Chatto & Windus.

——(1971). *The Doors of Perception and Heaven and Hell*. Harmondsworth: Penguin Books.

——(1972). Visionary Experience. In J. White (ed.), *The Highest State of Consciousness*. New York: Anchor Books.

ILLIUS, B. (1992). The Concept of Nihue among the Shipibo-Conibo of Eastern Peru. In E. J. Langdon and G. Baer (eds.), *Portals of Power: Shamanism in South America*. Albuquerque, NM: University of New Mexico Press.

JACOBS, B. (1987). How Hallucinogenic Drugs Work. *American Scientist*, 75: 386–92.

JACOBS, L. (ed.) (1990). *The Jewish Mystics*. London: Kyle Cathie.

JAMES, W. (1882). On Some Hegelisms. *Mind*, 7: 186–208.

——(1890/1950). *The Principles of Psychology*. New York: Holt, Rinehart & Winston.

——(1929). *The Varieties of Religious Experience*. New York: Modern Library.

JARVIK, K. (1970). Drugs, Hallucinations and Memory. In W. Keup (ed.), *Origins and Mechanisms of Hallucinations*. New York: Plenum Press.

JAYNES, J. (1976). *The Origin of Consciousness in the Breakdown of the Bicameral Mind*. Boston: Houghton-Mifflin.

JOHNSON, M., and RAYE, C. (1981). Reality Monitoring. *Psychological Review*, 88: 67–85.

JOHNSON-LAIRD, P. N. (1983). *Mental Models*. Cambridge, MA: Harvard University Press.

JUNG, C. G. (1951/1957). On Synchronicity. In H. Read, M. Fordham, and G. Adler (eds.), *The Collected Works of C. G. Jung*. New York: Pantheon Books, viii.

——(1954/1969). Psychological Aspects of the Mother Archetype. In H. Read, M. Fordham, and G. Adler (eds.), *The Archetypes and the Collective Unconscious*. Princeton, NJ: Princeton University Press.

JUNG, C. G. (1964). *Man and his Symbols*. London: Aldus.

——(1969). *The Archetypes and the Collective Unconscious*. Princeton, NJ: Princeton University Press.

——(1976). *The Visions Seminars*. Zurich: Spring.

KANT, I. (1781/1953). *Critique of Pure Reason*. New York: Macmillan.

KAPLAN, A. (ed. and trans.) (1997). *Sefer Yetzirah: The Book of Creation*. Rev. edn. York Beach, ME: S. Weiser.

KARSTEN, R. (1935). *The Head-Hunters of Western Amazonas*. Helsingforst: Societas Scientiarum Fennica.

KATZ, F., and DOBKIN DE RIOS, M. (1971). Hallucinogenic Music: An Analysis of the Role of Whistling in Peruvian Ayahuasca Healing Sessions. *Journal of American Folklore*, 84: 320–7.

KATZ, J. (1972). *Semantic Theory*. New York: Harper & Row.

——(1981). *Language and Other Abstract Objects*. Totowa, NJ: Rowman & Littlefield.

KATZ, S. T. (ed.) (1992). *Mysticism and Language*. Oxford: Oxford University Press.

KATZ, S. T. (1978). Language, Epistemology, and Mysticism. In S. T. Katz (ed.), *Mysticism and Philosophical Analysis*. London: Sheldon Press.

KENSINGER, K. (1973). Banisteriopsis Usage among the Peruvian Cashinahua. In M. J. Harner (ed.), *Hallucinogens and Shamanism*. Oxford: Oxford University Press.

KEUP, W. (ed.) (1970). *Origins and Mechanisms of Hallucinations*. New York: Plenum Press.

KIVY, P. (1987). Platonism in Music: Another Kind of Defence. *American Philosophical Quarterly*, 24: 245–52.

KLÜVER, H. (1928). *Mescal: The 'Divine' Plant and its Psychological Effects*. London: Kegan, Paul, Trench, Trubner & Co.

——(1966). *Mescal and Mechanisms of Hallucinations*. Chicago: University of Chicago Press.

KOCH-GRÜNBERG, T. (1921). *Zwei Jahre Bei den Indianern Nordwest-Brasiliens*. Stuttgart: Strecker & Schröder.

KOLERS, P., and SMYTHE, W. (1984). Symbol Manipulation: Alternatives to the Computational View of Mind. *Journal of Verbal Learning and Verbal Behaviour*, 23: 289–314.

KUSEL, H. (1965). Ayahuasca Drinkers among the Chama Indians of Northeast Peru. *Psychedelic Review*, 6: 58–66.

LA BARRE, W. (1972). *The Ghost Dance: Origins of Religion*. New York: Delta Books.

LABERGE, S. (1985). *Lucid Dreaming*. Los Angeles: J. P. Tarcher.

LAGROU, E. (1991). Uma Etnografia da Cultura Kaxinawá: Entre a Cobra e o Inca. Master's thesis, Federal University of Santa Catarina.

——(1995). Compulsão visual: Resenha de 'Visual Compulsion' de Peter Gow. In *Antropologia em Primeria Mão—PPGAS*. Florianópolis: Editora UFSC.

——(1996). Xamanismo e representação entre os Kaxinawá. In E. J. Langdon (ed.), *Xamanismo no Brasil: Novas Perspectivas*. Florianópolis: Editora UFSC.

——(1998). Cashinahua Cosmovision: A Perspectival Approach to Identity and Alterity. Unpublished Doctoral dissertation, University of St. Andrews.

——(1999). Cashinahua Drawings and their Role in the Vision Quest. In *Visions That the Plants Gave Us*. Richard F. Brush Gallery. Canton, NY: St. Lawrence University.

LAMB, B. (1971). *Wizard of the Upper Amazon: The Story of Manuel Córdoba-Ríos*. Boston: Houghton-Mifflin.

——(1985). *Río Tigre and Beyond: The Amazon Jungle Medicine of Manuel Córdoba-Ríos*. Berkeley, Calif.: North Atlantic Books.

LANGDON, E. J. (1979*a*). Yagé among the Siona: Cultural Patterns in Visions. In *Spirits, Shamans and Stars*. The Hague: Mouton Publishers.

——(1979*b*). The Siona Hallucinogenic Ritual: Its Meaning and Power. In J. Morgan (ed.), *Understanding Religion and Culture: Anthropological and Theoretical Perspectives*. Washington, DC: University Press of America.

——(1992*a*). Dau: Shamanic Power in Siona Religion and Medicine. In E. J. Langdon and G. Baer (eds.), *Portals of Power: Shamanism in South America*. Albuquerque, NM: University of New Mexico Press.

——(1992*b*). Yagé and Dau: Shamanic Power in Siona Religion. *Shaman's Drum*, 29: 33–9.

LANGDON, E. J., and BAER, G. (eds.) (1992). *Portals of Power: Shamanism in South America*. Albuquerque, NM: University of New Mexico Press.

LANGER, S. (1953). *Feeling and Form: A Theory of Art*. New York: C. Scribner's Sons.

——(1969). *Philosophy in a New Key: A Study in the Symbolism of Reason, Rite, and Art*. 3rd edn. Cambridge, MA: Harvard University Press.

LEACH, E. (1976). *Claude Lévi-Strauss*. Harmondsworth: Penguin.

LEARY, T., METZNER, R., and ALPERT, R. (1969). *The Psychedelic Experience: A Manual Based on the Tibetan Book of the Dead*. London: Academic Press.

LEIBNIZ, G. (1714/1945). *The Monadology and Other Philosophical Writings*. Oxford: Oxford University Press.

LÉVI-STRAUSS, C. (1964). *Mythologiques I: Le Crut et le Cuit*. Paris: Plon.

LEWIS-WILLIAMS, J., and DOWSON, T. (1988). The Signs of All Times: Entoptic Phenomena in Upper Palaeolithic Art. *Current Anthropology*, 29: 201–45.

LLINÁS, R. R., and RIBARY, U. (1994). Perception as an Oneiric-Like State Modulated by the Senses. In C. Koch and J. L. Davis, (eds.), *Large-Scale Neuronal Theories of the Brain*. Cambridge, MA: MIT Press.

LOCKE, J. (1690/1964). *An Essay Concerning Human Understanding*. London: Collins.

LOVELOCK, J. (1979). *Gaia: A New Look at Life on Earth*. Oxford: Oxford University Press.

——(1988). *The Ages of Gaia*. New York: W. Norton.

LUDWIG, A. (1969). Altered States of Consciousness. In C. Tart (ed.), *Altered States of Consciousness*. New York: Wiley & Sons.

LUNA, L. E. (1984*a*). The Concept of Plants as Teachers among Four Mestizo Shamans of Iquitos, Northeastern Peru. *Journal of Ethnopharmacology*, 11: 135–56.

——(1984*b*). The Healing Practices of a Peruvian Shaman. *Journal of Ethnopharmacology*, 11: 123–33.

——(1986*a*). Bibliografía sobre el Ayahuasca. *America Indígena*, 46: 235–45.

——(1986*b*). *Vegetalismo Shamanism Among the Mestizo Population of the Peruvian Amazon*. Stockholm: Almqvist & Wiksell.

——(1992). Icaros: Magic Melodies among the Mestizo Shamans of the Peruvian Amazon. In E. J. Longdon and G. Baer (eds.), *Portals of Power: Shamanism in South America*. Albuquerque, NM: University of New Mexico Press.

LUNA, L. E., and AMARINGO, P. (1993). *Ayahuasca Visions*. Berkeley: North Atlantic Books.

LUNA, L. E., and WHITE, S. (eds.) (2000). *Ayahuasca Reader: Encounters with the Amazon's Sacred Vine*. Santa Fe: Synergetic Press.

LYONS, J. (1977). *Semantics I*. Cambridge: Cambridge University Press.

MABIT, J. (1988). *L'Hallucination par l'ayahuasca chez les guerisseurs de la Haute-Amazone peruvienne (Tarapoto)*. Lima: Institut Français D'Études Andines.

——(1992). Consideraciones acerca del brebaje ayahuasca y perspectivas terapeuticas. *Revista de Neuro-Psiquiatria*, 55: 118–31.

——(1996). Ayahuasca and Shamanism in Addiction Therapy: A Report from the Peruvian Amazon. *Bulletin of the Multidisciplinary Association for Psychedelic Studies*, 6: 24–31.

MCGOVERN, W. (1927). *Jungle Paths and Inca Ruins*. New York: Grosset & Dunlap.

MCKENNA, D., CALLAWAY, J., and GROB, C. (1998). The Scientific Investigation of Ayahuasca: A Review of Past and Current Research. *The Heffter Review of Psychedelic Research*, 1: 65–77.

MCKENNA, T. (1991). *The Archaic Revival: Speculation on Psychedelic Mushrooms, the Amazon, Virtual Reality, UFOs, Evolution, Shamanism, the Rebirth of the Goddess and the End of History*. New York: HarperSanFrancisco.

——(1992). *Food of the Gods: The Search for the Original Tree of Knowledge*. New York: Bantam Books.

MACRAE, E. (1992). *Guiado Pela Lua: Xamanismo e Uso Ritual da Ayahuasca no Culto do Santo Daime*. São Paulo: Editora Brasilense.

MADER, E. (1999). *Metamorfosis del Poder: Persona, Mito y Visión en la Sociedad Shuar y Achuar*. Quito: Abya-Yala.

MALLIN, S. (ed. and trans.) (1979). *R. Yehuda Leove ben Bezalel (The Maharal of Prague)'s Book of Divine Power: Introductions*. 2nd rev. edn. Jerusalem: Ber Aryeh International-Gavriel Beer Books.

MALLOL DE RECASENS, M. (1963). Cuatro representaciones de la imagenes allucinatorias originadas por la toma de yage. *Revista Colombiana de Folclor*, 3: 57–79.

MARKS, L. E. (1978). *The Unity of the Senses*. New York: Academic Press.

MASLOW, A. (1968). *Toward a Psychology of Being*. New York: Van Nostrand Reinhold.

——(1970). *Religions, Values, and Peak Experiences*. New York: Viking Press.

——(1971). *The Farther Reaches of Human Nature*. New York: Viking Press.

MASTERS, R., and HOUSTON, J. (1966). *The Varieties of Psychedelic Experience*. New York: Holt, Rinehart & Winston.

MAYHEW, C. (1961). Untitled chapter. In D. Ebin (ed.), *The Drug Experience*, pp. 293–300. New York: Evergreen Black Cat.

MERKUR, D. (1998). *The Ecstatic Imagination: Psychedelic Experiences and the Psychoanalysis of Self-Actualization*. Albany, NY: State University of New York Press.

——(1999). *Mystical Moments and Unitive Thinking*. Albany, NY: State University of New York.

——(2000). *The Mystery of the Manna: The Psychedelic Sacrament of the Bible*. Rochester, Vt.: Inner Traditions International.

MERLEAU-PONTY, M. (1962). *The Phenomenology of Perception*. London: Routledge & Kegan Paul.

MERRELL-WOLFF, F. (1995). *Transformations in Consciousness: The Metaphysics and Epistemology*. Albany, NY: SUNY Press.

METZNER, R. (ed.) (1968). *The Ecstatic Adventure*. New York: Macmillan.

——(ed.) (1999). *Ayahuasca: Hallucinogens, Consciousness, and the Spirit of Nature*. New York: Thunder's Mouth Press.

MEYER-BAER, K. (1970). *Music of the Spheres and the Dance of Death: Studies in Musical Iconography*. Princeton: Princeton University Press.

MEYERRATKEN, U., and SALEM, N. (1997). *Daime: Brasiliens Kult der heilended Kraftpflanzen*. Munich: Knaur.

MICHAELS, C., and CARELLO, C. (1981). *Direct Perception*. Englewood Cliffs, NJ: Prentice Hall.

MICHAUX, H. (1972). *Misérable Miracle: Le Mescaline*. Paris: Gallimard.

MINTZ, S., and ALPERT, M. (1972). Imagery Vividness, Reality Testing, and Schizophrenic Hallucinations. *Journal of Abnormal Psychology*, 19: 310–16.

MONTAGUE, R. (1974). *Formal Philosophy: Selected Papers of R. Montague*, ed. R. Thomason. New Haven: Yale University Press.

MONTEIRO DA SILVA, C. (1983). O Palacio de Juramidam. Santo Daime, um ritual de transcendencia e despoluição. Master's thesis, Federal University of Pernambuco. Unpublished.

MOODY, R. (1975). *Life After Life*. Atlanta: Mockingbird Books.

MORRIS, C. (1938). Foundations of the Theory of Signs. In O. Neurath (ed.), *International Encyclopedia of United Science*. Chicago: Chicago University Press, i. 77–138.

MORTON, C. (1931). Notes on Yagé, a Drug Plant of Southern Colombia. *Journal of the Washington Academy of Sciences*, 21: 485–8.

MURPHY, J. (1976). Psychiatric Labeling in Cross-Cultural Perspective. *Science*, 191: 1019–28.

NAKAZAWA, R. (1996). *Medicina Tradicional Amazonica en el Tratamiento del Abuso de Drogas*. Tarapoto: Centro Takiwasi.

NARANJO, C. (1973*a*). Psychological Aspects of the Yaje Experience in an Experimental Setting. In M. J. Harner (ed.), *Hallucinogens and Shamanism*. Oxford: Oxford University Press.

—— (1973*b*). *The Healing Journey: New Approaches to Consciousness*. New York: Pantheon Books.

—— (1979). Psychotropic Properties of the Harmala Alkaloids. In D. Efron, B. Holmstedt, and N. Kline (eds.), *Ethnopharmacologic Search for Psychoactive Drugs*. New York: Raven Press.

—— (1983). *Ayahuasca, Etnomedicina y Mitologia*. Quito: Ediciones Libri Mundi.

—— (1987). Ayahuasca Imagery and the Therapeutic Property of the Harmala Alkaloids. *Journal of Mental Imagery*, 11: 131–6.

NARANJO, P. (1986). El ayahuasca en la arqueología ecuatoriana. *America Indígena*, 46: 117–27.

NARBY, J. (1998). *The Cosmic Serpent: DNA and the Origins of Knowledge*. New York: Jeremy P. Tarcher.

NEMIAH, J. (1989). Dissociative Disorders (Hysterical Neuroses, Dissociative Type). In H. Kaplan and B. Sadock (eds.), *Comprehensive Textbook of Psychiatry*, 5th edn. Baltimore, Md.: Williams & Wilkins.

NOLL, R. (1985). Mental Imagery Cultivation as a Cultural Phenomenon: The Role of Visions in Shamanism. *Current Anthropology*, 26: 443–61.

ODEBERG, H. (trans.) (1973). *The Hebrew Book of Enoch*. New York: Ktav.

ORNSTEIN, R. (1969). *On the Experience of Time*. Harmondsworth: Penguin.

ORTONY, A. (ed.) (1979). *Metaphor and Thought*. Cambridge: Cambridge University Press.

Oster, G. (1970). Phosphenes. *Scientific American*, 222: 83–8.

Ott, J. (1986). Carved 'Disembodied Eyes' of Teotihuacan. In R. G. Wasson, S. Kramrisch, J. Ott, and C. A. P. Ruck (eds.), *Persephone's Quest: Entheogens and the Origins of Religion*. New Haven: Yale University Press.

——(1993). *Pharmacotheon: Entheogenic Drugs, their Plant Sources and History*. Kennewick, WA: Natural Products.

Ott, J. (1994). *Ayahuasca Analogues—Pangean Entheogens*. Kennewick, WA: Natural Products.

——(1996). Entheogens II: On Entheology and Ethnobotany. *Journal of Psychoactive Drugs*, 28: 205–9.

Otto, R. (1957). *The Idea of the Holy: An Inquiry into the Non-Rational Factor in the Idea of the Divine and its Relation to the Rational*. Oxford: Oxford University Press.

Overing, J. (1985*a*). There is No End of Evil: The Guilty Innocents and their Fallible God. In D. Parkin (ed.), *The Anthropology of Evil*. London: Basil Blackwell.

——(1985*b*). Today I Shall Call Him 'Mummy': Multiple Worlds and Classificatory Confusion. In J. Overing (ed.), *Reason and Morality*. London: Routledge.

Pahnke, W. N. (1967). LSD and Religious Experience. In R. C. Debold and R. C. Leaf (eds.), *LSD, Man and Society*. Middletown, CT: Wesleyan University Press.

——(1972). Drugs and Mysticism. In J. White (ed.), *The Highest State of Consciousness*. Garden City, NY: Anchor Books.

Pahnke, W. N., and Richards, W. (1966). Implications of LSD and Experimental Mysticism. *Journal of Religion and Health*, 5: 175–208.

Parsons, W. B. (1999). *The Enigma of the Oceanic Feeling: Revisioning the Psychoanalytic Theory of Mysticism*. Oxford: Oxford University Press.

Payaguaje, F. (1983). *El Bebedor de Yaje*. Quito: Libri Mundi.

Peat, F. D. (1987). *Synchronicity: The Bridge Between Matter and Mind*. New York: Bantam Books.

Pellegrini, L. (1994). Jornada ao Coracao do Daime. *Planeta*, 265: 45–9.

Penrose, R. (1990). *The Emperor's New Mind: Concerning Computers, Minds, and the Laws of Physics*. London: Vintage.

——(1994). *Shadows of the Mind: A Search for the Missing Science of Consciousness*. Oxford: Oxford University Press.

Peters, L., and Price-Williams, D. (1980). Towards an Experiential Analysis of Shamanism. *American Ethologist*, 73: 397–418.

Piaget, J. (1983). Piaget's Theory. In P. H. Mussen (ed.), *Carmichael's Manual of Child Psychology*. 3rd edn. New York: John Wiley.

Phillips, S. (1995). *Classical Indian Metaphysics*. Chicago: Open Court.

Pinch, G. (1994). *Magic in Ancient Egypt*. London: British Museum Press.

Plato. (1914). *Phaedrus*. Loeb Classical Library, 1. London: Heinemann.

——(1935). *The Republic*. Loeb Classical Library, 2. London: Heinemann.

Polanyi, M. (1968). Life's Irreducible Structure: Live Mechanisms and Information in DNA are Boundary Conditions with a Sequence of Boundaries Above Them. *Science*, 160: 1308–12.

Polari, A. (1984). *O Livro das Mirações*. Rio de Janeiro: Editora Record.

——(1992). *O Guia da Floresta*. Rio de Janeiro: Nova Era.

——(1998). *O Evangelho Segundo Sebastião Mota*. Céu do Mapiá: CEFLURIS.

Popper, K. R. (1972). *Objective Knowledge*. Oxford: Clarendon Press.

Prance, G. (1970). Notes on the Use of the Plant Hallucinogens in Amazonian Brazil. *Economic Botany*, 24: 62–8.

Progoff, I. (1973). *Jung, Synchronicity, and Human Destiny*. New York: Dell Publishing Co.

Putnam, H. (1973). Reductionism and the Nature of Psychology. *Cognition*, 8: 263–368.

Pylyshyn, Z. W. (1979). Do Mental Events have Durations? *The Behavioural and Brain Sciences*, 2: 277–8.

—— (1984). *Computation and Cognition*. Cambridge, MA: MIT Press.

Quinn, E., and Penrose, R. (1965). *Picasso at Work: An Intimate Photographic Study*. New York: Doubleday.

Radden, J. (1996). *Divided Minds and Successive Selves: Ethical Issues in Disorders of Identity and Personality*. Cambridge, MA: MIT Press.

—— (1998). Pathologically Divided Minds: Synchronic Unity and Models of Self. *Journal of Consciousness Studies*, 5: 658–72.

Ramírez de Jara, M. C., and Pinzón Castaño, C. E. (1992). Sibundoy Shamanism and Popular Culture in Colombia. In E. J. Langdon and G. Baer (eds.), *Portals of Power: Shamanism in South America*. Albuquerque, NM: University of New Mexico Press.

Reber, A. (1984). *The Penguin Dictionary of Psychology*. Harmondsworth: Penguin Books.

Reed, E. (1991). James Gibson's Ecological Approach to Cognition. In A. Still and A. Costall (eds.), *Against Cognitivism: Alternative Foundations for Cognitive Psychology*. Hemel Hempstead: Harvester Wheatsheaf.

Reed, G. (1972). *The Psychology of Anomalous Experience: A Cognitive Approach*. London: Hutchinson University Library.

Reichel-Dolmatoff, G. (1960). Notas etnográficas sobre los indios del Chocó. *Revista Colombiana de Antropología*, 9: 75–158.

—— (1971). *Amazonian Cosmos: The Sexual and Religious Symbolism of the Tukano Indians*. Chicago: University of Chicago Press.

—— (1975). *The Shaman and the Jaguar: A Study of Narcotic Drugs among the Indians in Colombia*. Philadelphia: Temple University Press.

—— (1978*a*). *Beyond the Milky Way: Hallucinatory Imagery of the Tukano Indians*. Los Angeles: UCLA Latin America Centre.

—— (1978*b*). Drug Induced Optical Sensations and their Relationship to Applied Art among Some Colombian Indians. In M. Greenglagh and V. Megaw (eds.), *Art in Society: Studies in Style, Culture and Aesthetics*. New York: St Martin's Press.

—— (1981). Brain and Mind in Desana Shamanism. *Journal of Latin American Lore*, 7: 73–98.

—— (1990). The Cultural Context of an Aboriginal Hallucinogen: Banisteriopsis Caapi. In P. T. Furst (ed.), *Flesh of the Gods: The Ritual Use of Hallucinogens*. Prospect Heights, IL: Waveland Press.

—— (1991). *Indians of Colombia: Experience and Cognition*. Bogota: Villegas Editores.

—— (1996). *The Forest Within: The World-View of the Tukano Amazonian Indians*. Foxhole, Devon: Themis Books.

—— (1997). *Chamanes de la Selva Pluvial*. Foxhole, Devon: Themis Books.

Reinburg, P. (1921). Contribution à l'étude des biossons toxiques des indiens du nord-ouest de l'Amazonie, l'ayahuasca, le yajé, le huanto. *Journal de la Société Ameriquane*, 13: 25–54, 197–216.

Reinburg, P. (1965). *Bebidas Toxicas de los Indios del Amazonas*. Lima: Universidad Nacional Mayor de San Marcos.

Richman, G. (1990/1). The Santo Daime Doctrine: An Interview with Alex Polari de Alverga. *Shaman's Drum*, 22: 30–41.

Rivier, L., and Lindgren, J. E. (1972). Ayahuasca, the South American Hallucinogenic Drink: An Ethnobotanical and Chemical Investigation. *Economic Botany*, 26: 101–29.

Roberts, T. (1983). New Learning. In L. Grinspoon and J. Bakalar (eds.), *Psychedelic Reflections*. New York: Human Sciences Press.

Rosch, E. (1978). Principles of Categorization. In E. Rosch and B. Lloyd (eds.), *Cognition and Categorization*. Hillsdale, NJ: Lawrence Erlbaum.

Rosch, E., Mervis, C., Gray, W., Johnson, D., and Boyes-Braem, P. (1976). Basic Objects in Natural Categories. *Cognitive Psychology*, 7: 573–605.

Rouget, G. (1985). *Music and Trance*. Chicago: University of Chicago Press.

Rouhier, A. (1924). Le Yagé: Plante télépathique. *Paris Médical*, 52: 341.

Ruck, C., Bigwood, J., Staples, R., Wasson, R., and Ott, J. (1979). Entheogens. *Journal of Psychedelic Drugs*, 11: 145–6.

Russell, B. (1967). *The Problems of Philosophy*. Oxford: Oxford University Press.

Sachs, J. (1967). Recognition Memory for Syntactic and Semantic Aspects of Connected Discourse. *Perception and Psychophysics*, 2: 437–42.

Saramago, J. (1989/1996). *The History of the Siege of Lisbon*. New York: Harcourt Brace.

Savage, C. (1975). The Continuity of Perceptual and Cognitive Experiences. In R. Siegel and L. West (eds.), *Hallucinations: Behaviour, Experience, and Theory*. New York: J. Wiley.

Scharfetter, C. (1980). *General Psychopathology: An Introduction*. Cambridge: Cambridge University Press.

Scharfstein, B. (1973). *Mystical Experience*. Oxford: Blackwell.

Schirn, M. (1998). *The Philosophy of Mathematics Today*. Oxford: Clarendon Press.

Schleiffer, H. (ed.) (1973). *Sacred Narcotic Plants of the New World Indians*. New York: Hafner Press.

Schultes, R. E. (1954). A New Narcotic Snuff from the Northwest Amazon. In *Botanical Museum Leaflets (Harvard University)*, 16: 241–60.

——(1972). An Overview of Hallucinogens in the Western Hemisphere. In P. T. Furst (ed.), *Flesh of the Gods: The Ritual Use of Hallucinogens*. Prospect Heights, IL: Waveland Press.

——(1982). The Beta-Carboline Hallucinogens of South America. *Journal of Psychoactive Drugs*, 14: 205–20.

——(1986). El desarrollo histórico de la identificatión de las malpigiáceas empleadas como allucinógenos. *America Indígena*, 46: 319–30.

Schultes, R. E., and Hofmann, A. (1979). *Plants of the Gods: Origins of Hallucinogenic Use*. London: McGraw-Hill.

——(1980). *The Botany and Chemistry of Hallucinogens*. Springfield, IL: Charles C. Thomas.

Schultes, R. E., and Winkelman, M. (1995). The Principal American Hallucinogenic Plants and their Bioactive and Therapeutic Properties. *Yearbook of Cross-Cultural Medicine and Psychotherapy*, 6: 205–39.

Searle, J. R. (1969). *Speech Acts: An Essay in the Philosophy of Language*. Cambridge: Cambridge University Press.

——(1980). Minds, Brains and Programs. *The Behavioural and Brain Sciences*, 3: 417–57.

——(1992). *The Rediscovery of the Mind*. Cambridge, MA: MIT Press.

Sena Araújo, W. (1999). *Navegando sobre as Ondas do Daime: História, Cosmologia e Ritual da Barquinha*. Campinas: Editoria da Unicamp.

Shanon, B. (1982). Colour Associates to Semantic Linear Orders. *Psychological Research*, 44: 75–83.

——(1983). Descartes' Puzzle: An Organismic Approach. *Cognition and Brain Theory*, 6: 185–95.

——(1984). The Case for Introspection. *Cognition and Brain Theory*, 7: 167–80.

——(1987). The Non-Abstractness of Mental Representations. *New Ideas in Psychology*, 5: 117–26.

——(1988*a*). Semantic Representations of Meaning—a Critique. *Psychological Bulletin*, 104: 70–83.

——(1988*b*). Channels of Thought. *Discourse Processes*, 11: 221–42.

——(1989*a*). Thought Sequences. *The European Journal of Cognitive Psychology*, 1: 129–59.

——(1989*b*). *Mental Images—a Phenomenological Investigation*. Working Paper no. 28. The Goldie Rotman Centre for Cognitive Science in Education, Jerusalem.

——(1989*c*). A Squish in Hebrew Morphology. *Theoretical Linguistics*, 6: 211–19.

——(1990*a*). Consciousness. *Journal of Mind and Behaviour*, 11: 137–52.

——(1990*b*). What is Context? *Journal for the Theory of Social Behaviour*, 20: 158–66.

——(1991). Cognitive Psychology and Modern Physics—Some Analogues. *The European Journal of Cognitive Psychology*, 3: 201–34.

——(1992). Metaphor: From Fixedness and Selection to Differentiation and Creation. *Poetics Today*, 13: 659–85.

——(1993*a*). *The Representational and the Presentational: An Essay on Cognition and the Study of Mind*. London: Harvester-Wheatsheaf.

——(1993*b*). Why Are We (at Least Sometimes) Conscious of Our Thoughts? *Pragmatics and Cognition*, 1: 25–49.

——(1998*a*). The Intrinsic Temporality of Human Cognition. *Behavioural and Brain Sciences*, 21: 650–1.

——(1998*b*). What is the Function of Consciousness? *Journal of Consciousness Studies*, 5: 295–308.

——(1998*c*). Ayahuasca Visions: A Comparative Cognitive Investigation. *Yearbook of Ethnomedicine and the Study of Consciousness*, 7: 227–50.

——(2001). Altered Temporality. *Journal of Consciousness Studies*, 8: 35–58.

——(2002*a*). Entheogens. *Journal of Consciousness Studies*, 9: 85–94.

——(2002*b*). Ayahuasca Visualizations. *Journal of Consciousness Studies*, 9: 3–30.

Shanon, B., and Eifermann, R. (1984). Dream-Reporting Discourse. *Text*, 4: 369–79.

Shapiro, S. (1997). *Philosophy of Mathematics: Structure and Ontology*. Oxford: Oxford University Press.

Shaw, R., and Bransford, J. (1977). Introduction: Psychological Approaches to the Problem of Knowledge. In R. Shaw and J. Bransford (eds.), *Perceiving, Acting, and Knowing: Toward an Ecological Psychology*. Hillsdale, NJ: Lawrence Erlbaum.

Sherzer, J. (1986). El arte verbal de los cantos chamanisticos Cuna. *America Indígena*, 46: 49–60.

SIEGEL, R. (1977). Hallucinations. *Scientific American*, 237: 132–40.

SIEGEL, R. (1980). The Psychology of Life after Death. *American Psychologist*, 35: 911–31.

SIEGEL, R., and JARVIK, M. (1975). Drug Induced Hallucinations in Animals and Man. In R. Siegel and L. West (eds.), *Hallucinations: Behaviour, Experience, and Theory*. New York: J. Wiley.

SIEGEL, R., and WEST, L. (eds.) (1975). *Hallucinations: Behaviour, Experience, and Theory*. New York: J. Wiley.

SISKIND, J. (1973*a*). Visions and Cures among the Sharanahua. In M. J. Harner (ed.), *Hallucinogens and Shamanism*. Oxford: Oxford University Press.

—— (1973*b*). *To Hunt in the Morning*. New York: Oxford University Press.

SLADE, P., and BENTALL, R. (1988). *Sensory Deception: A Scientific Analysis of Hallucination*. Baltimore: The Johns Hopkins University Press.

SMITH, A., and TART, C. (1998). Cosmic Consciousness Experience and Psychedelic Experiences: A First Person Comparison. *Journal of Consciousness Studies*, 1: 97–107.

SMITH, H. (1964). Do Drugs Have Religious Import? *The Journal of Philosophy*, 61: 517–30.

—— (2000). *Cleansing the Doors of Perception: The Religious Significance of Entheogenic Plants and Chemicals*. New York: Jeremy P. Tarcher/Putnam.

SOGYAL RINPOCHE (1993). *The Tibetean Book of Living and Dying*. New York: HarperSanFrancisco.

SPINELLA, M. (2001). *The Psychopharmacology of Herbal Medicine: Plant Drugs that Alter Mind, Brain, and Behaviour* Cambridge, MA: MIT Press.

SPINOZA, B. (1670/1967). *Ethics*. London: Everyman's Library.

—— (1852/1985). Short Treatise on God, Man, and his Well-Being. In E. Curley (ed.), *The Collected Works of Spinoza*. Princeton: Princeton University Press.

SPITZER, M., THIMM, M., HOLZMANN, K., HEIMANN, H., GOUZOULIS-MAYFRANK, E., KISCHKA. U., and SCHNEIDER, F. (1996). Increased Activation of Indirect Semantic Associations under Psilocybin. *Biological Psychiatry*, 39: 1055–7.

STACE, W. (1961). *Mysticism and Philosophy*. London: Macmillan Press.

STAFFORD, P. (1993). *Psychedelics Encyclopedia*. Los Angeles: J. P. Tarcher.

STEINER, R. (1969). *Cosmic Memory*. New York: The Rudolph Steiner Publications.

STERLING, E. E. (1997). Law Enforcement Against Entheogens: Is It Religious Persecution? In R. Forte (ed.), *Entheogens and the Future of Religion*. San Francisco: Council on Spiritual Practices.

STOLBA, K. (1990). *The Development of Western Music: A History*. Dubuque, IA: William C. Brown.

STRASSMAN, R. J. (1994). Human Psychopharmacology of LSD, Dimethyltryptamine and Related Compounds. In A. Pletscher and D. Ladewig (eds.), *50 Years of LSD: Current Status and Perspectives on Hallucinogens*. New York: Parthenon.

—— (1996). Human Psychopharmacology of N,N-dimethyltryptamine. *Behavioural Brain Research*, 73: 121–4.

STRASSMAN, R. J. (2001). *DMT the Spirit Molecule*. Rochester, VT: Park Street Press.

SUDNOW, D. (1980). *Talk's Body: A Meditation Between Two Keyboards*. Harmondsworth: Penguin.

SWEDENBORG, E. (1854). *Of Heaven and Hell: Concerning Heaven and its Wonders and Concerning Hell and Things Heard and Seen*. Boston: Otis Clapp.

TART, C. (ed.) (1969). *Altered States of Consciousness: A Book of Readings*. New York: John Wiley.

——(1972). States of Consciousness and State-Specific Sciences. *Science*, 176: 1203–10.

——(ed.) (1992). *Transpersonal Psychologies: Perspectives on the Mind from Seven Great Spiritual Traditions*. San Francisco: Harper.

TAUSSIG, M. (1987). *Shamanism, Colonialism, and the Wild Man: A Study in Terror and Healing*. Chicago: University of Chicago Press.

TEILHARD DE CHARDIN, P. (1965). *The Phenomenon of Man*. New York: Harper & Row.

TREISMAN, M. (1963). Temporal Discrimination and the Indifference Interval: Implications for a Model of the 'Internal Clock'. *Psychological Monographs: General and Applied*, 77: 1–40.

TURVEY, M., SHAW, R., REED, E., and MACE, W. (1981). Ecological Laws of Perceiving and Acting: In Reply to Fodor and Pylyshyn. *Cognition*, 3: 237–304.

TYLER, C. (1978). Some New Entoptic Phenomena. *Vision Research*, 18: 1633–9.

UNDERHILL, E. (1955). *Mysticism: A Study in the Nature and Development of Man's Spiritual Consciousness*. New York: Meridian.

VAUGHAN, F. (1983). Perception and Knowledge: Reflections on Psychological and Spiritual Learning in the Psychedelic Experience. In L. Grinspoon and J. Bakalar (eds.), *Psychedelic Reflections*. New York: Human Sciences Press.

VILLAVICENCIO, M. (1858). *Geographia de la Republica de Ecuador*. New York: Robert Craighead.

VILLOLDO, A., and JENDRESEN, E. (1994). *Island of the Sun: Mastering the Inca Medicine Wheel*. Rochester, Vt.: Inner Traditions International.

VIVEIROS DE CASTRO, E. (1992). *From the Enemy's Point of View: Humanity and Divinity in an Amazonian Society*. Chicago, IL: University of Chicago Press.

VYGOTSKY, L. (1978). *Mind in Society*. Cambridge, Mass.: Harvard University Press.

WADE, J. (1996). *Changes of Mind: A Holonomic Theory of the Evolution of Consciousness*. Albany, NY: State University of New York Press.

WAINWRIGHT, W. (1981). *Mysticism: A Study of its Nature, Cognitive Value and Moral Implications*. Milwaukee, WI: University of Wisconsin Press.

WAISBARD, S. R. (1958/9). Les Indiens Shamas de l'Ucayali et du Tamaya. *Ethnographie: Societe d'ethnographie de paris*, 53: 19–74.

——(1969). *Connaissances des Amazonies: Chez les chasseurs de têtes d'Amazonie*. Paris: Sociéte Continentale D'Éditions Modernes Illustrés.

WALSH, R. (1990). *The Spirit of Shamanism*. Los Angeles: Jeremy P. Tarcher.

——(1993). The Transpersonal Movement: A History and State of the Art. *Journal of Transpersonal Psychology*, 25: 123–39.

WALSH, R., and VAUGHAN, F. (eds.) (1980). *Beyond Ego: Transpersonal Dimensions in Psychology*. Los Angeles: Jeremy P. Tarcher.

WASSÉN, S., and HOLMSTEDT, B. (1963). The Use of Parica, an Ethnological and Pharmacological Review. *Ethnos*, 1: 5–45.

WASSON, R. G. (1961). Hallucinogenic Fungi of Mexico: An Inquiry into the Origins of the Religious Idea Among Primitive Peoples. *Botanical Museum Leaflets (Harvard University)*, 19: 137–62.

——(1968). *Soma: Divine Mushroom of Immortality*. New York: Harcourt, Brace & World.

WASSON, R. G., HOFMANN, A., and RUCK, C. A. P. (1978). *The Road to Eleusis: Unveiling the Secret of the Mysteries*. New York: Harcourt, Brace, Jovanovich.

Wasson, R. G., Kramrisch, S., Ott, J., and Ruck, C. A. P. (1986). *Persephone's Quest: Entheogens and the Origins of Religion*. New Haven: Yale University Press.
Watts, A. (1962). *The Joyous Cosmology: Adventures in the Chemistry of Consciousness*. New York: Pantheon Books.
Webster's Third New International Dictionary (1975). Springfield, Mass.: G. & C. Merriam.
Weiskopf, J. (1995). From Agony to Ecstasy: The Transformative Spirit of Yagé. *Shaman's Drum*, 39: 40–7.
Weiss, G. (1973). Shamanism and Priesthood in Light of the Campa Ayahuasca Ceremony. In M. J. Harner (ed.), *Hallucinogens and Shamanism*. Oxford: Oxford University Press.
Wertsch, J. V. (ed.) (1981). *The Concept of Activity in Soviet Psychology*. New York: Sharpe.
Wertsch, J. V. (1985). *Culture, Communication and Cognition: Vygotskian Perspectives*. Cambridge: Cambridge University Press.
West, L. (1962). A General Theory of Hallucinations and Dreams. In L. West (ed.), *Hallucinations*. New York: Grune & Stratton.
——(1975). A Clinical and Theoretical Overview of Hallucinatory Phenomena. In R. Siegel and L. West (eds.), *Hallucinations: Behaviour, Experience, and Theory*. New York: J. Wiley.
——(ed.). (1962). *Hallucinations*. New York: Grune & Stratton.
Whitehead, A. (1978). *Process and Reality*. New York: The Free Press.
Wilber, K. (1977). *The Spectrum of Consciousness*. Wheaton, IL: Quest.
——(1983). *Up from Eden*. London: Routledge.
——(1990). *Eye to Eye: The Quest for the New Paradigm*. (Expanded edn). Boston: Shambhala.
——(1996). *A Brief History of Everything*. Boston and London: Shambhala.
Wilson, A. (1949). Ayahuasca, Peyotl, Yagé. *Proceedings of the Society of Psychical Research*, 48: 353–62.
Wittgenstein, L. (1922). *Tractatus Logico Philosophicus*. London: Routledge & Kegan Paul.
——(1953). *Philosophical Investigations*. Oxford: Blackwell.
——(1958). *The Blue and Brown Books*. Oxford: Blackwell.
——(1980). *Remarks on the Philosophy of Psychology*. Oxford: Basil Blackwell, i.
Wolf, F. A. (1992). *The Eagle Quest: A Physicist's Search for Truth in the Heart of the Shamanic World*. New York: Touchstone.
Wolfson, E. R. (1994). *Through a Speculum that Shines: Vision and Imagination in Medieval Jewish Mysticism*. Princeton: Princeton University Press.
Zaehner, R. (1957). *Mysticism, Sacred and Profane: An Inquiry into Some Varieties of the Religious Experience*. Oxford: Oxford University Press.
——(1972). *Zen, Drugs and Mysticism*. New York: Vintage Press.
Zakay, D. (1993). Relative and Absolute Duration Judgements under Prospective and Retrospective Paradigms. *Perception and Psychophysics*, 54: 656–64.
Zakay, D., and Block, R. (1997). Temporal Cognition. *Current Directions in Psychological Science*, 6: 12–16.
Zinberg, N. (1984). *Drugs: Set and Setting*. New Haven: Yale University Press.

INDEX

Notes

1. Most references are to Ayahuasca and visions experienced, therefore these words are mainly omitted as qualifiers
2. Page numbers for chapters are **emboldened**